# Studies in Fuzziness and Soft Con

Volume 363

**Series editor**

Janusz Kacprzyk, Polish Academy of Sciences, Warsaw, Poland
e-mail: kacprzyk@ibspan.waw.pl

The series "Studies in Fuzziness and Soft Computing" contains publications on various topics in the area of soft computing, which include fuzzy sets, rough sets, neural networks, evolutionary computation, probabilistic and evidential reasoning, multi-valued logic, and related fields. The publications within "Studies in Fuzziness and Soft Computing" are primarily monographs and edited volumes. They cover significant recent developments in the field, both of a foundational and applicable character. An important feature of the series is its short publication time and world-wide distribution. This permits a rapid and broad dissemination of research results.

More information about this series at http://www.springer.com/series/2941

Sunil Mathew · John N. Mordeson
Davender S. Malik

# Fuzzy Graph Theory

Sunil Mathew
Bio-Computational and Cancer Research
Laboratory, Department of Mathematics
National Institute of Technology Calicut
Kozhikode, Kerala
India

Davender S. Malik
Department of Mathematics
Creighton University
Omaha, NE
USA

John N. Mordeson
Department of Mathematics
Creighton University
Omaha, NE
USA

ISSN 1434-9922 ISSN 1860-0808 (electronic)
Studies in Fuzziness and Soft Computing
ISBN 978-3-319-89070-8 ISBN 978-3-319-71407-3 (eBook)
https://doi.org/10.1007/978-3-319-71407-3


Softcover re-print of the Hardcover 1st edition 2018


Printed on acid-free paper

This Springer imprint is published by Springer Nature
The registered company is Springer International Publishing AG
The registered company address is: Gewerbestrasse 11, 6330 Cham, Switzerland

*Sunil Mathew would like to dedicate this book to his father.*

*John N. Mordeson and Davender S. Malik would like to dedicated this book to those who have been victims of human trafficking.*

# Foreword

Fuzzy logic, based on fuzzy sets, was introduced by Lotfi A. Zadeh along the mid 1960s and 1970s, and meant one of the most important, creative, and fruitful concepts introduced in science and technology during the second half of twentieth century, as it is shown by the many theoretic results and applications that are based on fuzzy sets. Its' introduction has, indeed, all the typical characteristics of creativity since, and by the first time, it deals with the imprecision that permeates natural language, and that logicians and scientists in general did left aside, and even abhorred it from a long time by over valuating the precision that is necessary for science. Nevertheless, if fuzzy logic tries to model the reasoning with imprecise concepts, it is not imprecise in itself; fuzzy logic is not fuzzy.

A proper fuzzy set is nothing else than the collective generated in the language by an imprecise word, and whose 'meaning extent' can be measured by its membership functions, each one being designed accordingly with the current context-dependent and purpose-driven meaning of the corresponding word that generates the fuzzy set. Linguistic collectives, or fuzzy sets, are cloudy entities placed in natural language where they are well anchored, since people manage them with ease and jointly with their linguistic labels; they are classical sets whenever the word is precisely used in the universe of discourse. If all this corresponds with philosophically viewing, à la Wittgenstein, the meaning of words as its contextual use, for the scientific analysis, or domestication of language, imprecision cannot be avoided, and advancing toward the mathematical modeling of natural language and ordinary reasoning is necessary for reaching machines that can actually 'think' like people, and for advancing into 'Computing with Words and Perceptions' as Zadeh advocates since the last nineties.

In addition, and once Zadeh did prove that a membership function whose maximum in the unit interval is 1 can be seen as a possibility distribution, this theoretic fact opened a window for using these functions to also measure the nonrandom uncertainty that exists in language, and for both precise and imprecise words.

Zadeh's trial to scientifically domesticating imprecision and nonrandom uncertainty was actually a novelty that did not receive immediate adhesion by scientists and that, without any doubt, aroused from the today disappeared field of cybernetics which is the ascent of the successful current information technologies. As Zadeh likes to say, fuzzy logic is a matter of degree, but it should not be forgotten that applying fuzzy logic is also a matter of design; that all the intervening imprecise terms should be carefully designed according with their conceptual use, and that for this to count with mathematical models of relations, connectives, hedges, antonyms, negations, etc., is very important.

Zadeh, a very creative engineer and scientist, not only introduced the basic ideas of fuzzy sets and fuzzy logic, but pushed ahead with his own work its application to many fields, ranging from expert systems and control engineering technology to economy or medicine; as something not very usual, during his large life he is contemplating the fertility of his ideas in fields different to those in which he did initially expected its usefulness. Fertility, and mainly in other fields, is what Karl Menger qualified as the success of a scientific concept.

As many creative new ideas, fuzzy logic was received with some content by many scientists; for example, I remember how an important mathematician, by my part a loved person and still in the first eighties, asked me if I was expecting to 'resolve the world' with just the unit interval! These scientists were closed in their own world and forgot that before Zadeh, John von Neumann did ask for the introduction of mathematical analysis into the study of 'automata,' or that the great geometer Karl Menger did introduce the probabilistic 'hazy sets' that can be seen as a precedent of fuzzy sets. They even forgot that, in language and reasoning, too much precision often conducts to a loosing of meaning.

The introduction of mathematical analysis in the study of language was done in fuzzy logic after the first papers of Zadeh, and thanks to the work of some mathematicians. Indeed, and from the very beginning of fuzzy logic, several (then young) mathematicians did enter into the new field and contributed to its theoretic development with the study of some of its ideas like it was, for instance, the analysis of the several ways of mathematically representing the basic 'If/then' conditionals that are relevant for the modeling of the important rule-based systems, that experts established for linguistically describing the behavior of some dynamical systems. This task was continued by mathematicians along the world, and today it is easy to recognize their work in, for instance, the mathematic contributions made in the study of t-norms, negations and antonyms, aggregation functions, implication functions, indistinguishability (or fuzzy equivalence) relations, fuzzy integrals, fuzzy graphs.

In my own case, after being mathematically educated in a strict 'bourbakism,' typical of the fifties and sixties of the twentieth Century, and coming from a postdoctoral study of probabilistic metric spaces, my interest in fuzzy logic was a partial consequence of feeling that mathematics, and at least its teaching, was going too far from the real world problems when most of the knowledge fixed in its corpus historically comes from questions posed in engineering, natural sciences, financial sciences, social sciences, etc. That the secession between mathematics and real world was not actually good neither for mathematics, nor for real world problems; at the end, and as Galileo said, the book of nature is written in the language of mathematics, and at least from Newton, the scientific study of the world cannot be made without mathematics. Natural language and ordinary reasoning are also in the world, and rationality (the old Greek 'Logos') deserves to be scientifically studied.

Along the years, each time I am more convinced that if something called 'fuzzy mathematics' has no sense, what is full of sense is to conduct studies on new mathematical concepts that can, eventually, facilitate some new instruments for the study of both linguistic imprecision and nonrandom uncertainty; instruments that can help penetrating into the 'logos.' This book is a nice example of this kind of mathematical contributions that flourished around Zadeh's initial ideas, namely, fuzzy graphs, that were introduced by the late Prof. Azriel Rosenfeld as early as in 1975.

If directed graphs grounded on a classical set and with graduate arcs are not only well known, but widely used in many applications, graphs defined on a fuzzy set in its ground have interesting applications in the many cases in which elements in such ground are affected by some imprecision allowing its graduation. Fuzzy graphs translate into fuzzy sets the graduate relational concept of a numerically valued graph and reduce them when the fuzzy set is but a crisp one; they still deserve more applications than those that have been done and, for instance, in the mathematical modeling of linguistic graded relationships between imprecise statements.

This book by Sunil Mathew, John N. Mordeson, and Devender S. Malik can contribute to both help continuing the mathematical development of fuzzy graph theory and to expand more its use in the applications; actually, the book comes to complement what is currently available in the literature concerning fuzzy graphs, like the two precedent books coauthored by John N. Mordeson.

Were only for this fact the book already deserves to be welcomed, but, in addition, it not only contains most of all that has been published on fuzzy graphs, including the important contributions made by its authors, but presents all in a mathematically well organized and clearly understandable form.

The three authors should be congratulated by the excellent presentation of the theory of fuzzy graphs they reached with this book that, I am sure, will receive careful attention and will be appreciated by both the specialists, as well as by those aiming at applying the big collection of ideas on fuzzy graphs it contains.

Oviedo (Asturias), Spain Enric Trillas

# Preface

Lotfi Zadeh introduced the concept of a fuzzy subset of a set in 1965 as a way to represent uncertainty. His ideas have motivated the interest of researchers worldwide. One such researcher was Azriel Rosenfeld. He was one of the fathers of fuzzy graph theory. His development of the concept of a fuzzy graph provides the motivation of this book and the research it contains.

The book deals with current ideas in fuzzy graphs. It is not an attempt to provide an exhaustive study. There are individual topics in fuzzy graphs that would provide enough material for an entire book in them. Still it covers most of the major developments in fuzzy graph theory during the period 1975–2017. The book should be of interest to research mathematicians, computer scientists, and social scientists. It is the first volume of a two volume set. The second volume focuses on the application of fuzzy graph theory to the problem of human trafficking.

Some of the material in this book has appeared in [127]. We include it here since the development of the book rests on it.

We provide in Chap. 1 only the very basics of fuzzy set theory needed to understand the book. We assume the reader is familiar with basic notions of mathematics including set theory. Since this book is designed primarily for researchers with a knowledge of fuzzy set theory, we only provide a few concepts from fuzzy sets and relations mainly to set forth our notation to be used in the book.

In Chap. 2, we present basic concepts of fuzzy graphs which are needed later in the chapter and in the remainder of the book. For example, we introduce and present basic results on paths, connectedness, forests, trees, and fuzzy cutsets. Other basic concepts include bridges, cutsets, and blocks. We examine the connection between cycles and fuzzy trees. We present deeper results on blocks and in fact give a characterization of blocks in fuzzy graphs. We examine special types of cycles such as strong cycles and locamin cycles. We then present results on important types of fuzzy graphs such as fuzzy line graphs and fuzzy interval graphs. The study of fuzzy interval graphs includes the fuzzy analog of Marczewski's theorem, the Gilmore and Hoffman characterization, and the Fulkerson and Gross characterization. The chapter is concluded with the development of certain operations on fuzzy graphs such as the Cartesian cross product, the composition, union, and join of two fuzzy graphs.

In Chap. 3, we focus on the connectivity of fuzzy graphs. We describe various types of edges in fuzzy graphs with respect to connectivity properties and characterize different fuzzy graph structures. We then consider vertex connectivity and edge connectivity. We provide generalized versions of connectivity parameters introduced by Yeh and Bang in 1975. Menger's theorem is very celebrated result in graph theory. We present a version of Menger's theorem for fuzzy graphs.

Chapter 4 develops further results involving blocks. An application involving undirected network of roads is given. Attention is then turned to critical blocks and block graphs. Connectivity-transitive and cyclically transitive fuzzy graphs are examined next.

Chapter 5 starts by considering connectedness and acyclic levels. A new measure of connectivity of fuzzy graphs, called cycle connectivity, and two different types of bridges, called bonds and cutbonds, are discussed. Various metrics are also examined. Attention is also given to detour distance in fuzzy graphs.

In Chap. 6, the notion of a sequence in fuzzy graphs is introduced. Most of the fuzzy graph structures are characterized using different types of sequences. The notion of saturation in fuzzy graphs is also introduced. The chapter concludes with a study of strong intervals and strong gates in fuzzy graphs.

In Chap. 7, we present the work on interval-valued fuzzy graphs that is mostly due to Akram. It includes results concerning the operations, Cartesian product, composition, union, and join of fuzzy interval graphs. Other results deal with isomorphisms, complete, and self-complementary interval-valued fuzzy graphs. The important work of Craine on fuzzy interval graphs appears in Chap. 2.

In Chap. 8, we present the work on bipolar fuzzy graphs that is mostly due to Akram. This work includes results on operations of bipolar fuzzy graphs similar to the results in Chaps. 2 and 7. Results concerning isomorphisms of bipolar fuzzy graphs as well as results concerning strong and regular bipolar fuzzy graphs are provided. The chapter concludes with the work by Mathew and others on connectivity concepts of bipolar fuzzy graphs.

The authors are thankful to everyone who supported this project. It is our hope that this book will help students and researchers all over the globe to learn and to apply fuzzy graph theory. We welcome all suggestions and comments from everyone so that we can improve this book as a useful resource for students, teachers, scientists, and engineers for addressing the challenges of today's world.

**Acknowledgements** We thank Dr. George Haddix and his wife Sue for their generous support of our endeavors involving mathematics of uncertainty. We also thank the journal New Mathematics and Natural Computation for allowing us to reuse some of our work.

Kozhikode, India Sunil Mathew
Omaha, USA John N. Mordeson
Omaha, USA Davender S. Malik

# Contents

# About the Authors

**Dr. Sunil Mathew** is currently a Faculty Member in the Department of Mathematics, NIT Calicut, India. He has acquired his masters from St. Joseph's College Devagiri, Calicut, and Ph.D. from National Institute of Technology Calicut in the area of Fuzzy Graph Theory. He has published more than 75 research papers and written two books. He is a member of several academic bodies and associations. He is editor and reviewer of several international journals. He has an experience of 20 years in teaching and research. His current research topics include fuzzy graph theory, bio-computational modeling, graph theory, fractal geometry, and chaos.

**Dr. John N. Mordeson** is Professor Emeritus of Mathematics at Creighton University. He received his B.S., M.S., and Ph.D. from Iowa State University. He is a Member of Phi Kappa Phi. He is the President of the Society for Mathematics of Uncertainty. He has published 15 books and 200 journal articles. He is on the editorial board of numerous journals. He has served as an external examiner of Ph.D. candidates from India, South Africa, Bulgaria, and Pakistan. He has refereed for numerous journals and granting agencies. He is particularly interested in applying mathematics of uncertainty to combat the problem of human trafficking.

**Dr. Davender S. Malik** is a Professor of Mathematics at Creighton University. He received his Ph.D. from Ohio University and has published more than 55 papers and 18 books on abstract algebra, applied mathematics, graph theory, fuzzy automata theory and languages, fuzzy logic and its applications, programming, data structures, and discrete mathematics.

# Chapter 1
# Fuzzy Sets and Relations

The notion of a fuzzy graph was initially introduced by Kauffman in [91]. However, the development of fuzzy graph theory is due to the ground setting papers of Rosenfeld [154] and Yeh and Bang [186]. In Rosenfeld's paper, basic structural and connectivity concepts were presented while Yeh and Bang introduced different connectivity parameters and discussed their application. Rosenfeld obtained the fuzzy analogs of several graph-theoretic concepts like bridges, paths, cycles, trees, and connectedness. Most of the theoretical development of fuzzy graph theory is based on Rosenfeld's initial work.

Fuzzy graph theory is finding more and more applications. Applications can be found in cluster analysis, pattern classification, data base theory, social sciences, neural networks, decision analysis, group structure, portfolio management, and many other areas [128].

## 1.1 Fuzzy Sets

Probability theory, considered as the only theory to deal with uncertainty till the middle of the 20th century was based on the well known YES or NO logic, known as the two valued logic of Aristotle. Aristotelian logic provides the information, whether an element belongs to a set or not. Most of the human inventions like the switch, computer, vehicles, and so on are based on this logic. In 1965, Lotfi A Zadeh, [190] in his seminal paper introduced a new type of set called a fuzzy set and a new logic later known as fuzzy logic. Zadeh's line of thought was different. Instead of YES or NO, regarding the existence of an element in a set, he used the degree of membership, which allows an element to exist in a set with partial grades of memberships. The applications of fuzzy logic are profound and widespread. Artificial intelligence, electronics, transportation, robotics and pattern recognition are some of the major areas that use fuzzy logic.

There are a large number of books available in both theory and applications of fuzzy sets and logic. In the first chapter, we present only the basic concepts from



S. Mathew et al., *Fuzzy Graph Theory*, Studies in Fuzziness
and Soft Computing 363, https://doi.org/10.1007/978-3-319-71407-3_1

fuzzy set theory needed for this book. We assume that the reader is knowledgeable on the basics of set theory. However, we first must set our notation.

$\mathbb{Z}$ denotes the integers

$\mathbb{R}$ denotes the set of all real numbers

$\mathbb{N}$ denotes the set of positive integers or natural numbers

We let $\wedge$ denote minimum or infimum and $\vee$ denote maximum or supremum.

Let $A$ and $B$ be subsets of a universal set $U$. We write $A \subseteq B$ if $A$ is a subset of $B$ or equivalently $B \supseteq A$ if $B$ contains $A$. We write $A \subset B$ if $A \subseteq B$ and there exists $x \in B$ such that $x \notin A$. The intersection of $A$ and $B$ is denoted by $A \cap B$ and the union of $A$ and $B$ is denoted by $A \cup B$. We let $B \backslash A$ denote the set difference of $A$ in $B$. Thus, $B \backslash A = \{x \in B \mid x \notin A\}$. If $B = U$, then we write $A^c$ for $U \backslash A$ and call $A^c$ the **complement** of $A$.We let $A \times B$ denote the Cartesian cross product of $A$ and $B$.

We let $\emptyset$ denote the empty set.

**Definition 1.1.1** Let $X$ be a set. A **fuzzy subset** of $X$ is a function from $X$ into the closed interval [0, 1].

We can interpret a fuzzy subset $\mu$ of a set $X$ as giving the membership degree of every element of $X$ in some "subset" of $X$, i.e, given in some descriptive manner. For example, $A$ might be the set of all young people in a set of people $X$. Of course, $A$ is not been well-defined here but we might let $\mu$ assign to every member of $X$ a value $t$ from [0, 1] in such away that $t$ represents the membership degree of $x$ in $A$, i.e., a measure of "youngness." We shall use the term **fuzzy set**, hereafter to denote fuzzy subset of a set, if there is no confusion regarding the underlying set. We let $\mathfrak{FP}(X)$ denote the set of all fuzzy subsets of $X$. We call $\mathfrak{FP}(X)$ the **fuzzy power set** of $X$.

A basic concept of fuzzy set theory is the $t$-cut of a fuzzy subset $\mu$ of a set $X$, where $t \in [0, 1]$. The **$t$-cut** or **$t$-level set** of a fuzzy subset $\mu$ of $X$ is defined to be $\mu^t = \{x \in X \mid \mu(x) \geq t\}$, where $t \in [0, 1]$. The **strong $t$-cut** of $\mu$ is defined as $\mu^{t+} = \{x \in X \mid \mu(x) > t\}$. The support of $\mu$ is defined to be $\text{Supp}(\mu) = \{x \in X \mid \mu(x) > 0\}$. We write $\mu^*$ for $\text{Supp}(\mu)$ at times. Clearly a $t$-cut of a fuzzy set is a crisp set and the support is indeed a strong $t$-cut. The 1-cut is usually termed as the **core** of the fuzzy set $\mu$. The height $h(\mu)$ of fuzzy set $\mu$ is the largest membership value obtained by an element in the fuzzy set. A fuzzy set is said to be **normal** if $h(\mu) = 1$ and **subnormal** if $h(\mu) < 1$. The height of $\mu$ may be defined alternately as the supremum of all $t$ such that $\mu^t \neq \emptyset$.

*Example 1.1.2* Consider the fuzzy subset $\mu$ of $\mathbb{R}$ given in Fig. 1.1. Here $\mu^* = (0, 3)$, $\mu^1 = [1, 2]$. Thus, the core of the fuzzy set $\mu$ is $[1, 2]$. Also, $h(\mu) = 1$ and hence it is a normal fuzzy set.

*Example 1.1.3* Consider the set $X = \{1, 2, 3, \ldots, 10\}$. Define the fuzzy subset $\mu$ of $X$, representing 'real numbers near to 5' as $\{(1, 0.1), (2, 0.3), (3, 0.6), (4, 0.8), (5, 1), (6, 0.8), (7, 0.6), (8, 0.3), (9, 0.1), (10, 0)\}$. Because $\mu(5) = 1$, $\mu$ is a normal fuzzy subset of $X$. Also, note that $10 \notin \mu^*$. Now $\mu^{0.5} = \mu^{0.5+} = \{3, 4, 5, 6, 7\}$, $\mu^{0.6} = \{3, 4, 5, 6, 7\}$ and $\mu^{0.6+} = \{4, 5, 6\}$.

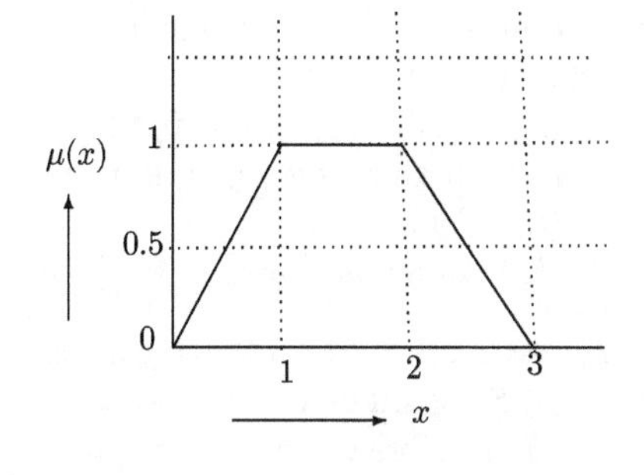


**Fig. 1.1** A fuzzy subset of the real line

**Definition 1.1.4** A fuzzy subset $\mu$ of a set $X$ is said to be **convex** if $\mu(\lambda x_1 + (1-\lambda)x_2) \geq \mu(x_1) \wedge \mu(x_2)$ for all $x_1, x_2 \in X$ and for all $\lambda \in [0, 1]$.

Clearly a fuzzy set is convex if all its $t$-cuts are convex.

**Definition 1.1.5** For a finite fuzzy subset $\mu$ of $X$, the **cardinality** of $\mu$ is defined as $|\mu| = \sum_{x \in X} \mu(x)$. The **relative cardinality** of $\mu$ is defined as $\|\mu\| = |\mu| \, / \, |X|$.

We next define some set theoretical operations for fuzzy sets. Let $\mu$ and $\nu$ be fuzzy subsets of a set $X$. We write $\mu \subseteq \nu$ if for all $x \in X$, $\mu(x) \leq \nu(x)$. If $\mu \subseteq \nu$ and there exists $x \in X$ such that $\mu(x) < \nu(x)$, we write $\mu \subset \nu$. We define $\mu \cap \nu$ by for all $x \in X$, $(\mu \cap \nu)(x) = \mu(x) \wedge \nu(x)$. We define $\mu \cup \nu$ by for all $x \in X$, $(\mu \cup \nu)(x) = \mu(x) \vee \nu(x)$.

The notion of intersection for fuzzy sets can also be defined by use of a variety of $t$-norms.

**Definition 1.1.6** A function $i : [0, 1] \times [0, 1] \to [0, 1]$ is called a $t$**-norm** if it satisfies the following conditions:

$(i)$ For all $x \in [0, 1]$, $i(1, x) = x$.
$(ii)$ For all $x, y \in [0, 1]$, $i(x, y) = i(y, x)$.
$(iii)$ For all $x, y, z \in [0, 1]$, $i(x, i(y, z)) = i(i(x, y), z)$.
$(iv)$ For all $w, x, y, z \in [0, 1]$, $w \leq x$ and $y \leq z$ implies $i(w, y) \leq i(x, z)$.

*Example 1.1.7* The following are examples of $t$-norms: for all $x, y \in [0, 1]$,

$(i)$ $i(x, y) = \begin{cases} x \wedge y & \text{if } x \vee y = 1, \\ 0 & \text{otherwise} \end{cases}$
$(ii)$ $i(x, y) = 0 \vee (x + y - 1)$
$(iii)$ $i(x, y) = \frac{xy}{2-(x+y-xy)}$
$(iv)$ $i(x, y) = xy$
$(v)$ $i(x, y) = x \wedge y$.

The $t$-norm in $(v)$ is often called the **standard intersection** for fuzzy subsets.

The notion of union for fuzzy sets can be defined by a variety of $t$-conorms as defined next.

**Definition 1.1.8** A function $u : [0,1] \times [0,1] \to [0,1]$ is called a ***t*-conorm** if it satisfies the following conditions:

$(i)$ For all $x \in [0,1]$, $u(0,x) = x$.

$(ii)$ For all $x, y \in [0,1]$, $u(x,y) = u(y,x)$.

$(iii)$ For all $x, y, z \in [0,1]$, $u(x, u(y,z)) = u(u(x,y), z)$.

$(iv)$ For all $w, x, y, z \in [0,1]$, $w \leq x$ and $y \leq z$ implies $u(w,y) \leq u(x,z)$.

*Example 1.1.9* The following are examples of $t$-conorms:

$(i)$ Standard union: $u(x,y) = x \vee y$

$(ii)$ Algebraic sum: $u(x,y) = x + y - xy$

$(iii)$ Bounded sum: $u(x,y) = 1 \wedge (x+y)$

$(iv)$ Drastic union: $u(x,y) = \begin{cases} x & \text{if } y = 0, \\ y & \text{if } x = 0, \\ 1 & \text{otherwise.} \end{cases}$

Next we introduce the notion of a complement.

**Definition 1.1.10** A function $c : [0,1] \to [0,1]$ is called a **fuzzy complement** if the following conditions hold:

$(i)$ $c(0) = 1$ and $c(1) = 0$.

$(ii)$ For all $x, y \in [0,1]$, $x \leq y$ implies $c(x) \geq c(y)$.

A desirous property for a fuzzy complement $c$ to possess is continuity. Another is that it be **involutive**, i.e., for all $x \in [0,1]$, $c(c(x)) = x$. An example of a fuzzy complement is the standard complement, i.e., $c(x) = 1 - x$ for all $x \in [0,1]$.

In classical set theory, the operations of intersection and union are dual with respect to complement in the sense that they satisfy De Morgan's laws,

$$(A \cap B)^c = A^c \cup B^c \text{ and } (A \cup B)^c = A^c \cap B^c$$

for subsets $A$ and $B$ of some universe.

For fuzzy subsets, De Morgan's laws become

$$c(i(a,b)) = u(c(a), c(b)) \text{ and } c(u(a,b)) = i(c(a), c(b)).$$

It can be easily shown that if $c$ is the standard complement, then the standard intersection and the standard union are duals with respect to $c$ as are many other pairs.

*Example 1.1.11* Standard union and intersection are illustrated in this example. Consider the fuzzy subsets $\sigma$ and $\mu$ of $\{1,2,3,4,5,6,7,8\}$ defined as follows.

$$\sigma = \{(1,0.1), (2,0.5), (3,0.8), (4,1), (5,0.8), (6,0.5), (7,0.1), (8,0.0)\}$$

and

$$\mu = \{(1,0.0), (2,0.0), (3,0.2), (4,0.4), (5,0.6), (6,0.8), (7,1), (8,1)\}.$$

Then

$$\sigma \cup \mu = \{(1, 0.1), (2, 0.5), (3, 0.8), (4, 1), (5, 0.8), (6, 0.8), (7, 1), (8, 1)\}$$

and

$$\sigma \cap \mu = \{(3, 0.2), (4, 0.4), (5, 0.6), (6, 0.5), (7, 0.1)\}.$$

## 1.2 Fuzzy Relations

As crisp relations represent the association between elements of two or more sets, a fuzzy relation gives the extent of relationship between elements between two fuzzy sets. Zadeh [190] introduced fuzzy relations in 1965. Later, Zadeh [192], Kaufman, [91] and Rosenfeld [154] developed significant results. There are several applications for fuzzy relations. We have only a theoretical discussion about fuzzy relations in this section. We provide a formal definition below. Most of the contents of this section are based on Rosenfeld's work in 1975 [154].

If $S$ represents a set, a fuzzy relation $\mu$ on $S$ is a fuzzy subset of $S \times S$. In symbols, $\mu : S \times S \to [0, 1]$ such that $0 \leq \mu(x, y) \leq 1$ for all $(x, y) \in S \times S$. When $\mu$ takes the values 0 and 1 alone, it becomes the characteristic function of a relation on $S$. If $R$ is a subset of $S$ and $P$ is a relation on $S$, then $P$ becomes a relation on $R$ only if $(x, y) \in P$ implies $x \in R$ and $y \in R$. If $\zeta$ and $\eta$ are the characteristic functions of $R$ and $P$ respectively, then $\eta(x, y) = 1$ implies $\zeta(x) = \zeta(y) = 1$ for all $x, y \in R$. This is equivalent to the expression $\eta(x, y) \leq \zeta(x) \wedge \zeta(y)$ for all $x, y \in R$. Motivated by this, we have the definition of a fuzzy relation on a fuzzy subset as follows.

**Definition 1.2.1** Let $\sigma$ be a fuzzy subset of a set $S$ and $\mu$ a fuzzy relation on $S$. Then $\mu$ is called a **fuzzy relation** on $\sigma$ if $\mu(x, y) \leq \sigma(x) \wedge \sigma(y)$ for all $x, y \in S$.

**Definition 1.2.2** If $S$ and $T$ are two sets and $\sigma$ and $\tau$ are fuzzy subsets of $S$ and $T$, respectively, then a fuzzy relation $\mu$ from the fuzzy subset $\sigma$ into the fuzzy subset $\tau$ is a fuzzy subset $\mu$ of $S \times T$ such that $\mu(x, y) \leq \sigma(x) \wedge \tau(y)$ for all $x \in S$ and $y \in T$.

It is interesting to see that for $\mu$ to be a fuzzy relation, the degree of membership of a pair of elements never exceeds the degree of membership of either of the elements. Later, while defining a fuzzy graph, this inequality allows us to organize the flow through an edge of a fuzzy graph in such a way that, it never exceeds the capacities of its end vertices. Also, $\mu^\alpha$ is a relation from $\sigma^\alpha$ into $\tau^\alpha$ for all $\alpha \in [0, 1]$ and as a consequence, $\mu^*$ becomes a relation from $\sigma^*$ into $\tau^*$.

In Definition 1.2.2, if $\sigma(x) = 1$ for all $x \in S$ and $\tau(y) = 1$ for all $y \in T$, then $\mu$ is called a fuzzy relation from $S$ into $T$. Similarly, if $\sigma(x) = 1$ for all $x \in S$ in Definition 1.2.1, $\mu$ is said to be a fuzzy relation on $S$.

**Definition 1.2.3** If $\sigma$ is a fuzzy subset of a set $S$, the **strongest fuzzy relation** on $\sigma$ is the fuzzy relation $\mu_\sigma$ defined by $\mu_\sigma(x, y) = \sigma(x) \wedge \sigma(y)$ for all $x, y \in S$.

**Definition 1.2.4** For a fuzzy relation $\mu$ on $S$, the **weakest fuzzy subset** of $S$, on which $\mu$ is a fuzzy relation is $\sigma_\mu$, defined by $\sigma_\mu(x) = \vee_{y \in S}(\mu(x, y) \vee \mu(y, x))$ for all $x \in S$.

**Definition 1.2.5** Let $\mu : S \times T \to [0, 1]$ be a fuzzy relation from a fuzzy subset $\sigma$ of $S$ into a fuzzy subset $\tau$ of $T$ and $\nu : T \times U \to [0, 1]$ is a fuzzy relation from the fuzzy subset $\rho$ of $T$ into a fuzzy subset $\eta$ of $U$. Define $\mu \circ \nu : S \times U \to [0, 1]$ by $(\mu \circ \nu)(x, z) = \vee\{\mu(x, y) \wedge \nu(y, z) \mid y \in T\}$ for all $x \in S, z \in U$. Then $\mu \circ \nu$ is called the **max–min composition** of $\sigma$ and $\tau$.

The composition of any two fuzzy relations as in Definition 1.2.5 is always a fuzzy relation. But in the next result, we only consider two fuzzy relations defined on the same fuzzy set.

**Proposition 1.2.6** *If $\mu$ and $\nu$ are fuzzy relations on a fuzzy set $\sigma$, then $\mu \circ \nu$ is a fuzzy relation on $\sigma$.*

*Proof* Let $S$ be a set and $\sigma$ be a fuzzy subset of $S$. Because $\mu$ and $\nu$ are fuzzy relations on $\sigma$, $\mu(x, y) \leq \sigma(x) \wedge \sigma(y)$ and $\nu(y, z) \leq \sigma(y) \wedge \sigma(z)$ for all $x, y, z \in S$. Thus, $\mu(x, y) \wedge \nu(y, z) \leq \sigma(x) \wedge \sigma(y) \wedge \sigma(z) \leq \sigma(x) \wedge \sigma(z)$ for all $y \in S$ and hence, $(\mu \circ \nu)(x, z) = \vee_{y \in S}(\mu(x, y) \wedge \nu(y, z)) \leq \sigma(x) \wedge \sigma(z)$ for all $x, z \in S$. ■

Max–min composition is similar to matrix multiplication, where addition is replaced by $\vee$ and multiplication by $\wedge$. We can easily show that the composition of fuzzy relations is associative. So if we denote $\mu \circ \mu$ by $\mu^2$, higher powers of the fuzzy relation $\mu^2, \mu^3$, and so on, can be easily defined. Define $\mu^\infty(x, y) = \vee\{\mu^k(x, y) \mid k = 1, 2, \ldots\}$ for all $x, y \in S$. Also, define $\mu^0(x, y) = 0$ if $x \neq y$ and $\mu^0(x, x) = \mu(x, x)$ otherwise.

**Proposition 1.2.7** *If $\mu$ and $\nu$ are two fuzzy relations on a finite set $S$, then for all $t \in [0, 1]$, we have $(\mu \circ \nu)^t = \mu^t \circ \nu^t$.*

*Proof* Let $(x, z) \in (\mu \circ \nu)^t$. Then $(\mu \circ \nu)(x, z) \geq t$. By definition, $\mu(x, y) \wedge \nu(y, z) \geq t$ for some $y \in S$. Therefore, $\mu(x, y) \geq t$ and $\nu(y, z) \geq t$, which implies $(x, y) \in \mu^t$ and $(y, z) \in \nu^t$. Thus, $(x, z) \in \mu^t \circ \nu^t$ by the definition of composition of functions. Hence, $(\mu \circ \nu)^t \subseteq \mu^t \circ \nu^t$. Similarly, $\mu^t \circ \nu^t \subseteq (\mu \circ \nu)^t$. Thus, $(\mu \circ \nu)^t = \mu^t \circ \nu^t$. ■

**Proposition 1.2.8** *Suppose $\mu, \nu, \lambda, \rho$ are fuzzy relations defined on a fuzzy subset $\sigma$ of $S$. If $\mu \subseteq \nu$ and $\lambda \subseteq \rho$, then $\mu \circ \lambda \subseteq \nu \circ \rho$.*

*Proof* We have $(\mu \circ \lambda)(x, z) = \vee_{y \in S}(\mu(x, y) \wedge \lambda(y, z)) \leq \vee_{y \in S}(\nu(x, y) \wedge \rho(y, z)) = (\nu \circ \rho)(x, z)$ for all $x, z \in S$. ■

Note that there are several types of composition of fuzzy relations that are available in the literature [154]. Next we have a unary operation on a fuzzy relation.

**Definition 1.2.9** Let $\mu$ be a fuzzy relation defined on a fuzzy subset $\sigma$ of a set $S$. Then the **compliment** $\mu^c$ of $\mu$ is defined as $\mu^c(x, y) = 1 - \mu(x, y)$ for all $x, y \in S$.

**Definition 1.2.10** Let $\mu : S \times T \to [0, 1]$ be a fuzzy relation from a fuzzy subset $\sigma$ of $S$ into a fuzzy subset $\nu$ of $T$. Then $\mu^{-1} : T \times S \to [0, 1]$, the **inverse** of $\mu$ from $\nu$ into $\sigma$ is defined as $\mu^{-1}(y, x) = \mu(x, y)$ for all $(y, x) \in T \times S$.

Some of the properties of fuzzy relations are given in the following result. Their proofs are omitted as they are obvious.

**Theorem 1.2.11** *Let $\tau$, $\pi$, $\rho$ and $\nu$ be a fuzzy relations on a fuzzy subset $\sigma$ of a set $S$. Then the following properties hold.*
$(i)$ $\tau \cup \pi = \pi \cup \tau$.
$(ii)$ $\tau \cap \pi = \pi \cup \tau$.
$(iii)$ $(\tau^c)^c = \tau$.
$(iv)$ $\pi \cup (\rho \cup \nu) = (\pi \cup \rho) \cup \nu$.
$(v)$ $\pi \cap (\rho \cap \nu) = (\pi \cap \rho) \cap \nu$.
$(vi)$ $\pi \circ (\rho \circ \nu) = (\pi \circ \rho) \circ \nu$.
$(vii)$ $\pi \cap (\rho \cup \nu) = (\pi \cap \rho) \cup (\pi \cap \nu)$.
$(viii)$ $\pi \cup (\rho \cap \nu) = (\pi \cup \rho) \cap (\pi \cup \nu)$.
$(ix)$ $(\tau \cup \pi)^c = \pi^c \cap \tau^c$.
$(x)$ $(\tau \cap \pi)^c = \pi^c \cup \tau^c$.
$(xi)$ *For every* $t \in [0, 1]$, $(\tau \cup \pi)^t = \tau^t \cup \pi^t$.
$(xii)$ *For every* $t \in [0, 1]$, $(\tau \cap \pi)^t = \tau^t \cap \pi^t$.
$(xiii)$ *If* $\tau \subseteq \rho$ *and* $\pi \subseteq \nu$, *then* $\tau \cup \pi \subseteq \rho \cup \nu$.
$(xiv)$ *If* $\tau \subseteq \rho$ *and* $\pi \subseteq \nu$, *then* $\tau \cap \pi \subseteq \rho \cap \nu$.

**Definition 1.2.12** Let $\mu$ be a fuzzy relation on $\sigma$, where $\sigma$ is a fuzzy subset of a set $S$. Then $\mu$ is said to be **reflexive** if $\mu(x, x) = \sigma(x)$ for all $x \in S$.

When $\mu$ is a reflexive fuzzy relation on $\sigma$, it is not hard to see that $\mu(x, y) \leq \sigma(x) = \mu(x, x)$ and $\mu(y, x) \leq \sigma(x) = \mu(x, x)$ for all $x, y \in S$. In other words, when we express a fuzzy relation in a matrix form, the elements of any row or any column will be less than or equal to the diagonal element belonging to that row or column. Sometimes we say $\mu$ is reflexive on $\sigma$. Next we have some interesting properties of reflexive fuzzy relations.

**Theorem 1.2.13** *Let $\mu$ and $\nu$ be fuzzy relations on a fuzzy subset $\sigma$ of a set $S$. If $\mu$ is reflexive, then $\nu \subseteq \nu \circ \mu$ and $\nu \subseteq \mu \circ \nu$.*

*Proof* Let $x, z \in S$. Then $(\mu \circ \nu)(x, z) = \vee\{\mu(x, y) \wedge \nu(y, z) \mid y \in S\} \geq \mu(x, x) \wedge \nu(x, z) = \sigma(x) \wedge \nu(x, z)$. But $\nu(x, z) \leq \sigma(x) \wedge \sigma(z)$. Therefore, $\sigma(x) \wedge \nu(x, z) = \nu(x, z)$. Thus, $\nu \subseteq \mu \circ \nu$. Similarly, we can prove that $\nu \subseteq \nu \circ \mu$. ■

**Corollary 1.2.14** *If $\mu$ is reflexive, then $\mu \subseteq \mu^2$.*

**Corollary 1.2.15** *If $\mu$ is reflexive, then $\mu^0 \subseteq \mu \subseteq \mu^2 \subseteq \mu^3 \subseteq \cdots \subseteq \mu^\infty$.*

The proofs of Corollaries 1.2.14 and 1.2.15 can be obtained by taking $\nu$ as $\mu$, $\mu^2$, and so on in the proof of Theorem 1.2.13.

**Theorem 1.2.16** *Let $\mu$ be a fuzzy relation on a fuzzy subset $\sigma$ of a set $S$. If $\mu$ is reflexive, $\mu^0(x, x) = \mu(x, x) = \mu^2(x, x) = \mu^3(x, x) = \cdots = \mu^\infty(x, x) = \sigma(x)$ for all $x \in S$.*

*Proof* We have $\mu(x, x) = \sigma(x)$ for all $x \in S$. Assume that the result is true for $k = n$. That is, $\mu^n(x, x) = \sigma(x)$, for all $x \in S$. Now, for all $x \in S$, we have $\mu^{n+1}(x, x) = \vee\{\mu(x, y) \wedge \mu^n(y, x) \mid y \in S \leq \vee\{\sigma(x) \wedge \sigma(x) \mid y \in S\} = \sigma(x)$. Also, $\mu^{n+1}(x, x) = \vee\{\mu(x, y) \wedge \mu^n(y, x) \mid y \in S\} \geq \mu(x, x) \wedge \mu^n(x, x) = \sigma(x)$. Thus, $\mu^{n+1}(x, x) = \sigma(x)$ for all $x \in S$. ■

**Theorem 1.2.17** *If $\mu$ and $\nu$ reflexive fuzzy relation on $\sigma$, then $\mu \circ \nu$ and $\nu \circ \mu$ are also reflexive.*

*Proof* $(\mu \circ \nu)(x, x) = \vee\{\mu(x, y) \wedge \nu(y, x) \mid y \in S\} \leq \vee\{\sigma(x) \wedge \sigma(x) \mid y \in S\} = \sigma(x)$ and $(\mu \circ \nu)(x, x) = \vee\{\mu(x, y) \wedge \nu(y, x) \mid y \in S\} \geq \mu(x, x) \wedge \nu(x, x) = \sigma(x) \wedge \sigma(x) = \sigma(x)$. The proof that $\nu \circ \mu$ is reflexive is similar. ■

**Theorem 1.2.18** *If $\mu$ is reflexive on $\sigma$, then $\sigma^t$ is reflexive on $\sigma^t$ for all $t \in [0, 1]$.*

*Proof* Suppose $\mu$ is reflexive. Let $x \in \sigma^t$. Then $\mu(x, x) = \sigma(x) \geq t$ and thus $(x, x) \in \mu^t$. ■

**Definition 1.2.19** Let $\mu$ be a fuzzy relation on $\sigma$, where $\sigma$ is a fuzzy subset of a set $S$. Then $\mu$ is said to be **symmetric** if $\mu(x, y) = \mu(y, x)$ for all $x, y \in S$.

From the definition, it follows that if $\mu$ is symmetric, then the matrix representation of $\mu$ is symmetric.

**Theorem 1.2.20** *Let $\mu$ and $\nu$ be fuzzy relations on a fuzzy subset $\sigma$ of a set $S$. Then the following properties hold.*

*$(i)$ If $\mu$ and $\nu$ are symmetric, then $\mu \circ \nu$ is symmetric if and only if $\mu \circ \nu = \nu \circ \mu$.*
*$(ii)$ If $\mu$ is symmetric, then every power of $\mu$ also is symmetric.*
*$(iii)$ If $\mu$ is symmetric, then $\mu^t$ is a symmetric relation on $\sigma^t$ for all $t \in [0, 1]$.*

*Proof* $(i)$ $(\mu \circ \nu)(x, z) = (\mu \circ \nu)(z, x) \Leftrightarrow \vee\{\mu(x, y) \wedge \nu(y, z) \mid y \in S\} = \vee\{\mu(z, y) \wedge \nu(y, x) \mid y \in S\} \Leftrightarrow \vee\{\mu(x, y) \wedge \nu(y, z) \mid y \in S\} = \vee\{\nu(y, x) \wedge \mu(z, y) \mid y \in S\} \Leftrightarrow \mu \circ \nu = \nu \circ \mu$ since $\mu$ and $\nu$ are symmetric.

$(ii)$ Assume that $\mu^n$ is symmetric for $n \in \mathbb{N}$. Then $\mu^{n+1}(x, z) = \vee\{\mu(x, y) \wedge \mu^n(y, z) \mid y \in S\} = \vee\{\mu(y, x) \wedge \mu^n(z, y) \mid y \in S\} = \vee\{\mu^n(z, y) \wedge \mu(y, x) \mid y \in S\} = \mu^{n+1}(z, x)$

$(iii)$ Let $0 \leq t \leq 1$. Suppose $(x, z) \in \mu^t$. Then $\mu(x, z) \geq t$. Because $\mu$ is symmetric, $\mu(z, x) \geq t$. Thus, $(z, x) \in \mu^t$. ■

**Definition 1.2.21** Let $\mu$ be a fuzzy relation on $\sigma$, where $\sigma$ is a fuzzy subset of a set $S$. Then $\mu$ is said to be **transitive** if $\mu^2 \subseteq \mu$.

From the definition, it is clear that for any fuzzy relation $\mu$, $\mu^\infty$ is a transitive fuzzy relation.

**Theorem 1.2.22** *Let $\nu$, $\mu$, and $\tau$ be fuzzy relations on a fuzzy subset $\sigma$ of a set S. Then the following properties hold.*

*$(i)$ If $\mu$ is transitive and $\nu \subseteq \mu$, $\tau \subseteq \mu$, then $\nu \circ \tau \subseteq \mu$.*

*$(ii)$ If $\mu$ is transitive, then so is every power of $\mu$.*

*$(iii)$ If $\mu$ is transitive, $\tau$ is reflexive and $\tau \subseteq \mu$, then $\mu \circ \tau = \tau \circ \mu = \mu$.*

*$(iv)$ If $\mu$ is reflexive and transitive, then $\mu^2 = \mu$.*

*$(v)$ If $\mu$ is reflexive and transitive, then $\mu^0 \subseteq \mu = \mu^2 = \mu^3 = \cdots = \mu^\infty$.*

*Proof* $(i)$ $(\nu \circ \tau)(x, z) = \vee\{\nu(x, y) \wedge \tau(y, z) \mid y \in S\} \leq \vee\{\mu(x, y) \wedge \mu(y, z) \mid y \in S\} = \mu^2(x, z) \leq \mu(x, z)$. Hence, $\nu \circ \tau \subseteq \mu$.

$(ii)$ Assume that $\mu^n$ is transitive. Then $\mu^n \circ \mu^n \subseteq \mu^n$ and $\mu^{n+1} \circ \mu^{n+1} = \mu^{2n+2} = \mu^{2n} \circ \mu^2 \subseteq \mu^n \circ \mu = \mu^{n+1}$.

$(iii)$ In $(i)$, take $\nu$ to be $\mu$. Then $\mu \circ \tau \subseteq \mu$. Also, $(\mu \circ \tau)(x, z) = \vee\{\mu(x, y) \wedge \tau(y, z) \mid y \in S\} \geq \mu(x, z) \wedge \tau(z, z) = \mu(x, z) \wedge \sigma(z) = \mu(x, z)$. That is, $\mu \circ \tau \supseteq \mu$ and hence $\mu \circ \tau = \mu$. Similarly, we can prove that $\tau \circ \mu = \mu$.

$(iv)$ Follows from $(iii)$.

$(v)$ By $(iv)$, $\mu = \mu^2$. Assume that $\mu^n = \mu^{n+1}$ for $n > 1$. Then $\mu^n \circ \mu = \mu^{n+1} \circ \mu$. Hence, $\mu^{n+1} = \mu^{n+2}$. ■

**Theorem 1.2.23** *Let $\nu$, $\mu$, and $\tau$ be fuzzy relations on a fuzzy subset $\sigma$ of a set S. Then the following properties hold.*

*$(i)$ If $\mu$ and $\tau$ are transitive and $\mu \circ \tau = \tau \circ \mu$, then $\mu \circ \tau$ is transitive.*

*$(ii)$ If $\mu$ is symmetric and transitive, then $\mu(x, y) \leq \mu(x, x)$ and $\mu(y, x) \leq \mu(x, x)$ for all $x, y \in S$.*

*$(iii)$ If $\mu$ is transitive, then for any $t \in [0, 1]$, $\mu^t$ is a transitive relation on $\sigma^t$.*

*Proof* $(i)$ $(\mu \circ \tau) \circ (\mu \circ \tau) = \mu \circ (\tau \circ \mu) \circ \tau = \mu \circ (\mu \circ \tau) \circ \tau = \mu^2 \circ \tau^2 \subseteq \mu \circ \tau$. Thus, $\mu \circ \tau$ is transitive.

$(ii)$ Because $\mu$ is transitive, $\mu \circ \mu \subseteq \mu$. Hence, $(\mu \circ \mu)(x, x) \leq \mu(x, x)$. That is, $\vee\{\mu(x, y) \wedge \mu(y, x) \mid y \in S\} \leq \mu(x, x)$. Because $\mu$ is symmetric, $\vee\{\mu(x, y) \wedge \mu(x, y) \mid y \in S\} \leq \mu(x, x)$. Thus, $\mu(x, y) \leq \mu(x, x)$. Because $\mu$ is symmetric, $\mu(y, x) \leq \mu(x, x)$.

$(iii)$ Let $0 \leq t \leq 1$. Let $(x, y), (y, z) \in \mu^t$. Then $\mu(x, y) \geq t$ and $\mu(y, z) \geq t$. Therefore, $\mu(x, z) = \vee\{\mu(x, z) \wedge \mu(w, z) \mid w \in S\} \geq \mu(x, y) \wedge \mu(y, z) \geq t$. Thus, $(x, z) \in \mu^t$. ■

**Definition 1.2.24** A fuzzy relation $\mu$ on a fuzzy subset $\sigma$ of a set $S$ is said to be a **fuzzy equivalence relation** if it is reflexive, symmetric and transitive.

Fuzzy equivalence relations have several applications including those in pattern classification given in [128]. Next we discuss similarity relations introduced by Zadeh in [192]. We present the work of Tamura, Higuchi and Tanaka [173].

**Definition 1.2.25** Let $\mu$ be a fuzzy relation on a set $S$. Then

$(i)$ $\mu$ is $\epsilon$-**reflexive** if for all $x \in S$, $\mu(x, x) \geq \epsilon$, where $\epsilon \in [0, 1]$.

$(ii)$ $\mu$ is **irreflexive** if for all $x \in S$, $\mu(x, x) = 0$.

$(iii)$ $\mu$ is **weakly reflexive** if for all $x, y \in S$ and for all $\epsilon \in [0, 1]$, $\mu(x, y) = \epsilon \Rightarrow \mu(x, x) \geq \epsilon$.

Note that if $\epsilon = 1$, then $\epsilon$-reflexive relation coincides with a reflexive relation.

**Lemma 1.2.26** *If $\mu$ is a fuzzy relation from S into T, then the fuzzy relation $\mu \circ \mu^{-1}$ is weakly reflexive and symmetric.*

*Proof* Now, $(\mu \circ \mu^{-1})(x, x') = \vee\{\mu(x, y) \wedge \mu^{-1}(y, x') \mid y \in T\} \leq \vee\{\mu(x, y) \wedge \mu(x, y) \mid y \in T\} = \vee\{\mu(x, y) \wedge \mu^{-1}(y, x) \mid y \in T\} = (\mu \circ \mu^{-1})(x, x)$. Hence, $\mu \circ \mu^{-1}$ is weakly reflexive.

Also, $(\mu \circ \mu^{-1})(x, x') = \vee\{\mu(x, y) \wedge \mu^{-1}(y, x') \mid y \in T\} = \vee\{\mu^{-1}(y, x) \wedge \mu(x', y) \mid y \in T\} = \vee\{\mu(x', y) \wedge \mu^{-1}(y, x) \mid y \in T\} = (\mu \circ \mu^{-1})(x', x)$. Hence, $\mu \circ \mu^{-1}$ is symmetric. ■

If $\mu$ is a weakly reflexive and symmetric fuzzy relation on a set $S$, we can define a family of non fuzzy subsets $F^{\mu}$ as follows.

$F^{\mu} = \{K \subseteq S \mid (\exists\, 0 < \epsilon \leq 1)(\forall x \in S)[x \in K \Leftrightarrow (\forall x' \in K)\ [\mu(x, x') \geq \epsilon]]\}$

Thus, if we let $F^{\mu}_{\epsilon} = \{K \subseteq S \mid (\forall x \in S)[x \in K \Leftrightarrow (\forall x' \in K)\ [\mu(x, x') \geq \epsilon]]\}$, then $\epsilon_1 \leq \epsilon_2 \Rightarrow F^{\mu}_{\epsilon_2} \preceq F^{\mu}_{\epsilon_1}$, where $\preceq$ denotes a covering relation. That is, every element in $F^{\mu}_{\epsilon_2}$ is a subset of an element in $F^{\mu}_{\epsilon_1}$.

**Definition 1.2.27** Let $\mu$ be a weakly reflexive and symmetric fuzzy relation on a set $S$. A subset $J$ of $S$ is called $\epsilon$- **complete** with respect to $\mu$ if for all $x, x' \in J$, $\mu(x, x') \geq \epsilon$. A **maximal** $\epsilon$-**complete** set is one which is not properly contained in any other $\epsilon$-complete set.

**Lemma 1.2.28** *$F^{\mu}$ is the family of all maximal $\epsilon$ -complete sets with respect to $\mu$ for $0 \leq \epsilon \leq 1$.*

*Proof* Let $K \in F^{\mu}$ and $x, x'' \in K$. Then there exists $0 < \epsilon \leq 1$ such that for all $x' \in K$, $\mu(x, x') \geq \epsilon$. Thus, $\mu(x, x'') \geq \epsilon$. Hence, $K$ is $\epsilon$-complete. Let $J$ be a subset of $S$ such that $K \subseteq J$ and $J$ is $\epsilon$-complete. Let $x \in J$. Because $J$ is $\epsilon$-complete, for all $x' \in K$, $\mu(x, x') \geq \epsilon$. Because $K \in F^{\mu}$, $x \in K$. Thus, $J \subseteq K$. Hence, $K$ is maximal. Now, let $K$ be a maximal $\epsilon$ -complete set. Let $x \in S$. Then clearly $x \in K \Leftrightarrow$ for all $x' \in K$, $\mu(x, x') \geq \epsilon$. Thus, $K \in F^{\mu}$. ■

**Lemma 1.2.29** *Let $\mu$ be a weakly reflexive and symmetric fuzzy relation on a set S. If $\mu(x, x') > 0$, then there is some $\epsilon$-complete set $K \in F^{\mu}$ such that $\{x, x'\} \subseteq K$.*

*Proof* If $x = x'$, then $\{x\}$ is clearly $\epsilon$-complete for $\epsilon = \mu(x, x)$. Suppose $x \neq x'$. Then because $\mu(x, x') = \mu(x', x)$, by symmetry, and $\mu(x, x) \geq \mu(x, x')$ and $\mu(x', x') \geq \mu(x, x')$ by weak reflexivity, we can see that $\{x, x'\}$ is $\epsilon$ -complete, where $\epsilon = \mu(x, x')$. Denote by $C_{\epsilon}$, the family of all $\epsilon$-complete sets $C$ which contain $\{x, x'\}$. Then $C_{\epsilon}$ is not empty because $\{x, x'\} \in C_{\epsilon}$. It follows by Zorn's lemma that $C_{\epsilon}$ has a maximal element $K$. This element is also maximal in the family of all $\epsilon$-complete sets because any set including $K$ must also include $\{x, x'\}$. Hence, $K \in F^{\mu}$ by Lemma 1.2.28. ■

**Lemma 1.2.30** *If $\chi_{\phi}$ represents the characteristic function of $\phi$ in $S \times S$, and if $\mu \neq \chi_{\phi}$ is a weakly reflexive and symmetric fuzzy relation on S, then there exists a set T and a fuzzy relation $\nu$ from S into T such that $\mu = \nu \circ \nu^{-1}$.*

*Proof* Let $T$ denotes the set $\{K^* \mid K \in F^{\mu}\}$. Define a fuzzy relation $\nu$ from $S$ to $T$ as follows

$$\nu(x, K^*) = \begin{cases} t & \text{if } x \in K \text{ and } t \text{ is the largest number such that } K \in F_t^{\mu} \\ 0 & \text{otherwise} \end{cases}$$

If $\mu(x, x') = t > 0$, then by Lemma 1.2.29, there is a $t$ -complete set $K \in F^{\mu}$ such that $\{x, x'\} \subseteq K$. Because $(\nu o \nu^{-1})(x, x') = \vee_{K^*}[\nu(x, K^*) \wedge \nu(x'^*)] \geq t = \mu(x, x')$, we conclude that $\mu \subseteq \nu \circ \nu^{-1}$.

Suppose now that $(\nu \circ \nu^{-1})(x, x') = s$. Then there exists $K^* \in F_s$ such that $\nu(x, K^*) = \nu(x'^*)$. This means that $\{x, x'\} \subseteq K$ and hence $\mu(x, x') \geq s$. ($s = (\nu \circ \nu^{-1})(x, x') = \vee_{K^*}[\nu(x, K^*) \wedge \nu(K^*, x')] = \vee_{K^*}[\nu(x, K^*) \wedge \nu(x'^*)]$. Thus, there exists $K^*$ such that either $\nu(x, K^*) = s$ and $\nu(x'^*) \geq s$ or $\nu(x, K^*) \geq s$ and $\nu(x'^*) = s$. Now, $s$ is the largest such that $K \in F_s^{\mu}$. Hence, $\nu(x, K^*) = s$ and $\nu(x'^*) = s$.) Therefore, $\nu \circ \nu^{-1} \subseteq \mu$. ■

**Theorem 1.2.31** *A fuzzy relation $\mu \neq \chi_{\phi}$ on a set $S$ is weakly reflexive and symmetric if and only if there is a set $T$ and a fuzzy relation $\nu$ from $S$ into $T$ such that $\mu = \nu \circ \nu^{-1}$.*

The proof of Theorem 1.2.31 follows from Lemmas 1.2.26 and 1.2.30.

From here on we use the notation $\phi_{\mu}$ to denote the fuzzy relation in Theorem 1.2.31.

**Definition 1.2.32** A **cover** $\mathcal{C}$ on a set $S$ is a family of subsets $S_i, i \in I$ of $S$ such that $\bigcup_{i \in I} S_i = S$, where $I$ is a nonempty index set.

**Definition 1.2.33** Let $\mu$ be a fuzzy relation from $S$ into $T$. For $\epsilon \in [0, 1]$, we have

$(i)$ $\mu$ is $\epsilon$-**determinate** if for each $x \in S$, there exists at most one $y \in T$ such that $\mu(x, y) \geq \epsilon$;

$(ii)$ $\mu$ is $\epsilon$-**productive** if for each $x \in S$, there exists at least one $y \in T$ such that $\mu(x, y) \geq \epsilon$;

$(iii)$ $\mu$ is an $\epsilon$-**function** if it is both $\epsilon$-determinate and $\epsilon$-productive.

**Lemma 1.2.34** *If $\mu$ is an $\epsilon$-reflexive fuzzy relation on $S$, then $\phi_{\mu}$ is $\epsilon$-productive and for each $\epsilon' \leq \epsilon$, $F_{\epsilon'}^{\mu}$ is a cover of $S$.*

*Proof* Let $0 < \epsilon' \leq \epsilon$. Because for each $x \in S$, $\mu(x, x) \geq \epsilon$, and because $\{x\}$ is $\epsilon$-complete, there is some $K$ in $F_{\epsilon'}^{\mu}$ such that $x \in K$. Hence, $F_{\epsilon'}^{\mu}$ is a cover of $X$. Also, by definition of $\phi_{\mu}$, $x \in K$ implies that $\phi_{\mu}(x, K^*) \geq \epsilon$, which implies that $\phi_{\mu}$ is $\epsilon$ -productive. ■

When $\epsilon = 1$, we use the terms determinate, productive and function for $\epsilon$-determinate, $\epsilon$-productive and $\epsilon$-function, respectively.

**Corollary 1.2.35** *If $\mu$ is reflexive, then $\phi_{\mu}$ is productive and each $F_{\epsilon}^{\mu}$, $(0 < \epsilon \leq 1)$ is a cover of $S$.*

**Corollary 1.2.36** *$\mu$ is a reflexive and symmetric relation on S if and only if there is a set T and a productive fuzzy relation $\nu$ from S into T such that $\mu = \nu \circ \nu^{-1}$.*

The proof of Corollary 1.2.36 follows directly from Theorem 1.2.31 and Corollary 1.2.35.

**Lemma 1.2.37** *Let $\mu$ be a weakly reflexive, symmetric and transitive fuzzy relation on S and let $\phi_\mu^\epsilon$ denotes the relation $\phi_\mu$ whose range is restricted to $F_\epsilon^\mu$. That is, $\phi_\mu^\epsilon$ equals $\phi_\mu$ on $S \times \{K^* \mid K \in F_\epsilon^\mu\}$. Then for each $0 < \epsilon \leq 1$, $\phi_\mu^\epsilon$ is $\epsilon$-determinate and the elements of $F_\epsilon^\mu$ are pairwise disjoint.*

*Proof* Let $K$ and $K'$ be two elements of $F_\epsilon^\mu$ and assume that $K \cap K' \neq \emptyset$. For any $q_1 \in K \cap K'$, we have $\mu(q, q_1) \geq \epsilon$ for all $q \in K$ and $\mu(q_1, q') \geq \epsilon$ for all $q' \in K'$. Because $\mu$ is transitive, we have $\mu(q, q') \geq \epsilon$ for all $q \in K$ and $q' \in K'$. Because $\mu$ is weakly reflexive and symmetric, we can conclude that $K \cup K'$ is $\epsilon$-complete. But because $K$ and $K'$ are maximal $\epsilon$-complete, we must conclude that $K = K'$. Hence, $K \neq K'$ implies $K \cap K' = \emptyset$. Suppose $x \in K$, where $K \in F_\epsilon^\mu$. Then $\phi_\mu(x, K^*) \geq \epsilon$, and because $x$ cannot belong to any other sets in $F^\mu$, $\phi_\mu^\epsilon$ is $\epsilon$-determinate. ■

**Definition 1.2.38** A **similarity relation** $\mu$ on $S$ is a fuzzy relation on $S$, which is reflexive, symmetric and transitive. $\mu$ is called an $\epsilon$-similarity relation if it is $\epsilon$-reflexive for some $0 < \epsilon \leq 1$, symmetric and transitive.

Clearly, a similarity relation on a set $S$ is a fuzzy equivalence relation on $S$.

**Corollary 1.2.39** *If $\mu$ is a similarity relation on S, then for each $0 < \epsilon \leq 1$, $F_\epsilon^\mu$ is a partition of S.*

The proof of Corollary 1.2.39 follows from the fact that reflexivity implies weak reflexivity, Lemmas 1.2.34 and 1.2.37.

We conclude this section with the following theorem, which is a characterization for similarity relations.

**Theorem 1.2.40** *A relation $\mu$ is an $\epsilon$-similarity* $(0 < \epsilon \leq 1)$ *relation on a set S if and only if there is another set T and an $\epsilon$-function $\nu$ from S into T such that $\mu = \nu \circ \nu^{-1}$.*

The proof of Theorem 1.2.40 follows from Theorem 1.2.31 and Corollary 1.2.39.

# Chapter 2
# Fuzzy Graphs

A graph represents a particular relationship between elements of a set $V$. It gives an idea about the extent of the relationship between any two elements of $V$. We can solve this problem by using a weighted graph if proper weights are known. But in most of the situations, the weights may not be known, and the relationships are 'fuzzy' in a natural sense. Hence, a fuzzy relation can deal with the situation in a better way. As an example, if $V$ represents certain locations and a network of roads is to be constructed between elements of $V$, then the costs of construction of the links are fuzzy. But the costs can be compared, to some extent using the terrain and local factors and can be modeled as fuzzy relations. Thus, fuzzy graph models are more helpful and realistic in natural situations.

Kaufman [91] gave the first definition of a fuzzy graph. But it was Rosenfeld [154] and Yeh and Bang [186] who laid the foundations for fuzzy graph theory. Rosenfeld introduced fuzzy analogs of several basic graph-theoretic concepts, including subgraphs, paths, connectedness, cliques, bridges, cutvertices, forests, and trees. Yeh and Bang independently introduced many connectivity concepts including vertex and edge connectivity in fuzzy graphs and applied fuzzy graphs for the first time in clustering of data.

In this chapter, we discuss fundamentals of fuzzy graph theory. We provide formal definitions, basic concepts, and properties of fuzzy graphs. For simplicity, we consider only undirected fuzzy graphs, unless otherwise specified. Thus, the edges of the fuzzy graph are unordered pairs of vertices.

## 2.1 Definitions and Basic Properties

Let $V$ be a nonempty set. Define the relation $\sim$ on $V \times V$ by for all $(x, y), (u, v) \in V \times V$, $(x, y) \sim (u, v)$ if and only if $x = u$ and $y = v$ or $x = v$ and $y = u$. Then it is easily shown that $\sim$ is an equivalence relation on $V \times V$. For all $x, y \in V$, let



S. Mathew et al., *Fuzzy Graph Theory*, Studies in Fuzziness
and Soft Computing 363, https://doi.org/10.1007/978-3-319-71407-3_2

$[(x, y)]$ denote the equivalence class of $(x, y)$ with respect to $\sim$ . Then $[(x, y)] = \{(x, y), (y, x)\}$. Let $\mathcal{E}_V = \{[(x, y)] \mid x, y \in V, x \neq y\}$. For simplicity, we often write $\mathcal{E}$ for $\mathcal{E}_V$ when $V$ is understood. Let $E \subseteq \mathcal{E}$. A graph is a pair $(V, E)$. The elements of $V$ are thought of as vertices of the graph and the elements of $E$ as the edges. For $x, y \in V$, we let $xy$ denote $[(x, y)]$. Then clearly $xy = yx$. We note that graph $(V, E)$ has no loops or parallel edges.

**Definition 2.1.1** A **fuzzy graph** $G = (V, \sigma, \mu)$ is a triple consisting of a nonempty set $V$ together with a pair of functions $\sigma : V \to [0, 1]$ and $\mu : \mathcal{E} \to [0, 1]$ such that for all $x, y \in V$, $\mu(xy) \leq \sigma(x) \wedge \sigma(y)$.

The fuzzy set $\sigma$ is called the **fuzzy vertex set** of $G$ and $\mu$ the **fuzzy edge set** of $G$. Clearly $\mu$ is a fuzzy relation on $\sigma$. We consider $V$ as a finite set, unless otherwise specified. For notational convenience, we use simply $G$ or $(\sigma, \mu)$ to represent the fuzzy graph $G = (V, \sigma, \mu)$. Also, $\sigma^*$ and $\mu^*$, respectively, represent the supports of $\sigma$ and $\mu$, also denoted by Supp($\sigma$) and Supp($\mu$).

*Example 2.1.2* Let $V = \{a, b, c\}$. Define the fuzzy set $\sigma$ on $V$ as $\sigma(a) = 0.5$, $\sigma(b) = 1$ and $\sigma(c) = 0.8$. Define a fuzzy set $\mu$ of $\mathcal{E}$ such that $\mu(ab) = 0.5$, $\mu(bc) = 0.7$ and $\mu(ac) = 0.1$. Then $\mu(xy) \leq \sigma(x) \wedge \sigma(y)$ for all $x, y \in V$. Thus, $G = (\sigma, \mu)$ is a fuzzy graph. If we redefine $\mu(ab) = 0.6$, then it is no longer a fuzzy graph.

It follows from the Definition 2.1.1 that any unweighted graph $(V, E)$ is trivially a fuzzy graph with $\sigma(x) = 1$ for all $x \in V$ and $\mu(xy) = 0$ or 1 for all $x, y \in V$. Also, we write $(V, \mu)$ to denote a fuzzy graph with $\sigma(x) = 1$ for all $x \in V$.

**Definition 2.1.3** Let $G = (V, \sigma, \mu)$ be a fuzzy graph. Then a fuzzy graph $H = (V, \tau, \nu)$ is called a **partial fuzzy subgraph** of $G$ if $\tau \subseteq \sigma$ and $\nu \subseteq \mu$. Similarly, the fuzzy graph $H = (P, \tau, \nu)$ is called a **fuzzy subgraph** of $G$ induced by $P$ if $P \subseteq V$, $\tau(x) = \sigma(x)$ for all $x \in P$ and $\nu(xy) = \mu(xy)$ for all $x, y \in P$. We write $\langle P \rangle$ to denote the fuzzy subgraph **induced** by $P$.

*Example 2.1.4* Let $G = (\tau, \nu)$, where $\tau^* = \{a, b, c\}$ and $\mu^* = \{ab, bc\}$ with $\tau(a) = 0.4$, $\tau(b) = 0.8$, $\tau(c) = 0.5$, $\nu(ab) = 0.3$ and $\nu(bc) = 0.2$. Then clearly $G$ is a partial fuzzy subgraph of the fuzzy graph in Example 2.1.2. Also, if $P = \{a, b\}$ and $H = (\tau, \nu)$, where $\tau(a) = 0.5$, $\tau(b) = 1$ and $\nu(ab) = 0.5$, then $H$ is the induced fuzzy subgraph of $G$ in Example 2.1.2, induced by $P$.

**Definition 2.1.5** Let $G = (\sigma, \mu)$ be a fuzzy graph. Then a partial fuzzy subgraph $(\tau, \nu)$ of $G$ is said to **span** $G$ if $\sigma = \tau$. In this case, we call $(\tau, \nu)$ a **spanning fuzzy subgraph** of $(\sigma, \mu)$.

In fact a fuzzy subgraph $H = (\tau, \nu)$ of a fuzzy graph $G = (\sigma, \mu)$ induced by a subset $P$ of $V$ is a particular partial fuzzy subgraph of $G$. Take $\tau(x) = \sigma(x)$ for all $x \in P$ and 0 for all $x \notin P$. Similarly, take $\nu(xy) = \mu(xy)$ if $xy$ is in a set of edges involving elements from $P$, and 0 otherwise.

**Definition 2.1.6** Let $G = (V, \sigma, \mu)$ be a fuzzy graph. Let $0 \leq t \leq 1$. Let $\sigma^t = \{x \in \sigma^* \mid \sigma(x) \geq t\}$ and $\mu^t = \{ uv \in \mu^* \mid \mu(uv) \geq t\}$.

Clearly, $\mu^t \subseteq \{xy \mid \sigma(x) \geq t, \sigma(y) \geq t\}$ and hence $H = (\sigma^t, \mu^t)$ is a graph with vertex set $\sigma^t$ and edge set $\mu^t$. $H$ is called the **threshold graph** of the fuzzy graph $G$, corresponding to $t$.

**Proposition 2.1.7** *Let $G = (\sigma, \mu)$ be a fuzzy graph and $0 \leq s < t \leq 1$. Then the threshold graph $(\sigma^t, \mu^t)$ is a subgraph of the threshold graph $(\sigma^s, \mu^s)$. Also, if $H = (\nu, \tau)$ is a partial fuzzy subgraph of $G$ and $t \in [0, 1]$, then $(\nu^t, \tau^t)$ is a subgraph of $(\sigma^t, \mu^t)$.*

## 2.2 Connectivity in Fuzzy Graphs

We mainly discuss the concepts of fuzzy cutvertices and fuzzy bridges in this section. Most of the results are due to Sunitha and Vijayakumar [167, 168]. Also, Theorem 2.2.1, by Rosenfeld gives a very strong characterization for a fuzzy bridge.

A **path** $P$ in a fuzzy graph $G = (\sigma, \mu)$ is a sequence of distinct vertices $x_0, x_1, \ldots, x_n$ (except possibly $x_0$ and $x_n$) such that $\mu(x_{i-1}x_i) > 0, i = 1, \ldots, n$. Here $n$ is called the **length** of the path. The consecutive pairs are called the **edges** of the path. The **diameter** of $x, y \in V$, written $\text{diam}(x, y)$, is the length of the longest path joining $x$ to $y$. The **strength** of $P$ is defined to be $\wedge_{i=1}^{n}\mu(x_{i-1}x_i)$. In words, the strength of a path is defined to be the weight of the weakest edge. We denote the strength of a path $P$ by $d(P)$ or $s(P)$. The strength of connectedness between two vertices $x$ and $y$ is defined as the maximum of the strengths of all paths between $x$ and $y$ and is denoted by $\mu^\infty(x, y)$ or $CONN_G(x, y)$. A strongest path joining any two vertices $x, y$ has strength $\mu^\infty(x, y)$. It can be shown that if $(\tau, \nu)$ is a partial fuzzy subgraph of $(\sigma, \mu)$, then $\nu^\infty \subseteq \mu^\infty$. We call $P$ a **cycle** if $x_0 = x_n$ and $n \geq 3$. Two vertices that are joined by a path are called **connected**. It follows that this notion of connectedness is an equivalence relation. The equivalence classes of vertices under this equivalence relation are called **connected components** of the given fuzzy graph. They are just its maximal connected partial fuzzy subgraphs.

Let $G = (\sigma, \mu)$ be a fuzzy graph, let $x, y$ be two distinct vertices and let $G'$ be the partial fuzzy subgraph of $G$ obtained by deleting the edge $xy$. That is, $G' = (\sigma, \mu')$, where $\mu'(xy) = 0$ and $\mu' = \mu$ for all other pairs. We call $xy$ a **fuzzy bridge** in $G$ if $\mu'^\infty(u, v) < \mu^\infty(u, v)$ for some $u, v$ in $\sigma^*$. In words, the deletion of the edge $xy$ reduces the strength of connectedness between some pair of vertices in $G$. Thus, $xy$ is a fuzzy bridge if and only if there exists vertices $u, v$ such that $xy$ is an edge of every strongest path from $u$ to $v$.

**Theorem 2.2.1** ([154]) *Let $G = (\sigma, \mu)$ be a fuzzy graph. Then the following statements are equivalent.*

*(i) $xy$ is a fuzzy bridge.*
*(ii) $\mu'^{\infty}(x, y) < \mu(xy)$.*
*(iii) $xy$ is not the weakest edge of any cycle.*

*Proof* $(ii) \Rightarrow (i)$ If $xy$ is not a fuzzy bridge, then $\mu'^{\infty}(x, y) = \mu^{\infty}(x, y) \geq \mu(xy)$.

$(i) \Rightarrow (iii)$ If $xy$ is the weakest edge of a cycle, then any path $P$ involving edge $xy$ can be converted into a path $P'$ not involving $xy$ but at least as strong as $P$, by using the rest of the cycle as a path from $x$ to $y$. Thus, $xy$ cannot be a fuzzy bridge.

$(iii) \Rightarrow (ii)$ If $\mu'^{\infty}(x, y) \geq \mu(xy)$, then there is a path from $x$ to $y$ not involving $xy$, that has strength $\geq \mu(xy)$, and this path together with $xy$ forms a cycle of $G$ in which $xy$ is a weakest edge. ■

Let $w$ be any vertex and let $G'$ be the partial fuzzy subgraph of $G$ obtained by deleting the vertex $w$. That is, $G' = (\sigma', \mu')$ is the partial fuzzy subgraph of $G$ such that $\sigma'(w) = 0$, $\sigma = \sigma'$ for all other vertices, $\mu'(wz) = 0$ for all vertices $z$, and $\mu' = \mu$ for all other edges. We call $w$ a **fuzzy cutvertex** in $G$ if $\mu'^{\infty}(u, v) < \mu^{\infty}(u, v)$ for some $u, v$ in $V$ such that $u \neq w \neq v$. In words, $w$ is a fuzzy cutvertex if deleting the vertex $w$ reduces the strength of connectedness between some other pair of vertices. Hence, $w$ is a fuzzy cutvertex if and only if there exists $u, v$ distinct from $w$ such that $w$ is on every strongest path from $u$ to $v$. $G'$ is called **nonseparable** or a **block** if it has no fuzzy cutvertices. Although in a fuzzy graph, a block may have fuzzy bridges, this cannot happen for crisp graphs. Sometimes we refer to a block in a fuzzy graph as a fuzzy block.

A **maximum spanning tree** of a connected fuzzy graph $(\sigma, \mu)$ is a fuzzy spanning subgraph $T = (\sigma, \nu)$ of $G$, which is a tree, such that $\mu^{\infty}(u, v)$ is the strength of the unique strongest $u - v$ path in $T$ for all $u, v \in G$. We next characterize fuzzy cutvertices and fuzzy bridges of fuzzy graphs using maximum spanning trees.

**Theorem 2.2.2** ([168]) *A vertex $w$ of a fuzzy graph $G = (\sigma, \mu)$ is a fuzzy cutvertex if and only if $w$ is an internal vertex of every maximum spanning tree of $G$.*

*Proof* Let $G = (\sigma, \mu)$ be a fuzzy graph and $w$ be a fuzzy cutvertex of $G$. Then there exists $u, v$ distinct from $w$ such that $w$ is on every strongest $u - v$ path. Now, each maximum spanning tree of $G$ contains a unique strongest $u - v$ path and hence $w$ is an internal vertex of every maximum spanning tree of $G$.

Conversely, let $w$ be an internal vertex of every maximum spanning tree. Let $T$ be a maximum spanning tree and let $uw$ and $wv$ be edges in $T$. Note that the path $u, w, v$ is a strongest $u - v$ path in $T$. If possible assume that $w$ is not a fuzzy cutvertex. Then between every pair of vertices $u, v$, there exists at least one strongest $u - v$ path not containing $w$. Consider one such $u - v$ path $P$ which clearly contains edges not in $T$. Now, without loss of generality, let $\mu^{\infty}(u, v) = \mu(uw)$ in $T$. Then edges in $P$ have strength $\geq \mu(uw)$. Removal of $uw$ and adding $P$ in $T$ will result in another maximum spanning tree of $G$, of which $w$ is an end vertex, which contradicts our assumption. ■

From Theorem 2.2.2, it can be seen that the end vertices of a maximum spanning tree $T$ of $G$ cannot be fuzzy cutvertices of $G$. Thus, we have the following corollary.

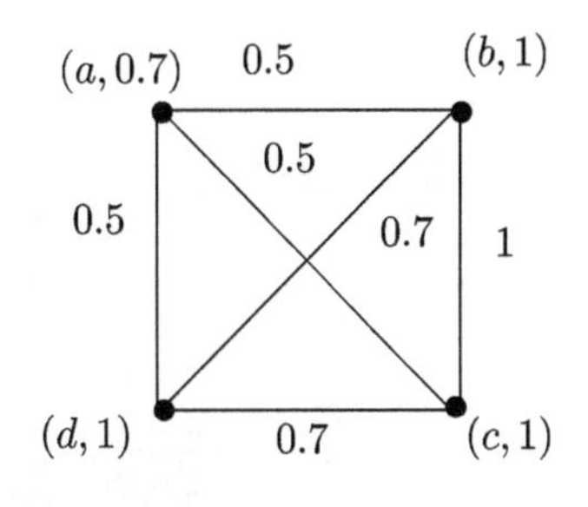


**Fig. 2.1** A noncomplete fuzzy graph satisfying $\mu^\infty = \mu$

**Corollary 2.2.3** *Every fuzzy graph G has at least two vertices which are not fuzzy cutvertices.*

**Corollary 2.2.4** *An edge $uv$ of a fuzzy graph $G = (\sigma, \mu)$ is a fuzzy bridge if and only if $uv$ is in every maximum spanning tree of G.*

*Proof* Let $uv$ be a fuzzy bridge of $G$. Then the edge $uv$ is the unique strongest $u - v$ path and hence is in every maximum spanning tree of $G$.

Conversely, let $uv$ be in every maximum spanning tree of $G$ and assume that $uv$ is not a fuzzy bridge. Then $uv$ is the weakest edge of some cycle in $G$ and $\mu^\infty(u, v) \geq \mu(uv)$, which implies that there is at least one maximum spanning tree of $G$ not containing $uv$. ■

In 1989, Bhutani [40] introduced the concept of a complete fuzzy graph as follows.

A **complete fuzzy graph** (**CFG**) is a fuzzy graph $G = (\sigma, \mu)$ such that $\mu(uv) = \sigma(u) \wedge \sigma(v)$ for all $u, v \in \sigma^*$. If $G = (\sigma, \mu)$ is a complete fuzzy graph, then $\mu^\infty = \mu$ and $G$ has no fuzzy cutvertices.

*Example 2.2.5* Let $V = \{a, b, c, d\}$ and $X = \{ab, bc, cd, da, ac, bd\}$. Let $\sigma(a) = 0.7$ and $\sigma(b) = \sigma(c) = \sigma(d) = 1$. Let $\mu$ be the fuzzy subset of $X$ defined by $\mu(ab) = \mu(da) = \mu(ac) = 0.5$, $\mu(bc) = 1$ and $\mu(cd) = \mu(bd) = 0.7$, Then $\mu^\infty = \mu$, and $G$ has no fuzzy cutvertices, but $G$ is not complete (Fig. 2.1).

*Example 2.2.6* We show that a complete fuzzy graph may have a bridge. Let $V = \{a, b, c, d\}$ and $X = \{ab, bc, cd, da, ac, bd\}$. Let $\sigma(a) = 0.7$, $\sigma(b) = 0.8$, $\sigma(c) = 1$ and $\sigma(d) = 0.6$. Let $\mu$ be the fuzzy subset of $X$ defined by $\mu(ab) = \mu(ac) = 0.7$, $\mu(bd) = \mu(cd) = \mu(ad) = 0.6$ and $\mu(bc) = 0.8$. Clearly, $(\sigma, \mu)$ is complete and $bc$ is a fuzzy bridge (Fig. 2.2).

**Theorem 2.2.7** *If $G = (\sigma, \mu)$ is a complete fuzzy graph, then for any edge $uv \in \mu^*$, $\mu^\infty(u, v) = \mu(uv)$.*

*Proof* By definition, $\mu^2(u, v) = \vee_{z \in \sigma^*}\{\mu(uz) \wedge \mu(zv)\} = \vee\{\sigma(u) \wedge \sigma(v) \wedge \sigma(z)\} = \sigma(u) \wedge \sigma(v) = \mu(uv)$.

Similarly, $\mu^3(u, v) = \mu(uv)$ and in the same way one can show that $\mu^k(u, v) = \mu(uv)$ for all positive integers $k$. Thus,

$\mu^\infty(u, v) = \sup\{\mu^k(u, v) \mid \text{for all integers } k \geq 1\} = \mu(uv)$. ■

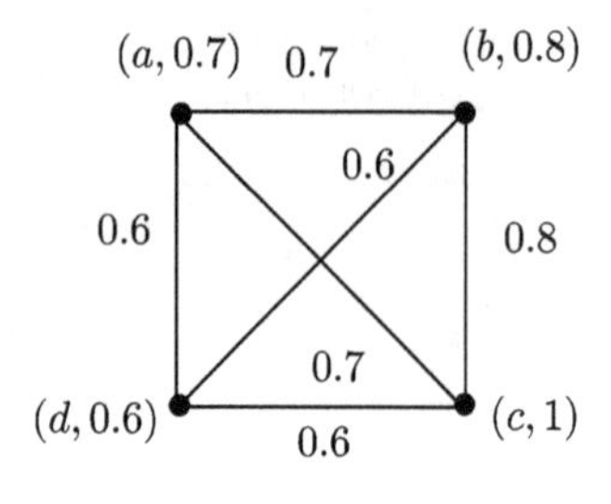


**Fig. 2.2** A complete fuzzy graph with a fuzzy bridge

**Corollary 2.2.8** *A complete fuzzy graph has no fuzzy cutvertices.*

Theorem 2.2.7 says that every edge $uv$ in a CFG is a strongest $u - v$ path. Also, a CFG can have at most one fuzzy bridge (Theorem 4, [168]), even though it has no fuzzy cutvertices. This bridge can be easily located as seen from the following theorem.

**Theorem 2.2.9** *Let $G = (\sigma, \mu)$ be a CFG with $|\sigma^*| = n$. Then $G$ has a fuzzy bridge if and only if there exists an increasing sequence $\{t_1, t_2, \ldots, t_n\}$ such that $t_{n-2} < t_{n-1} \leq t_n$, where $t_i = \sigma(u_i)$ for $i = 1, 2, \ldots, n$. Also, the edge $u_{n-1}u_n$ is the fuzzy bridge of $G$.*

*Proof* Assume that $G = (\sigma, \mu)$ is a complete fuzzy graph and that $G$ has a fuzzy bridge $uv$. Now, $\mu(uv) = \sigma(u) \wedge \sigma(v)$. Without loss of generality let, $\sigma(u) \leq \sigma(v)$, so that $\mu(uv) = \sigma(u)$. Note that $uv$ is not a weakest edge of any cycle in $G$. It is required to prove that $\sigma(u) > \sigma(w)$ for all $w \neq v$. On the contrary, assume that there is at least one vertex $w \neq v$ such that $\sigma(u) \leq \sigma(w)$. Now, consider the cycle $C : u, v, w, u$. Then $\mu(uv) = \mu(uw) = \sigma(u)$ and $\mu(vw) = \sigma(v)$ if $\sigma(u) = \sigma(v)$ or $\sigma(u) < \sigma(v) \leq \sigma(w)$ and $\mu(vw) = \sigma(w)$ if $\sigma(u) < \sigma(w) < \sigma(v)$. In either case, the edge $uv$ becomes the weakest edge of a cycle and by Theorem 2.2.1, $uv$ cannot be a fuzzy bridge, a contradiction.

Conversely, let $t_1 \leq t_2 \leq \cdots \leq t_{n-2} \leq t_{n-1} \leq t_n$ and $t_i = \sigma(u_i)$ for all $i$.

**Claim**. Edge $u_{n-1}u_n$ is the fuzzy bridge of $G$.

We have, $\mu(u_{n-1}u_n) = \sigma(u_{n-1}) \wedge \sigma(u_n) = \sigma(u_{n-1})$ and by hypothesis, all other edges of $G$ will have strength strictly less than that of $\sigma(u_{n-1})$. Thus, the edge $u_{n-1}u_n$ is not the weakest edge of any cycle in $G$ and by Theorem 2.2.1, is a fuzzy bridge. ■

We next present more connectivity properties of fuzzy cutvertices and fuzzy bridges.

**Theorem 2.2.10** *Let $G = (\sigma, \mu)$ be a fuzzy graph such that $(\sigma^*, \mu^*)$ is a cycle. Then a vertex of $G$ is a fuzzy cutvertex if and only if it is a common vertex of two fuzzy bridges.*

*Proof* Let $w$ be a fuzzy cutvertex of $G$. Then there exists $u$ and $v$, other than $w$ such that $w$ is on every strongest $u$-$v$ path. Because $G^* = (\sigma^*, \mu^*)$ is a cycle, there exists only one strongest $u$-$v$ path containing $w$ and all its edges are fuzzy bridges. Thus, $w$ is a common vertex of two fuzzy bridges.

Conversely, let $w$ be a common vertex of two fuzzy bridges $uw$ and $wv$. Then both $uw$ and $wv$ are not weakest edges of $G$. Also, the path from $u$ to $v$ not containing edges $uw$ and $wv$ has strength less than $\mu(uw) \wedge \mu(wv)$. Hence, the strongest $u$-$v$ path is the path $u, w, v$ and $\mu^{\infty}(u, v) = \mu(uw) \wedge \mu(wv)$. Thus, $w$ is a fuzzy cutvertex. ■

**Theorem 2.2.11** *If $w$ is a common vertex of at least two fuzzy bridges, then $w$ is a fuzzy cutvertex.*

*Proof* Let $u_1 w$ and $wu_2$ be two fuzzy bridges. Then there exists $u, v$ such that $u_1 w$ is on every strongest $u$-$v$ path. If $w$ is distinct from $u$ and $v$, then it follows that $w$ is a fuzzy cutvertex. Next suppose one of $v, u$ is $w$ so that $u_1 w$ is on every strongest $u$-$w$ path or $wu_2$ is on every strongest $w$-$v$ path. Suppose that $w$ is not a fuzzy cutvertex. Then between every two vertices there exists at least one strongest path not containing $w$. In particular, there exists at least one strongest path $P$ joining $u_1$ and $u_2$, not containing $w$. This path together with $u_1 w$ and $wu_2$ forms a cycle.

We now consider two cases. First, suppose that $u_1, w, u_2$ is not a strongest path. Then clearly one of $u_1 w, wu_2$ or both become weakest edges of a cycle, which contradicts that $u_1 w$ and $wu_2$ are fuzzy bridges.

Second, suppose that $u_1, w, u_2$ is also a strongest path joining $u_1$ to $u_2$. Then $\mu^{\infty}(u_1, u_2) = \mu(u_1 w) \wedge \mu(wu_2)$, the strength of $P$. Thus, edges of $P$ are at least as strong as $\mu(u_1 w)$ and $\mu(wu_2)$, which implies that $u_1 w, wu_2$ are both weakest edges of a cycle, which again is a contradiction. ■

*Example 2.2.12* This example shows that the condition in Theorem 2.2.11 is not necessary. Let $V = \{a, b, c, d\}$ and $X = \{ab, bc, cd, da, ac, db\}$. Let $\sigma(x) = 1$ for all $x \in V$ and let $\mu$ be the fuzzy subset of $X$ defined by $\mu(ac) = \mu(bd) = 0.9$, $\mu(da) = \mu(cd) = 0.3$ and $\mu(ab) = \mu(bc) = 0.8$. Clearly, $b$ is a fuzzy cutvertex. However, $ac$ and $db$ are the only fuzzy bridges.

*Example 2.2.13* Consider the fuzzy graph $G = (V, X)$, where $V = \{a, b, c, d\}$. Let $X = \{ab, bc, cd, ad\}$. Let $\sigma(s) = 1$ for all $s \in V$ and let $\mu$ be the fuzzy subset of $X$ defined by $\mu(ab) = \mu(cd) = 0.2$, and $\mu(bc) = \mu(ad) = 0.1$. Note that $ab$ and $cd$ are fuzzy bridges and no vertex is a fuzzy cutvertex. This is a significant difference from the crisp graph theory.

The fuzzy graphs in Examples 2.2.12 and 2.2.13 are given in Fig. 2.3.

**Theorem 2.2.14** *If $uv$ is a fuzzy bridge, then $\mu^{\infty}(u, v) = \mu(uv)$.*

*Proof* Suppose that $uv$ is a fuzzy bridge and that $\mu^{\infty}(u, v) > \mu(uv)$. Then there exists a strongest $u$–$v$ path with strength greater than $\mu(uv)$ and all edges of this strongest path have strength greater than $\mu(uv)$. Now, this path together with the edge $uv$ forms a cycle in which $uv$ is the weakest edge, contradicting that $uv$ is a fuzzy bridge. ■

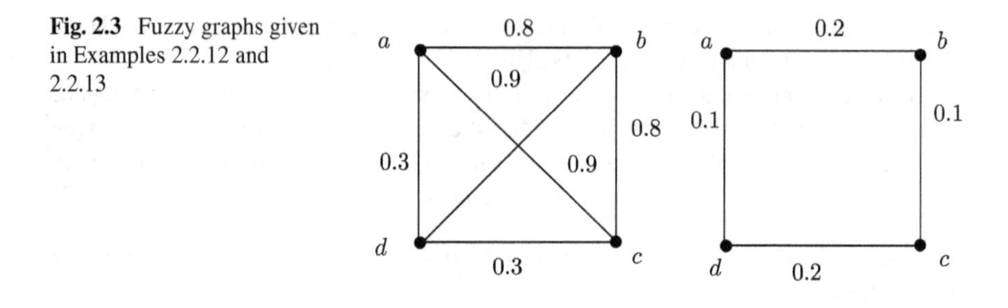


**Fig. 2.3** Fuzzy graphs given in Examples 2.2.12 and 2.2.13

## 2.3 Forests and Trees

Rosenfeld first studied the concepts of fuzzy forests and fuzzy trees in [154]. The results in this section are by Rosenfeld [154], Mordeson and Nair [127] and Sunitha and Vijayakumar [167, 168]. The structure of fuzzy trees is significantly different from that of trees. In fact, a fuzzy tree can contain cycles in the classical sense. There are several characterizations for fuzzy trees in this book.

A crisp graph that has no cycles is called **acyclic** or a **forest**. A connected forest is a **tree**. A fuzzy graph is called a **forest** if the graph consisting of its nonzero edges is a forest, and a **tree** if this graph is also connected. If $G = (\sigma, \mu)$ is a fuzzy graph, we call $G$ a **fuzzy forest** if it has a partial fuzzy spanning subgraph $F = (\sigma, \nu)$, which is a forest, where for all edges $xy$ not in $F$, i.e., such that $\nu(xy) = 0$, we have $\mu(xy) < \nu^{\infty}(x, y)$. In words, if $xy$ is in $G$, but is not in $F$, there is a path in $F$ between $x$ and $y$ whose strength is greater than $\mu(xy)$. Clearly, a forest is a fuzzy forest.

**Theorem 2.3.1** *A fuzzy graph $G$ is a fuzzy forest if and only if in any cycle of $G$ there is an edge $xy$ such that $\mu(xy) < \mu'^{\infty}(x, y)$, where $G' = (\sigma, \mu')$ is the partial fuzzy subgraph obtained by deleting the edge $xy$ from $G$.*

*Proof* Suppose $xy$ is an edge, belonging to a cycle which has the property of the theorem and for which $xy$ is the smallest. (If there are no cycles, $G$ is a forest and we are done.) If we delete $xy$, the resulting partial fuzzy subgraph satisfies the path property of a fuzzy forest. If there are still cycles in this graph, we can repeat the process. Now, at each stage, no previously deleted edge is stronger than the edge being currently deleted. Thus, the path guaranteed by the property of the theorem involves only edges that have not yet been deleted. When no cycles remain, the resulting partial fuzzy subgraph is a forest $F$. Let $xy$ not be an edge of $F$. Then $xy$ is one of the edges that we deleted in the process of constructing $F$, and there is a path from $x$ to $y$ that is stronger than $\mu(xy)$ and that does not involve $xy$ nor any of the edges deleted prior to it. If this path involves edges that were deleted later, it can be diverted around them using a path of still stronger edges; if any of these were deleted later, the path can be further diverted; and so on. This process eventually stabilizes with a path consisting entirely of edges of $F$. Thus, $G$ is a fuzzy forest.

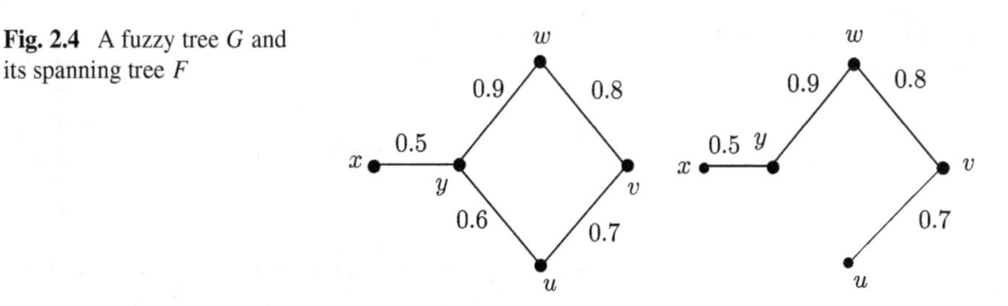


**Fig. 2.4** A fuzzy tree $G$ and its spanning tree $F$

Conversely, if $G$ is a fuzzy forest and $P$ is any cycle, then some edge $xy$ of $P$ is not in $F$. Thus, by definition of a fuzzy forest, we have $\mu(xy) < \nu^{\infty}(x, y) \leq \mu'^{\infty}(x, y)$. ■

We see that if $G$ is connected, then so is $F$ as determined by the construction in the first part of the proof. The tree $F$ thus constructed plays a very important role in the study of fuzzy trees. Also, Theorem 2.3.1 allows fuzzy forests to have cycles, as mentioned before.

**Proposition 2.3.2** *If there is at most one strongest path between any two vertices of $G$, then $G$ is a fuzzy forest.*

*Proof* Suppose $G$ is not a fuzzy forest. Then by the previous theorem, there is a cycle $P$ in $G$ such that $\mu(xy) \geq \mu'(xy)$ for all edges $xy$ of $P$. Thus, $xy$ is a strongest path from $x$ to $y$. If we choose $xy$ to be a weakest edge of $P$, it follows that the rest of $P$ also is a strongest path from $x$ to $y$, a contradiction. ■

We note that the converse of the previous proposition does not hold.

*Example 2.3.3* Consider the fuzzy graphs $G = (\sigma, \mu)$ and $F = (\tau, \nu)$ given in Fig. 2.4 with $V = \{x, y, u, v, w\}$. Define $\sigma, \tau : V \to [0, 1]$ by for all $z \in V$, $\sigma(z) = 1 = \tau(z)$. Define $\mu, \nu : V \times V \to [0, 1]$ as $\mu(xy) = 0.5$, $\mu(yw) = 0.9$, $\mu(wv) = 0.8$, $\mu(vu) = 0.7$, $\mu(yu) = 0.6$, $\nu(xy) = 0.5$, $\nu(yw) = 0.9$, $\nu(wv) = 0.8$, $\nu(uv) = 0.7$. It is clear that $G$ is a fuzzy forest. $F$ is the spanning tree of $G$. But note that both $x, xy, y, yu, u, uv, v$ and $x, xy, y, yw, w, wv, v$ are strongest $x - v$ paths in $G$.

**Proposition 2.3.4** *If $G = (\sigma, \mu)$ is a fuzzy forest, then the edges of $F = (\tau, \nu)$ are precisely the bridges of $G$.*

*Proof* An edge $xy$ not in $F$ cannot be a bridge because $\mu(xy) < \nu^{\infty}(x, y) \leq \mu'^{\infty}(x, y)$. Suppose that $xy$ is an edge in $F$. If it were not a bridge, we would have a path $P$ from $x$ to $y$, not involving $xy$, of strength greater than or equal to $\mu(xy)$. This path must involve edges not in $F$ because $F$ is a forest and has no cycles. However, by definition, any such edge $u_i v_i$ can be replaced by a path $P_i$ in $F$ of strength greater than $\mu(u_i v_i)$. Now, $P_i$ cannot involve $xy$ because all its edges are strictly stronger than $\mu(u_i v_i) \geq \mu(xy)$. Thus, by replacing each $u_i v_i$ by $P_i$, we can construct a path in $F$ from $x$ to $y$ that does not involve $xy$, giving a cycle in $F$, a contradiction. ■

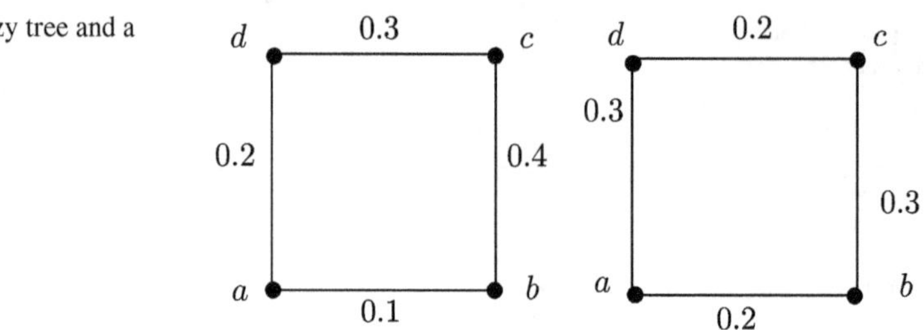


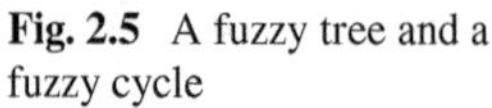
**Fig. 2.5** A fuzzy tree and a fuzzy cycle

**Definition 2.3.5** Let $G = (\sigma, \mu)$ be a fuzzy graph. Then

(i) $G$ is called a **tree** if $(\text{Supp}(\sigma), \text{Supp}(\mu))$ is a tree.
(ii) $G$ is called a **fuzzy tree** if $G$ has a fuzzy spanning subgraph $F = (\sigma, \nu)$, which is a tree, such that for all $uv \in \text{Supp}(\mu)\backslash\text{Supp}(\nu)$, $\mu(uv) < \nu^{\infty}(u, v)$. That is, there exists a path in $(\sigma, \nu)$ between $u$ and $v$ whose strength is greater than $\mu(uv)$.

**Definition 2.3.6** Let $G = (\sigma, \mu)$ be a fuzzy graph. Then

(i) $G$ is called a **cycle** if $(\text{Supp}(\sigma), \text{Supp}(\mu))$ is a cycle.
(ii) $G$ is called a **fuzzy cycle** if $(\text{Supp}(\sigma), \text{Supp}(\mu))$ is a cycle and $\nexists$ unique $xy \in \text{Supp}(\mu)$ such that $\mu(xy) = \wedge\{\mu(uv) \mid uv \in \text{Supp}(\mu)\}$.

*Example 2.3.7* Let $V = \{a, b, c, d\}$ and $X = \{ab, ac, ad, bc, cd, db\}$. Let $\sigma(x) = 1$ for all $x \in V$ and let $\mu$ be the fuzzy subset of $X$ defined by $\mu(ab) = 0.9$, $\mu(bc) = \mu(cd) = 0.7$, $\mu(bd) = 0.3$. Then $(\sigma, \mu)$ is neither a fuzzy cycle nor a fuzzy tree.

*Example 2.3.8* Let $V = \{a, b, c, d\}$ and $X = \{ab, ac, ad, bc, bd, cd\}$. Let $\sigma(x) = 1$ for all $x \in V$ and $\mu$, $\mu'$ be fuzzy subsets of $X$ defined by $\mu(ab) = 0.1$, $\mu(bc) = 0.4$, $\mu(cd) = 0.3$, $\mu(ad) = 0.2$ and $\mu'(ab) = 0.2$, $\mu'(bc) = 0.3$, $\mu'(cd) = 0.2$, $\mu'(ad) = 0.3$. Then $(\sigma, \mu)$ is a fuzzy tree, but not a tree and not a fuzzy cycle while $(\sigma, \mu')$ is a fuzzy cycle, but not a fuzzy tree (Fig. 2.5).

**Theorem 2.3.9** ([127]) *Let $G = (\sigma, \mu)$ be a cycle. Then $G$ is a fuzzy cycle if and only if $G$ is not a fuzzy tree.*

*Proof* Suppose that $G$ is a fuzzy cycle. Then there exists edges $x_1y_1, x_2y_2 \in \text{Supp}(\mu)$ such that $\mu(x_1y_1) = \mu(x_2y_2) = \wedge\{\mu(uv) \mid uv \in \text{Supp}(\mu)\}$. If $(\sigma, \nu)$ is any spanning tree of $(\sigma, \mu)$, then $\text{Supp}(\mu)\backslash \text{Supp}(\nu) = \{uv\}$ for some $u, v \in V$ because $(\sigma, \mu)$ is a cycle. Hence, $\nexists$ a path in $(\sigma, \nu)$ between $u$ and $v$ of greater strength than $\mu(uv)$. Thus, $(\sigma, \mu)$ is not a fuzzy tree.

Conversely, suppose that $(\sigma, \mu)$ is not a fuzzy tree. Because $(\sigma, \mu)$ is a cycle, we have for all $uv \in \text{Supp}(\mu)$, $(\sigma, \nu)$ is a fuzzy spanning subgraph of $(\sigma, \mu)$, which is a tree, and $\nu^{\infty}(u, v) \leq \mu(uv)$, where $\nu(uv) = 0$ and $\nu(xy) = \mu(xy)$ for all $xy \in \text{Supp}(\mu)\backslash\{uv\}$. Hence, $\mu$ does not attain $\wedge\{\mu(xy) \mid xy \in \text{Supp}(\mu)\}$ uniquely. Thus, $(\sigma, \mu)$ is a fuzzy cycle. ■

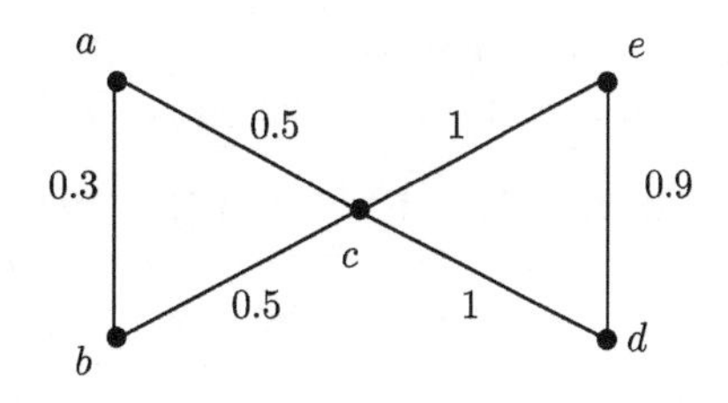


**Fig. 2.6** Fuzzy tree in Example 2.3.11

**Theorem 2.3.10** *Let $G = (\sigma, \mu)$ be a fuzzy graph. If there exists $t \in (0, 1]$ such that $(Supp(\sigma), \mu^t)$ is a tree, then $G$ is a fuzzy tree. Conversely, if $G$ is a cycle and $G$ is a fuzzy tree, then there exists $t \in (0, 1]$ such that $(Supp(\sigma), \mu^t)$ is a tree.*

*Proof* Suppose that there exists $t \in (0, 1]$ such that $(\mathrm{Supp}(\sigma), \mu^t)$ is a tree. Let $\nu$ be the fuzzy subset of $V \times V$ such that $\nu = \mu$ on $\mu^t$ and $\nu(xy) = 0$ if $xy \in E \backslash \mu^t$. Then $(\sigma, \nu)$ is a spanning fuzzy subgraph of $(\sigma, \mu)$ such that $(\sigma, \nu)$ is a fuzzy tree because $(\mathrm{Supp}(\sigma), \mathrm{Supp}(\nu))$ is a tree. Suppose that $uv \in E$ and $uv \notin \mu^t$. Then $\exists$ a path between $u$ and $v$ of strength $\geq t > \mu(uv)$. Thus, $(\sigma, \mu)$ is a fuzzy tree. For the converse, we note that because $(\sigma, \mu)$ is a cycle and a fuzzy tree, $\exists$ unique $xy \in \mathrm{Supp}(\mu)$ such that $\mu(xy) = \wedge\{\mu(uv) \mid uv \in \mathrm{Supp}(\mu)\}$. Let $t$ be such that $\mu(xy) < t \leq \wedge\{\mu(uv) \mid uv \in \mathrm{Supp}(\mu)\backslash\{xy\}\}$. Then $(\mathrm{Supp}(\sigma), \mu^t)$ is a tree. ■

*Example 2.3.11* Let $V = \{a, b, c, d, e\}$ and $X = \{ab, bc, ac, cd, de, ec\}$. Let $\sigma(x) = 1$ for all $x \in V$ and let $\mu$ be the fuzzy subset of $X$ defined by $\mu(ab) = 0.3$, $\mu(bc) = \mu(ac) = 0.5$, $\mu(ec) = \mu(cd) = 1$, $\mu(de) = 0.9$. Then $\nexists t \in (0, 1]$ such that $(\mathrm{Supp}(\sigma), \mu^t)$ is a tree. However, $(\sigma, \mu)$ is a fuzzy tree (see Fig. 2.6).

**Theorem 2.3.12** *If $G = (\sigma, \mu)$ is a fuzzy tree and $(\sigma^*, \mu^*)$ is not a tree, then there exists at least one edge $uv$ in $Supp(\mu)$ for which $\mu(uv) < \mu^\infty(u, v)$.*

*Proof* If $G$ is a fuzzy tree, then by definition there exists a fuzzy spanning subgraph $F = (\sigma, \nu)$, which is a tree and $\mu(uv) < \nu^\infty(u, v)$ for all edges $uv$ not in $F$. Also, $\nu^\infty(u, v) \leq \mu^\infty(u, v)$. Thus, $\mu(uv) < \mu^\infty(u, v)$ for all $uv$ not in $F$ and by hypothesis there exists at least one edge $uv$ not in $F$. ■

The rest of the results in this section are from [167, 168].

**Theorem 2.3.13** *Let $G = (\sigma, \mu)$ be a connected fuzzy graph with no fuzzy cycle. Then $G$ is a fuzzy tree.*

*Proof* If $G^*$ has no cycles, then $G^*$ is a tree and $G$ is a fuzzy tree. So assume that $G$ has cycles and by hypothesis no cycle is a fuzzy cycle. That is, every cycle in $G$ will have exactly one weakest edge in it. Remove the weakest edge say $e$ in a cycle $C$ of $G$. If there are still cycles in the resulting fuzzy graph, repeat the process, which will eventually results in a fuzzy subgraph, which is a tree, and which is the required spanning subgraph $F$. ■

When we delete a bridge, different fuzzy graph structures behave differently. Consider the case of fuzzy trees in the following theorem.

**Theorem 2.3.14** *If $G$ is a fuzzy tree, then the removal of any fuzzy bridge reduces the strength of connectedness between its end vertices and also between some other pair of vertices.*

*Proof* Let $G = (\sigma, \mu)$ be a fuzzy tree and let $uv$ be a fuzzy bridge of $G$. Then by Proposition 2.3.4 $uv$ is an edge of the maximum spanning tree $T$ of $G$ and $T$ contains unique strongest paths joining every pair of vertices. So removal of $uv$ reduces the strength of connectedness between some other pair of vertices $x, y$, where $x$ is adjacent to $u$ and $y$ is adjacent to $v$ if $uv$ is an internal edge of $T$, and $u = x$ or $v = y$ otherwise. ■

Note that when $G^*$ is $K_2$, its unique edge is a fuzzy bridge and its removal reduces the strength of connectedness between its end vertices alone.

**Theorem 2.3.15** *If $G = (\sigma, \mu)$ is a fuzzy tree, then $G$ is not complete.*

*Proof* If possible, let $G$ be a complete fuzzy graph. Then $\mu^\infty(u, v) = \mu(uv)$ for all $u, v$. Now, $G$ being a tree, $\mu(uv) < \nu^\infty(u, v)$ for all $u, v$ not in $F$. Thus, $\mu^\infty(u, v) < \nu^\infty(u, v)$, which is impossible. ■

**Theorem 2.3.16** *If $G$ is a fuzzy tree, then the internal vertices of $F$ are fuzzy cutvertices of $G$.*

*Proof* Let $w$ be any vertex in $G$, which is not an end vertex of $F$. Then $w$ is the common vertex of at least two edges in $F$, which are fuzzy bridges of $G$ and by Theorem 2.2.11, $w$ is a fuzzy cutvertex. Also, if $w$ is an end vertex of $F$, then $w$ is not a fuzzy cutvertex; else there would exist $u, v$ distinct from $w$ such that $w$ is on every $u$-$v$ path and one such path certainly lies in $F$. But because $w$ is an end vertex of $F$, this is not possible. ■

**Corollary 2.3.17** *A fuzzy cutvertex of a fuzzy tree is the common vertex of at least two fuzzy bridges.*

**Theorem 2.3.18** ([167]) *Let $G = (\sigma, \mu)$ be a fuzzy graph. Then $G$ is a fuzzy tree if and only if the following conditions are equivalent for all $u, v \in V$.*

*(i)* *$uv$ is a fuzzy bridge.*
*(ii)* *$\mu^\infty(u, v) = \mu(uv)$.*

*Proof* Let $G = (\sigma, \mu)$ be a fuzzy tree and suppose that $uv$ is a fuzzy bridge. Then $\mu^\infty(u, v) = \mu(uv)$ by Theorem 2.2.14. Now, let $uv$ be an edge in $G$ such that $\mu^\infty(u, v) = \mu(uv)$. If $G^*$ is a tree, then clearly $uv$ is a fuzzy bridge; otherwise, it follows from Theorem 2.3.12 that $uv$ is in $F$ and $uv$ is a bridge.

Conversely, assume that $(i)$ and $(ii)$ are equivalent. Construct a maximum spanning tree $T = (\sigma, \nu)$ for $G$ [38]. If $uv$ is in $T$, by an algorithm in [38], $\mu^{\infty}(u, v) = \mu(uv)$ and hence $uv$ is a fuzzy bridge. Now, these are the only fuzzy bridges for $G$; for, if possible, let $u'v'$ be a fuzzy bridge of $G$, which is not in $T$. Consider a cycle $C$ consisting of $u'v'$ and the unique $u'$-$v'$ path in $T$. Now, edges of this $u'$-$v'$ path are fuzzy bridges and so they are not weakest edges of $C$ and thus $u'v'$ must be the weakest edge of $C$ and cannot be a fuzzy bridge.

Moreover, for all edges $u'v'$ not in $T$, we have $\mu'(u'v') < \nu'(u'v')$; for if possible let $\mu(u'v') \geq \nu^{\infty}(u', v')$. But $\nu^{\infty}(u', v') < \mu^{\infty}(u', v')$, were strict inequality holds because $u'v'$ is not a fuzzy bridge. Hence, $\nu^{\infty}(u', v') < \mu^{\infty}(u', v')$, which gives a contradiction because $\nu^{\infty}(u', v')$ is the strength of the unique $u'$-$v'$ path in $T$ and by the algorithm in [36], $\mu^{\infty}(u', v') = \nu^{\infty}(u', v')$. Thus, $T$ is the required spanning subgraph $F$, which is a tree and hence $G$ is a fuzzy tree. ■

From the previous theorem, it follows that the spanning fuzzy subgraph of a fuzzy tree is unique. Also, it follows that $F$ is nothing but the maximum fuzzy spanning tree of $G$. Thus, we have the following theorem.

**Theorem 2.3.19** *A fuzzy graph is a fuzzy tree if and only if it has a unique maximum fuzzy spanning tree.*

If $G$ is a fuzzy graph such that $G^*$ is not a tree and $T$ is the maximum fuzzy spanning tree of $G$, then there is at least one edge in $G$ which is not a fuzzy bridge. Also, edges not in $T$ are not fuzzy bridges of $G$. So we have the following result.

**Theorem 2.3.20** *If $G = (\sigma, \mu)$ is a fuzzy graph with $Supp(\sigma) = V$ and $|V| = p$, then $G$ has at most $p - 1$ fuzzy bridges.*

If $G$ is a fuzzy graph with $T$ as its unique maximum fuzzy spanning tree, then end vertices of $T$ are not fuzzy cutvertices $G$. Thus, every fuzzy graph will have at least two vertices which are not fuzzy cutvertices.

## 2.4 Fuzzy Cut Sets

This section is based on [127], a work by Mordeson and Nair in 1996. We begin by some topics in graph theory, which can be found in [83]. When $G$ is a graph, one can associate $G$ with two vector spaces over the field of scalars $\mathbb{Z}_2 = \{0, 1\}$, where addition and multiplication are modulo 2. Note that for $1 \in \mathbb{Z}_2$, $1 + 1 = 0$. Let $V(G) = \{v_1, v_2, \ldots, v_n\}$ and edge set $E(G) = \{e_1, e_2, \ldots, e_m\}$. A **0-chain** of $G$ is a formal linear combination $\sum \epsilon_i v_i$ of vertices and a **1-chain** is a formal linear combination of edges $\sum \epsilon_i e_i$, where $\epsilon_i \in \mathbb{Z}_2$. The **boundary operator** $\partial$ is a linear function which maps 1-chains to 0-chains such that if $e = xy$, then $\partial(e) = x + y$. The **coboundary operator** $\delta$ is a linear function which maps 0-chains to 1-chains such that $\delta(v) = \sum \epsilon_i e_i$ whenever $e_i$ is incident with $v$.

*Example 2.4.1* Let $G = (V, E)$, where $V = \{v_1, v_2, \ldots, v_6\}$ and $E(G) = \{e_1, e_2, \ldots, e_9\}$, where $e_1 = v_1v_2$, $e_2 = v_1v_3$, $e_3 = v_2v_3$, $e_4 = v_2v_4$, $e_5 = v_2v_5$, $e_6 = v_3v_5$, $e_7 = v_3v_6$, $e_8 = v_4v_5$ and $e_9 = v_5v_6$. The 1-chain $\gamma_1 = e_1 + e_2 + e_4 + e_9$ has boundary

$$\begin{aligned}\partial(\gamma_1) &= (v_1 + v_2) + (v_1 + v_3) + (v_2 + v_4) + (v_5 + v_6)\\ &= v_3 + v_4 + v_5 + v_6.\end{aligned}$$

The 0-chain $\gamma_0 = v_3 + v_4 + v_5 + v_6$ has coboundary

$$\begin{aligned}\delta(\gamma_0) &= (e_2 + e_3 + e_6 + e_7) + (e_4 + e_8) + (e_5 + e_6 + e_8 + e_9) + (e_7 + e_9)\\ &= e_2 + e_3 + e_4 + e_5.\end{aligned}$$

A 1-chain with boundary 0 is called a **cycle vector** of $G$ which can be visualized as a set of edge disjoint cycles. The collection of all cycle vectors is called the **cycle space** of $G$ and it is clearly a vector space over $\mathbb{Z}_2$. A cut set of a connected graph is a collection of edges whose removal results in a disconnected graph. A **cocycle** is a minimal cutset. A **coboundary** of $G$ is the coboundary of some 0-chain in $G$. The coboundary of a subset of $V$ is the set of all edges joining a point in this subset to a point not in the subset. Hence, every coboundary is a cutset. Because any minimal cutset is a coboundary, a cocycle is just a minimal nonzero coboundary. The collection of all coboundaries of $G$ is a vector space over $\mathbb{Z}_2$ and is called the **cocycle space** of $G$. A basis of this space which consists entirely of cocycles is called a **cocycle basis** for $G$.

Let $G$ be a connected graph. A **chord** of a spanning tree $T$ of $G$ is an edge of $G$ which is not in $T$. The subgraph of $G$ consisting of $T$ and any chord of $T$ has only one cycle. The set $C(T)$ of cycles obtained in this way is independent. Every cycle $C$ depends on the set $C(T)$ because $C$ is the symmetric difference of the cycles determined by the chords of $T$ which lie in $C$. The **cycle rank** $m(G)$ is defined to be the number of cycles in a basis for the cycle space of $G$. Thus, we have the following result.

**Theorem 2.4.2** *The cycle rank of a connected graph G is equal to the number of chords of any spanning tree in G.*

Similar results can be derived for the cocycle space. Assume that $G$ is a connected graph. The **cotree** $T'$ of a spanning tree $T$ of $G$ is the spanning subgraph of $G$ containing exactly those edges which are not in $T$. A cotree of $G$ is the cotree of some spanning tree $T$. The edges of $G$ which are not in $T'$ are called its **twigs**. The subgraph of $G$ consisting of $T'$ and any one of its twigs contains exactly one cocycle. The collection of cocycles obtained by adding twigs to $T'$, one at a time is a basis for the cocycle space of $G$. The cocycle rank $m'(G)$ is the number of cocycles in a basis for the cocycle space of $G$.

**Theorem 2.4.3** *The cocycle rank of a connected graph G is the number of twigs in any spanning tree T of G.*

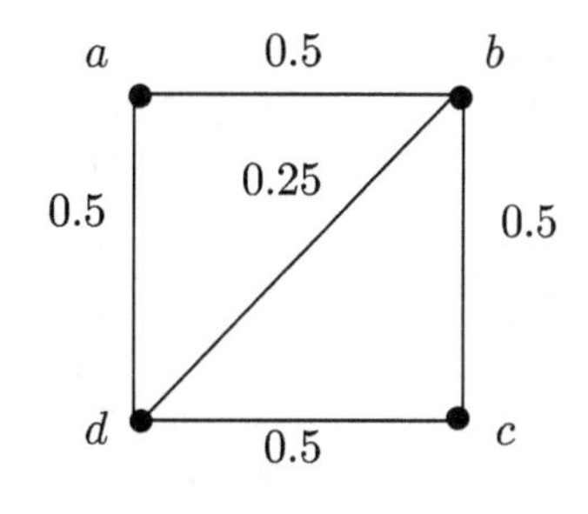

**Fig. 2.7** Fuzzy graph having no fuzzy bridges

**Definition 2.4.4** Let $G = (\sigma, \mu)$ be a fuzzy graph. Let $x \in V$ and let $t \in [0, 1]$. Define the fuzzy subset $x_t$ of $V$ by for all $y \in V$, $x_t(y) = 0$ if $y \neq x$ and $x_t(y) = t$ if $y = x$. Then $x_t$ is called a **fuzzy singleton** in $V$. If $xy \in E$, then $xy_{\mu(xy)}$ denotes a fuzzy singleton in $E$.

**Definition 2.4.5** Let $G = (\sigma, \mu)$ be a fuzzy graph and let $S$ be a subset of Supp$(\mu)$. Then

(i) $\{xy_{\mu(xy)} \mid xy \in S\}$ is called a **cut set** of $(\sigma, \mu)$ if $S$ is a cut set of (Supp$(\sigma)$, Supp$(\mu)$).
(ii) $\{xy_{\mu(xy)} \mid xy \in S\}$ is called a **fuzzy cut set** of $(\sigma, \mu)$ if $\exists\, u, v \in$ Supp$(\sigma)$ such that $\mu'^{\infty}(u, v) < \mu^{\infty}(u, v)$, where $\mu'$ is the fuzzy subset of $E$ defined by $\mu' = \mu$ on Supp$(\mu)$ and $\mu'(xy) = 0$ for all $xy \in S$.

When $S$ is a singleton set, a cut set is called a **bridge** and a fuzzy cut set is a **fuzzy bridge**.

*Example 2.4.6* In this example, we show there is a fuzzy graph $(\sigma, \mu)$ that has no fuzzy bridges and $\mu$ is not a constant function. Let $V = \{a, b, c, d\}$ and $x = \{ab, bc, cd, da, bd\}$. Let $\sigma(x) = 1$ for all $x \in V$, $\mu(ab) = \mu(bc) = \mu(cd) = \mu(da) = 1$ and $\mu(bd) = 0.25$. Then $\mu$ is not a constant, but $(\sigma, \mu)$ does not have a fuzzy bridge because the strength of connectedness between any pair of vertices of $(\sigma, \mu)$ remains 0.5 even after the removal of an edge as seen from Fig. 2.7.

**Theorem 2.4.7** *Let $G = (\sigma, \mu)$ be a fuzzy graph. Let $V = \{v_1, \ldots, v_n\}$ and $C = \{v_1v_2, v_2v_3, \ldots, v_{n-1}v_n, v_nv_1\}$, $n \geq 3$.*

*(i) Suppose that $C \subseteq$ Supp$(\mu)$ and that for all $v_jv_k \in$ Supp$(\mu)\backslash C$, $\mu(v_jv_k) < \vee\{\mu(v_iv_{i+1}) \mid i = 1, \ldots, n\}$, where $v_{n+1} = v_1$. Then either $\mu$ is a constant function on $C$ or $G$ has a fuzzy bridge.*
*(ii) If $\emptyset \neq$ Supp$(\mu) \subset C$, then $G$ has a fuzzy bridge.*

*Proof* $(i)$ Suppose $\mu$ is not constant on $C$. Let $v_hv_{h+1} \in C$ be such that $\mu(v_hv_{h+1}) = \vee\{\mu(v_iv_{i+1}) \mid i = 1, \ldots, n\}$. Because $\mu$ is not constant on $C$, the strength of the path $C\backslash\{v_hv_{h+1}\}$ between $v_h$ and $v_{h+1}$ is strictly less than $\mu(v_hv_{h+1})$. The strength of any other path $P$ between $v_h$ and $v_{h+1}$ is also strictly less than $\mu(v_hv_{h+1})$ because $P$ must contain an edge from Supp$(\mu)\backslash C$. Thus, $v_hv_{h+1\mu(v_hv_{h+1})}$ is a fuzzy bridge.
$(ii)$ The result here is immediate. ■

**Theorem 2.4.8** *Let $G = (\sigma, \mu)$ be a fuzzy graph. Suppose that the dimension of the cycle space of $(Supp(\sigma), Supp(\mu))$ is 1. Then G does not have a fuzzy bridge if and only if G is a cycle and $\mu$ is a constant function.*

*Proof* Suppose it is not the case that $(\sigma, \mu)$ is a cycle and $\mu$ is a constant function. If $(\sigma, \mu)$ is not a cycle, then there exists $xy \in \text{Supp}(\mu)$ which is not part of a cycle. Then $xy_{\mu(xy)}$ is a bridge and hence a fuzzy bridge. Suppose that $(\sigma, \mu)$ is a cycle, but $\mu$ is not a constant function. Let $xy \in \text{Supp}(\mu)$ be such that $\mu(xy)$ is maximal. Then $xy_{\mu(xy)}$ is a fuzzy bridge.

Conversely, suppose that $(\sigma, \mu)$ is a cycle and $\mu$ is not a constant function. Then the deletion of an edge $v_i v_{i+1}$ yields a unique path between $v_i$ and $v_{i+1}$ of strength equal to $\mu(v_i v_{i+1})$. Thus, $v_i v_{i+1\mu(v_i v_{i+1})}$ is not a fuzzy bridge. ■

Several other concepts like fuzzy chords, fuzzy cotrees, and fuzzy twigs can also be found in [127].

## 2.5 Bridges, Cutsets, and Blocks

In 1985, Delgado, Verdegay, and Vila [62], defined the notions of connectedness, fuzzy cycles, and fuzzy trees differently than Rosenfeld. They used the notion of level sets to define these terms. They pointed out some valid reasons for their definitions. For example, they noted that a fuzzy graph may have different degrees of connectedness and that two fuzzy graphs may share the property that neither is connected, but there is a $t$-cut of one which is connected while no $t$-cut of the other is connected.

Later in 2002, Mordeson and Yao [131] studied this further and obtained several new results on connectedness by levels. The work in this section is from [131]. Connectivity analysis by levels is important in any interconnection network. The structural properties of finite fuzzy graphs provide tools for the solutions of Operations Research problems. In this section, several connectedness properties of various types of fuzzy graph structures are discussed. Level graphs are used to define different variants.

**Definition 2.5.1** Let $d(\mu) = \wedge\{\mu(xy) \mid xy \in \mu^*\}$ and $h(\mu) = \vee\{\mu(xy) \mid xy \in \mu^*\}$. Then $d(\mu)$ is called the **depth** of $\mu$ and $h(\mu)$ is called the **height** of $\mu$.

Note that $d(\mu)$ and $h(\mu)$ are undefined in Definition 2.5.1 if $\mu^* = \emptyset$.

**Definition 2.5.2** Let $xy \in \mu^*$. Then

(i) $xy$ is called a **bridge** if $xy$ is a bridge of $(\sigma^*, \mu^*)$.
(ii) $xy$ is called a **fuzzy bridge** if $\mu'^{\infty}(u, v) < \mu^{\infty}(u, v)$ for some $uv \in \mu^*$, where $\mu'$ is $\mu$ restricted to $E \setminus \{xy\}$.
(iii) $xy$ is called a **weak fuzzy bridge** if $\exists\, t \in (0, h(\mu)]$ such that $xy$ is a bridge for $G^t$.

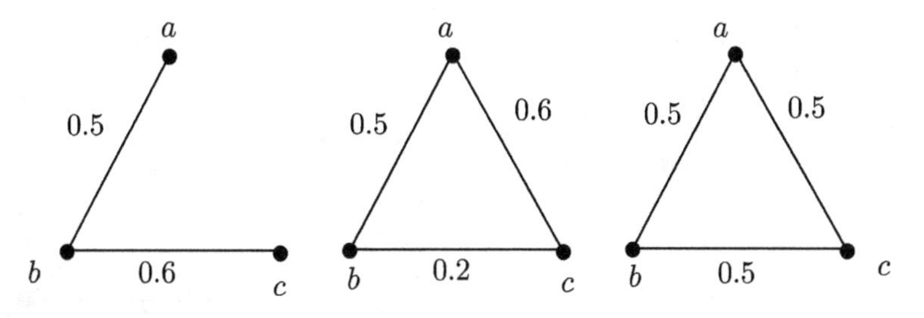


**Fig. 2.8** Different types of bridges in a fuzzy graph

(iv) $xy$ is called a **partial fuzzy bridge** if $xy$ is a bridge for $G^t$ for all $t \in (d(\mu), h(\mu)] \cup \{h(\mu)\}$.

(v) $xy$ is called a **full fuzzy bridge** if $xy$ is a bridge for $G^t$ for all $t \in (0, h(\mu)]$.

We note that in $(iv)$ of Definition 2.5.2, $(d(\mu), h(\mu)] = \emptyset$ if $d(\mu) = h(\mu)$.

*Example 2.5.3* Let $V = \{a, b, c\}$. Define the fuzzy subsets $\sigma$ of $V$ and $\mu$ of $E = \{ab, bc\}$ as follows: $\sigma(a) = \sigma(b) = \sigma(c) = 1$ and $\mu(ab) = 0.5$, $\mu(bc) = 0.6$. Then $d(\mu) = 0.5$ and $h(\mu) = 0.6$. For $0 < t \leq 0.5$, $G^t = (V, \{ab, bc\})$ and for $0.5 < t \leq 0.6$, $G^t = (V, \{bc\})$. Hence, $bc$ is a full fuzzy bridge and $ab$ is a weak fuzzy bridge, but not a partial fuzzy bridge. Both $ab$ and $bc$ are bridges and fuzzy bridges.

*Example 2.5.4* Let $V = \{a, b, c\}$. Define the fuzzy sets $\sigma$ of $V$ and $\mu$ of $E = \{ab, bc, ac\}$ as follows: $\sigma(a) = \sigma(b) = \sigma(c) = 1$, $\mu(ab) = 0.5$, $\mu(ac) = 0.6$, and $\mu(bc) = 0.2$. Then $d(\mu) = 0.2$ and $h(\mu) = 0.6$. For $0 < t \leq 0.2$, $G^t = (V, \{ab, bc, ac\})$, for $0.2 < t \leq 0.5$, $G^t = (V, \{ab, ac\})$, and for $0.5 < t \leq 0.6$, $G^t = (V, \{ac\})$. Then $ac$ is a fuzzy bridge and a partial fuzzy bridge, but not a full fuzzy bridge and not a bridge. The edge $bc$ is not any of the five types of bridges.

*Example 2.5.5* Let $V = \{a, b, c\}$. Define the fuzzy subsets $\sigma$ of $V$ and $\mu$ of $E = \{ab, bc, ac\}$ as follows: $\sigma(a) = \sigma(b) = \sigma(c) = 1$ and $\mu(ab) = \mu(bc) = \mu(ac) = 0.5$. Then $G$ has no bridges of any of the five types.

The fuzzy graphs of Examples 2.5.3–2.5.5 are given in Fig. 2.8.

*Example 2.5.6* Let $V = \{a, b, c, d\}$. Define the fuzzy subsets $\sigma$ of $V$ and $\mu$ of $E = \{ab, bc, cd, ac\}$ as follows: $\sigma(a) = \sigma(b) = \sigma(c) = \sigma(d) = 1$ and $\mu(ab) = 0.3 = \mu(bc)$, $\mu(ac) = 0.9 = \mu(cd)$. Then $d(\mu) = 0.3$ and $h(\mu) = 0.8$. For $0 < t \leq 0.3$, $G^t = (V, \{ab, bc, cd, ac\})$ and for $0.3 < t \leq 0.8$, $G^t = (V, \{ac, cd\})$. Hence, $cd$ is a full fuzzy bridge and $ac$ is a partial fuzzy bridge, but not a full fuzzy bridge (Fig. 2.9).

**Proposition 2.5.7** *$xy$ is a full fuzzy bridge if and only if $xy$ is a bridge for $G^*$ and $\mu(xy) = h(\mu)$.*

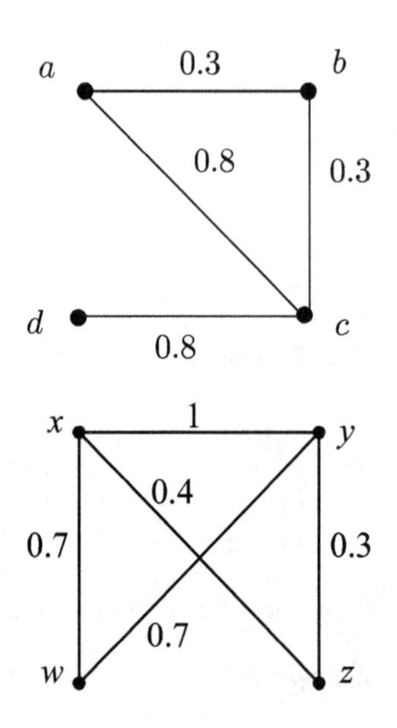


**Fig. 2.9** Fuzzy graph with a full fuzzy bridge

**Fig. 2.10** Fuzzy graph in Example 2.5.9

*Proof* Suppose $xy$ is a full fuzzy bridge. Then $xy$ is a bridge for $G^t$ for all $t \in (0, h(\mu)]$. Hence, $xy \in \mu^{h(\mu)}$ and so $\mu(xy) = h(\mu)$. Because $xy$ is a bridge for $G^t$ for all $t \in (0, h(\mu)]$, it follows that $xy$ is a bridge for $G^*$ because $\sigma^* = \sigma^{d(\mu)}$ and $\mu^* = \mu^{d(\mu)}$. Conversely, suppose that $xy$ is a bridge for $G^*$ and $\mu(xy) = h(\mu)$. Then $xy \in \mu^t$ for all $t \in (0, h(\mu)]$. Thus, because $xy$ is also a bridge for $G^*$, $xy$ is a bridge for $G^t$ for all $t \in (0, h(\mu)]$, because each $G^t$ is a subgraph of $G^*$. Hence, $xy$ is a full fuzzy bridge. ■

**Proposition 2.5.8** *Suppose that $xy$ is not contained in a cycle of $G^*$. Then the following conditions are equivalent.*

*(i)* $\mu(xy) = h(\mu)$.
*(ii)* *$xy$ is a partial fuzzy bridge.*
*(iii)* *$xy$ is a full fuzzy bridge.*

*Proof* Because $xy$ is not contained in a cycle of $G^*$, $xy$ is a bridge of $G^*$. Hence, by Proposition 2.5.7, (i) $\Leftrightarrow$ (iii). Clearly, (iii) $\Rightarrow$ (ii). Suppose that (ii) holds. Then $xy$ is a bridge for $G^t$ for all $t \in (d(\mu), h(\mu)]$ and so $xy \in \mu^{h(\mu)}$. Hence, $\mu(xy) = h(\mu)$, i.e., (i) holds. ■

*Example 2.5.9* Consider the fuzzy graph $G = (\sigma, \mu)$ with $\sigma^* = \{x, y, z, w\}$ and $\mu^* = \{xz, xw, wy, yz\}$ (Fig. 2.10). Define the fuzzy subsets $\sigma$ and $\mu$ as follows: $\sigma(x) = \sigma(y) = \sigma(z) = \sigma(w) = 1$, $\mu(xy) = 1$, $\mu(xz) = 0.4$, $\mu(xw) = 0.7$, $\mu(wy) = 0.7$, and $\mu(yz) = 0.3$. Then $d(\mu) = 0.3$ and $h(\mu) = 1$. For $0 < t \leq 0.3$, $G^t = (V, \{xy, xz, xw, wy, yz\})$. For $0.3 < t \leq 0.4$, $G^t = (V, \{xy, xz, xw, wy\})$, for $0.4 < t \leq 0.7$, $G^t = (V, \{xy, xw, yz\})$, and for $0.7 < t \leq 1$, $G^t = (V, \{xy\})$. Then $xy$ is in a cycle of $G^*$, $xy$ is not a partial fuzzy bridge, and $\mu(xy) = h(\mu)$. Also, $xy$ is a weak fuzzy bridge and a fuzzy bridge, but not a bridge (Fig. 2.10).

**Proposition 2.5.10** *If $xy$ is a bridge, then $xy$ is a weak fuzzy bridge and a fuzzy bridge.*

*Proof* $xy$ is a bridge $\Leftrightarrow$ $xy$ is a bridge for $G^*$ $\Leftrightarrow$ $xy$ is a bridge for $G^{d(\mu)}$, because $G^* = G^{d(\mu)}$ $\Rightarrow$ $xy$ is a weak fuzzy bridge. $xy$ is a bridge implies that its removal disconnects $G^*$ and so $xy$ is a fuzzy bridge. ■

**Theorem 2.5.11** *$xy$ is a fuzzy bridge if and only if $xy$ is a weak fuzzy bridge.*

*Proof* Suppose $xy$ is a weak fuzzy bridge. Then $\exists\ t \in (0, h(\mu)]$ such that $xy$ is a bridge for $G^t$. Hence, the removal of $xy$ disconnects $G^t$. Thus, any path from $x$ to $y$ in $G$ has an edge $uv$ with $\mu(uv) < t$. Hence, the removal of $xy$ results in $\mu'^{\infty}(x, y) < t \leq \mu^{\infty}(x, y)$. Thus, $xy$ is a fuzzy bridge. Conversely, suppose $xy$ is a fuzzy bridge. Then $\exists\ u, v$ such that removal of $xy$ results $\mu'^{\infty}(u, v) < \mu^{\infty}(u, v)$. Hence, $xy$ is on every strongest path connecting $u$ and $v$ and in fact, $\mu(xy)$ is greater than or equal to this value. Thus, there does not exist a path (other than $xy$) connecting $x$ and $y$ in $G^{\mu(xy)}$, else this other path without $xy$ would be of strength $\geq \mu(xy)$ and would be part of a strongest path connecting $u$ and $v$, contrary to the fact $xy$ is on every such path. Hence, $xy$ is a bridge of $G^{\mu(xy)}$ and $0 < \mu(xy) \leq h(\mu)$. Thus, $\mu(xy)$ is a desired $t$. ■

**Definition 2.5.12** Let $x \in V$.

(i) $x$ is called a **cutvertex** if $x$ is a cutvertex of $G^*$.
(ii) $x$ is called a **fuzzy cutvertex** if $\exists\ u, v \in V \setminus \{x\}$ such that $\mu'^{\infty}(u, v) < \mu^{\infty}(u, v)$, where $\mu'$ is $\mu$ restricted to $E \setminus \{xz, zx\} \mid z \in V\}$.
(iii) $x$ is called a **weak fuzzy cutvertex** if $\exists\ t \in (0, h(\mu)]$ such that $x$ is a cutvertex for $G^t$.
(iv) $x$ is called a **partial fuzzy cutvertex** if $x$ is a cutvertex for $G^t$ for all $t \in (d(\mu), h(\mu))] \cup \{h(\mu)\}$.
(v) $x$ is called a **full fuzzy cutvertex** if $x$ is a cutvertex for $G^t$ for all $t \in (0, h(\mu)]$.

*Example 2.5.13* Consider the fuzzy graph $G = (\sigma, \mu)$ with $\sigma^* = \{x, y, z\}$ and $\mu^* = \{xy, xz, yz\}$. Let the fuzzy subsets $\sigma$ and $\mu$ be defined as $\sigma(x) = \sigma(y) = \sigma(z) = 1$, $\mu(xy) = 0.5$, $\mu(xz) = 0.4$, and $\mu(yz) = 0.3$. Then $d(\mu) = 0.3$ and $h(\mu) = 0.5$. For $0 < t \leq 0.3$, $G^t = (V, \{xy, xz, yz\})$, for $0.3 < t \leq 0.4$, $G^t = (V, \{xy, xz\})$, and for $0.4 < t \leq 0.5$, $G^t = (V, \{xy\})$. Thus, $x$ is a fuzzy cutvertex and a weak fuzzy cutvertex, but neither a cutvertex nor a partial cutvertex.

*Example 2.5.14* Consider the fuzzy graph $G = (V, \sigma, \mu)$ with $V = \{x, y, z\}$. Define the fuzzy subsets $\sigma$ of $V$ and $\mu$ of $E = \{xy, xz, yz\}$ as follows: $\sigma(x) = \sigma(y) = \sigma(z) = 1$ and $\mu(xy) = \mu(xz) = 0.7$ and $\mu(yz) = 0.1$. Then $d(\mu) = 0.1$ and $h(\mu) = 0.7$. For $0 < t \leq 0.1$, $G^t = (V, \{xy, xz, yz\})$ and for $0.1 < t \leq 0.7$, $G^t = (V, \{xy, xz\})$. Thus, $x$ is a fuzzy cutvertex and a partial fuzzy cutvertex, but neither a cutvertex nor a full fuzzy cutvertex.

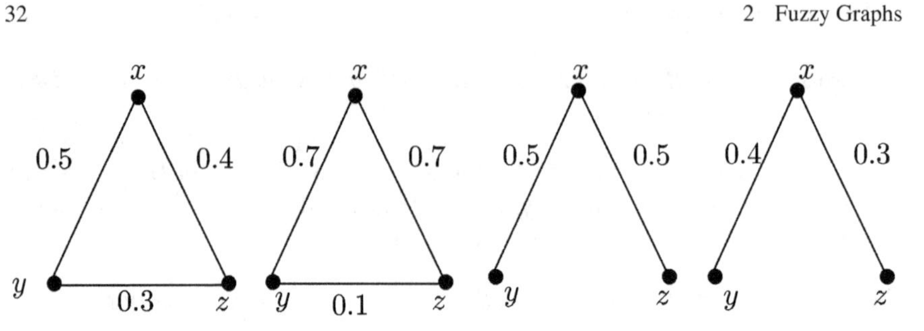


**Fig. 2.11** Fuzzy Graphs in Examples 2.5.13–2.5.16

*Example 2.5.15* Let $V = \{x, y, z\}$. Define the fuzzy subsets $\sigma$ of $V$ and $\mu$ of $E = \{xy, xz\}$ as follows: $\sigma(x) = \sigma(y) = \sigma(z) = 1$ and $\mu(xy) = \mu(xz) = 0.5$. Then $d(\mu) = h(\mu) = 0.5$. For $0 < t \leq 0.5$, $G^t = (V, \{xy, xz\})$. Thus, $x$ is a full fuzzy cutvertex, a fuzzy cutvertex, and a cutvertex.

*Example 2.5.16* Let $V = \{x, y, z\}$. Define the fuzzy subsets $\sigma$ of $V$ and $\mu$ of $E = \{xy, xz\}$ as follows: $\sigma(x) = \sigma(y) = \sigma(z) = 1$ and $\mu(xy) = 0.4$ and $\mu(xz) = 0.3$. Then $d(\mu) = 0.3$ and $h(\mu) = 0.4$. For $0 < t \leq 0.3$, $G^t = (V, \{xy, xz\})$ and for $0.3 < t \leq 0.4$, $G^t = (V, \{xy\})$. Thus, $x$ is a cutvertex, a fuzzy cutvertex, and a weak fuzzy cutvertex, but not a partial fuzzy cutvertex.

The fuzzy graphs in Examples 2.5.13–2.5.16 are given in Fig. 2.11.

**Definition 2.5.17** (i) $G$ is called a **block** if $G^*$ is a block.

(ii) $G$ is called a **fuzzy block** if it has no fuzzy cutvertices.

(iii) $G$ is called a **weak fuzzy block** if $\exists\, t \in (0, h(\mu)]$ such that $G^t$ is a block.

(iv) $G$ is called a **partial fuzzy block** if $G^t$ is a block for all $t \in (d(\mu), h(\mu)] \cup \{h(\mu)\}$.

(v) $G$ is called a **full fuzzy block** if $G^t$ is a block for all $t \in (0, h(\mu)]$.

*Example 2.5.18* Consider the fuzzy graph $G = (\sigma, \mu)$ with $\sigma^* = \{x, y, z\}$ and $\mu^* = \{xy, yz, xz\}$. The fuzzy subsets $\sigma$ and $\mu$ are defined as follows: $\sigma(x) = \sigma(y) = \sigma(z) = 1$ and $\mu(xy) = \mu(yz) = 0.6$, and $\mu(xz) = 0.7$. Then $d(\mu) = 0.6$ and $h(\mu) = 0.7$. For $0 < t \leq 0.6$, $G^t = (V, \{xy, yz, xz\})$ and for $0.6 < t \leq 0.7$, $G^t = (V, \{xz\})$. Thus, $G$ is a block, a fuzzy block, and a weak fuzzy block. $G$ is not a partial fuzzy block because $G^t$ is not a block for $0.5 < t \leq 0.9$; it is not connected.

*Example 2.5.19* Consider the fuzzy graph $G = (\sigma, \mu)$ with $\sigma^* = \{x, y, z\}$ and $\mu^* = \{xy, yz, xz\}$. The fuzzy subsets $\sigma$ and $\mu$ are defined as follows: $\sigma(x) = \sigma(y) = \sigma(z) = 1$ and $\mu(xy) = \mu(xz) = 0.8$ and $\mu(yz) = 0.7$. Then $d(\mu) = 0.7$ and $h(\mu) = 0.8$. For $0 < t \leq 0.7$, $G^t = (V, \{xy, xz, yz\})$ and for $0.7 < t \leq 0.8$, $G^t = (V, \{xy, xz\})$. Thus, $G$ is a block and a weak fuzzy block because $G$ is a block for $0 < t \leq 0.7$. However, $G$ is not a fuzzy block because $x$ is a fuzzy cutvertex of $G$. Also, $G$ is not a partial fuzzy block because $x$ is a cutvertex for $G^t$ for $0.7 < t \leq 0.8$.

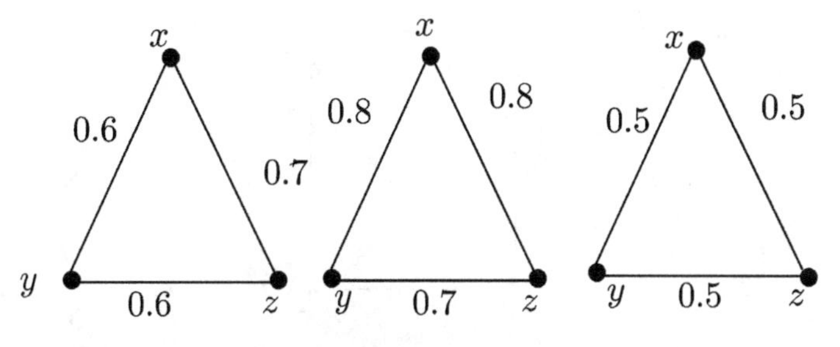


**Fig. 2.12** Examples given in Examples 2.5.18–2.5.20

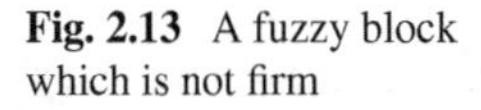

**Fig. 2.13** A fuzzy block which is not firm

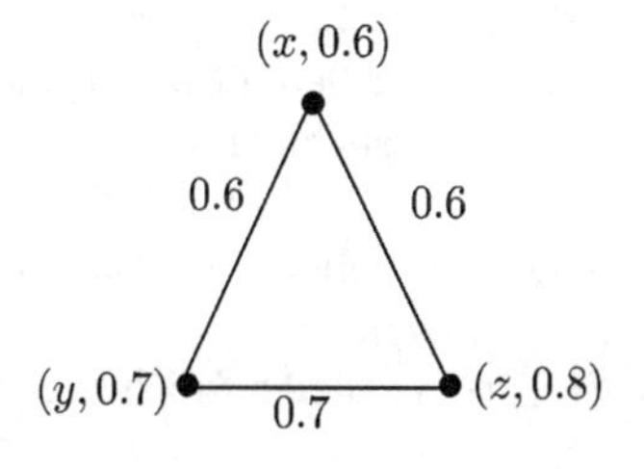


*Example 2.5.20* Consider the fuzzy graph $G = (\sigma, \mu)$ with $\sigma^* = \{x, y, z\}$ and $\mu^* = \{xy, yz, xz\}$. The fuzzy subsets $\sigma$ and $\mu$ are defined as follows: $\sigma(x) = \sigma(y) = \sigma(z) = 1$ and $\mu(xy) = \mu(xz) = \mu(yz) = 0.5$. Then $d(\mu) = h(\mu) = 0.5$. For $0 < t \leq 0.5$, $G^t = (V, \{xy, xz, yz\})$. Thus, $G$ is a block, a fuzzy block and a full fuzzy block.

The fuzzy graphs of Examples 2.5.18–2.5.20 are given in Fig. 2.12.

**Definition 2.5.21** $G$ is said to be **firm** if $\wedge\{\sigma(x) \mid x \in V\} \geq \vee(\mu(xy) \mid xy \in \mu^*\}$.

To this point all examples of fuzzy graphs except Fig. 2.2 in Example 2.2.6 have been firm.

*Example 2.5.22* Consider the fuzzy graph $G = (\sigma, \mu)$ with $\sigma^* = \{x, y, z\}$ and $\mu^* = \{xy, yz, xz\}$. The fuzzy subsets $\sigma$ and $\mu$ are defined as follows: $\sigma(x) = 0.6$, $\sigma(y) = 0.7$, $\sigma(z) = 0.8$, and $\mu(xy) = \mu(xz) = 0.6$, and $\mu(yz) = 0.7$. Then $d(\mu) = 0.6$ and $h(\mu) = 0.7$. For $0 < t \leq 0.6$, $G^t = (V, \{xy, xz, yz\})$ and for $0.6 < t \leq 0.7$, $G^t = (V, \{yz\})\})$. Thus, $G$ is a block, a fuzzy block, and a full fuzzy block. We note that $G$ is not firm (Fig. 2.13).

## 2.6 Cycles and Trees

In this section, we discuss the connectedness properties of cycles and trees in fuzzy graphs by levels. This is a continuation of results from [131].

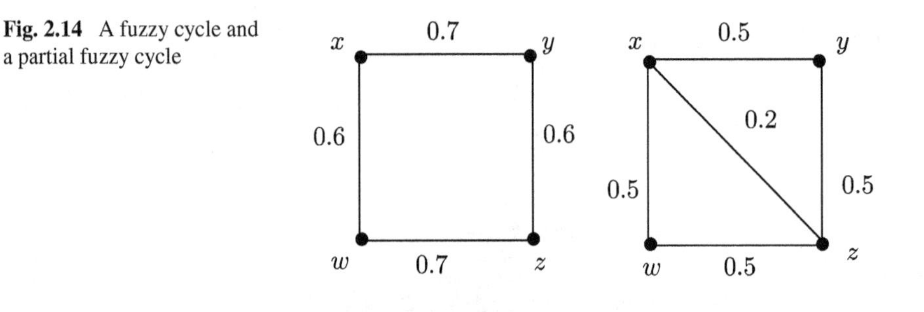


**Fig. 2.14** A fuzzy cycle and a partial fuzzy cycle

**Definition 2.6.1** (i) $G$ is called a **cycle** if $G^*$ is a cycle.
(ii) $G$ is called a **fuzzy cycle** if $G^*$ is a cycle and there does not exist unique $xy \in \mu^*$ such that $\mu(xy) = \wedge\{\mu(uv) \mid uv \in \mu^*\}$.
(iii) $G$ is called a **weak fuzzy cycle** if $\exists\, t \in (0, h(\mu)]$ such that $G^t$ is a cycle.
(iv) $G$ is called a **partial fuzzy cycle** if $G^t$ is a cycle for all $t \in (d(\mu), h(\mu)] \cup \{h(\mu)\}$.
(v) $G$ is called a **full fuzzy cycle** if $G^t$ is a cycle for all $t \in (0, h(\mu)]$.

*Example 2.6.2* Consider $V = \{x, y, z, w\}$. Let $\sigma$ be the fuzzy subset of $V$ and $\mu$ the fuzzy subset of $E = \{xy, wz, xw, yz\}$ defined as follows: $\sigma(x) = \sigma(y) = \sigma(z) = \sigma(w) = 1$, $\mu(xy) = \mu(wz) = 0.7$ and $\mu(xw) = \mu(yz) = 0.6$. Then $d(\mu) = 0.6$ and $h(\mu) = 0.7$. For $0 < t \leq 0.6$, $G^t = (V, \{xy, xw, yz, wz\})$ and for $0.6 < t \leq 0.7$, $G^t = (V, \{xy, wz\})$. Thus, $G$ is a fuzzy cycle and a weak fuzzy cycle, but $G$ is not a partial fuzzy cycle.

*Example 2.6.3* Let $V = \{x, y, z, w\}$. Let $\sigma$ be the fuzzy subset of $V$ and $\mu$ the fuzzy subset of $E = \{xy, yz, zw, wx\}$ defined as follows: $\sigma(x) = \sigma(y) = \sigma(z) = \sigma(w) = 1$, $\mu(xy) = \mu(yz) = \mu(zw) = \mu(wx) = 0.5$, and $\mu(x, z) = 0.2$. Then $G$ is not a cycle. Now, $d(\mu) = 0.2$ and $h(\mu) = 0.5$. For $0 < t \leq 0.2$, $G^t = (V, \{xy, yz, zw, wx, xz\})$ which is not a cycle and for $0.2 < t \leq 0.5$, $G^t = (V, \{xy, yz, zw, wx\})$ which is a cycle. Thus, $G$ is a partial fuzzy cycle, but not a full fuzzy cycle.

Fuzzy graphs in Examples 2.6.2 and 2.6.3 are given in Fig. 2.14.

**Proposition 2.6.4** *Suppose G is a cycle. Then G is a partial fuzzy cycle if and only if G is a full fuzzy cycle.*

*Proof* Suppose $G$ is a partial fuzzy cycle. Let $t \in (0, d(\mu)]$. Then $G^t = G^*$ and $G^*$ is given to be a cycle. Hence, $G$ is a full fuzzy cycle. ■

**Proposition 2.6.5** *G is a full fuzzy cycle if and only if G is a cycle and $\mu$ is constant on $\mu^*$.*

*Proof* Suppose $G$ is a full fuzzy cycle. Then $G^* = G^{d(\mu)}$ is a cycle. Suppose $\exists\, t_1$ and $t_2 \in \text{Im}(\mu)$ with $0 < t_1 < t_2$. Then $\exists\, xy \in \mu^*$ such that $\mu(xy) = t_1$. Hence, $xy \notin \mu^{t_2}$. Thus, $G^{t_2}$ is not a cycle, a contradiction. Hence, $\mu$ is constant on $\mu^*$. The converse is immediate. ■

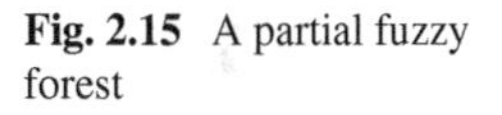

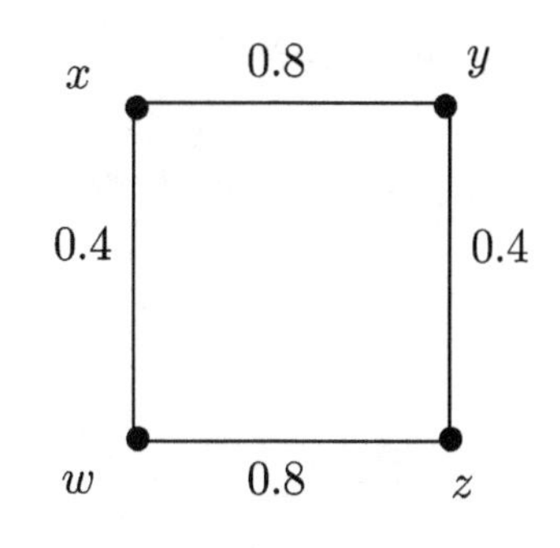


**Fig. 2.15** A partial fuzzy forest

**Corollary 2.6.6** *If G is a full fuzzy cycle, then G is a fuzzy cycle.*

**Proposition 2.6.7** *G is a partial fuzzy cycle if and only if $G^{h(\mu)}$ is a cycle and $|Im(\mu)\backslash\{0\}| \leq 2$.*

*Proof* Suppose $G$ is a partial fuzzy cycle. Then clearly $G^{h(\mu)}$ is a cycle and in fact $G^t$ is a cycle for all $t \in (d(\mu), h(\mu)] \cup \{h(\mu)\}$. Suppose $|\text{Im}(\mu)\backslash\{0\}| > 2$. Then $\exists\, t$ such that $0 < d(\mu) < t < h(\mu)$. Hence, $\exists\, xy \in \mu^*$ such that $\mu(xy) = t$. Thus, $xy \notin \mu^{h(\mu)}$ and so $G^{h(\mu)}$ is not a cycle, a contradiction. Conversely, suppose $G^{h(\mu)}$ is a cycle and $|\text{Im}(\mu)\backslash\{0\}| \leq 2$. If $|\text{Im}(\mu)\backslash\{0\}| = 1$, then $G$ is a full fuzzy cycle by Proposition 2.6.5. Suppose $|\text{Im}(\mu)\backslash\{0\}| = 2$. Then $\text{Im}(\mu)\backslash\{0\} = \{d(\mu), h(\mu)\}$. Because $G^t = G^{h(\mu)}$ for $d(\mu) < t \leq h(\mu)$, it follows that $G$ is a partial fuzzy tree. ∎

**Definition 2.6.8** (i) $G$ is called a **forest** (**tree**) if $G^*$ is a forest (*tree*).

(ii) $G$ is called a **fuzzy forest** ***(tree)*** if $G$ has a fuzzy spanning subgraph $(\sigma, \nu)$ which is a forest (tree) such that for all $uv \in \mu^* \setminus \nu^*$, $\mu(uv) < \nu^{\infty}(uv)$.

(iii) $G$ is called a **weak fuzzy forest** (**tree**) if $\exists\, t \in (0, h(\mu)]$ such that $G^t$ is a forest (tree).

(iv) $G$ is called a **partial fuzzy forest** (**tree**) if $G^t$ is a forest (tree) for all $t \in (d(\mu), h(\mu] \cup \{h(\mu)\}$.

(v) $G$ is called a **full fuzzy forest** (**tree**) if $G^t$ is a forest (tree) for all $t \in (0, h(\mu)]$.

The definition of a weak fuzzy forest in Definition 2.6.8(iii) is equivalent to the definition of a fuzzy graph being acyclic by $t$-cuts in Chap. 5. We will show that the definition of a full fuzzy forest here and the one in Chap. 5 are equivalent, but that this is not the case for the notion of a full fuzzy tree.

*Example 2.6.9* Consider the fuzzy graph $G = (\sigma, \mu)$ with $\sigma^* = \{x, y, z, w\}$ and $\mu^* = \{xw, yz, xy, wz\}$. The fuzzy subsets $\sigma$ and $\mu$ are defined as follows: $\sigma(x) = \sigma(y) = \sigma(z) = \sigma(w) = 1$ and $\mu(xw) = \mu(yz) = 0.4$, $\mu(xy) = \mu(wz) = 0.8$. Then $d(\mu) = 0.4$ and $h(\mu) = 0.8$. For $0 < t \leq 0.4$, $G^t = (V, \{xw, yz, xy, wz\})$ and for $0.4 < t \leq 0.8$, $G^t = (V, \{xy, wz\})$. Hence, $G$ is a partial fuzzy forest, but is neither a fuzzy forest nor a full fuzzy forest (Fig. 2.15).

**Proposition 2.6.10** *G is a full fuzzy forest if and only if G is a forest.*

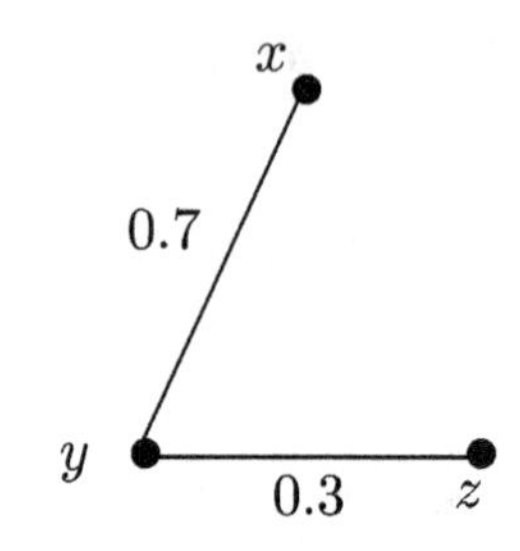


**Fig. 2.16** A full fuzzy forest

*Proof* Suppose $G$ is a full fuzzy forest. Then $G^* = G^{d(\mu)}$ is a forest. Conversely, suppose $G$ is a forest. Then $G^*$ is a forest and hence so must be $G^t$ for all $t \in (0, h(\mu)]$ because each such $G^t$ is a subgraph of $G^*$. ■

*Example 2.6.11* Consider the fuzzy graph $G = (\sigma, \mu)$ with $\sigma^* = \{x, y, z\}$ and $\mu^* = \{xy, yz\}$. The fuzzy subsets $\sigma$ and $\mu$ are defined as follows: $\sigma(x) = \sigma(y) = \sigma(z) = 1$, $\mu(xy) = 0.7$ and $\mu(yz) = 0.3$. Then $d(\mu) = 0.3$ and $h(\mu) = 0.7$. For $0 < t \leq 0.3$, $G^t = (V, \{xy, yz\})$ and for $0.3 < t \leq 0.7$, $G^t = (V, \{xy\})$. Hence, $G$ is a forest (and a full fuzzy forest) without being a constant on $\mu^*$ (Fig. 2.16). Note that $G^{h(\mu)}$ has more connected components than $G^*$.

**Proposition 2.6.12** *$G$ is a weak fuzzy forest if and only if $G$ does not contain a cycle whose edges are of strength $h(\mu)$.*

*Proof* Suppose $G$ contains a cycle whose edges are of strength $h(\mu)$. Then $G^t$, $t \in (0, h(\mu)]$, contains this cycle and so is not a forest. Thus, $G$ is not a weak fuzzy forest. Conversely, suppose $G$ does not contain a cycle all of whose edges are of strength $h(\mu)$. Then $G^{h(\mu)}$ does not contain a cycle and so is a forest. ■

**Corollary 2.6.13** *If $G$ is a fuzzy forest, then $G$ is a weak fuzzy forest.*

*Proof* $G$ cannot have a cycle all of whose edges are of strength $h(\mu)$, else it could not have a fuzzy spanning forest with the property that for all $uv \in \mu^* \setminus \nu^*$, $\mu(uv) < \nu^\infty(u, v)$. ■

**Theorem 2.6.14** *$G$ is a forest and $\mu$ is a constant on $\mu^*$ if and only if $G$ is a full fuzzy forest, $G^*$ and $G^{h(\mu)}$ have the same number of connected components, and $G$ is firm.*

*Proof* Suppose that $G$ is a forest and $\mu$ is constant on $\mu^*$. Then for all $t \in (0, h(\mu)]$, $G^t = G^*$ and so $G$ is a full fuzzy forest and $G^*$ and $G^{h(\mu)}$ have the same number of connected components. Clearly, $G$ is firm because $\mu$ is a constant on $\mu^*$. Conversely, suppose $G$ is a full fuzzy forest, $G^*$ and $G^{h(\mu)}$ have the same number of connected components, and $G$ is firm. Suppose $\exists\, t_1, t_2 \in \text{Im}(\mu)$ such that $0 < t_1 < t_2$. Then $\exists\, xy \in \mu^*$ such that $\mu(xy) = t_1$. Now, $xy \in \mu^{t_1}$, $xy \notin \mu^{t_2}$. Hence, $G^{t_2}$ has more connected components than $G^{t_1}$ because $G$ is firm, i.e., no vertices were lost. Thus, $G^{h(\mu)}$ has more connected components than $G^*$, a contradiction. ■

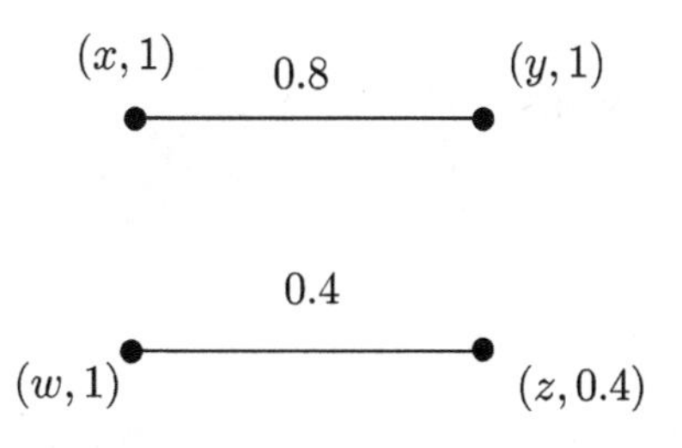


**Fig. 2.17** A full fuzzy forest

*Example 2.6.15* Consider $V = \{x, y, z, w\}$. Define the fuzzy subsets $\sigma$ of $V$ and $\mu$ of $E = \{xy, zw\}$ as follows: $\sigma(x) = \sigma(y) = \sigma(z) = 1$, $\sigma(w) = 0.4$ and $\mu(xy) = 0.8$, $\mu(zw) = 0.4$. Then $d(\mu) = 0.4$ and $h(\mu) = 0.8$. For $0 < t \leq 0.4$, $G^t = (V, \{xy, zw\})$ and for $0.4 < t \leq 0.8$, $G^t = (\{x, y, w\}, \{xy\})$. Thus, $G^*$ and $G^{h(\mu)}$ are forests with the same number of connected components. $G$ is a full fuzzy forest (Fig. 2.17), $\mu$ is not constant on $\mu^*$, and $G$ is not firm.

**Corollary 2.6.16** *G is a tree and $\mu$ is constant on $\mu^*$ if and only if G is a full fuzzy tree and G is firm.*

*Proof* Suppose $G$ is a full fuzzy tree and $G$ is firm. Because $G^t$ is a tree for all $t \in (0, h(\mu)]$, $G^*$ is a tree and so $G^{h(\mu)}$ and $G^*$ have the same number of connected components. The desired result now follows from Theorem 2.6.14. ■

*Example 2.6.17* Consider the fuzzy graph $G = (\sigma, \mu)$ with $\sigma^* = \{x, y, z\}$ and $\mu^* = \{xy, yz\}$. The fuzzy subsets $\sigma$ and $\mu$ are defined as follows: $\sigma(x) = \sigma(y) = 1$, $\sigma(z) = 0.6$ and $\mu(xy) = 0.8$, $\mu(yz) = 0.6$. Then $d(\mu) = 0.6$ and $h(\mu) = 0.8$. For $0 < t \leq 0.6$, $G^t = (V, \{(x, y), (y, z)\})$ and for $0.6 < t \leq 0.8$, $G^t = (\{x, y\}, \{xy\})$. Thus, $G$ is a tree, $G$ is a full fuzzy tree, and $G^*$ and $G^{h(\mu)}$ have the same number of connected components. However, $G$ is not firm and $\mu$ is not constant on $\mu^*$.

*Example 2.6.18* Let $V = \{x, y, z\}$. Define the fuzzy subsets $\sigma$ of $V$ and $\mu$ of $E = \{xy, xz, yz\}$ as follows: $\sigma(x) = \sigma(y) = 1$, $\sigma(z) = 0.7$ and $\mu(xy) = 0.8$, $\mu(xz) = \mu(yz) = 0.7$. Then $d(\mu) = 0.7$ and $h(\mu) = 0.8$. For $0 < t \leq 0.7$, $G^t = (V, \{xy, xz, yz\})$ and for $0.7 < t \leq 0.8$, $G^t = (\{x, y\}, \{xy\})$. Thus, $G$ is a partial fuzzy tree, but not a full fuzzy tree. $G$ is not a fuzzy tree. Hence, it is not the case that if $G$ is a weak fuzzy tree, then $G$ is a fuzzy tree. $G$ is not firm.

The fuzzy graphs in Examples 2.6.17 and 2.6.18 are given in Fig. 2.18.

**Definition 2.6.19** For all $t \in (0, 1]$ define $\sigma^{(t)} : \sigma^t \to [0, 1]$ and $\mu^{(t)} : \mu^t \to [0, 1]$ by $\sigma^{(t)}(x) = \sigma(x)$ for all $x \in \sigma^t$; $\sigma^{(t)}(x) = 0$ otherwise, and $\mu^{(t)}(xy) = \mu(xy)$ for all $xy \in \mu^t$ and $\mu^{(t)}(xy) = 0$ otherwise. Let $G^{(t)} = (\sigma^{(t)}, \mu^{(t)})$ for all $t \in (0, 1]$.

**Proposition 2.6.20** *Suppose that G is firm. If G is a weak fuzzy tree, then G is a fuzzy tree.*

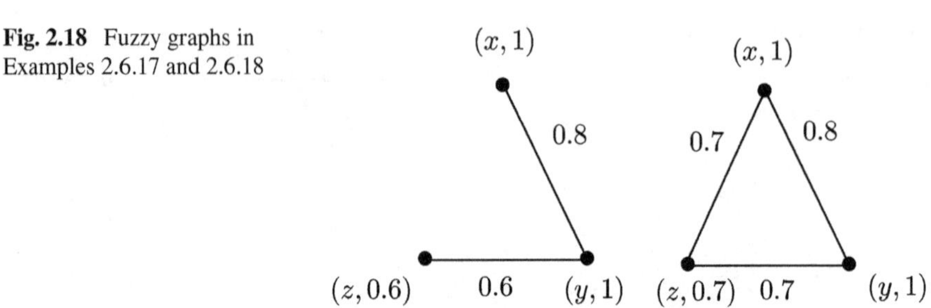


**Fig. 2.18** Fuzzy graphs in Examples 2.6.17 and 2.6.18

*Proof* There exist $t \in (0, h(\mu)]$ such that $G^t$ is a tree. Because $G$ is firm, $G^{(t)}$ is a fuzzy spanning subgraph of $G$ which is a tree. If $uv$ is in $\mu^* \setminus \mu^t$, then $\mu(uv) < t$ and so it follows that $G$ is a fuzzy tree. ■

*Example 2.6.21* Consider the fuzzy graph $G = (\sigma, \mu)$ with $\sigma^* = \{x, y, z, w\}$ and $\mu^* = \{xy, yz, xz, zw, xw\}$. The fuzzy subsets $\sigma$ and $\mu$ are defined as follows: $\sigma(x) = \sigma(y) = \sigma(z) = \sigma(w) = 1$ and $\mu(xy) = \mu(yz) = 0.8$, $\mu(xz) = \mu(zw) = 0.5$, $\mu(xw) = 0.2$. Then $d(\mu) = 0.2$ and $h(\mu) = 0.8$. For $0 < t \leq 0.2$, $G^t = (V, \{xy, yz, zw, wx, xz\})$, for $0.2 < t \leq 0.5$, $G^t = (V, \{xy, yz, zw, xz)\})$, and for $0.5 < t \leq 0.8$, $G^t = (V, \{xy, yz\})$. We see that $G$ is not a weak fuzzy tree. However, it is a fuzzy tree because $(\sigma, \nu)$ is a fuzzy spanning subgraph of $G$, which is a tree, where $\nu(xy) = \nu(yz) = 0.8$ and $\nu(zw) = 0.5$.

*Example 2.6.22* Consider $V = \{x, y, z, w\}$. Let $\sigma$ be the fuzzy subset of $V$ and $\mu$ the fuzzy subset of $E = \{xy, yz, zw\}$ defined as follows: $\sigma(x) = \sigma(y) = \sigma(z) = \sigma(w) = 1$ and $\mu(xy) = \mu(yz) = 0.8$, $\mu(zw) = 0.6$. Then $G$ is a tree, a fuzzy tree, a weak fuzzy tree, but not a partial fuzzy tree (if we were to define, $\mu(w) = 0.6$, then $G$ would be a full fuzzy tree, but not firm).

*Example 2.6.23* Let $V = \{x, y, z\}$. Let $\sigma$ be the fuzzy subset of $V$ and $\mu$ the fuzzy subset of $E = \{xy, yz, yz\}$ defined as follows: $\sigma(x) = \sigma(y) = \sigma(z) = 1$ and $\mu(xy) = \mu(xz) = 0.8$, $\mu(yz) = 0.2$. Then $G$ is a fuzzy tree, but not a tree. $G$ is a partial fuzzy tree, but not a full fuzzy tree.

The fuzzy graphs in Examples 2.6.21–2.6.23 are given in Fig. 2.19.

**Definition 2.6.24** (i) $G$ is called **connected** if $G^*$ is connected.
(ii) $G$ is called **fuzzy connected** if $G$ is a fuzzy block.
(iii) $G$ is called **weakly connected** if $\exists\, t \in (0, h(\mu)]$ such that $G^t$ is connected.
(iv) $G$ is called **partially connected** if $G^t$ is connected for all $t \in (d(\mu), h\mu)] \cup \{h(\mu)\}$.
(v) $G$ is called **fully connected** if $G^t$ is connected for all $t \in (0, h(\mu)]$.

*Example 2.6.25* Consider the fuzzy graph $G = (\sigma, \mu)$ with $\sigma^* = \{x, y, z, w\}$ and $\mu^* = \{xy, zw\}$. The fuzzy subsets $\sigma$ and $\mu$ are defined as follows: $\sigma(x) = \sigma(y) = 1$, $\sigma(z) = \sigma(w) = 0.6$ and $\mu(xy) = 0.8$ and $\mu(zw) = 0.6$. Then $G$ is not connected. $G$ is partially connected, but not fully connected. We see that $G$ is not firm (Fig. 2.20).

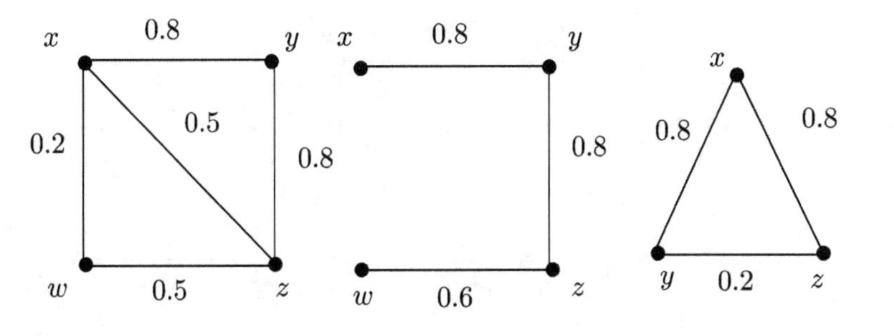


Fig. 2.19 Fuzzy graphs in Examples 2.6.21–2.6.23

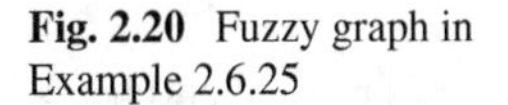

**Fig. 2.20** Fuzzy graph in Example 2.6.25

$(x, 1)$ 0.8 $(y, 1)$

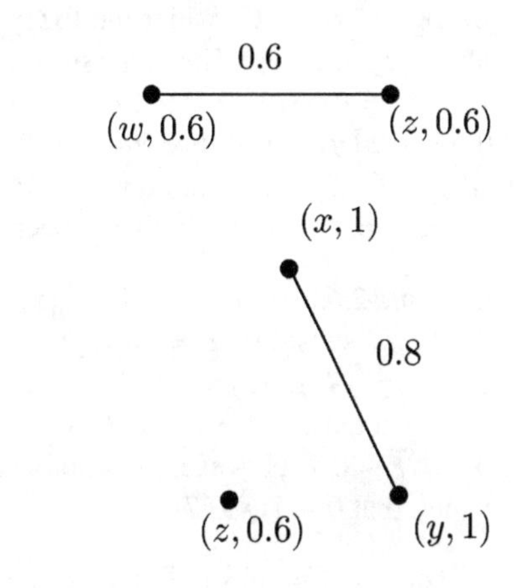


**Fig. 2.21** Fuzzy graph in Example 2.6.26

*Example 2.6.26* Consider $V = \{x, y, z\}$. Let $\sigma$ be the fuzzy subset of $V$ and $\mu$ be the fuzzy subset of $E = \{xy\}$ defined as follows: $\sigma(x) = \sigma(y) = 1, \sigma(z) = 0.6$ and $\mu(xy) = 0.8$. Then $G$ is not connected. $G$ is partially connected, but not fully connected. Note that $G$ is not firm (Fig. 2.21).

**Proposition 2.6.27** *If $G$ is connected, then $G$ is weakly connected. Conversely, if $G$ is firm and weakly connected, then $G$ is connected.*

*Proof* $G$ connected implies $G^*$ is connected. Now, $G^* = G^{d(\mu)}$ and so $G$ is weakly connected. Conversely, if $G^t$ is connected for some $t \in (0, h(\mu)]$, then $G^*$ is connected because $G$ is firm. ■

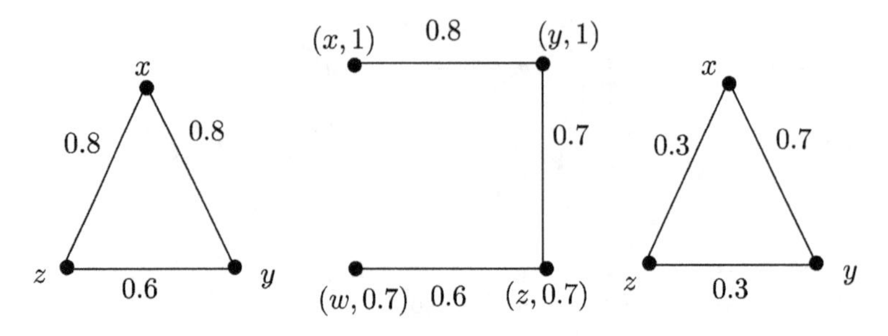


**Fig. 2.22** Fuzzy graphs in Examples 2.6.28–2.6.30

*Example 2.6.28* Consider the fuzzy graph $G = (\sigma, \mu)$ with $\sigma^* = \{x, y, z\}$ and $\mu^* = \{xy, yz, xz\}$. The fuzzy subsets $\sigma$ and $\mu$ are defined as follows: $\sigma(x) = \sigma(y) = \sigma(z) = 1$ and $\mu(xy) = \mu(xz) = 0.8$, $\mu(yz) = 0.6$. Then $G$ is fully connected, but $\mu$ is not constant on $\mu^*$.

*Example 2.6.29* Consider the fuzzy graph $G = (\sigma, \mu)$ with $\sigma^* = \{x, y, z, w\}$ and $\mu^* = \{xy, yz, zw\}$. The fuzzy subsets $\sigma$ and $\mu$ are defined as follows: $\sigma(x) = \sigma(y) = 1$, $\sigma(z) = \sigma(w) = 0.7$ and $\mu(xy) = 0.8$, $\mu(yz) = 0.7$, $\mu(zw) = 0.6$. Then $d(\mu) = 0.6$ and $h(\mu) = 0.8$. For $0 < t \leq 0.6$, $G^t = (V, \{xy, yz, zw\})$, for $0.6 < t \leq 0.7$, $G^t = (V, \{xy, yz\})$, and for $0.7 < t \leq 0.8$, $G^t = (\{x, y\}, \{xy\})$. Thus, $G$ is weakly connected, but not partially connected. $G$ is connected but $G$ is not firm.

*Example 2.6.30* Consider $V = \{x, y, z\}$. Let $\sigma$ be the fuzzy subset of $V$ and $\mu$ be the fuzzy subset of $E = \{xy, yz, xz\}$ be defined as: $\sigma(x) = \sigma(y) = \sigma(z) = 1$ and $\mu(xy) = 0.7$, $\mu(yz) = \mu(xz) = 0.3$. Then $G$ is weakly fuzzy connected because $G^t$ is connected for $0 < t \leq 0.3$. $G$ is a weak fuzzy forest because $G^t$ is a forest for $0.3 < t \leq 0.7$. However $G$ is not a weak fuzzy tree because $G^t$ is not a tree for any $t$ such that $0 < t \leq 0.7$.

The fuzzy graphs in Examples 2.6.28–2.6.30 are given in Fig. 2.22.

**Proposition 2.6.31** *(i) If $G$ is a weak fuzzy tree, then $G$ is weakly connected and $G$ is a weak fuzzy forest. Conversely, if $\exists\, t_1, t_2 \in (0, h(\mu)]$ with $t_1 < t_2$ such that $G^{t_1}$ is a forest and $G^{t_2}$ is connected, then $G$ is a weak fuzzy tree.*
*(ii) $G$ is a tree if and only if $G$ is a forest and $G$ is connected.*
*(iii) $G$ is a partial fuzzy tree if and only if $G$ is a partial fuzzy forest and $G$ is partially connected.*
*(iv) $G$ is a full fuzzy tree if and only if $G$ is a full fuzzy forest and $G$ is fully connected.*

*Proof* (i) If $G^t$ is a tree for some $t \in (0, h(\mu)]$, then $G^t$ is connected and is a forest. For the converse, we note that $G^{t_2}$ must also be a forest. Because also $G^{t_2}$ is connected, $G^{t_2}$ is a tree.

(ii), (iii), (iv): Immediate. ■

**Proposition 2.6.32** *$G$ is firm if and only if $G^{(t)}$ is firm for all $t \in (0, h(\mu)]$.*

*Proof* Suppose $G$ is firm. Let $t \in (0, h(\mu)]$. Let $xy \in \mu^t$. Then $t \leq \mu(xy) \leq \wedge \{\sigma(x) \mid x \in \sigma^*\} \leq \wedge\{\sigma(x) \mid x \in \sigma^t\}$. Hence, $\vee\{\mu(xy) \mid x, y \in \mu^t\} \leq \wedge(\sigma(x) \mid x \in \sigma^t\}$. Thus, if we note that $\mu^{(t)*} = \mu^t$ and $\sigma^{(t)*} = \sigma^t$, we see that $G^{(t)}$ is firm. Conversely, suppose $G^{(t)}$ is firm for all $t \in (0, h(\mu)]$. Let $\wedge\{\sigma(x) \mid x \in \sigma^*\} = t_0$. Then $t_0 > 0$. Now, $\vee(\mu(xy) \mid xy \in \mu^{t_0}\} \leq t_0$ because $G^{(t_0)}$ is firm and $\sigma^* = \sigma^{t_0} = \sigma^{(t_0)*}$. Let $xy \in \mu^* \setminus \mu^{t_0}$. Then $\mu(xy) < t_0$. Thus, $\vee\{\mu(xy) \mid xy \in \mu^*\} \leq t_0 = \wedge\{\sigma(x) \mid x \in \sigma^*\}$. Hence, $G$ is firm. ■

## 2.7 Blocks in Fuzzy Graphs

The definition of a nonseparable fuzzy graph first appeared in Rosenfeld's classic paper in 1975 [154]. But a formal study of blocks in fuzzy graphs was made by Sunitha and Vijayakumar in 2005 [168]. Later Mathew and Sunitha [110] studied blocks further in 2010 and characterized a class of blocks in fuzzy graphs. In graph theory, a graph without cutvertices is called a **block** (or **nonseparable**). This concept is generalized to fuzzy graph theory. A fuzzy graph is said to be a block if it has no fuzzy cutvertices. It is clear that a block in fuzzy graphs has no cutvertices. Thus, a fuzzy block is trivially a block in the classical sense, but the converse is not true. In contrast to the conventional concept of a block in graphs, the study of blocks in fuzzy graphs is challenging due to the complexity of its cutvertices. Note that the cutvertices of a fuzzy graph are those vertices which reduce the strength of connectedness between some pair of vertices rather than the total disconnection of the fuzzy graph on its removal from the fuzzy graph.

Rosenfeld [154] observed that a block may have a fuzzy bridge. Sunitha and Vijayakumar [168] identified that a fuzzy graph can have more than one fuzzy bridge as seen from the example below.

*Example 2.7.1* Let $V = \{u, v, w, x, y\}$. Let $\sigma$ be the fuzzy subset of $V$ and $\mu$ be the fuzzy subset of $V \times V$ defined as follows. $\sigma(u) = \sigma(v) = \sigma(w) = \sigma(x) = \sigma(y) = 1$ and $\mu(uv) = \mu(xy) = 0.9$, $\mu(vy) = \mu(uy) = \mu(ux) = 0.5$ and $\mu(wx) = \mu(wy) = 0.3$. It can be verified easily that $G = (\sigma, \mu)$ is a block. But note that both $uv$ and $xy$ are fuzzy bridges (Fig. 2.23).

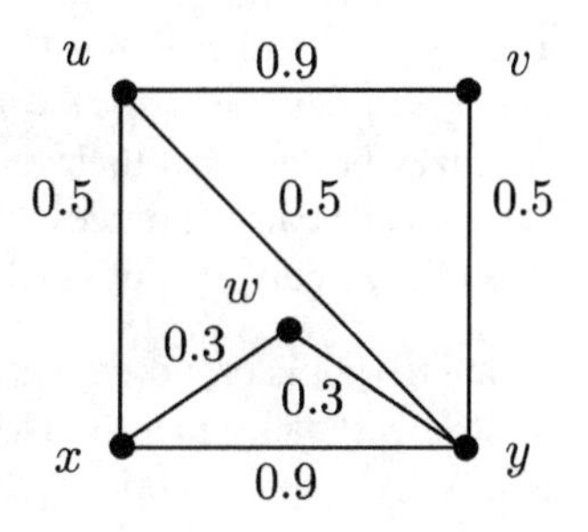


**Fig. 2.23** A block with two fuzzy bridges

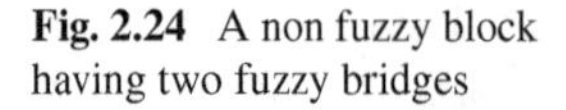

**Fig. 2.24** A non fuzzy block having two fuzzy bridges

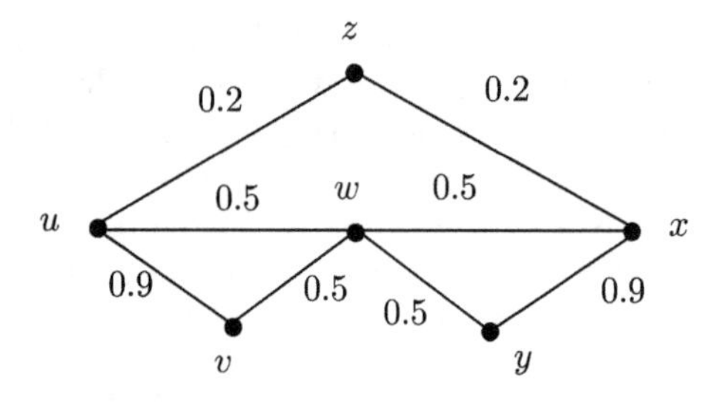


It is obvious from the definition that no two fuzzy bridges in a block can have a common vertex. A complete fuzzy graph is clearly a block. As pointed out in Theorem 2.3.14, the removal of a fuzzy bridge from a fuzzy tree reduces the strength of connectedness between some pair of vertices other than its end vertices. But, the situation is different in blocks as seen from the following theorem.

**Theorem 2.7.2** *If $G = (\sigma, \mu)$ is a block with at least one fuzzy bridge, then removal of any fuzzy bridge reduces the strength of connectedness between its end vertices alone.*

*Proof* Let $G = (\sigma, \mu)$ be a block and $uv$ be a fuzzy bridge of $G$. Assume on the contrary that removal of $uv$ reduces the strength of connectedness between some other pair of vertices $u_l$ and $v_l$.

**Case 1**: Both $u_l$ and $v_l$ are distinct from $u$ and $v$.

Without loss of generality let $u_l \neq u$ and $v_l \neq v$. By assumption, every strongest $u_l - v_l$ path contains the edge $uv$. Thus, clearly removal of either $u$ or $v$ reduces the strength of connectedness between $u_l$ and $v_l$, which shows that $u$ and $v$ are fuzzy cutvertices of $G$, contradicting that $G$ is a block.

**Case 2**: One of $u$ or $v$ is $u_l$ or $v_l$.

Let $v_l = v$ and $u_l \neq u$. Then as before removal of $v$ reduces the strength of connectedness between $u_l$ and $v_l$ showing that $v$ is a fuzzy cutvertex of $G$ and similarly if $u_l = u$ and $v_l \neq v$, then $u$ becomes a fuzzy cutvertex, both contradict the hypothesis that $G$ is a block. Thus, the only possibility is that $u_l = u$ and $v_l = v$ and hence the theorem. ■

The condition in Theorem 2.7.2 is not sufficient as seen from Example 2.7.3.

*Example 2.7.3* Let $V = \{u, v, w, x, y, z\}$. Let $\sigma$ be a fuzzy subset of $V$ and $\mu$ be a fuzzy subset of $\mathcal{E}$ defined as $\sigma(s) = 1$ for all $s \in V$ and $\mu(uv) = \mu(xy) = 0.9$, $\mu(vw) = \mu(yw) = \mu(uw) = \mu(xw) = 0.5$, $\mu(xz) = \mu(uz) = 0.2$. In $G$, $uv$ and $xy$ are fuzzy bridges, but their removal does not reduce the strength of connectedness between any pair of vertices other than their endvertices. Clearly, $G$ is not a block as $w$ is a fuzzy cutvertex of $G$ (See Fig. 2.24).

In [154], it is proposed that if every pair of vertices in a fuzzy graph $G$ are joined by strongest paths, then $G$ is a block and the converse is not true. Also, if an edge $uv$ of a fuzzy graph is a bridge, then it is the unique strongest $u - v$ path. Sunitha and

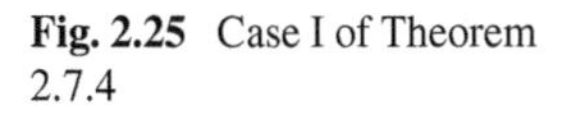
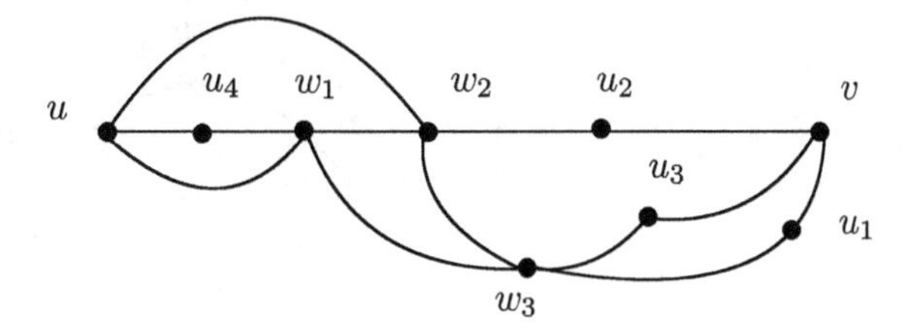


**Fig. 2.25** Case I of Theorem 2.7.4

Vijayakumar [168] proved that the converse of Rosenfeld's observation is true only for blocks having no fuzzy bridges. Hence, we have the following result.

**Theorem 2.7.4** ([168]) *The following statements are equivalent for a fuzzy graph* $G = (\sigma, \mu)$.

(i) *G is a block.*
(ii) *Any two vertices u and v such that uv is not a fuzzy bridge are joined by two internally disjoint strongest paths.*
(iii) *For every three distinct vertices of G, there is a strongest path joining any two of them not containing the third.*

*Proof* $(i) \Rightarrow (ii)$ Let $G = (\sigma, \mu)$ be a block. Let $u$ and $v$ be any two vertices such that $\mu(uv) \geq 0$ and $uv$ is not a fuzzy bridge. If there exists a unique strongest $u - v$ path of length greater than or equal to 2, then the vertices on this path other than $u$ and $v$ are fuzzy cutvertices of $G$. Hence, there exist more than one strongest $u - v$ paths. If these strongest $u - v$ paths are internally disjoint, then we are done. Note that all strongest $u - v$ paths do not have a common vertex, if so, that vertex becomes a fuzzy cutvertex. So consider the following cases.

**Case 1**:

Let $P_1 : u - w_2 - w_3 - u_1 - v$, $P_2 : u - u_4 - w_1 - w_2 - u_2 - v$ and $P_3 : u - w_1 - w_3 - u_3 - v$ be strongest $u - v$ paths. Let $w_2$ be the last common vertex of $P_1$ and $P_2$ (Fig. 2.25). Then $u - w_2$ subpath in $P_1$ together with $w_2 - u_2 - v$ subpath in $P_2$ is a path (say) $P$ disjoint from $P_3$.

**Claim**: $P$ is a strongest $u - v$ path.

Let $e_1, e_2$ and $e_3$ be weakest edges in $P_1, P_2$ and $P_3$, respectively, and let $\mu(e_1) = \mu(e_2) = \mu(e_3) = \mu^\infty(u, v)$. Then $e_1$ should be in $u - w_2$ subpath of $P_1$ or $e_2$ should be in $w_2 - u_2 - v$ subpath of $P_2$; for if not, then strength of $P > \mu^\infty(u, v)$, contradiction. Hence, $P$ is a strongest $u - v$ path.

**Case 2**:

Let $P_1 : u - u_1 - w_1 - w_2 - v$, $P_2 : u - w_1 - w_3 - u_2 - v$ and $P_3 : u - w_2 - w_3 - v$ be strongest $u - v$ paths. Let $w_2$ be the first common vertex of $P_1$ and $P_3$. Then $u - w_2$ subpath in $P_3$ together with $w_2 - v$ subpath in $P_1$ is a path disjoint from $P_2$ (Fig. 2.26). As in Case 1, it can be proved that $P$ is a strongest $u - v$ path.

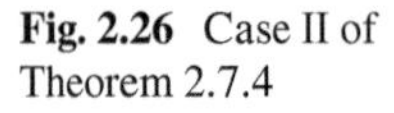

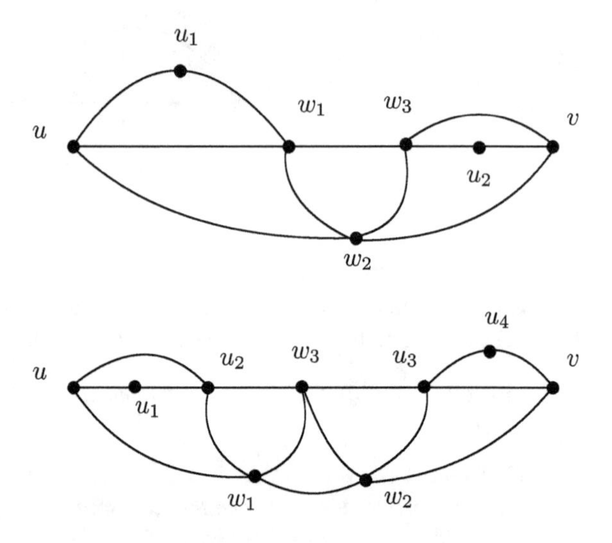


**Fig. 2.26** Case II of Theorem 2.7.4

**Fig. 2.27** Case III of Theorem 2.7.4

**Case 3**:

Let $P_1 : u - u_2 - w_1 - w_2 - u_3 - u_4 - v$, $P_2 : u - u_1 - u_2 - w_3 - u_3 - v$ and $P_3 : u - w_1 - w_3 - w_2 - v$ be strongest $u - v$ paths. Let $w_1$ and $w_2$ be the first and last common vertices of $P_1$ and $P_3$, respectively (Fig. 2.27). Then $u - w_1$ subpath in $P_3$ and $w_1 - w_2$ subpath in $P_1$ together with $w_2 - v$ subpath in $P_3$ will give a strongest $u - v$ path disjoint from $P_2$.

$(ii) \Rightarrow (iii)$

Let $u \neq v \neq w$ be any three vertices of $G$. Choose any two (say) $u$ and $v$. If edge $uv$ is a fuzzy bridge, then it is the strongest $u - v$ path and $(iii)$ holds. So assume $uv$ is not a fuzzy bridge. Now, by $(ii)$, there exist two internally disjoint strongest $u - v$ paths and hence $w$ cannot be in both.

$(iii) \Rightarrow (i)$ If possible let $w$ be a fuzzy cutvertex of $G$. Then by definition there exist $u, v$ different from $w$ such that $w$ is on every strongest $u - v$ path. But this contradicts $(iii)$. ■

A fuzzy analogue of the characterization of blocks in graphs given in [83] with all six conditions is not possible. But in any fuzzy graph there exists a strongest path between every pair of vertices. We will discuss a characterization using strongest strong paths in this section.

An edge $xy$ is said to be a strong if its membership value is at least as great as the connectedness of its end vertices when the edge is deleted. That is, if $\mu(xy) \geq \mu'^{\infty}(x, y)$ or $\mu(xy) \geq CONN_{G-xy}(x, y)$. A detailed discussion of strong edges will be made in Chap. 3. From the following example, it can be seen that a block can contain edges which are even not strong.

*Example 2.7.5* Let $G = (\sigma, \mu)$ with $\sigma^* = \{u, v, w, x\}$, $\sigma(s) = 1$ for all $s \in \sigma^*$ and $\mu(uv) = \mu(xu) = \mu(vw) = \mu(wx) = 0.9$, $\mu(vx) = 0.2$. Then $G$ is a block. Here $vx$ is the unique weakest edge of the cycle $uvxu$ which is not strong (Fig. 2.28).

Fig. 2.28 A block with a non strong edge

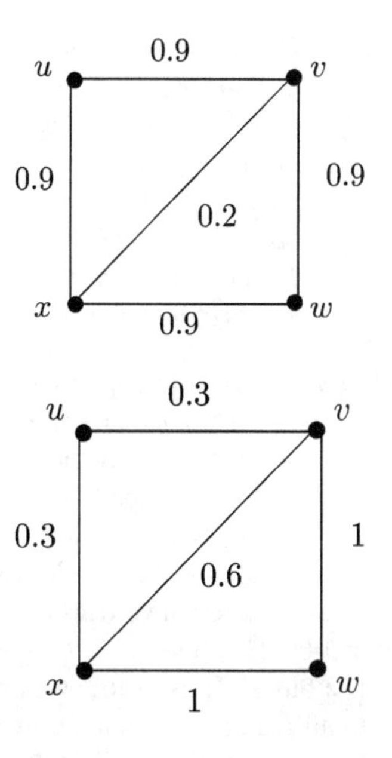


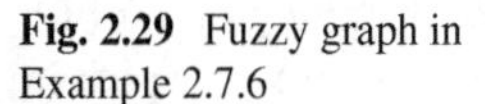
Fig. 2.29 Fuzzy graph in Example 2.7.6

A path in a fuzzy graph $G$ is strong if all its edges are strong. Recall that an $x - y$ path $P$ in a fuzzy graph $G$ is said to be a strongest $x - y$ path if $d(P) = CONN_G(x, y)$. In a block, a strongest path need not be strong and a strong path need not be strongest. But there exists a strongest strong path between any two vertices of $G$.

*Example 2.7.6* Let $G = (\sigma, \mu)$ with $\sigma^* = \{u, v, w, x\}$, $\sigma(s) = 1$ for all $s \in \sigma^*$ and $\mu(uv) = 0.3 = \mu(xu)$, $\mu(vw) = \mu(wx) = 1$, $\mu(vx) = 0.6$. Edge $xv$ is not strong because $0.6 = \mu(xv) < CONN_{G-xv}(x, v) = 1$. Thus, $P : uxv$ is not a strong path even though it is a strongest $u - v$ path. $d(P) = 0.3 = CONN_G(u, v)$. Also, the $x - v$ path $Q : xuv$ is not a strongest $x - v$ path even if it is a strong $x - v$ path (Fig. 2.29).

**Definition 2.7.7** A cycle in a fuzzy graph $G$ is called a **strong cycle** if all its edges are strong.

*Example 2.7.8* Consider the fuzzy graph in Example 2.7.6. The cycle $uvwxu$ is a strong cycle whereas $vwxv$ is not.

Next we have a characterization of blocks having no fuzzy bridges.

**Theorem 2.7.9** ([110]) *Let $G = (\sigma, \mu)$ be a fuzzy graph with at least three vertices and having no fuzzy bridges. Then the following statements are equivalent.*

*(i) $G$ is a block.*
*(ii) For any two vertices $x$, $y$ of $G$, there exists a cycle containing the vertices $x$ and $y$ which is formed by two strongest strong $x - y$ paths.*

*(iii)* *For each vertex u and each strong edge $vw$ of G, there exists a cycle containing the vertex u and the edge $vw$ which is formed by two strongest strong $u - v$ paths or $u - w$ paths.*

*(iv)* *For each pair of strong edges $xy$ and $uv$ of G, there exists a cycle containing the edges $xy$ and $uv$ which is formed by two strongest strong $x - u$ or $y - u$ paths.*

*(v)* *For every three distinct vertices of G there exists a strongest strong path joining any two of them not containing the third.*

*Proof* $(i) \Rightarrow (ii)$ Suppose $G$ is a block. Consider a maximum spanning tree $T$ of $G$. Clearly, every edge in a maximum spanning tree is strong. Also, every $x - y$ path in $T$ is a strongest $x - y$ path in $G$. Thus, between any two vertices of $G$ there exists a strongest strong path. Let $P$ be a strongest strong $x - y$ path in $G$. Assume that $P$ is a unique $x - y$ path in $G$. Then $P$ should belong to all maximum spanning trees. Also, note that the length of $P$ is at least two because $G$ has no fuzzy bridges. Thus, all internal vertices of $P$ are internal vertices of every maximum spanning tree and by Theorem 2.2.2, they are all fuzzy cutvertices contradicting the fact that $G$ is a block. Thus, it follows that the strongest strong $x - y$ path $P$ does not belong to all maximum spanning trees. Hence, there exists a maximum spanning tree say $T_1$ not containing $P$. Let $P_1$ be a strongest strong $x - y$ path in $T_1$. This strongest strong path $P_1$ together with $P$ gives a cycle in $G$ containing the vertices $x$ and $y$ as required. Note that $P$ and $P_1$ should be internally disjoint because otherwise the common vertices of $P$ and $P_1$ become fuzzy cutvertices of $G$.

$(ii) \Rightarrow (iii)$

Let $u$ be a vertex and $vw$, a strong edge of $G$. Let $C_1$ be the cycle containing $u$ and $v$ satisfying the conditions in $(ii)$ and $C_2$ be the cycle containing $u$ and $w$ satisfying the conditions in $(ii)$. If $w$ is a neighbor of $v$ in $C_1$ or $v$ is a neighbor of $w$ in $C_2$ then we are done.

So suppose that $vw$ is neither in $C_1$ nor in $C_2$. Let $P_1$ and $P_2$ be the strongest strong $u - v$ paths in $C_1$ and $Q$ be a strongest strong $u - w$ path in $C_2$. Let $z$ be the vertex in $Q$ before $w$ and nearest to it at which $Q$ meets $P_1$ or $P_2$ (note that $z$ can be the vertex $u$ itself). Without loss of generality suppose that $P_2$ is the $u - v$ path which meets $Q$ at $z$. Let $P$ be the union of $w - z$ sub path of $Q$ and $z - u$ sub path of $P_2$. Then let $C = P_1 \cup vw \cup P$.

**Claim**: $C$ is the required cycle.

Let $xy$ be an edge in $P_1$ such that $s(P_1) = s(P_2) = \mu(xy)$. Then three cases arise.

**Case 1**: $\mu(vw) > \mu(xy)$.

**Sub Claim 1**: $s(P) = \mu(xy)$.

We have, $s(P)$ cannot exceed $\mu(xy)$, for otherwise $P \cup vw$ will become a strong $u - v$ path having strength more than the strongest $u - v$ path $P_1$, a contradiction. Therefore, $s(P) \leq \mu(xy)$. Because $s(P_2) = \mu(xy)$, the strength of the $u - z$ sub path of $P_2$ is greater than or equal to $\mu(xy)$. Hence, if $s(P) < \mu(xy)$, then the strength of $z - w$ sub path of $Q < \mu(xy)$ and thus $s(Q) < \mu(xy)$. Thus, we have $P_1 \cup vw$ is a strong $u - w$ path which is stronger than the strongest $u - w$ path $Q$, a contradiction. Hence, the only possibility is that $s(P) = \mu(xy)$.

Now, we have two strongest strong paths between $u$ and $v$ namely $P_1$ and $P \cup vw$ whose union gives the required cycle containing the vertex $u$ and the edge $vw$.

**Case 2**: $\mu(vw) = \mu(xy)$.

If $s(P) < \mu(xy)$, then as in Case-1, $s(Q) < \mu(xy)$ and hence $P_1 \cup vw$ becomes a strong $u - w$ path which is stronger than the strongest $u - w$ path $Q$, a contradiction. Thus, $s(P) \geq \mu(xy)$ and hence we have two strongest strong $u - v$ paths namely $P_1$ and $P \cup vw$ whose union gives the required cycle.

**Case 3**: $\mu(vw) < \mu(xy)$.

**Sub Claim 2**: $s(P) = \mu(vw)$.

If $s(P) > \mu(vw)$, then all edges in $P$ and $P_1$ have strength more than $\mu(vw)$ and thus $vw$ becomes the unique weakest edge of the cycle $P_1 \cup vw \cup P$, contradicting our assumption that $vw$ is a strong edge. If $s(P) < \mu(vw)$, then the strength of $z - w$ sub path of $Q < \mu(vw)$ because the strength of the $u - z$ sub path of $P_2 \geq \mu(xy) > \mu(vw)$. Therefore, $s(Q) < \mu(vw) < \mu(xy) = s(P_1)$ and hence $P_1 \cup vw$ is a strong path having strength more than that of $Q$, which is a contradiction to the fact that $Q$ is a strongest strong $u - w$ path. Thus, $s(P) = \mu(vw)$.

**Sub Claim 3**: $P$ is a strongest $u - w$ path.

To prove Sub Claim 3, it is sufficient to prove that $s(Q) = \mu(vw)$ because $Q$ is a strongest $u - w$ path. Clearly, $s(Q) \geq \mu(vw)$. If $s(Q) < \mu(vw)$, then $P$ will become a strong $u - w$ path which is stronger than the strongest $u - w$ path $Q$, a contradiction. If $s(Q) > \mu(vw)$ because $\mu(vw) < \mu(xy)$, all edges in $P_1 \cup Q$ have strength more than $\mu(vw)$ and hence $vw$ becomes the unique weakest edge of the cycle $P_1 \cup vw \cup Q$ contradicting that $vw$ is a strong edge. Thus, $s(Q) = \mu(vw)$ and the sub Claim 3 is proved.

Now, we have two strongest strong $u - w$ paths namely $P$ and $P_1 \cup vw$ whose union gives the required cycle containing the vertex $u$ and edge $vw$. Thus, in all the three cases the claim is proved.

$(iii) \Rightarrow (iv)$

Let $xy$ and $uv$ be any two given strong edges of $G$. Let $C_1$ be a cycle containing the vertex $x$ and the strong edge $uv$ and let $C_2$ be a cycle containing the vertex $u$ and the edge $xy$. If $y$ is a neighbor of $x$ in $C_1$ or $v$ is a neighbor of $u$ in $C_2$, we are done. Suppose not. That is, $xy$ is not in $C_1$ and $uv$ is not in $C_2$. Without loss of generality suppose that $C_1$ is the union of two strongest strong $x - u$ paths $P_1$ and $P_2$ with $P_2$ containing the strong edge $uv$ and $C_2$ is the union of two strongest strong $u - y$ paths. Let $Q$ be the strongest strong $u - y$ path in $C_2$ not containing the edge $xy$. Let $z$ be the vertex nearest to $u$ at which $Q$ meets $P_1$ or $P_2$. Without loss of generality let $Q$ meets $P_2$ at $z$. Now, let $P$ be the union of $y - z$ sub path of $Q$ and $z - u$ sub path (containing the strong edge $vu$) of $P_2$.

**Claim**: $P_1 \cup P \cup xy$ is the required cycle.

Let $x'y'$ be an edge in $P_1$ such that $s(P_1) = \mu(x'y')$. Then three cases arise.

**Case 1**: $\mu(xy > \mu(x'y')$.

**Sub Claim 1**: $s(Q) = \mu(x'y')$.

If $s(Q) < \mu(x'y')$, then $P_1 \cup xy$ is a strong $u - y$ path having strength more than the strength of $Q$, a contradiction to the assumption that $Q$ is a strongest strong $u - y$ path. If $s(Q) > \mu(x'y')$, then the $u - x$ path $Q \cup yx$ has strength greater than

$\mu(x'y')$ because, $\mu(xy) > \mu(x'y')$, which contradicts the fact that $P_1$ is a strongest strong $u - x$ path. Therefore, only possibility is that $s(Q) = \mu(x'y')$ and hence Sub Claim 1 is proved.

Thus, we have strength of the $y - z$ sub path of $Q \geq \mu(x'y')$. Also, the strength of $z - u$ sub path of $P_2 \geq \mu(x'y')$, because $s(P_2) = \mu(x'y')$. Now, if both these sub paths are of strength greater than $\mu(x'y')$, then $s(P)$ is greater than $\mu(x'y')$, which contradicts the fact that $Q$ is a strongest strong $u - y$ path. Thus, at least one of this sub paths should have strength equal to $\mu(x'y')$ and $s(P)$ is equal to $\mu(x'y')$. Hence, we have two strongest strong $x - u$ paths namely $P_1$ and $P \cup xy$ whose union gives the required cycle containing the edges $xy$ and $uv$.

**Case 2**: $\mu(xy) = \mu(x'y')$.

Because the strength of the $y - u$ path passing through $x$ and $P_1$ is $\mu(xy)$, we have the strength of $Q$ is at least $\mu(xy)$. Also, because $s(P_2)$ being $\mu(xy)$, the $z - u$ sub path of $P_2$ has strength at least $\mu(xy)$. Thus, $s(P)$ is at least $\mu(xy)$ and hence we have two strongest strong $x - u$ paths namely $P_1$ and $P \cup xy$ whose union gives the required cycle containing the edges $xy$ and $uv$.

**Case 3**: $\mu(xy) < \mu(x'y')$.

**Sub Claim 2**: $s(Q) = \mu(xy)$.

Clearly, $s(Q) \leq \mu(xy)$, for otherwise all edges in $P_1 \cup Q$ have strength more than $\mu(xy)$ and hence $xy$ becomes the unique weakest edge of the cycle $P_1 \cup Q \cup xy$, contradicting our assumption that $xy$ is a strong edge. Now, if $s(Q) < \mu(xy)$ we have a strong path from $x$ to $u$ namely $xy \cup P_1$ which is stronger than the strongest $y - u$ path $Q$, a contradiction. Thus, $s(Q) = \mu(xy)$.

**Sub Claim 3**: The strength of $y - z$ sub path of $Q$ is precisely $\mu(xy)$.

Because $s(Q) = \mu(xy)$, the strength of $y - z$ sub path of $Q$ is at least $\mu(xy)$, but if the strength of the $y - z$ sub path of $Q$ is greater than $\mu(xy)$, then the $x - z$ sub path of $P_2$ has strength $\geq \mu(x'y') > \mu(xy)$ and so we have all the edges in the two sub paths have strength greater than $\mu(xy)$ and $xy$ becomes the unique weakest edge of the cycle formed by the edge $xy$, the $x - z$ sub path of $P_2$ and the $y - z$ sub path of $Q$, which contradicts the assumption that $xy$ is a strong edge.

Also, because $P_2$ is a strongest strong $x - u$ path, $s(P_2) = \mu(x'y')$. Therefore, the strength of $z - u$ sub path of $P_2 \geq \mu(x'y') > \mu(xy)$. Thus, the strength of the $y - u$ path $P$ is $\mu(xy)$ and hence it is a strongest strong $y - u$ path in $G$. Also, note that the strength of the $y - u$ path $P_1 \cup xy$ is $\mu(xy)$ because $\mu(x'y') > \mu(xy)$. Hence, it follows that $P \cup P_1 \cup xy$ is a cycle containing the edges $xy$ and $uv$ formed by the union of two strongest strong $y - u$ paths as required.

$(iv) \Rightarrow (v)$

Let $x$, $u$ and $w$ be any three distinct vertices of $G$. Let $P$ be a strongest strong $x - u$ path in $G$ with strength $\alpha$ (say). Let $y$ be a strong neighbor of $x$ and $v$ a strong neighbor of $u$ in $P$. Then $xy$ and $uv$ are strong edges of $G$. By (iv), there exists a cycle $C$ containing the edges $xy$ and $uv$ formed by two strongest strong $x - u$ paths or $y - u$ paths.

**Case 1**: $C$ is the union of two strongest strong $x - u$ paths.

In this case, because there exist two internally disjoint strongest strong $x - u$ paths, at least one of them will not contain $w$.

**Case 2**: $C$ is the union of two strongest strong $y-u$ paths.

Let $P_1$ be the $y-u$ strongest strong path containing the edge $uv$ and let $C-P_1$ be the other path containing edge $xy$. Because $P$ is a strongest strong $x-u$ path, containing $y$ with strength $\alpha$, we have $\mu(xy)\geq\alpha$. Also, we have $s(P_1)\geq\alpha$ for otherwise the $y-u$ sub path of $P$ will become a strong path stronger than the strongest. But $s(P_1)$ cannot exceed $\alpha$ because $C-P_1$ is also a strongest strong $y-u$ path which passes through the edge $xy$ with $\mu(xy)=\alpha$. Thus, only possibility is that $s(P_1)=\alpha$. In this case, $s(P_1\cup xy)$ is also $\alpha$ and hence $P_1\cup xy$ becomes a strongest strong $x-u$ path. Also, the strength of $C-\{P_1\cup xy\}$ cannot exceed $\alpha$ for otherwise it is a contradiction to the fact that $P$ is a strongest strong $x-u$ path. That is,

$$s(C-\{P_1\cup xy\})\leq\alpha \tag{2.1}$$

Also, because $C-P_1$ is a strongest strong $y-u$ path, it follows that

$$s(C-\{P_1\cup xy\})\geq\alpha. \tag{2.2}$$

From (2.1) and (2.2), $s(C-\{P_1\cup xy\})=\alpha$. Thus, we have two internally disjoint strongest strong $x-u$ paths namely $P_1\cup xy$ and $C-\{P_1\cup xy\}$ with at least one of them not containing the vertex $w$.

$(v)\Rightarrow(i)$

Assume $(v)$. Let $w$ be a vertex in $G$. By $(v)$, between any two vertices $u$ and $v$ other than $w$ there exists a strongest strong $u-v$ path not containing $w$. Thus, $w$ is not in all strongest paths between any pair of vertices and hence is not a fuzzy cutvertex. Hence, $G$ is a block. ■

The remaining statements of the characterization of blocks in graphs given in Harary [83] cannot be extended to fuzzy graphs as seen from the following example.

*Example 2.7.10* Let $G=(\sigma,\mu)$ with $\sigma^*=\{u,v,w,x,y,z\}$ and $\mu(uv)=\mu(vw)=\mu(wz)=\mu(zu)=0.8$, $\mu(ux)=\mu(xw)=\mu(uy)=\mu(yw)=0.1$. $G$ is a block. But there is no strongest strong $v-w$ path containing the vertex $x$ and no strongest strong $v-w$ path containing the strong edge $ux$ (Fig. 2.30).

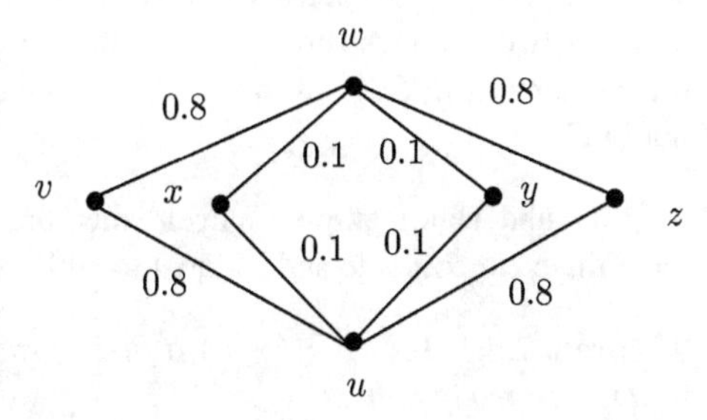


**Fig. 2.30** A fuzzy block different from block

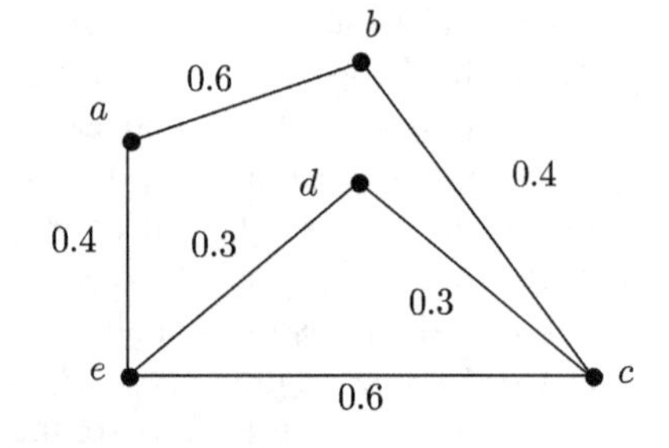


**Fig. 2.31** Strongest strong cycles in a fuzzy graph

## 2.8 Strongest Strong Cycles and $\theta$-Fuzzy Graphs

In the previous section, we observed that blocks in fuzzy graphs cannot be fully characterized even by cycles formed by two strongest strong paths. When the underlying structure of a fuzzy graph is a cycle, we can see that it is a block only when it is strong and is the union of two different strongest paths. A cycle is called a **locamin cycle** if every vertex of the cycle lies on a weakest edge.

**Definition 2.8.1** The **strength of a cycle** $C$ in a fuzzy graph $G$ is defined as the weight of a weakest edge in $C$.

**Definition 2.8.2** A cycle $C$ in a fuzzy graph $G$ is said to be a **strongest strong cycle (SSC)** if $C$ is the union of two strongest strong $u - v$ paths for every pair of vertices $u$ and $v$ in $C$ except when $uv$ is a fuzzy bridge of $G$ in $C$.

Note that in Definition 2.8.2, it is possible that edge $uv$ can be a fuzzy bridge of $G$. But the condition that $C$ is the union of two strongest strong $u - v$ paths can be relaxed for those vertices which are the end vertices of fuzzy bridges of $G$ which are in $C$. Also, $CONN_G(x, y) = CONN_C(x, y)$ for all vertices $x, y$ in $C$. The concept of SSC is illustrated below.

*Example 2.8.3* Let $G = (\sigma, \mu)$ with $\sigma^* = \{a, b, c, d, e\}$, $\sigma(x) = 1$ for all $x \in \sigma^*$, $\mu(ab)=\mu(ce) = 0.6$, $\mu(ae)=\mu(bc) = 0.4$ and $\mu(cd) = \mu(de)=0.3$ (Fig. 2.31). Here $ab$ and $ce$ are the fuzzy bridges of $G$. $C_1 = a, b, c, e, a$ and $C_2 = e, c, d, e$ are strongest strong cycles while $C_3 = a, b, c, d, e, a$ is not, because $C_3$ is not a union of two strongest strong $c - e$ paths. Here, $CONN_G(c, e) = 0.6$. But none of the $c - e$ paths in $C_3$ is strongest. Also, note that $ce$ is a fuzzy bridge of $G$ which is not in $C_3$.

If the underlying graph is a cycle, then the concepts of strongest strong cycle and locamin cycle coincide and is equal to a block as seen from the next theorem.

**Theorem 2.8.4** *Let $G = (\sigma, \mu)$ a fuzzy graph such that $G^*$ is a cycle. Then the following are equivalent.*

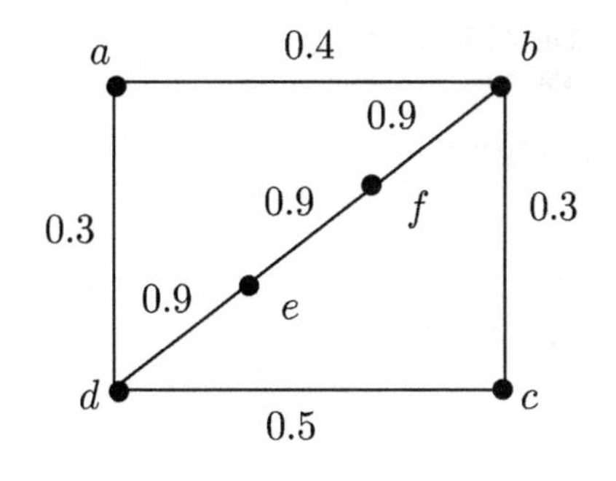


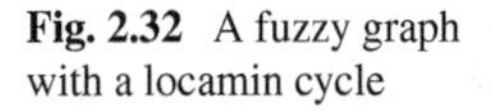
**Fig. 2.32** A fuzzy graph with a locamin cycle

*(i) $G$ is a block.*
*(ii) $G$ is an SSC.*
*(iii) $G$ is a locamin cycle.*

*Proof* $(i) \Rightarrow (ii)$ Suppose that $G$ is a block, where $G^*$ is a cycle. Then by Theorem 2.7.4, any two vertices $u$ and $v$ such that $uv$ is not a fuzzy bridge are joined by two internally disjoint strongest paths. Clearly every edge in $G$ is strong, otherwise $G$ will have exactly one non strong edge, whose removal from $G$ results in a tree, with all internal vertices fuzzy cutvertices, contradicting the assumption that $G$ is a fuzzy block. Thus, $G$ is the union of two strongest strong $u - v$ paths for every pair of vertices $u$ and $v$ in $G$ except when $uv$ is a fuzzy bridge of $G$. Thus, $G$ is an SSC.

$(ii) \Rightarrow (iii)$ Suppose that $G$ is an SSC. If possible suppose that $G$ is not locamin. Then there exists some vertex $w$ such that $w$ is not on a weakest edge of $G$. Let $uw$ and $wv$ be the two edges incident on $w$, which are not weakest edges. This implies that the path $uwv$ is the unique strongest $u - v$ path in $G$, contradiction to the assumption that $G$ is an SSC.

$(iii) \Rightarrow (i)$ Because we consider only simple fuzzy graphs, $G$ will have at least three edges and the proof follows from Theorem 2.7.4. ■

Generally in a fuzzy graph, a locamin cycle need not be an SSC and an SSC need not be a locamin cycle. (See Examples 2.8.5 and 2.8.6).

*Example 2.8.5* Let $G = (\sigma, \mu)$ with $\sigma^* = \{a, b, c, d, e, f\}$, $\sigma(x) = 1$ for all $x \in \sigma^*$ and $\mu(ab) = 0.4$, $\mu(bc) = \mu(da) = 0.3$, $\mu(cd) = 0.5$, $\mu(de) = \mu(ef) = \mu(fb) = 0.9$ (Fig. 2.32). Here, $C_1 = a, b, c, d, a$ is a locamin cycle but it is not an SSC because there do not exist two strongest $b - d$ paths in $C_1$. Note that $CONN_G(b, d) = 1$ (strength of the path $b, f, e, d$).

*Example 2.8.6* Let $G = (\sigma, \mu)$ with $\sigma^* = \{a, b, c, d, e, f\}$, $\sigma(x) = 1$ for all $x \in \sigma^*$ and $\mu(ab) = \mu(bc) = \mu(ca) = \mu(cd) = \mu(de) = \mu(ef) = \mu(fb) = 0.8$, $\mu(da) = 0.5$ (See Fig. 2.33). Here $G$ contains no locamin cycles but there are several strongest strong cycles. Note that in $G$, any cycle not containing the weakest edge $da$ is a strongest strong cycle.

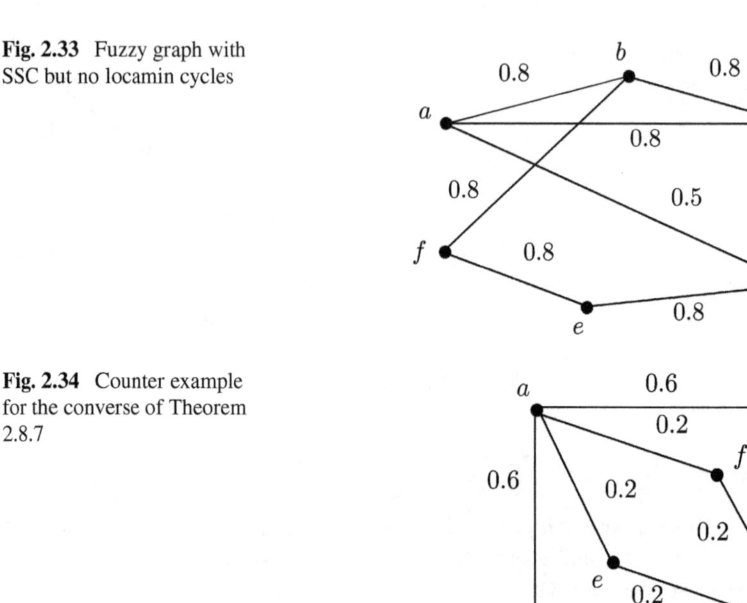


**Fig. 2.33** Fuzzy graph with SSC but no locamin cycles

**Fig. 2.34** Counter example for the converse of Theorem 2.8.7

Next we discuss a sufficient condition for a fuzzy graph to be a block.

**Theorem 2.8.7** *If any two vertices of a fuzzy graph $G$ lie on a common SSC, then $G$ is a block.*

*Proof* Let $G = (\sigma, \mu)$ be a fuzzy graph satisfying the condition of the theorem. Clearly $G$ is connected. Let $w$ be a vertex in $G$. For any two vertices $x$ and $y$ such that $x \neq w \neq y$, there exists an SSC containing $x$ and $y$. That is, there exist two internally disjoint strongest $x - y$ paths in $G$. At most one of these paths can contain the vertex $w$ and hence $w$ cannot be a fuzzy cutvertex of $G$. Because $w$ is arbitrary, it follows that $G$ is a block. ■

The converse of the above result is not true in general as seen from the next example, but is true for a sub family of fuzzy graphs which will be discussed soon.

*Example 2.8.8* Let $G = (\sigma, \mu)$ with $\sigma^* = \{a, b, c, d, e, f\}$, $\sigma(x) = 1$ for all $x \in \sigma^*$ and $\mu(ab) = \mu(bc) = \mu(cd) = \mu(da) = 0.6$, $\mu(ae) = \mu(ec) = \mu(cf) = \mu(fa) = 0.2$ (Fig. 2.34). Here the vertices $b$ and $e$ do not belong to a common SSC, but $G$ is a block.

A fuzzy graph is said to be **edge disjoint** when no two cycles share a common edge. When the fuzzy graph is edge disjoint, the blocks can be easily characterized as in the following theorem.

**Theorem 2.8.9** *Let $G = (\sigma, \mu)$ be an edge disjoint fuzzy graph with at least three vertices. Then the following are equivalent.*

*(i) G is a block.*
*(ii) G is an SSC.*
*(iii) G is a locamin cycle.*

*Proof* $(i) \Rightarrow (ii)$ Suppose $G$ is a block. In view of Theorem 2.8.4, it is sufficient to prove that $G^*$ is a cycle. Suppose $G^*$ is not a cycle. Because $G$ is a block, each vertex must be on a cycle. Thus, $G$ is a union of more than one cycle with a unique vertex in common. Let $w$ be this common vertex of intersection. Then clearly $w$ is a cutvertex, a contradiction to $(i)$.

$(ii) \Rightarrow (iii)$ Let $G$ be an edge disjoint fuzzy graph such that $G$ is an SSC. To prove $G$ is locamin. Because $G$ is an SSC, $G^*$ is a cycle and hence by Theorem 2.8.4, $G$ is a locamin cycle.

$(iii) \Rightarrow (i)$ Let $G$ be a locamin cycle. Then $G^*$ is a cycle and hence by Theorem 2.8.4, $G$ is a block. ■

The relevance of the above theorem is that any connected edge disjoint fuzzy graph with more than one cycle will always have a fuzzy cutvertex. The following are a set of necessary conditions for a fuzzy graph to be a block.

**Theorem 2.8.10** *If* $G = (\sigma, \mu)$ *is a block, then the following conditions hold and are equivalent.*

*(i) Every two vertices of G lie on a common strong cycle.*
*(ii) Each vertex and a strong edge of G lie on a common strong cycle.*
*(iii) Any two strong edges of G lie on a common strong cycle.*
*(iv) For any two given vertices and a strong edge in G, there exists a strong path joining the vertices containing the edge.*
*(v) For every three distinct vertices of G, there exist strong paths joining any two of them containing the third.*
*(vi) For every three vertices of G, there exist strong paths joining any two of them which does not contain the third.*

*Proof* $(i)$ Suppose that $G$ is a block. Let $u$ and $v$ be any two vertices in $G$ such that there exists a unique strong path between $u$ and $v$. Now, two cases arise. (1) $uv$ is a strong edge. (2) $uv$ is either weakest edge of a cycle or there exists a $u - v$ path of length more than one in $G$.

**Case 1**: $uv$ is a strong edge.

Because $uv$ is not on any strong cycle, $uv$ is an edge in every maximum spanning tree of $G$ and hence it is a fuzzy bridge. If $u$ is an end vertex in all maximum spanning trees, then clearly $u$ is a fuzzy end vertex of $G$ and hence $v$ is a fuzzy cutvertex of $G$ or vice versa, contradicting our assumption that $G$ is a block.

Now, suppose that $u$ is an end vertex in some maximal spanning tree $T_1$ and $v$ is an end vertex in some maximal spanning tree $T_2$. Let $u'$ be a strong neighbor of $u$ in $T_2$. Because $u$ is an end vertex and $v$ is an internal vertex in $T_1$, there exists a strong path $P$ in $T_1$ from $u$ to $u'$ passing through $v$. The path $P$ together with the strong edge $uu'$ forms a strong cycle in $G$, a contradiction.

**Case 2**: Either $uv$ is the weakest edge of a cycle or there exists a strong $u - v$ path of length more than one in $G$.

If $uv$ is a weakest edge of a cycle, then there exists a strong path between $u$ and $v$. Because there is a unique strong $uv$ path $P$ in $G$, $P$ belongs to all maximum spanning trees. Thus, all internal vertices in $P$ are internal vertices in all the maximum spanning trees and hence all of them are fuzzy cutvertices in $G$, contradiction to the assumption that $G$ is a block. Thus, condition $(i)$ holds. Next we show that each of the given conditions are equivalent.

$(i) \Rightarrow (ii)$ Suppose that every two vertices of $G$ lie on a common strong cycle. To prove that a given vertex and a strong edge lie on a common strong cycle. Let $u$ be a vertex and $vw$ be an edge in $G$. Let $C$ be a strong cycle containing $u$ and $v$. If $w$ is a neighbor of $v$ in $C$, then there is nothing to prove. Now, suppose that $w$ is not a neighbor of $v$ in $C$. Let $C_1$ be a strong cycle containing $u$ and $w$. Let $P_1$ and $P_2$ be the strong $u - v$ paths in $C$ and $P_1'$ and $P_2'$ the strong $u - w$ paths in $C_1$. Let $x_1$ be the vertex at which $P_1'$ leaves $P_1$. Then clearly $u \ldots (P_1) \ldots x_1 \ldots (P_1') \ldots wv \ldots (P_2)u$ is a strong cycle containing $u$ and $vw$. If $x = u$ then $u \ldots (P_1')) \ldots wv \ldots (P_2)u$ is the required cycle. If $x_1 = v$, then let $x_2$ be the vertex at which $P_2'$ leaves $P_2$. Then $u \ldots (P_1) \ldots vw \ldots (P_2') \ldots x_2 \ldots (P_2)u$ is the required strong cycle. If $x_2 = u$, then $u \ldots (P_2') \ldots wv \ldots (P_1) \ldots u$ is the required strong cycle. Because $P_1$ and $P_2$ are internally disjoint, both $x_1$ and $x_2$ cannot be the same as $v$.

$(ii) \Rightarrow (iii)$ Suppose that each vertex and strong edge lie on a common strong cycle. To prove any two strong edges lie on a common strong cycle. Let $uv$ and $xy$ be two strong edges of $G$. Let $P_1$ and $P_2$ be two internally disjoint strong paths between $v$ and $x$ and $Q_1$ and $Q_2$ be two internally disjoint strong paths between $u$ and $y$. If $P_1$, $P_2$, $Q_1$ and $Q_2$ are internally disjoint, then $uv \ldots (P_1) \ldots xy \ldots (Q_2) \ldots u$ is a strong cycle containing $uv$ and $xy$. If $Q_1$ and $Q_2$ intersect $P_1$ or $P_2$, then a strong cycle containing $uv$ and $xy$ can be extracted from the parts of the four cycles $P_1$, $P_2$, $Q_1$ and $Q_2$.

$(iii) \Rightarrow (iv)$ Let $x$ and $y$ be two vertices and let $uv$ be a strong edge in $G$. Let $x'$ be a strong neighbor of $x$ and $y'$ be a strong neighbor of $y$. Now, there exists a strong cycle $C_1$ containing $xx'$ and $uv$ and a strong cycle $C_2$ containing $yy'$ and $uv$. Now, $xx' \ldots (C_1) \ldots uv \ldots (C_2) \ldots y'y$ is a strong $x - y$ path containing the edge $uv$.

$(iv) \Rightarrow (v)$ Let $G$ be a block and $u$, $v$, $w$ be three distinct vertices of $G$. Let $v'$ be a strong neighbor of $v$. Then $u$ and $w$ are distinct vertices and $vv'$ is a strong edge of $G$. By (iv), there exists a strong path from $u$ to $w$ containing the edge $uv'$ (Even if $v' = u$ or $w$). Thus, we have a strong path between the two given vertices containing the third.

$(v) \Rightarrow (vi)$ Let $u, v, w$ be three distinct vertices of $G$. Let $P$ be a strong path between $u$ and $w$ containing $v$. Then clearly the $u - v$ strong sub path say $P'$ does not contain $w$.

$(vi) \Rightarrow (i)$ Let $u$ and $v$ be two given vertices. Let $w$ be a third vertex in $G$. Let $P_1$ be the strong path joining $u$ and $v$ not containing $w$. Let $P_2$ be the strong path joining $u$ and $w$ not containing $v$ and let $P_3$ be the strong path joining $v$ and $w$ not containing $u$. Then $P_1 \cup P_2 \cup P_3$ will contain a strong cycle containing $u$ and $v$. ■

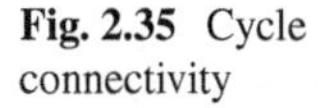
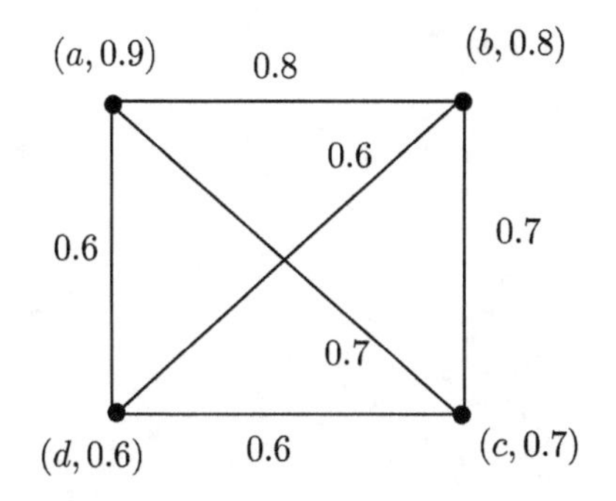


**Fig. 2.35** Cycle connectivity

So far, we have been trying to characterize fuzzy blocks as in classical graph theory. But, in a graph, the strength of every cycle is 1 whereas in fuzzy graphs, cycles of different strengths can pass through pairs of vertices. So a similar generalization is impossible. But we can find a subfamily where all the characterizations are valid. In this section, two new connectivity concepts in fuzzy graphs, namely $\theta$-evaluation and cycle connectivity and a new subfamily of fuzzy graphs called $\theta$-fuzzy graphs are discussed. Blocks in $\theta$-fuzzy graphs are fully characterized.

Consider the following definition of $\theta$-evaluation of a pair of vertices in a fuzzy graph.

**Definition 2.8.11** Let $G = (\sigma, \mu)$ be a fuzzy graph. Then for any two vertices $u$ and $v$ of $G$, there associated a set say $\theta(u, v)$ called the **$\theta$-evaluation** of $u$ and $v$ and is defined as $\theta(u, v) = \{\alpha \mid \alpha \in (0, 1]\}$, where $\alpha$ is the strength of a strong cycle passing through both $u$ and $v$.

Note that if there are no strong cycles passing both $u$ and $v$, then $\theta(u, v) = \emptyset$.

Using $\theta$-evaluation, a new measure of connectivity in graphs, namely cycle connectivity can be introduced.

**Definition 2.8.12** Let $G = (\sigma, \mu)$ be a fuzzy graph. Then $\vee\{\alpha \mid \alpha \in \theta(u, v), u, v \in \sigma^*\}$ is defined as the **cycle connectivity** between $u$ and $v$ in $G$ and denoted by $C^G_{u,v}$. If $\theta(u, v) = \emptyset$ for some pair of vertices $u$ and $v$, define the cycle connectivity between $u$ and $v$ to be 0.

$\theta$-evaluation and cycle connectivity are illustrated in the following example.

*Example 2.8.13* Let $G = (\sigma, \mu)$ with $\sigma^* = \{a, b, c, d\}$ with $\sigma(a) = 0.9$, $\sigma(b) = 0.8$, $\sigma(c) = 0.7$, $\sigma(d) = 0.6$ and $\mu(ab) = 0.8$, $\mu(bc) = \mu(ac) = 0.7$, $\mu(bd) = \mu(cd) = \mu(da) = 0.6$ (Fig. 2.35). Here $G$ is a complete fuzzy graph. $\theta\{a, c\} = \{0.6, 0.7\}$ and hence $C^G_{a,c} = 0.7$

Cycle connectivity is a measure of connectedness in a fuzzy graph and it is always less than or equal to the strength of connectedness between any two vertices $u$ and $v$. In a crisp graph the cycle connectivity between two vertices $u$ and $v$ is 1 if $u$ and $v$ belong to a common cycle and 0 otherwise.

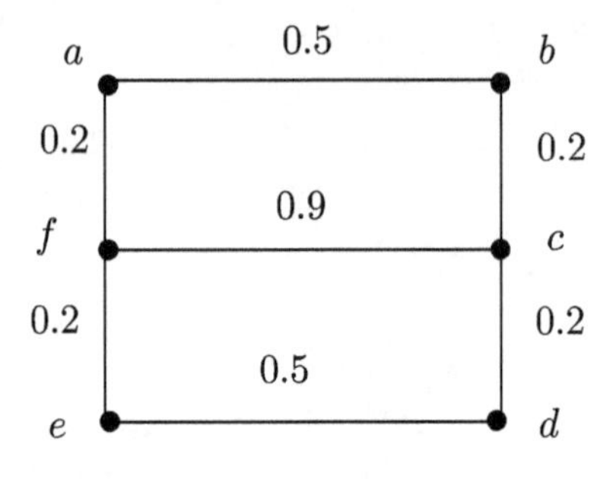


**Fig. 2.36** A $\theta$-fuzzy graph

A new subclass of fuzzy graphs called $\theta$-fuzzy graphs, with the property that every cycle passing through a particular pair of vertices have the same strength, is discussed below.

**Definition 2.8.14** Let $G = (\sigma, \mu)$ be a fuzzy graph. $G$ is said to be a **$\theta$-fuzzy graph** if $\theta$- evaluation of each pair of vertices in $G$ is either empty or a singleton set. In other words $G$ is called a $\theta$-fuzzy graph if for each pair of vertices $u$ and $v$, either there is no strong cycle passing through $u$ and $v$ or all strong cycles passing through $u$ and $v$ have the same strength.

Consider the following example.

*Example 2.8.15* Let $G = (\sigma, \mu)$ with $\sigma^* = \{a, b, c, d, e, f\}$, $\sigma(x) = 1$ for all $x \in \sigma^*$ and $\mu(ab) = \mu(de) = 0.5$, $\mu(bc) = \mu(cd) = \mu(ef) = \mu(fa) = 0.2$, $\mu(cf) = 0.9$ (Fig. 2.36). $G$ is clearly a $\theta$-fuzzy graph because all strong cycles in this graph have strength 0.1 and hence $\theta(u, v) = \{0.1\}$ for any two vertices $u$ and $v$.

**Proposition 2.8.16** *Edge disjoint fuzzy graphs are $\theta$-fuzzy graphs.*

*Proof* Let $G$ be an edge disjoint fuzzy graph. First we show that two distinct vertices cannot be in two different cycles of $G$. Because $C_1$ and $C_2$ share no edge in common, $C_1$ and $C_2$ intersect at $u$ and $v$ (say). Let $P_1$ and $P_2$ be the two $u - v$ paths in $C_1$ and $Q_1$ and $Q_2$ be the two $u - v$ paths in $C_2$. Then clearly $P_1 \cup Q_1$, $P_1 \cup Q_2$, $P_2 \cup Q_1$, $P_2 \cup Q_2$ are all cycles and some of them will definitely share edges with $C_1$ or $C_2$ which is not possible. Thus, between any pair of vertices $u$ and $v$ of $G$ there exists at most one strong cycle and hence it follows that $\theta$- evaluation of any two vertices in $G$ is either empty set or a singleton. Thus, $G$ is a $\theta$-fuzzy graph. ■

In a CFG, the cycle connectivity between vertices can be easily evaluated using the following result.

**Theorem 2.8.17** *Let $G = (\sigma, \mu)$ be a complete fuzzy graph. Then for any two vertices $u$ and $v$ in $G$, $C^G_{u,v} = \vee\{\wedge\{\sigma(u), \sigma(v), \sigma(w)\} \mid w \in \sigma^*\}$.*

*Proof* Let $G = (\sigma, \mu)$ be a CFG. By Theorem 2.2.7, all edges of $G$ are strong and hence all cycles in $G$ are strong cycles. Now, let $u, v \in \sigma^*$ and consider a strong cycle $C : u = u_1u_2u_3 \ldots u_n = vu$.

Then $s(C) = \wedge\{\mu(u_1u_2), \mu(u_2u_3), \ldots, \mu(u_{n-1}u_n), \mu(uv)\}$.

Because $G$ is complete, $\mu(xy) = \sigma(x) \wedge \sigma(y)$ for all $x$ and $y$ and hence $s(C) = \wedge\{\sigma(u_1), \sigma(u_2), \ldots, \sigma(u_n)\} = \sigma(u_l)$ for some $l \in \{1, 2, \ldots, n\}$. Thus, the strength of $C = \wedge\{\sigma(u), \sigma(v), \sigma(w)\}$ for some vertex $w \in \sigma^*$, and hence for any two vertices $u$ and $v$ in $G$, $C^G_{u,v} = \vee\{\wedge\{\sigma(u), \sigma(v), \sigma(w)\} \mid w \in \sigma^*\}$. ■

In other words, the cycle connectivity between $u$ and $v$ in a complete fuzzy graph is the maximum of strengths of all triangles passing through $u$ and $v$.

The next lemma is the key for the characterization of blocks in $\theta$-fuzzy graphs. It gives the relationship between strong paths and strongest paths in $\theta$-fuzzy graphs which are blocks.

**Lemma 2.8.18** *Let $G$ be a $\theta$-fuzzy graph which is a block. Then any strong $u - v$ path such that $uv$ is not a fuzzy bridge is a strongest $u - v$ path and hence any strong cycle in $G$ is a strongest strong cycle.*

*Proof* Let $G = (\sigma, \mu)$ be a $\theta$-fuzzy graph which is a block. Let $u, v \in \sigma^*$ be such that $uv$ is not a fuzzy bridge. Let $P$ be a strong $u - v$ path in $G$. If $P$ is not a strongest $u - v$ path, then because $G$ is a block, there exist two internally disjoint strongest strong $u - v$ paths say $P_1$ and $P_2$. Then $P_1 \cup P$ is a strong cycle with strength less than that of the cycle $P_1 \cup P_2$. Both these cycles pass through $u$ and $v$ and hence $\theta(u, v)$ is not a singleton or empty which is a contradiction to the fact that $G$ is a $\theta$-fuzzy graph. Thus, $P$ must be a strongest strong $u - v$ path.

To prove the second assertion of the lemma, let $C$ be a strong cycle in $G$. Let $u, v$ be two vertices in $C$ such that $uv$ is not a fuzzy bridge. Then by first part both these $u - v$ paths in $C$ are strongest $u - v$ paths. Thus, $G$ is a strongest strong cycle. ■

Thus, in a $\theta$-fuzzy graph which is a block, the concepts of strong path and strongest path coincide and as a result, the concepts of strong cycle and SSC also coincide. Thus, all the six necessary and sufficient conditions for blocks in graphs can be generalized to blocks in $\theta$- fuzzy graphs.

Next we have the awaited characterization for blocks in $\theta$-fuzzy graphs.

**Theorem 2.8.19** *Let $G = (\sigma, \mu)$ be a $\theta$-fuzzy graph. Then the following statements are equivalent.*

(i) *$G$ is a block.*
(ii) *Every pair of vertices of $G$ lie on a common strongest strong cycle.*
(iii) *Each vertex and a strong edge of $G$ lie on a common strongest strong cycle.*
(iv) *Any two strong edges of $G$ lie on a common strongest strong cycle.*
(v) *For any two given vertices $u$ and $v$ such that $uv$ is not a fuzzy bridge and a strong edge $xy$ in $G$, there exists a strongest strong $u - v$ path containing the edge $xy$.*

(vi) *For every three distinct vertices $u_i, i = 1, 2, 3$ of $G$ such that $u_i u_j, i \neq j$ is not a fuzzy bridge, there exist strongest strong paths joining any two of them containing the third.*

(vii) *For every three distinct vertices $u_i, i = 1, 2, 3$ of $G$ such that $u_i u_j, i \neq j$ is not a fuzzy bridge, there exist strongest strong paths joining any two of them not containing the third.*

*Proof* Theorem 2.8.10 and Lemma 2.8.18 together give all the required implications except $(vii) \Rightarrow (i)$

To prove $(vii) \Rightarrow (i)$, note that for any vertex $w$ of $G$ and for every pair of vertices $x, y$, other than $w$, there exists a strongest $x - y$ path not containing the vertex $w$. That is, $w$ is not on every strongest $x - y$ path for any $x$ and $y$ and hence $w$ is not a fuzzy cutvertex. Because $w$ is arbitrary, it follows that $G$ is a block. ■

## 2.9 Fuzzy Line Graphs

The contents of this section are from [125]. The study of fuzzy line graphs was carried out by Mordeson in 1993. It was one of the first theoretical topics studied in fuzzy graphs after Rosenfeld's introductory paper.

The line graph, $L(G)$, of a (crisp) graph $G$ is the intersection graph of the set of edges of $G$. Hence, the vertices of $L(G)$ are the edges of $G$ with two vertices of $L(G)$ adjacent whenever the corresponding edges of $G$ are. We present the notion of a fuzzy line graph. Let $G = (V, X)$ and $G' = (V', X')$ be graphs.

**Definition 2.9.1** Let $(\sigma, \mu)$ and $(\sigma', \mu')$ be partial fuzzy subgraphs of $G$ and $G'$, respectively. Let $f$ be a one-to-one function of $V$ onto $V'$. Then

$(i)$ $f$ is called a (**weak**) **vertex-isomorphism** of $(\sigma, \mu)$ onto $(\sigma', \mu')$ if and only if for all $v \in V$, $(\sigma(v) \leq \sigma'(f(v))$ and $\text{Supp}(\sigma') = f(\text{Supp}(\sigma))$, $\sigma(v) = \sigma'(f(v))$;

$(ii)$ $f$ is called a (**weak**) **line-isomorphism** of $(\sigma, \mu)$ onto $(\sigma', \mu')$ if and only if for all $u, v \in V$, $\mu(uv) \leq \mu'(f(u)f(v))$ and $\text{Supp}(\mu') = \{(f(u)f(v)) \mid u, v \in \text{Supp}(\mu)\}$, $\mu(uv) = \mu'(f(u)f(v))$.

If $f$ is a (weak) vertex-isomorphism and a (weak) line-isomorphism of $(\sigma, \mu)$ onto $(\sigma', \mu')$, then $f$ is called a (**weak**) **isomorphism** of $(\sigma, \mu)$ onto $(\sigma', \mu')$. If $(\sigma, \mu)$ is isomorphic to $(\sigma', \mu')$, then we write $(\sigma, \mu) \simeq (\sigma', \mu')$.

Let $G = (V, X)$ be a graph, where $V = \{v_1, \ldots, v_n\}$. Let $S_i = \{v_i, x_{i1}, \ldots, x_{iq_i}\}$, where $x_{ij} \in X$ and $x_{ij}$ has $v_i$ as a vertex, $j = 1, \ldots, q_i$; $i = 1, \ldots, n$. Let $S = \{S_1, \ldots, S_n\}$. Let $T = \{S_i S_j \mid S_i, S_j \in S, S_i \cap S_j \neq \emptyset, i \neq j\}$. Then $\mathcal{I}(S) = (S, T)$ is an intersection graph and $G \simeq \mathcal{I}(S)$. Any partial fuzzy subgraph $(\iota, \gamma)$ of $\mathcal{I}(S)$ with $\text{Supp}(\gamma) = T$ is called a **fuzzy intersection graph**.

Let $(\sigma, \mu)$ be a partial fuzzy subgraph of $G$. Let $I(S)$ be the intersection graph described above. Define the fuzzy subsets $\iota, \gamma$ of $S$ and $T$, respectively as follows:

For all $S_i \in S$, $\iota(S_i) = \sigma(v_i)$;

For all $S_i S_j \in T$, $\gamma(S_i S_j) = \mu(v_i v_j)$.

**Proposition 2.9.2** *Let $(\sigma, \mu)$ be a partial fuzzy subgraph of $G$. Then*
*(i) $(\iota, \gamma)$ is a partial fuzzy subgraph of $\mathcal{I}(S)$;*
*(ii) $(\sigma, \mu) \simeq (\iota, \gamma)$.*

*Proof* $(i)$ $\gamma(S_i S_j) = \mu(v_i v_j) \leq \sigma(v_i) \wedge \sigma(v_j) = \iota(S_i) \wedge \iota(S_j)$.
$(ii)$ Define $f : V \to S$ by $f(v_i) = S_i, i = 1, \ldots, n$. Clearly $f$ is a one-to-one function of $V$ onto $S$. Now, $v_i v_j \in X$ if and only if $S_i S_j \in T$ and so $T = \{f(v_i) f(v_j) \mid v_i v_j \in X\}$. Also, $\iota(f(v_i)) = \iota(S_i) = \sigma(v_i)$ and $\gamma(f(v_i) f(v_j)) = \gamma(S_i S_j) = \mu(v_i v_j)$. Thus, $f$ is an isomorphism of $(\sigma, \mu)$ onto $(\iota, \gamma)$. ■

Let $\mathcal{I}(S)$ be the intersection graph of $(V, X)$. Let $(i, \gamma)$ be the fuzzy intersection graph of $\mathcal{I}(S)$ as defined above. We call $(\iota, \gamma)$ the *fuzzy intersection graph* of $(\sigma, \mu)$. The previous proposition shows that any fuzzy graph is isomorphic to a fuzzy intersection graph.

The line graph of $G$, $L(G)$, is by definition the intersection graph $\mathcal{I}(X)$. That is, $L(G) = (Z, W)$, where $Z = \{\{x\} \cup \{u_x, v_x\} \mid x \in X, u_x, v_x \in V, x = u_x v_x\}$ and $W = \{S_x S_y \mid S_x \cap S_y \neq \emptyset,\ x, y \in X,\ x \neq y\}$ and where $S_x = \{x\} \cup \{u_x, v_x\}, x \in X$. Let $(\sigma, \mu)$ be a partial fuzzy subgraph of $G$. Define the fuzzy subsets $\lambda, \omega$ of $Z, W$, respectively, as follows:

For all $S_x \in Z$, $\lambda(S_x) = \mu(x)$;
For all $S_x S_y \in W$, $\omega(S_x S_y) = \mu(x) \wedge \mu(y)$.

**Proposition 2.9.3** *$(\lambda, \omega)$ is a fuzzy subgraph of $L(G)$ and is called the* ***fuzzy line graph*** *corresponding to $(\sigma, \mu)$.*

*Proof* $\omega(S_x S_y) = \mu(x) \wedge \mu(y) = \lambda(S_x) \wedge \lambda(S_y)$. ■

Every cutpoint of $L(G)$ is a bridge of $G$ which is not a pendent edge, and conversely [83]. It is shown in [125] that the relationship between cutpoints in $L(G)$ and bridges in $G$ does not carry over to the fuzzy case.

Let $(\sigma, \mu)$ and $(\sigma', \mu')$ be partial fuzzy subgraphs of $G$ and $G'$, respectively. If $f$ is a weak isomorphism of $(\sigma, \mu)$ onto $(\sigma', \mu')$, then it can be shown that $f$ is an isomorphism of (Supp$(\sigma)$, Supp$(\mu)$) onto (Supp$(\sigma')$,Supp$(\mu')$). If $(\lambda, \omega)$ is the fuzzy line graph of $(\sigma, \mu)$, then it can also be shown that (Supp$(\lambda)$, Supp$(\omega)$) is the fuzzy line graph of (Supp$(\sigma)$, Supp$(\mu)$).

*Example 2.9.4* Let $G = (V, X)$, where $V = \{v_1, v_2, v_3, v_4\}$ and $X = \{x_1 = v_1 v_2, x_2 = v_1 v_3, x_3 = v_2 v_3, x_4 = v_3 v_4\}$. Let $\sigma(v_i) = 1, i = 1, 2, 3, 4$, $\mu(x_1) = \mu(x_3) = 1$ and $\mu(x_2) = \mu(x_4) = 1/2$. Then $\lambda(S_{x_1}) = \lambda(S_{x_3}) = 1$, $\lambda(S_{x_2}) = \lambda(S_{x_4}) = 1/2$ and $\omega(S_{x_1} S_{x_2}) = 1$, $\omega(S_{x_2} S_{x_3}) = \omega(S_{x_3} S_{x_4}) = \omega(S_{x_2} S_{x_4}) = 1/2$. If we delete $x_1$ from $G$, then the strength of connectedness between $v_1$ and $v_2$ before the deletion of $x_1$ is $1 = \mu(x_1)$. Thus, $x_1$ is a bridge of $(\sigma, \mu)$ (not an endline of $G$). However, the strength of connectedness between any pair of vertices $S_{x_2}$, $S_{x_3}$, $S_{x_4}$ is $1/2$ before and after the deletion of $S_{x_1}$. Thus, $S_{x_1}$ is not a cutvertex of $(\sigma, \mu)$.

**Proposition 2.9.5** *Let $(\sigma, \mu)$ and $(\sigma', \mu')$ be partial fuzzy subgraphs of $G$ and $G'$, respectively. If $f$ is a weak isomorphism of $(\sigma, \mu)$ onto $(\sigma', \mu')$, then $f$ is an isomorphism of (Supp$(\sigma)$, Supp$(\mu)$) onto (Supp$(\sigma')$, Supp$(\mu')$).*

*Proof* $v \in \text{Supp}(\sigma) \Leftrightarrow f(v) \in \text{Supp}(\sigma')$ and $uv \in \text{Supp}(\mu) \Leftrightarrow f(u)f(v) \in \text{Supp}(\mu')$. ■

**Proposition 2.9.6** *If $(\lambda, \omega)$ is the fuzzy line graph of the fuzzy graph $(\sigma, \mu)$, then $(Supp(\lambda), Supp(\omega))$ is the line graph of $(Supp(\sigma), Supp(\mu))$.*

*Proof* $(\sigma, \mu)$ is a partial fuzzy subgraph of $G$ and $(\lambda, \omega)$ is a partial fuzzy subgraph of $L(G)$. Now, $\lambda(S_x) = \mu(x)$ for all $x \in X$ and so $S_x \in \text{Supp}(\lambda) \Leftrightarrow x \in \text{Supp}(\mu)$. Also, $\omega(S_x S_y) = \mu(x) \wedge \mu(y)$ for all $S_x S_y \in W$ and so $\text{Supp}(\omega) = \{S_x S_y \mid S_x \cap S_y \neq \emptyset, x, y \in \text{Supp}(\mu), x \neq y\}$. ■

**Theorem 2.9.7** ([125]) *Let $(\lambda, \omega)$ be the fuzzy line graph corresponding to $(\sigma, \mu)$. Suppose that $(Supp(\sigma), Supp(\mu))$ is connected. Then the following properties hold.*

*(i) There exists a weak isomorphism of $(\sigma, \mu)$ onto $(\lambda, \omega)$ if and only if $(Supp(\sigma), Supp(\mu))$ is a cycle and $\sigma$ and $\mu$ are constant functions on $Supp(\sigma)$ and $Supp(\mu)$, respectively, taking on the same value;*
*(ii) If $f$ is a weak isomorphism of $(\sigma, \mu)$ onto $(\lambda, \omega)$, then $f$ is an isomorphism.*

*Proof* Suppose that $f$ is a weak isomorphism of $(\sigma, \mu)$ onto $(\lambda, \omega)$. By Proposition 2.9.5, we can see that $f$ is an isomorphism of $(\text{Supp}(\sigma), \text{Supp}(\mu))$ onto $(\text{Supp}(\lambda), \text{Supp}(\omega))$. By Proposition 2.9.6, $(\text{Supp}(\sigma), \text{Supp}(\mu))$ is a cycle. Let $\text{Supp}(\sigma) = \{v_1, v_2, \ldots, v_n\}$ and $\text{Supp}(\mu) = \{v_1v_2, v_2v_3, \ldots, v_nv_1\}$, where $v_1v_2 \ldots v_nv_1$ is a cycle. Let $\sigma(v_i) = s_i$ and $\mu(v_iv_{i+1} = r_i$, $i = 1, 2, \ldots, n$, where $v_{n+1} = v_1$. Then $s_{n+1} = s_1$ and

$$r_i \leq s_i \wedge s_{i+1}, \ i = 1, 2, \ldots, n. \tag{2.3}$$

Now, we have $\text{Supp}(\lambda) = \{S_{(v_i, v_{i+1})} \mid i = 1, 2, \ldots, n\}$ and $\text{Supp}(\omega) = \{(S_{(v_i, v_{i+1})}, S_{(v_{i+1}, v_{i+2})}) \mid i = 1, 2, \ldots, (n-1)\}$. Also, for $r_{n+1} = r_1$, $\lambda(S_{(v_i, v_{i+1})}) = \mu(v_i, v_{i+1}) = r_i$ and $\omega(S_{(v_i, v_{i+1})}, S_{(v_{i+1}, v_{i+2})}) = \mu(v_i, v_{i+1}) \wedge \mu(v_{i+1}, v_{i+2}) = r_i \wedge r_{i+1}$, $i = 1, 2, \ldots, n$, where $v_{n+2} = v_2$. Because $f$ is an isomorphism of $(\text{Supp}(\sigma), \text{Supp}(\mu))$ onto $(\text{Supp}(\lambda), \text{Supp}(\omega))$, $f$ maps $\text{Supp}(\sigma)$ onto $\text{Supp}(\lambda) = \{S_{(v_1, v_2)}, \ldots, S_{(v_n, v_1)}\}$. Also, $f$ preserves adjacency. Hence, $f$ induces a permutation $\pi$ of $\{1, 2, \ldots, n\}$ such that $f(v_i) = S_{(v_{\pi(i)}, v_{\pi(i)+1})}$ and

$$(v_i, v_{i+1}) \to (f(v_i, f(v_{i+1})) = (S_{(v_{\pi(i)}, v_{\pi(i)+1})}, S_{(v_{\pi(i+1)}, v_{\pi(i+1)+1})})$$

for $i = 1, 2, \ldots, (n-1)$. Now,

$$s_i = \sigma(v_i) \leq \lambda(f(v_i)) = \lambda(S_{(v_{\pi(i)}, v_{\pi(i)+1})}) = r_{\pi(i)}$$

and

$$\begin{aligned} r_i &= \mu(v_i, v_{i+1}) \leq \omega(f(v_i), f(v_{i+1})) \\ &= \omega(S_{(v_{\pi(i)}, v_{\pi(i)+1})}, S_{(v_{\pi(i+1)}, v_{\pi(i+1)+1})}) = \mu(v_{\pi(i)}, \end{aligned}$$

$$v_{\pi(i)+1}) \wedge \mu(v_{\pi(i+1)}, v_{\pi(i+1)+1}) = r_{\pi(i)} \wedge r_{\pi(i+1)}, i = 1, 2, \ldots, n.$$

That is,

$$s_i \le r_{\pi(i)} \text{ and } r_i \le r_{\pi(i)} \wedge r_{\pi(i+1)}, i = 1, 2, \ldots, n. \qquad (2.4)$$

By the second part of (2.4), we have that $r_i \le r_{\pi(i)}, i = 1, 2, \ldots, n$, and so $r_{\pi(i)} \le r_{\pi(\pi(i))}, i = 1, 2, \ldots, n$. Continuing we have that $r_i \le r_{\pi(i)} \le \cdots \le r_{\pi^j} \le r_i$ and so $r_i = r_{\pi(i)}, i = 1, 2, \ldots, n$, where $\pi^{j+1}$ is the identity map. By (2.4) again, we have $r_i \le r_{\pi(i+1)} = r_{i+1}, i = 1, 2, \ldots, n$, where $r_{n+1} = r_1$. Hence, by (2.3) and (2.4), $r_1 = r_2 = \cdots = r_n = s_1 = s_2 = \cdots = s_n$. Thus, we have not only proved the conclusion about $\sigma$ and $\mu$ being constant functions, but we have also shown that $(ii)$ holds.

Conversely, suppose that (Supp($\sigma$), Supp($\mu$)) is a cycle and for all $v \in$ Supp($\sigma$) and $x \in$ Supp($\mu$), $\sigma(v) = \mu(x)$. By Proposition 2.9.6, (Supp($\lambda$), Supp($\omega$)) is the line graph of (Supp($\sigma$), Supp($\mu$)). Because (Supp($\sigma$), Supp($\mu$)) is a cycle, (Supp($\sigma$), Supp($\mu$)) $\cong$ (Supp($\lambda$), Supp($\omega$)) by Theorem 8.2 of [83]. This isomorphism induces an isomorphism of $(\sigma, \mu)$ onto $(\lambda, \omega)$, because $\sigma(v) = \mu(x)$ for all $v \in V$ and $x \in X$ and so $\sigma = \mu = \lambda = \omega$ on their respective domains. ■

**Theorem 2.9.8** *Let $(\sigma, \mu)$ and $(\sigma', \mu')$ be partial fuzzy subgraphs of $G$ and $G'$, respectively, such that (Supp($\sigma$), Supp($\mu$)) and (Supp($\sigma'$), Supp ($\mu'$)) are connected. Let $(\lambda, \omega)$ and $(\lambda', \omega')$ be the line graphs corresponding to $(\sigma, \mu)$ and $(\sigma', \mu')$, respectively. Suppose that it is not the case that one of (Supp($\sigma$), Supp($\mu$)) and (Supp($\sigma'$), Supp($\mu'$)) is $K_3$ and the other is $K_{1,3}$. If $(\lambda, \omega) \simeq (\lambda', \omega')$, then $(\sigma, \mu)$ and $(\sigma', \mu')$ are line isomorphic.*

*Proof* Because $(\lambda, \omega) \simeq (\lambda', \omega')$, (Supp($\lambda$), Supp($\omega$)) $\simeq$ (Supp($\lambda'$), Supp($\omega'$)) by Proposition 2.9.5. Because (Supp($\lambda$), Supp($\omega$)) and (Supp($\lambda'$), Supp($\omega'$)) are line graphs of (Supp($\sigma$), Supp($\mu$)) and (Supp($\sigma'$), Supp($\mu'$)), respectively, by Proposition 2.9.6, we have that (Supp($\sigma$), Supp($\mu$)) $\simeq$ (Supp ($\sigma'$), Supp($\mu'$)) by Theorem 8.3 of [83]. Let $g$ denote the isomorphism of $(\lambda, \omega)$ onto $(\lambda', \omega')$ and $f$ the isomorphism of (Supp($\sigma$), Supp($\mu$)) onto (Supp($\sigma'$), Supp($\mu'$)). Then $\lambda(S_{uv}) = \lambda'(g(S_{uv}) = \lambda'(g(S_{f(u)f(v)})$, where the latter equality holds by the proof of Theorem 8.3 in [83] and so $\mu(uv) = \mu'(f(u)f(v))$. Hence, $(\sigma, \mu)$ and $(\sigma', \mu')$ are line isomorphic. ■

**Proposition 2.9.9** *Let $(\tau, \nu)$ be a partial fuzzy subgraph of $L(G)$. Then $(\tau, \nu)$ is a fuzzy line graph of some partial fuzzy subgraph of $G$ if and only if for all $S_x S_y \in W$, $\nu(S_x S_y) = \tau(x) \wedge \tau(y)$.*

*Proof* Suppose that $\nu(S_x S_y) = \tau(S_x) \wedge \tau(S_y)$ for all $S_x S_y \in W$. For all $x \in X$, define $\sigma(x) = \tau(S_x)$. Then $\nu(S_x S_y) = \tau(S_x) \wedge \tau(S_y) = \mu(x) \wedge \mu(y)$. Any $\sigma$ that yields the property $\mu(uv) \le \sigma(x) \wedge \sigma(y)$ will suffice, i.e., $\sigma(v) = 1$ for all $v \in V$. The converse is immediate. ■

**Theorem 2.9.10** *$(\sigma, \mu)$ is a fuzzy line graph if and only if (Supp($\sigma$), Supp($\mu$)) is a line graph and for all $uv \in$ Supp($\mu$), $\mu(uv) = \sigma(u) \wedge \sigma(v)$.*

*Proof* Suppose that $(\sigma, \mu)$ is a fuzzy line graph. Then the conclusion holds by Propositions 2.9.6 and 2.9.9. Conversely, suppose that (Supp($\sigma$), Supp($\mu$)) is a line graph and for all $uv \in$ Supp($\mu$), $\mu(uv) = \sigma(u) \wedge \sigma(v)$. Then the conclusion holds from Proposition 2.9.9. ■

## 2.10 Fuzzy Interval Graphs

The results in this section are due to the important work of Craine in [61]. In [61], it was shown that a fuzzy graph without loops is the intersection graph of some family of fuzzy sets. It was shown that the characterization of interval graphs by Gilmore and Hoffman naturally extends to fuzzy interval graphs while that of Fulkerson and Gross does not. We present these results here.

As stated in [61], Roberts [153] cites applications of interval graphs in archaeology, developmental psychology, ecological modeling, mathematical sociology and organization theory. These disciplines all have components that are ambiguously defined, require subjective evaluation, or are satisfied to differing degrees. These are extremely active areas of application of fuzzy methods. It is therefore valuable to explore the extent that intersection graph results can be extended using fuzzy set theory.

The intersection graph of a family (perhaps with repeated members) of sets is a graph with a vertex representing each member of the family and an edge connecting two vertices if and only if the two sets have nonempty intersection. Generally loops are suppressed. If the family is composed of intervals or is the edge set of a hypergraph, then the intersection graph is called an *interval graph* or a *line graph*, respectively.

McAllister [119] used a different approach in defining a fuzzy intersection graph. However, his approach did not yield the usual definition of an intersection graph when applied to families of crisp sets. A different approach is taken in [61]. Each definition and theorem is a natural generalization of the crisp theory.

The $t$-norm minimum is used to define the fuzzy intersection graph of a family of fuzzy sets. We present a proof of a fuzzy analog of Marczewski's theorem, [61]. The proof shows that every fuzzy graph without loops is the intersection graph of some family of fuzzy sets. We also show that the natural generalization of the Fulkerson and Gross characterization of interval graphs fails, [61]. We then present a natural generalization of the Gilmore and Hoffman characterization.

Results characterizing a fuzzy property in terms of cut level set properties are significant, in that such theorems demonstrate the extent to which the crisp theory can be generalized. To accomplish this here, we provide the sequence of crisp cut level graphs given in [26]. Define the fundamental sequence of a fuzzy graph $G = (\sigma, \mu)$ to be the ordered set

$$\text{fg}(G) = \{\sigma(x) > 0 \mid x \in X\} \cup \{\mu(xy) > 0 \mid (x, y) \in X \times X\},$$

where the decreasing order inherited from the real interval [0, 1] is used.

The first element listed in fg($G$) is the maximal vertex strength while the last element listed is the minimal nonzero edge strength.

### 2.10.1 *Fuzzy Intersection Graphs*

We now define a fuzzy intersection graph and prove some basic results. Let $X$ be a set. Define the function $h : \mathcal{FP}(X) \to [0, 1]$ by for all $\alpha \in \mathcal{FP}(X)$, $h(\alpha) = \vee\{\alpha(x) \mid x \in X\}$. Then $h(\alpha)$ is called the **height** of $\alpha$.

**Definition 2.10.1** Let $\mathcal{F} = \{\alpha_1, \ldots, \alpha_n\}$ be a finite family of fuzzy sets defined on a set $X$ and consider $\mathcal{F}$ a crisp vertex set. The **fuzzy intersection graph** of $\mathcal{F}$ is the fuzzy graph Int($\mathcal{F}$) $= (\sigma, \mu)$, where $\sigma : \mathcal{F} \to [0, 1]$ is defined by $\sigma(\alpha_i) = h(\alpha_i)$ and $\mu : \mathcal{E}_{\mathcal{F}} \to [0, 1]$ is defined by

$$\mu(\alpha_i\alpha_j) = \begin{cases} h(\alpha_i \wedge \alpha_j) & \text{if } i \neq j \\ 0 & \text{if } i = j, \end{cases}$$

where we recall that $\mathcal{E}_{\mathcal{F}} = \{[(\alpha_i, \alpha_j)] \mid (a_i, \alpha_j) \in \mathcal{F} \times \mathcal{F}\}$.

The purpose of requiring $\mu(\alpha_i, \alpha_j) = 0$ for $i = j$, is to preclude loops. We note that an edge $\alpha_i\alpha_j$ has zero strength if and only if $\alpha_i \wedge \alpha_j$ is the zero function or $i = j$.

Let $\mathcal{F}$ be a family of sets and $c \in [0, 1]$. Define $\mathcal{F}^c = \{\alpha^c \mid \alpha \in \mathcal{F}\}$, where $\alpha^c$ is the $c$-level set of $\alpha$. Let $G = (\sigma, \mu)$ be a fuzzy graph. Let $G^c$ denote $(\sigma_c, \mu_c)$.

If $\mathcal{F} = \{\alpha_1, \ldots, \alpha_n\}$ is a family of fuzzy sets and $c \in [0, 1]$, then Int($\mathcal{F}^c$) $=$ (Int($\mathcal{F}$))$^c$. The graph Int($\mathcal{F}^c$) has a vertex representing $\alpha_i \in \mathcal{F}$ if and only if $h(\alpha_i) > c$. The pair $\{(\alpha_i)^c, (\alpha_j)^c\}$ is an edge of Int($\mathcal{F}^c$) if and only if $i \neq j$ and $h(\alpha_i \wedge \alpha_j) \geq c$. These conditions also characterize the graph (Int($\mathcal{F}$))$^c$. In particular, if $\mathcal{F}$ is a family of crisp subsets of $X$, then the fuzzy intersection graph and crisp intersection graph definitions coincide.

**Theorem 2.10.2** [61] (Fuzzy analog of Marczewski's theorem [105]) *If $G = (\sigma, \mu)$ is a fuzzy graph without loops, then for some family of fuzzy sets $\mathcal{F}$, $G = Int(\mathcal{F}^c)$.*

*Proof* Let $G = (\sigma, \mu)$ be a fuzzy graph on $V$. For each $x \in V$ define the anti-reflexive, symmetric fuzzy subset $\alpha_x : \mathcal{E}_V \to [0, 1]$ by for all $y, z \in V$,

$$\alpha_x(yz) = \begin{cases} \sigma(x) & \text{if } y = x \text{ and } z = x \\ \mu(xz) & \text{if } y = x \text{ and } z \neq x \\ \mu(yx) & \text{if } y \neq x \text{ and } z = x \\ 0 & \text{if } y \neq x \text{ and } z \neq x. \end{cases}$$

We show that $G$ is the fuzzy intersection graph of $F = \{\alpha_x \mid x \in V\}$. By definition, $\alpha_x(x, x) = \sigma(x) \geq \mu(xy)$ and so $h(\alpha_x) = \sigma(x)$ as required. For $x \neq y$ a nonzero value of $(\alpha_x \cap \alpha_y)(zw) = \alpha_x(zw) \wedge \alpha_y(zw)$ occurs only if $x = z$ and $y = w$(or $y = z$ and $x = w$). Thus, $h(\alpha_x \cap \alpha_y) = (\alpha_x \cap \alpha_y)(xy) = \mu(xy)$ and the desired result holds. ■

### *2.10.2 Fuzzy Interval Graphs*

The families of sets most often considered in connection with intersection graphs are families of intervals of a linearly ordered set. This class of interval graphs is central to many applications. In this section, we define fuzzy interval graphs and examine some of their basic properties.

In both the crisp and fuzzy cases, distinct families of sets can have the same intersection graph. In particular, the intersection properties of a finite family of real intervals (fuzzy numbers) can be characterized by a family of intervals (fuzzy intervals) defined on a finite set. Therefore, as is common in interval graph theory [120], we restrict our attention to intervals (fuzzy intervals) with finite support.

We generalize two characterizations of (crisp) interval graphs. Theorem 2.10.8 gives the Fulkerson and Gross characterization [74] and Theorem 2.10.16 provides the Gilmore and Hoffman characterization [78]. Both theorems make use of relationships between the finite number of points which define the intervals and the cliques of the corresponding interval graph.

Recall a clique is a maximal (with respect to set inclusion) complete subgraph. We adopt the convention of naming a clique by its vertex set. Clearly, if a vertex $z$ is not a member of a clique $K$, then there exists an $x \in K$ such that $xz$ is not an edge of $G$. We generalize this concept in Definition 2.10.7.

**Definition 2.10.3** Let $X$ be a linearly ordered set. A. **fuzzy interval** $\mathcal{I}$ on $X$ is a normal, convex fuzzy subset of $X$. That is, there exists an $x \in X$ with $\mathcal{I}(x) = 1$ and the ordering $w \leq y \leq z$ implies that $\mathcal{I}(y) \geq \mathcal{I}(w) \wedge \mathcal{I}(z)$. A **fuzzy number** is a real fuzzy interval. A **fuzzy interval graph** is the fuzzy intersection graph of a finite family of fuzzy intervals.

By normality of the fuzzy intervals, the vertex set of a fuzzy interval graph is crisp.

**Theorem 2.10.4** ([61]) *Let* $G = Int(\mathcal{F})$ *be a fuzzy interval graph. Then for all* $c \in (0, 1]$, *the level graph* $G^c$ *is an interval graph.*

*Proof* Let $G = \text{Int}(\mathcal{F})$ for a family of fuzzy intervals $\mathcal{F} = \{\alpha_1, \ldots, \alpha_n\}$. For all $c \in (0, 1]$, convexity implies that each $(\alpha_i)^c \in F^c$ is a crisp interval. Now, $G = (\text{Int}(\mathcal{F}))^c = \text{Int}(\mathcal{F}^c)$ and $G^c$ is an interval graph. ■

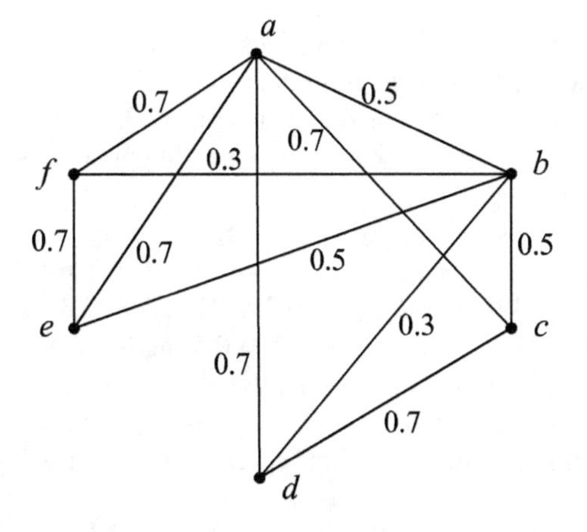


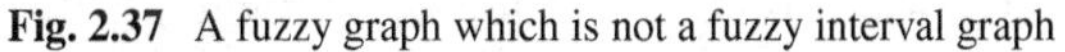

Fig. 2.37 A fuzzy graph which is not a fuzzy interval graph

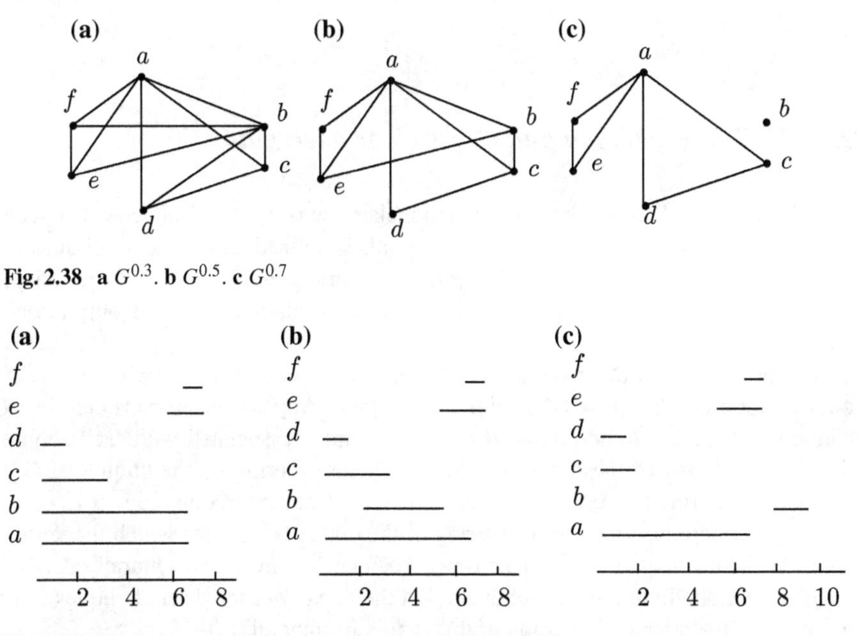


Fig. 2.38 a $G^{0.3}$. b $G^{0.5}$. c $G^{0.7}$

Fig. 2.39 Interval representation of the fuzzy graph in Fig. 2.37

*Example 2.10.5* The converse of the above result is not true. Consider the fuzzy graph given in Fig. 2.37.

The level graphs of $G$ are given in Fig. 2.38 and its interval representation in Fig. 2.39. $G$ is not a fuzzy interval graph.

Consider $G^{0.7}$ in Fig. 2.38c. It has an interval representation. Let $S_a = \{a\} \cup \{af, ae, ad, ac\}$, $S_b = \{b\}$, $S_c = \{c\} \cup \{ac, cd\}$, $S_d = \{d\} \cup \{cd, ad\}$, $S_e = \{e\} \cup \{ef, ae\}$ and $S_f = \{f\} \cup \{af, ef\}$. Let $\{S_a, S_b, S_c, S_d, S_e, S_f\}$ be the vertex set.

We can see that the intervals in Fig. 2.39c gives an approximate representation. Similarly, Fig. 2.39b is an approximate representation of $G^{0.5}$ and Fig. 2.39a is that of $G^{0.3}$. Suppose that $G = \text{Int}(\mathcal{F})$, where the fuzzy interval $v \in \text{Int}(\mathcal{F})$ corresponds to vertex $v$ of $G$. Because $h(c \cap e) = 0$, we can assume that Supp($c$) lies strictly to the left of Supp($e$). By Interval Graph Theorem, there exists $x_1$ such that $x_1 \in a^{0.7} \cap c^{0.7} \cap d^{0.7}$ because $\{a, c, d\}$ defines a clique of $G^{r_1}$. Therefore, $a(x_1) \wedge c(x_1) \wedge d(x_1) \geq 0.7$. Similarly, there exists an $x_5$ such that $a(x_5) \wedge e(x_5) \wedge f(x_5) \geq 0.7$. Now, $h(b \cap d) = 0.3$ and $h(b \cap f) = 0.3$ implies $b(x_1) \leq 0.3$ and $b(x_5) \leq 0.3$, respectively.

Continuing $h(b \cap c) = 0.5$ and $h(b \cap e) = 0.5$ imply there exist $x_2$ and $x_4$ with $b(x_2) \geq 0.5$ and $b(x_4) \geq 0.5$. By the normality of $b$ there exists $x_3$ such that $b(x_3) = 1$. By the convexity of the fuzzy intervals and the assumption that Supp($c$) lies strictly to the left of Supp($e$), the ordering of these points must be $x_1 \leq x_2 \leq x_3 \leq x_4 \leq x_5$, with $x_2 < x_4$.

Because $a(x_1) \geq 0.7$, $a(x_5) \geq 0.7$ and $a$ is convex, it follows that $a(x_3) \geq 0.7$. Hence, $h(a \cap b) \geq 0.7$. This contradicts $h(a \cap b) = 0.6$. Hence, $G$ is not a fuzzy interval graph.

### 2.10.3 The Fulkerson and Gross Characterization

The Fulkerson and Gross characterization makes use of a correspondence between the set of points on which the family of intervals is defined and the set of cliques of the corresponding interval graph. We provide natural generalizations of the (crisp) definitions and then show that for fuzzy graphs this relationship holds only in one direction.

The proof of the Fulkerson and Gross theorem rests on the following ideas for a crisp graph $G^*$. Suppose $G^*$ is an interval graph. Any set of intervals defining a clique will have a common point. If one such point is associated with each clique, the linear ordering of these points induces a linear ordering on the cliques of $G^*$. Using this ordering the resulting vertex clique incidence matrix has convex rows.

Suppose there exists a linear ordering of the cliques of $G^*$ for which the vertex clique incidence matrix has convex rows. Then each convex row naturally defines the characteristic function of a subinterval of the linearly ordered set of cliques. The graph $G^*$ is the intersection graph of this family of intervals.

**Theorem 2.10.6** [74] (Fulkerson and Gross) *A graph G is an interval graph if and only if there exists a linear ordering of the cliques of G for which the vertex clique incidence matrix has convex rows.*

**Definition 2.10.7** Let $G = (\sigma, \mu)$ be a fuzzy graph. We say that a fuzzy subgraph $\mathcal{K}$ defines a **fuzzy clique** of $G$ if for each $t \in (0, 1]$, $\mathcal{K}^t$ induces a clique of $G^t$. We associate with $G$ a **vertex clique incidence matrix**, where the rows are indexed by the domain of $\sigma$, the column are indexed by the family of all cliques of $G$, and the $x$, $\mathcal{K}$ entry is $\mathcal{K}(x)$.

Suppose that $G$ is a fuzzy graph with $\text{fs}(G) = \{r_1, \ldots, r_n\}$ and let $\mathcal{K}$ be a fuzzy clique of $G$. The level sets of $\mathcal{K}$ define a sequence $\mathcal{K}^{r_1} \subseteq \cdots \subseteq \mathcal{K}^{r_n}$, where each $\mathcal{K}^{r_i}$ is a clique of $G^{r_i}$. Conversely, any sequence $K_1 \subseteq \cdots \subseteq K_n$, where each $K_i$ is a clique of $G^{r_i}$ defines a fuzzy clique $\mathcal{K}$, where $\mathcal{K}(x) = \vee\{r_i \mid x \in K_i\}$ Therefore, $K$ is a clique of the $t$-level graph $G^t$ if and only if $K = \mathcal{K}^t$ for some fuzzy clique $\mathcal{K}$.

**Theorem 2.10.8** [61] (Fuzzy analog of Fulkerson and Gross) *Let $G = (V, \mu)$ be a fuzzy graph. Then the row of any vertex clique incidence matrix of G defines a family of fuzzy subsets $\mathcal{F}$ for which $G = Int(\mathcal{F})$. Further, if there exists an ordering of the fuzzy cliques of G such that each row of the vertex clique incidence matrix is convex, then G is a fuzzy interval graph.*

*Proof* Let $I = \{\mathcal{K}_1, \ldots, \mathcal{K}_p\}$ be an ordered family of the fuzzy cliques of $G$ and let $M$ be the vertex clique incidence matrix where the columns are given in this ordering. For each $x \in V$, define the fuzzy subset $\mathcal{I}_x : I \to [0, 1]$ by $\mathcal{I}_x(\mathcal{K}_i) = \mathcal{K}_i(x)$ and let $\mathcal{F} = \{\mathcal{I}_x \mid x \in V\}$. Because each vertex $x$ has strength 1, $x$ is contained in the 1-level cut of some fuzzy clique $K_i$ in $I$. Therefore, $\mathcal{I}_x(\mathcal{K}_i) = \mathcal{K}_i(x) = 1$ and $\mathcal{I}_x$ is normal.

We must now show for $x \neq y \in V$ that $h(\mathcal{I}_x \cap \mathcal{I}_y) = \mu(xy)$. Also, assuming that each row is convex implies that each $\mathcal{I}_x$ is a fuzzy interval and that $G$ is a fuzzy interval graph. By definition, if $x \neq y$, then

$$\begin{aligned} h(\mathcal{I}_x \cap \mathcal{I}_y) &= \vee\{(\mathcal{I}_x \cap \mathcal{I}_y)(\mathcal{K}_i) \mid \mathcal{K}_i \in I\}h(\mathcal{I}_x \cap \mathcal{I}_y) \\ &= \vee\{(\mathcal{I}_x \cap \mathcal{I}_y)(\mathcal{K}_i) \mid \mathcal{K}_i \in I\} \\ &= \vee\{\mathcal{K}_i(x) \wedge \mathcal{K}_i(y) \mid \mathcal{K}_i \in I\} = \vee\{t \in [0, 1\} \mid \mathcal{K}_i \in I \\ &\quad \text{and } xy \text{ is an edge of } (\mathcal{K}_i)^t\}. \end{aligned}$$

The edge strength $\mu(xy) = t$ is the maximal value where $xy$ is an edge of $G^t$ so is contained in a clique of $G^t$. Thus, $h(I_x \cap I_y) = \mu(xy)$ as required. ■

*Example 2.10.9* The fuzzy graph $G$ given in Fig. 2.40 shows that the converse of Theorem 2.10.8 does not hold.

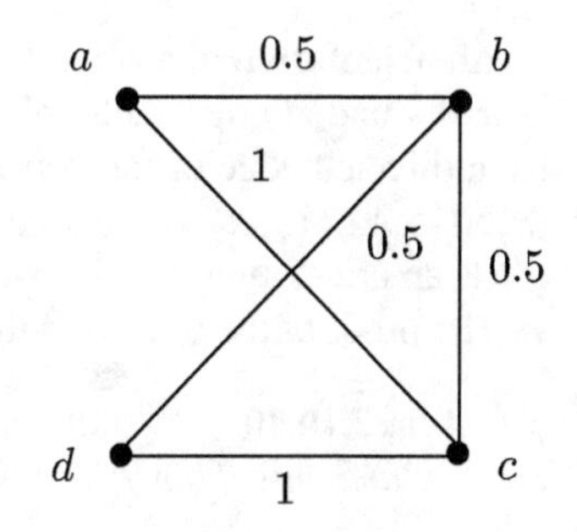


**Fig. 2.40** A fuzzy interval graph

Let the set $\mathcal{F}$ of fuzzy intervals be defined by the rows of the matrix $F$ given by

$$F = \begin{array}{c} \\ I_a \\ I_b \\ I_c \\ I_d \end{array} \begin{array}{c} \begin{array}{cccc} 1 & 2 & 3 & 4 \end{array} \\ \begin{bmatrix} 1 & 0.5 & 0 & 0 \\ 0.5 & 0.5 & 0.5 & 1 \\ 1 & 1 & 1 & 0.5 \\ 0 & 0 & 1 & 0.5 \end{bmatrix} \end{array}$$

Then $G = \text{Int}(\mathcal{F})$. A vertex clique incidence matrix $M$ is given below.

$$M = \begin{array}{c} \\ a \\ b \\ c \\ d \end{array} \begin{array}{c} \begin{array}{cccc} \mathcal{K}_1 & \mathcal{K}_2 & \mathcal{K}_3 & \mathcal{K}_4 \end{array} \\ \begin{bmatrix} 1 & 0.5 & 0 & 0 \\ 0.5 & 1 & 0.5 & 1 \\ 1 & 0.5 & 1 & 0.5 \\ 0 & 0 & 1 & 0.5 \end{bmatrix} \end{array}$$

We can verify by exhaustion that no ordering of the fuzzy cliques produces a vertex clique incidence matrix $M$ with convex rows.

## 2.10.4 *The Gilmore and Hoffman Characterization*

We begin with several graph theory definitions and state the Gilmore and Hoffman characterization. We then give corresponding fuzzy definitions, and conclude with the result that the Gilmore and Hoffman characterization generalizes exactly for fuzzy interval graphs.

Let $G = (X, E)$ be a connected graph. Recall that a **chord** of a spanning tree $T$ is an edge of $G$ which is not in $T$, and recall that a cycle of length $n$ in $G = (X, E)$ is a sequence $x_0, \ldots, x_n$ of distinct vertices, where $x_0 x_n \in E$ and $1 \leq i \leq n$ implies $x_{i-1}x_i \in E$. A graph is **chordal** (**triangulated**) if each cycle with $n > 4$ has a chord. Formally, if there exist integers $j \neq 0$ or $k \neq n$ with $0 \leq j < k - 1 \leq n$ and $x_j x_k \in E$.

An **orientation** of a graph $G = (X, E)$ is a directed graph $G_A = (X, A)$ that has $G$ as its underlying graph. We use the notation $xy$ for an edge of $G$, and $(x, y)$ for a directed edge of the corresponding orientation. That is, $xy \in E$ implies that $(x, y) \in A$ or $(y, x) \in A$ but not both. A graph $G$ is **transitively orientable** if there exists an orientation of $G$ for which $(u, v) \in A$ and $(v, w) \in A$ implies $(u, w) \in A$.

The proof of the following theorem can be found in [153].

**Theorem 2.10.10** [78] (Gilmore and Hoffman) *A graph $G = (V, X)$ is an interval graph if and only if it satisfies the following two conditions.*

*(i) Each subgraph of G induced by four vertices is chordal,*
*(ii) $G^c$ is transitively orientable.*

We now show that fuzzy interval graphs are chordal and have transitively orientable compliments.

**Definition 2.10.11** A **cycle of length** $n$ in a fuzzy graph is a sequence of distinct vertices $x_0, x_1, \ldots, x_n$ such that $\mu(x_0x_n) > 0$ and if $1 \leq i \leq n$, then $\mu(x_{i-1}x_i) > 0$. A fuzzy graph $G = (\sigma, \mu)$ is **chordal** if for each cycle with $n \geq 4$, there exist integers $j \neq 0$ or $k \neq n$ such that $0 \leq j < k - 1 \leq n$ and $\mu(x_jx_k) \geq \wedge\{\mu(x_{i-1}x_i) \mid i = 1, 2, \ldots, n\} \wedge \mu(x_0x_n)$.

It is easily shown that a fuzzy graph $G = (\sigma, \mu)$ is chordal if and only if for each $t \in (0, 1]$ the $t$-level graph of $G$ is chordal.

**Theorem 2.10.12** ([61]) *If $G$ is a fuzzy interval graph, then $G$ is chordal.*

*Proof* By Theorem 2.10.4, each cut level graph $G^t$ is an interval graph. As in the proof of Theorem 2.10.10, any interval graph is chordal. The result then follows from Definition 2.10.11. ∎

To avoid confusion when dealing with cut level graphs, we base an orientation of a fuzzy graph on an orientation of its underlying graph.

**Definition 2.10.13** Let $G = (\sigma, \mu)$ be a fuzzy graph with $\text{fs}(G) = \{r_1, \ldots, r_n\}$ and let $A$ be an orientation of $G^{r_n}$. Then the **orientation** of $G$ by $A$ is the fuzzy digraph $G_A = (\sigma, \mu_A)$, where

$$\mu_A((x, y)) = \begin{cases} \mu(xy) & \text{if } (x, y) \in A, \\ 0 & \text{if } (x, y) \notin A. \end{cases}$$

The fuzzy graph $G$ is called **transitively orientable** if there exists an orientation which is transitive, i.e., $\mu_A((x, y)) \wedge \mu_A((y, z)) \leq \mu_A((x, z))$ for all $x, y, z \in V$.

The $c$ level graph of $G_A$ has edge set $\{(x, y) \mid \mu_A((x, y)) \geq c\}$. Therefore, an orientation of a fuzzy graph induces consistent orientations on each member of the fundamental sequence of cut level graphs. Conversely, it is possible to have a sequence of transitively oriented subgraphs $G_1 \subseteq G_2 \subseteq G_3$, where the transitive orientation of $G_2$ does not induce a transitive orientation of $G_1$, and the transitive orientation of $G_2$ cannot be extended to a transitive orientation of $G_3$.

**Lemma 2.10.14** *Suppose that $G = Int(F)$ is a fuzzy interval graph. Then there exists an orientation $A$ that induces a transitive orientation of $G^c$.*

*Proof* Suppose $(\alpha, \beta)$ is a nontrivial edge of $G^c$. Then $h(\alpha \cap \beta) < r_1 = 1$ and $\alpha^{r_1}$ and $\beta^{r_1}$ are disjoint. We let $(\alpha, \beta) \in A$ if and only if $\alpha^{r_1}$ lies strictly to the left of $\beta^{r_1}$. Clearly $A$ is a well-defined and transitive orientation of $C^c$. ∎

*Example 2.10.15* The fuzzy graph in Example 2.10.5 (Fig. 2.37) is not a fuzzy interval graph because any orientation of $(d, e)$ shows that there is no transitive orientation of $G^c$. Figure 2.41 shows the cut level graphs of $G^c$ with $(d, e) \in A$. Note that $(G^1)^c = (G^c)^{1-r_2}$, $(G^{r_2})^c = (G^c)^{1-r_3}$ and $(G^{r_3})^c = (G^c)^1$ where $r_1 = 1$ here.

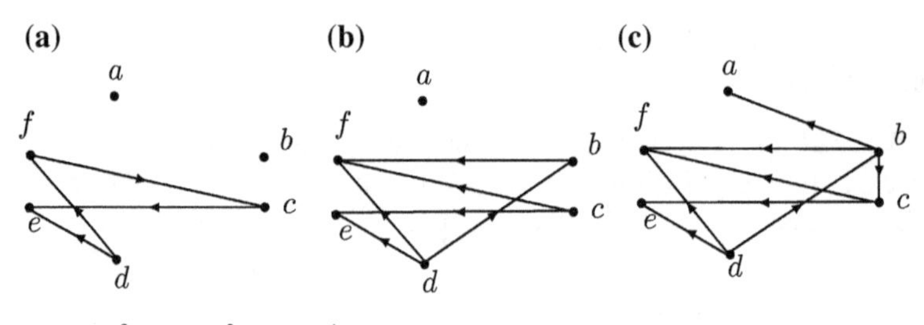


**Fig. 2.41** **a** $(G^{r3})^c$. **b** $(G^{r2})^c$. **c** $(G^{r1})^c$

**Theorem 2.10.16** [61] (Fuzzy analog of Gilmore and Hoffman characterization) *A fuzzy graph $G = (\sigma, \mu)$ is a fuzzy interval graph if and only if the following conditions hold:*

*(i) For all $x \in Supp(\sigma) = V$, $\sigma(x) = 1$ ($\sigma$ is a crisp set);*
*(ii) Each fuzzy subgraph of G induced by four vertices is chordal;*
*(iii) $G^c$ is transitively orientable.*

If $G$ is a fuzzy interval graph, the three conditions follow from Definitions 2.10.1, 2.10.3, Theorem 2.10.12 and Lemma 2.10.14, respectively.

The following discussion is from [61]. For the remainder of the section, we assume that each fuzzy subgraph of $G = (V, \mu)$ induced by four vertices is chordal and that $A$ is a transitive orientation of $G^c$. Because the proof that $G$ is a fuzzy interval graph is quite involved we first outline the proof. Details are given in Definition 2.10.17 through Lemma 2.10.23; the algorithm is applied in Example 2.10.24. For notational convenience we let $\mathcal{K}_{ij}$ denote the $r_j$ cut level set of the fuzzy set $\mathcal{K}_i$.

**Definition 2.10.17** Define the relation $<$ on the family of all fuzzy cliques of $G$ as follows. Suppose $\mathcal{K} < \mathcal{L}$ if and only if $K^t <_t L^t$, where $t$ is the smallest element of $fs(G)$ such that $K^t \neq L^t$.

The lexicographic ordering $<$ in the previous definition is clearly well defined, complete, and transitive. Therefore, $<$ defines a linear ordering on the family of all fuzzy cliques of $G$.

**Definition 2.10.18** Let $G$ satisfy the conditions of Theorem 2.10.16 and let $<$ be the relation of Definition 2.10.17. Let $t \in fs(G)$ and let $\mathcal{K} \neq \mathcal{L}$ be fuzzy cliques of $G$. We say $\mathcal{K}$ and $\mathcal{L}$ are **consistently ordered** by $<$ at level $t$ provided $\mathcal{K}^t <_t \mathcal{L}^t$ if and only if $\mathcal{K} < \mathcal{L}$. We say the linear ordering $<$ is **cut level consistent** if for each pair of distinct fuzzy cliques of $G$ and for each $t \in fs(G)$ the pair is consistently ordered by $<$ at level $t$.

By Theorem 2.10.8, the rows of any vertex clique incidence matrix of $G$ define a family of fuzzy subsets that has $G$ as its fuzzy intersection graph. If $<$ of Definition 2.10.17 is cut level consistent, then the rows of the vertex clique incidence matrix will be convex and the result follows from Theorem 2.10.8.

If $<$ is not level consistent, then some row is not convex. We modify this matrix in a "bottom up" construction using the notion of cut level consistent to determine which columns are modified or deleted from the vertex clique incidence matrix. We complete the proof by showing that in the modified matrix each row is normal and convex and that $G$ is the fuzzy intersection graph of the family of fuzzy intervals defined by the rows.

By the discussion following Definitions 2.10.11 and 2.10.13 each level graph $G^t$ is chordal and has a transitively orientable complement. Thus, each $G^t$ is an interval graph and there exists a linear ordering $<_t$ on the family of all cliques of $G^t$. We now establish definitions that will be used extensively in the discussion to follow.

If the linear ordering $<$ is cut level consistent then each row of the vertex clique incidence matrix is convex (and $G$ is a fuzzy interval graph by Theorem 2.10.8). We prove this statement by contrapositive. Assume there exists a row that is not convex. Suppose that there exists a vertex $x \in V$ and a sequence of fuzzy cliques $\mathcal{K} < \mathcal{L} < \mathcal{M}$ such that $\mathcal{L}(x) < \mathcal{K}(x) \wedge \mathcal{M}(x) = t$. Then $x \in \mathcal{K}^t$, $x \notin \mathcal{L}^t$ and $x \in \mathcal{M}^t$. As in Theorem 2.10.10 there exists $y \in \mathcal{L}^t$ such that $(x, y) \notin \mathcal{E}^t$. If $(x, y) \in A$, then $\mathcal{M}^t <_t \mathcal{L}^t$ with $\mathcal{L} < \mathcal{M}$. If $(y, x) \in A$ then $\mathcal{L}^t <_t \mathcal{K}^t$ with $\mathcal{K} < \mathcal{L}$. In either case the ordering $<$ is not cut level consistent.

By Example 2.10.9, there exist fuzzy interval graphs where no ordering of the fuzzy cliques is cut level consistent. We formalize a process that modifies or deletes "inconsistent" fuzzy cliques (matrix columns). The proof of the following lemma shows the "local" structure of noncut level consistent orderings. The lemma is also used to show the construction in Definition 2.10.20 is well defined.

**Lemma 2.10.19** *Suppose that $\mathcal{K}$ and $\mathcal{L}$ are fuzzy cliques of $G$ and that $s > t$. If $\mathcal{K}^s <_s \mathcal{L}^s$ and $\mathcal{L}^t <_t \mathcal{K}^t$, then there exists a clique $M$ of $G^t$ such that either*

*(i) $\mathcal{K}^s \subseteq M$ and $M <_t \mathcal{K}^t$ or*
*(ii) $\mathcal{L}^s \subseteq M$ and $L^t <_t M$.*

*Proof* We check all possible edge configurations. Recall the edge set of the graph $G^s$ is denoted by $\mathcal{E}^s$. Each case shares the general conditions shown in Fig. 2.42. By definition of $<_t$, there exist $x \in \mathcal{K}^t$ and $y \in \mathcal{L}^t$, with $xy \notin \mathcal{E}^t$ and $(x, y) \in A$. Similarly, there exist $x' \in \mathcal{K}^s$ and $y' \in \mathcal{L}^s$ with $x'y' \notin \mathcal{E}^s$ and $(y', x') \in A$. Then

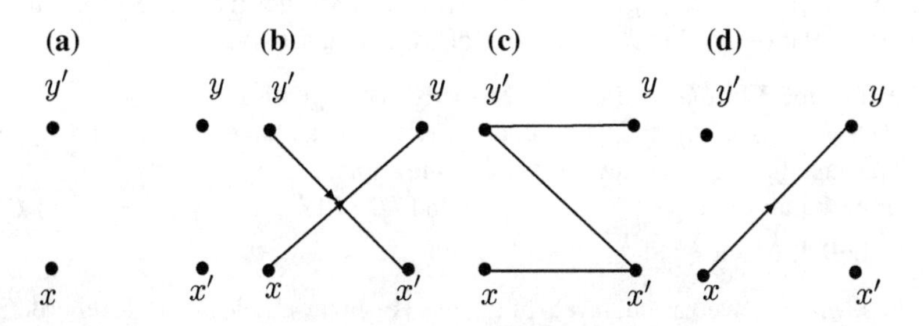


**Fig. 2.42** Basic conditions for inconsistent cut level orderings

$s > t$ implies $xy \notin \mathcal{E}^s$, $xx' \in \mathcal{E}^s$ (or $x = x'$) and $yy' \in \mathcal{E}^t$ (or $y = y'$). Because $<_t$ is well defined, $x'y' \in \mathcal{E}^t$ and either $xx' \notin \mathcal{E}^s$ or $yy' \notin \mathcal{E}^s$.

If $y'x \notin \mathcal{E}^t$, then $y'x \notin \mathcal{E}^s$ and transitivity requires $(y', x) \in A$ and $xx' \notin \mathcal{E}^s$ (so $x \notin \mathcal{K}^s$). We claim for each $x'' \in \mathcal{K}^s \subseteq \mathcal{K}^t$ that $y'x'' \in \mathcal{E}^t$. For $(y', x'') \notin \mathcal{E}^t$ with $\mathcal{L}^t <_t \mathcal{K}^t$ implies $(y', x'') \in A$. However, $\mathcal{E}^s \subseteq \mathcal{E}^t$ and $\mathcal{K}^s <_s \mathcal{L}^s$ imply $(x'', y') \in A$; a contradiction. Therefore, $\{y'\} \cup \mathcal{K}^s$ is a complete subgraph of $G^t$ and is contained in a clique $M$ of $G^t$. Because $x \notin M$ and $(y', x) \in A$, we have $M <_t \mathcal{K}^t$. Thus, property $(i)$ holds.

Similarly, if $yx' \notin \mathcal{E}^t$, we have that $(y, x') \in A$ and $yx' \notin \mathcal{E}^s$. By transitivity, $\{x'\} \cup \mathcal{L}^s$ is a complete subgraph of $G^t$ and hence is contained in a clique $M$ of $G^t$. Then $(y, x') \in A$ implies $\mathcal{L}^t <_t M$ and property $(ii)$ holds.

If $y'x \in \mathcal{E}^t$ and $yx' \in \mathcal{E}^t$, then we show that $\mathcal{K}^s \cup \mathcal{L}^s$ is a complete subgraph of $G^t$. We need only to show for each $x'' \in \mathcal{K}^s$ and $y'' \in \mathcal{L}^s$ that $y''x'' \in \mathcal{E}^t$. Again $x''y'' \notin \mathcal{E}^t$ and $\mathcal{L}^t <_t \mathcal{K}^t$ implies $(y'', x'') \in A$ and $y''x'' \notin \mathcal{E}^s$. However, $\mathcal{K}^t <_s \mathcal{L}^s$ implies $(x'', y'') \in A$; a contradiction.

Therefore, $\mathcal{K}^s \cup \mathcal{L}^s$ induces a complete subgraph of $G^t$ that is contained in some clique $M$ of $G^t$. If $M <_t \mathcal{L}^t <_t \mathcal{K}^t$, property $(i)$ holds. If $\mathcal{L}^t <_t M$ property $(ii)$ holds. ■

We now construct a directed graph $F$ and in turn a linearly ordered family of fuzzy subsets that define columns of an incidence matrix. These fuzzy subsets will either be fuzzy cliques of $G$ or modifications of fuzzy cliques. The graph theory analogy of a forest with trees allows a good visualization of "vertically growing" cut level sets which define the required fuzzy sets.

We use the fuzzy clique ordering $<$ to recursively construct a forest $F$ whose vertex set is the set of all cut level cliques of $G$ and whose edges connect cut levels of fuzzy sets. We recursively build the forest by "vertically" adding cut level cliques as vertices of $F$ and defining a set of edges between cut levels. In the recursion let $i$ range from 1 to $n - 1$.

**Definition 2.10.20** ([61]) Let $G = (V, \mu)$ with $fs(G) = \{r_1, r_2, \ldots, r_n\}$ be a chordal fuzzy graph and let $G^c$ be transitively oriented by $A$.

Level $r_n$: Linearly order the set of all cliques of $G^{r_n}$ by the relation $<_{r_n}$ of Definition 2.10.18. Each of these cliques of $G^{r_n}$ (vertices of $F$) represent the root of a tree in the forest.

Level $r_{n-i}$: Let $s = r_{n-i}$ and $t = r_{n-i+1}$. Linearly order the set of all cliques of $G^s$ by the relation $<_s$. Let $X^s$ be any set of edges that satisfy:

1. Each clique $\mathcal{K}^s$ of $G^s$ is a vertex of exactly one edge of $X^s$.
2. If $(\mathcal{K}^t, \mathcal{K}^s) \in X^s$ then $\mathcal{K}_t$ is a clique of $G^t$, $\mathcal{K}^s$ is a clique of $G^s$, and $\mathcal{K}^s \subseteq \mathcal{K}^t$. Thus, an edge joins two level sets of (some) fuzzy clique.
3. For each pair of edges $(\mathcal{K}^t, \mathcal{K}^s) \in X^s$ and $(\mathcal{L}^t, \mathcal{L}^s) \in X^s$ we have $\mathcal{K}^s <_s \mathcal{L}^s$ if and only if $\mathcal{K}^t <_t \mathcal{L}^t$ or $\mathcal{K}^t = \mathcal{L}^t$.

Thus, when viewed as cut levels of a family of fuzzy cliques, the $s$ level ordering is level consistent with the next "lower" level.

We continue with the discussion in [61]. We use Lemma 2.10.19 to demonstrate the existence of at least one such forest, and show in the last paragraph of this section that there may be a number of edge sets that satisfy these conditions. Let $\mathcal{K}^s$ be the minimal (with respect to $<_s$) clique of $G^s$. Clearly there exists a minimal (with respect to $<_t$) clique $\mathcal{K}^t$ of $G^t$ where $\mathcal{K}^s \subseteq \mathcal{K}^t$. Let $< \mathcal{K}^t, \mathcal{K}^s > \in X^s$.

Next let $\mathcal{L}^s$ be the successor of $\mathcal{K}^s$ (with respect to $<_s$) and let $\mathcal{L}^t$ be minimal (with respect to $<_t$) such that $\mathcal{L}^s \subseteq \mathcal{L}^t$ and $\mathcal{K}^t <_t \mathcal{L}^t$ or $\mathcal{K}^t = \mathcal{L}^t$. If $\mathcal{L}^t$ does not exist, let $L$ be maximal (with respect to $<_t$) with $\mathcal{L}^s \subseteq L$. Now, $\mathcal{K}^s <_s \mathcal{L}^s$ and $L <_t \mathcal{K}^t$ are the conditions of Lemma 2.10.19. However, property (1) contradicts the minimality of $\mathcal{K}^t$ and property (2) contradicts the maximality of $L$. Therefore, $\mathcal{L}^t$ exists and $(\mathcal{L}^t, \mathcal{L}^s) \in X^s$ is well defined.

We continue recursively to construct one edge for each clique of $G^s$. It may be that for some clique $M_t$ of $G^t$, there is no edge from $M_t$. We call such a clique a dead branch of $F$.

Combining the edge sets $F^{r_{n-i}}$ for $i \in \{1, \ldots, r-1\}$ defines a forest with edge set $\cup_{i=1}^{n-1} F^{r_{n-i}}$. As in Definition 2.10.18, we lexicographically order the set of paths from a root to a dead branch or a $r_1$ level clique. For notational convenience, we denote the $t$ level vertex of path $P_j$ by $P_{jt}$. To ensure convex rows in our (still undefined) incidence matrix, we add nonempty vertices "above" dead branches if "adjacent" cliques have nonempty intersection.

Suppose the path $P_j$ ends with a dead branch at the $t$ level. For each $s \in fs(G)$ with $s > t$, we continue the path $P_j$ through the new vertex $P_{js}$, where $x \in P_{js}$ if and only if there exist $i < j < k$ with $x \in P_{is} \cap P_{ks}$. We call this final forest $F$. Each path in $F$ has length $n$, and it is possible for a vertex $P_{js}$ to be the empty set.

We complete the construction by letting paths in $F$ define a linearly ordered family of fuzzy sets, say $I$. The fuzzy sets define columns of the *vertex forest matrix* of $(G, <)$; $G$ is the interval graph of the family of rows.

**Definition 2.10.21** Let $G$ satisfy the conditions of Theorem 2.10.16. $F$ be a forest for $G$ as defined in Definition 2.10.20, and $P_j$ be a path in $F$ of length $n$. Associated with $P_j$ define the fuzzy set $\mu_j \in I$ on the vertex set of $G$ by $\mu_j(x) = \vee\{s \in fs(G) \mid x$ is an element of the $s$ level vertex of $P_j\}$.

We construct the *vertex forest matrix* of $(G, <)$ indexing rows by the vertex set of $G$, columns by the (ordered) fuzzy sets of $I$ and defining the $x, \mu_i$ entry as $\mu_i(x)$. By construction each $\mu_j$ is either a fuzzy clique of $G$, or has a cut level set that is the intersection of two cut level cliques.

Let $\mathcal{F}$ denote the family of fuzzy sets defined by the rows of the vertex forest matrix. We now complete the proof of Theorem 2.10.16 by showing that each member of $\mathcal{F}$ is normal and convex (a fuzzy interval) and that $G = \text{Int}\,\mathcal{F}$.

**Lemma 2.10.22** *We assume the conditions and notation above. For each vertex $x$ of $G$, define $\mathcal{J}_x$ is a fuzzy interval.*

*Proof* Let $x$ be a vertex of $G$. Then $x$ is a vertex in some clique of $G^{r_1}$, say $K$. By Definitions 2.10.20 and 2.10.21, $K$ is the $r_1 = 1$ level cut of some fuzzy set $\mu$ in $I$. Therefore, $\mathcal{J}_x(\mu) = \mu(x) = 1$ and $\mathcal{J}_x$ is normal.

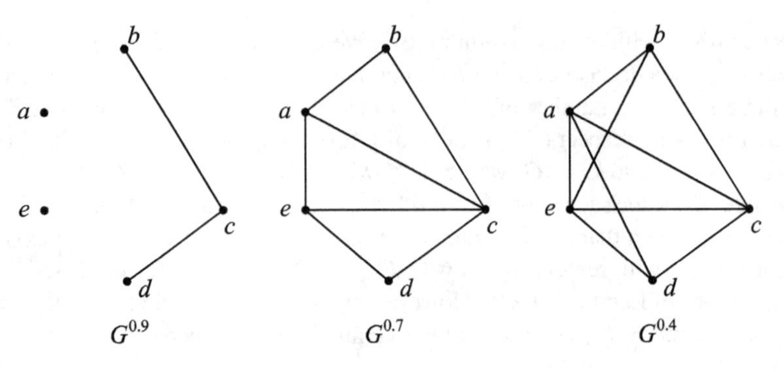


**Fig. 2.43** Level graphs $G^{0.9}$, $G^{0.7}$, $G^{0.4}$

Each $\mathcal{J}_x$ is convex if $i < j < k$ implies $\mathcal{J}_x(\mu_i) \wedge \mathcal{J}_x(\mu_k) \leq \mathcal{J}_x(\mu_j)$, or equivalently, if $\mu_i(x) \wedge \mu_k(x) \leq \mu_j(x)$. However, Definition 2.10.20 clearly provides these conditions. If $\mu_i$, $\mu_j$ and $\mu_k$ are all fuzzy cliques, the result follows immediately from the discussion after Definition 2.10.18. Otherwise, the result follows by definition of the fuzzy sets $\mu_i$, $\mu_j$ and $\mu_k$. ■

Now, we conclude proof of Theorem 2.10.16, by next lemma.

**Lemma 2.10.23** *Given the definitions and conditions of Theorem 2.10.16 through Lemma 2.10.22,* $\mathcal{G} = Int(\mathcal{F})$.

*Proof* There is a correspondence between the crisp vertex set $V$ and the family of fuzzy intervals $\mathcal{F}$. Let $x, y$ be distinct elements of $V$. We must show that $\mu(xy) = h(\mathcal{J}_x \cap \mathcal{J}_y)$. By definition, $h(\mathcal{J}_x \cap \mathcal{J}_y) = \vee\{\mathcal{J}_x(\mu_j) \wedge \mathcal{J}_y(\mu_j) \mid \mu_j \in I\} = \vee\{\mu_j(x) \wedge \mu_j(y) \mid \mu_j \in I\} = \vee\{s \in fs(G) \mid \{x, y\} \subseteq \mu_j^s\}$.

Because $\mu(xy) = t$ is the maximal value where $xy$ is an edge of $G^t$, $\mu(xy)$ is the maximal value where $xy$ is in a clique of $G^t$. By definition each clique of $G^t$ is the $t$ level set of some fuzzy set $\mu_j \in I$. Hence, $\mu(xy) = h(\mathcal{J}_x \cap \mathcal{J}_y)$ as required. ■

We provide an illustration for Theorem 2.10.16 in Example 2.10.24

*Example 2.10.24* Consider the fuzzy graph $G$ defined by the incidence matrix $G$ below, where $fs(G) = \{s, t, u\} = \{0.9, 0.7, 0.4\}$. Figure 2.43 shows the cut level graphs of $G$ and Fig. 2.44 shows transitive orientation $A$ of $G^c$.

$$G = \begin{array}{c} \\ a \\ b \\ c \\ d \\ e \end{array} \begin{array}{c} \begin{array}{ccccc} a & b & c & d & e \end{array} \\ \begin{bmatrix} 0 & 0.7 & 0.7 & 0.4 & 0.7 \\ 0.7 & 0 & 0.9 & 0 & 0.4 \\ 0.7 & 0.9 & 0 & 0.9 & 0.7 \\ 0.4 & 0 & 0.9 & 0.9 & 0.7 \\ 0.7 & 0.4 & 0.7 & 0.7 & 0 \end{bmatrix} \end{array}$$

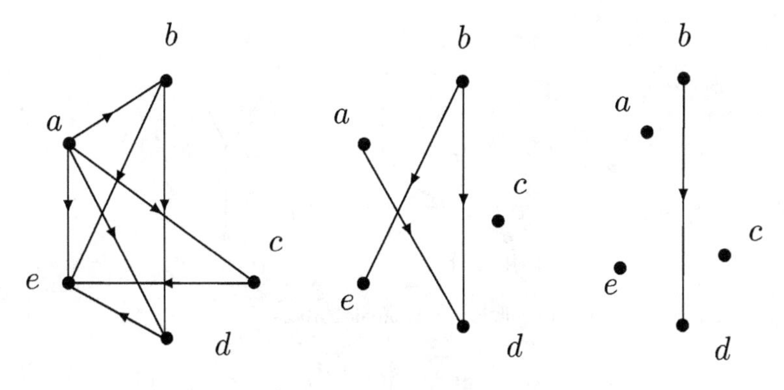


**Fig. 2.44** Transitive orientations of the complements

Using Definition 2.10.18, we linearly order the cut level cliques by: $s = 0.9$, $\{a\} <_s \{b, c\} <_s \{c, d\} <_s \{e\}$, $t = 0.7$, $\{a, b, c\} <_t \{a, c, e\} <_t \{c, d, e\}$, $u = 0.4$, $\{a, b, c, e\} <_u \{a, c, d, e\}$.

There are eight fuzzy cliques of $G$; subscripts indicate the order induced by Definition 2.10.18. The vertex clique incidence matrix $M$ for $G$ is given below. The only convex row is indexed by $d$. Thus, the fuzzy clique ordering is not cut level consistent.

$$M = \begin{array}{c} \\ a \\ b \\ c \\ d \\ e \end{array} \begin{array}{c} \begin{array}{cccccccc} \mathcal{K}_1 & \mathcal{K}_2 & \mathcal{K}_3 & \mathcal{K}_4 & \mathcal{K}_5 & \mathcal{K}_6 & \mathcal{K}_7 & \mathcal{K}_8 \end{array} \\ \begin{bmatrix} 0.9 & 0.7 & 0.9 & 0.7 & 0.9 & 0.7 & 0.4 & 0.4 \\ 0.7 & 0.9 & 0.4 & 0.4 & 0 & 0 & 0 & 0 \\ 0.7 & 0.9 & 0.7 & 0.7 & 0.7 & 0.7 & 0.9 & 0.7 \\ 0 & 0 & 0 & 0 & 0 & 0.4 & 0.9 & 0.7 \\ 0.4 & 0.4 & 0.7 & 0.9 & 0.7 & 0.9 & 0.7 & 0.9 \end{bmatrix} \end{array}$$

Following Definition 2.10.20, gives the forest $F$ of Fig. 2.45 with the incidence matrix $V$, given below.

$$V = \begin{array}{c} \\ a \\ b \\ c \\ d \\ e \end{array} \begin{array}{c} \begin{array}{ccccc} P_1 & P_2 & P_3 & P_4 & P_5 \end{array} \\ \begin{bmatrix} 0.9 & 0.7 & 0.7 & 0.4 & 0.4 \\ 0.7 & 0.9 & 0.4 & 0 & 0 \\ 0.7 & 0.9 & 0.9 & 0.9 & 0.7 \\ 0 & 0 & 0 & 0.9 & 0.7 \\ 0.4 & 0.4 & 0.7 & 0.7 & 0.9 \end{bmatrix} \end{array}$$

The paths $P_1$, $P_2$, $P_3$, $P_4$ and $P_5$ correspond, respectively, to the fuzzy cliques $\mathcal{K}_1$, $\mathcal{K}_2$, $\mathcal{K}_4$, $\mathcal{K}_7$ and $\mathcal{K}_8$. The clique $\{a, c, e\}$ is a dead branch; so $P_{3s} = P_{2s} \cap P_{4s} = \{c\}$. The path $P_3$ is a modification of $\mathcal{K}_4$, the fuzzy cliques $\mathcal{K}_3$, $\mathcal{K}_5$, and $\mathcal{K}_6$ are deleted.

The interval representation of a fuzzy graph $G$ is not in general unique. The construction heavily depends on the orientation of $G^c$. Different orientations can give different vertex interval matrices. Slight modifications in Definition 2.10.20 can

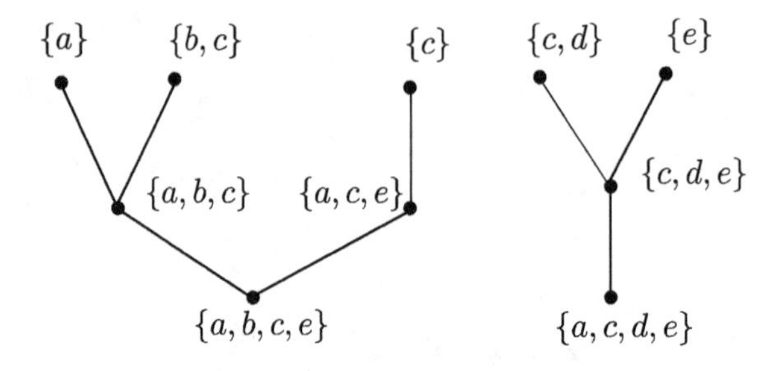


**Fig. 2.45** A fuzzy interval representation for Example 2.10.24

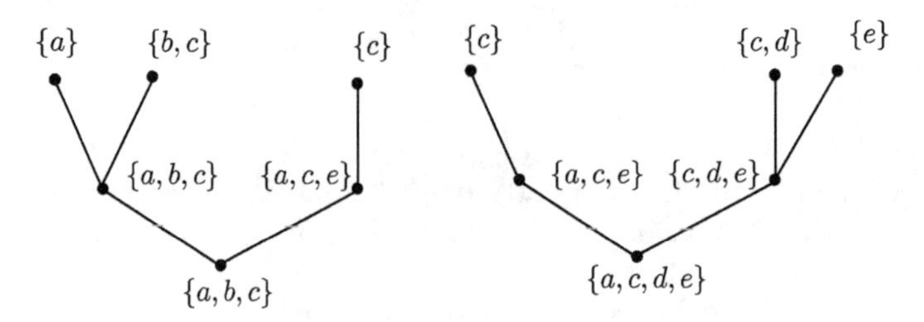


**Fig. 2.46** Alternate fuzzy interval representation for Example 2.10.24

produce different vertex interval matrices. A left right construction was followed in the example given, while a right to left also will work nicely. Also, in the example, it is specified that each cut level clique will be the terminal vertex of only one edge. One can relax this condition as long as cut level consistency is maintained. Figure 2.46 gives an alternate interval representation for the fuzzy graph given in Example 2.10.24.

## 2.11 Operations on Fuzzy Graphs

Fuzzy graph operations were first studied in [130] by Mordeson and Peng in 1994. Later Sunitha and Vijayakumar [169] investigated the properties of compliments of fuzzy graphs with respect to these operations in 2002. By a **partial fuzzy subgraph** of a graph $G = (V, X)$, we mean a partial fuzzy subgraph of $(\chi_V, \chi_X)$, where $\chi_V$ and $\chi_X$ denote the characteristic functions of $V$ and $X$, respectively. Let $(\sigma_i, \mu_i)$ be a partial fuzzy subgraph of the graph $G_i = (V_i, X_i), i = 1, 2$. The operations of Cartesian product, composition, union, and join on $(\sigma_1, \mu_1)$ and $(\sigma_2, \mu_2)$ are given in [130]. If the graph $G$ is formed from $G_1$ and $G_2$ by one of the these operations, necessary and sufficient conditions are given in [130] for an arbitrary partial fuzzy subgraph of $G$ to also be formed by the same operation from partial fuzzy subgraphs of $G_1$ and $G_2$. Recall that the Cartesian product $G = G_1 \times G_2$ of graphs $G_1 = (V_1, X_1)$ and $G_2 = (V_2, X_2)$ is given by $V = V_1 \times V_2$ and $X = \{(u, u_2)(u, v_2) \mid u \in$

$V_1, u_2v_2 \in X_2\} \cup \{(u_1, w)(v_1w) \mid w \in V_2, u_1v_1 \in X_1\}$. Let $\sigma_i$ be a fuzzy subset of $V_i$ and $\mu_i$ be a fuzzy subset of $X_i$, $i = 1, 2$. Define the fuzzy subsets $\sigma_1 \times \sigma_2$ of $V$ and $\mu_1\mu_2$ of $X$ as follows:

For all $(u_1, u_2) \in V$, $(\sigma_1 \times \sigma_2)(u_1, u_2) = \sigma_1(u_1) \wedge \sigma_2(u_2)$,

For all $u \in V_1$, for all $u_2v_2 \in X_2$, $\mu_1\mu_2((u, u_2)(u, v_2)) = \sigma_1(u) \wedge \mu_2(u_2v_2)$,

For all $w \in V_2$, for all $u_1v_1 \in X_1$, $\mu_1\mu_2((u_1, w)(v_1, w)) = \sigma_2(w) \wedge \mu_1(u_1v_1)$.

**Proposition 2.11.1** *Let $G$ be the Cartesian product of graphs $G_1$ and $G_2$. Let $(\sigma_i, \mu_i)$ is a partial fuzzy subgraph of $G_i$, $i = 1, 2$. Then $(\sigma_1 \times \sigma_2, \mu_1\mu_2)$ is a partial subgraph of $G$.*

*Proof* We have

$$\begin{aligned}\mu_1\mu_2((u, u_2)(u, u_2)) &= \sigma_1(u) \wedge \sigma_2(u_2v_2) \leq \sigma_1(u) \wedge (\sigma_2(u_2) \wedge \sigma_2(v_2)) \\ &= (\sigma_1(u) \wedge \sigma_2(u_2)) \wedge (\sigma_1(u) \wedge \sigma_2(u_2)) \\ &= (\sigma_1 \times \sigma_2)(u, u_2) \wedge (\sigma_1 \times \sigma_2)(u, v_2).\end{aligned}$$

Similarly, $\mu_1\mu_2((u_1, w)(v_1, w)) \leq (\sigma_1 \times \sigma_2)(u_1, w) \wedge (\sigma_1 \times \sigma_2)(v_1, w)$. ■

**Theorem 2.11.2** *Suppose that $G$ is a Cartesian product of two graphs $G_1$ and $G_2$. Let $(\sigma, \mu)$ be a partial fuzzy subgraph of $G$. Then $(\sigma, \mu)$ is a Cartesian product of a partial fuzzy subgraph of $G_1$ and a partial fuzzy subgraph of $G_2$ if and only if the following three equations have solutions for $x_i, y_j, z_{jk}$, and $w_{ih}$, where $V_1 = \{v_{11}, v_{12}, \ldots, v_{1n}\}$ and $V_2 = \{v_{21}, v_{22}, \ldots, v_{2m}\}$:*

*(i)* $x_i \wedge y_j = \sigma(v_{1i}, v_{2j}), i = 1, \ldots, n;\ j = 1, \ldots, m;$
*(ii)* $x_i \wedge z_{jk} = \mu((v_{1i}, v_{2j})(v_{1i}, v_{2k})), i = 1, \ldots, n;$ *$j, k$ such that* $v_{2j}v_{2k} \in X_2;$
*(iii)* $y_j \wedge w_{ik} = \mu((v_{1i}, v_{2j})(v_{1h}, v_{2j})), j = 1, \ldots, m;$ *$i, h$ such that* $v_{1i}v_{1h} \in X_1.$

*Proof* Suppose that a solution exists. Consider an arbitrary, but fixed $j, k$ in equations $(ii)$ and $i, h$ in equations $(iii)$. Let

$$\begin{aligned}\widehat{z}_{jk} &= \vee\{\mu((v_{1i}, v_{2j})(v_{1i}, v_{2k})) \mid i = 1, \ldots, n\}, \\ \widehat{w}_{ik} &= \vee\{\mu((v_{1i}, v_{2j})(v_{1h}, v_{2j})) \mid j = 1, \ldots, m\}.\end{aligned}$$

Set $J = \{(j, k) \mid j, k$ are such that $v_{2j}v_{2k} \in X_2\}$ and $I = \{(i, h) \mid i, h$ are such that $v_{1i}v_{1h} \in X_1\}$. Now, if $\{x_1, \ldots, x_n\} \cup \{z_{jk} \mid (j, k) \in J\} \cup \{w_{ih} \mid (i, h) \in I\}$ is any solution to $(i)$, $(ii)$, and $(iii)$, then $\{x_1, \ldots, x_n\} \cup \{\widehat{z}_{jk} \mid (j, k) \in J\} \cup \{\widehat{w}_{ih} \mid (i, h) \in I\}$ is also a solution and in fact $\widehat{z}_{jk}$ is a smallest possible $z_{jk}$ and $\widehat{w}_{ih}$ is a smallest $w_{ih}$. Fix such a solution and define the fuzzy subsets $\sigma_1, \sigma_2, \mu_1$, and $\mu_2$ of $V_1, V_2, X_1$, and $X_2$, respectively, as follows:

$$\begin{aligned}\sigma_1(v_{1i}) &= x_i \text{ for } i = 1, \ldots, n; \\ \sigma_2(v_{2j}) &= y_j \text{ for } j = 1, \ldots, m; \\ \mu_2(v_{2j}v_{2k}) &= \widehat{z}_{jk} \text{ for } j, k \text{ such that } v_{2j}v_{2k} \in X_2; \\ \mu_1(v_{1i}v_{1h}) &= \widehat{w}_{ih} \text{ for } i, h \text{ such that } v_{1i}v_{1h} \in X_1.\end{aligned}$$

For any fixed $j, k$,

$$\begin{aligned}\mu((v_{1i}, v_{2j})(v_{1i}, v_{2k})) &\leq \sigma(v_{1i}, v_{2j}) \wedge \sigma(v_{1i}, v_{2k}) \\ &= (\sigma_1(v_{1i}) \wedge \sigma_2(v_{2j}) \wedge (\sigma_1(v_{1i}) \wedge \sigma_2(v_{2k})) \\ &\leq \sigma_2(v_{2j}) \wedge \sigma_2(v_{2k}), i = 1, \ldots, n.\end{aligned}$$

Thus, $\widehat{z}_{jk} = \vee\{\mu((v_{1i}, v_{2j}(v_{1i}, v_{2k})) \mid i = 1, \ldots, n\} \leq \sigma_2(v_{2j}) \wedge \sigma_2(v_{2k})$. Hence, $\mu_2(v_{2j}v_{2k}) \leq \sigma_2(v_{2j}) \wedge \sigma_2(v_{2k})$. Thus, $(\sigma_2, \mu_2)$ is a partial fuzzy subgraph of $G_2$. Similarly, $(\sigma_1, \mu_1)$ is a partial fuzzy subgraph of $G_1$. Clearly, $\sigma = \sigma_1 \times \sigma_2$ and $\mu = \mu_1\mu_2$.

Conversely, suppose that $(\sigma, \mu)$ is the Cartesian product of partial fuzzy subgraphs $G_1$ and $G_2$. Then solutions to equations $(i)$, $(ii)$, and $(iii)$ exist by definition of Cartesian product. ∎

*Remark 2.11.3* Consider an arbitrary fixed solution to the equations $(i)$, $(ii)$ and $(iii)$ in the proof of Theorem 2.11.2 $(if\ one\ exists)$. Then

(i) Let $(j, k) \in J$ and let $I' = \{i_{jk} \in I \mid \hat{z}_{jk} = \mu((v_{1i_{jk}}, v_{2j})(v_{1i_{jk}}, v_{2k}))\}$ in Theorem 2.11.2. If $x_{i_{jk}} > \hat{z}_{jk}$ for some $i_{jk} \in I'$, then $z_{jk}$ is unique for these particular $x_1, x_2, \ldots, x_n$ and equals $\hat{z}_{jk}$; if $x_{i_{jk}} = \hat{z}_{jk}$ for all $i_{jk} \in I'$. Then $\hat{z}_{jk} \leq z_{jk} \leq 1$ for these particular $x_1, x_2, \ldots, x_n$.

(ii) Let $(i, h) \in I$ and let $J' = \{j_{ih} \in J \mid \hat{w}_{ih} = \mu((v_{ii}, v_{2j_{ih}})(v_{1h}, v_{2j_{ih}}))\}$ in Theorem 2.11.2. If $y_{j_{ih}} > \hat{w}_{ih}$ for some $j_{ih} \in J'$, then $w_{ih}$ is unique for these particular $y_1, y_2, \ldots, y_m$ and equals with $\hat{w}_{ih}$; if $y_{j_{ih}} = \hat{w}_{ih}$ for all $j_{ih} \in J'$, then $\hat{w}_{ih} \leq w_{ih} \leq 1$ for these particular $y_1, y_2, \ldots, y_m$.

*Example 2.11.4* Consider $V_1 = \{v_{11}, v_{12}\}$, $V_2 = \{v_{21}, v_{22}\}$, $X_1 = \{v_{11}v_{12}\}$ and $X_2 = \{v_{21}v_{22}\}$. If $\sigma((v_{11}, v_{21})) = 0.25$, $\sigma((v_{11}, v_{22})) = 0.5$, $\sigma((v_{12}, v_{21})) = 0.1$ and $\sigma((v_{12}, v_{22})) = 0.6$, then $(\sigma, \mu)$ is not a Cartesian product of partial fuzzy subgraphs of $G_1$ and $G_2$ for any choice of $\mu$, because equation $(i)$ in Theorem 2.11.2 is inconsistent; $x_1 \wedge y_1 = \sigma((v_{11}, v_{21})) = 0.25$, $x_1 \wedge y_2 = \sigma((v_{11}, v_{22})) = 0.5$, $x_2 \wedge y_1 = \sigma((v_{12}, v_{21})) = 0.1$, $x_2 \wedge y_2 = \sigma((v_{12}, v_{22})) = 0.6$, is impossible.

Note that examples satisfying Theorem 2.11.2(i) can be easily constructed, but either Theorem 2.11.2(ii) or Theorem 2.11.2(iii) may be inconsistent.

We now consider the composition of two fuzzy graphs. Let $G_1[G_2]$ denote the composition of graph $G_1 = (V_1, X_1)$ with graph $G_2 = (V_2, X_2)$. Then $G_1[G_2] = (V_1 \times V_2, X^0)$, where

$$\begin{aligned}X^0 = &\{(u, u_2)(u, v_2) \mid u \in V_1, u_2v_2 \in X_2\} \\ &\cup\{(u_1, w)(v_1, w) \mid w \in V_2, u_1v_1 \in X_1\} \\ &\cup\{(u_1, u_2)(v_1, v_2) \mid u_1v_1 \in X_1, u_2 \neq v_2\}.\end{aligned}$$

Let $\sigma_i$ be a fuzzy subset of $V_i$ and $\mu_i$ a fuzzy subset of $X_i$, $i = 1, 2$. Define the fuzzy subsets $\sigma_1 \circ \sigma_2$ and $\mu_1 \circ \mu_2$ of $V_1 \times V_2$ and $X^0$, respectively, as follows:

$$(\sigma_1 \circ \sigma_2)(u_1, u_2) = \sigma_1(u_1) \wedge \sigma_2(u_2) \text{ for all } (u_1, u_2) \in V_1 \times V_2,$$

$$(\mu_1 \circ \mu_2)((u, u_2)(u, v_2)) = \sigma_1(u) \wedge \mu_2(u_2v_2) \text{ for all } u \in V_1, \text{ for all } u_2v_2 \in X_2,$$

$$(\mu_1 \circ \mu_2)((u_1, w)(v_1, w)) = \sigma_2(w) \wedge \mu_1(u_1v_1) \text{ for all } w \in V_2, \text{ for all } u_1v_1 \in X_1,$$

$$(\mu_1 \circ \mu_2)((u_1, v_2)(v_1, v_2)) = \sigma_2(u_2) \wedge \sigma_2(v_2) \wedge \mu_1(u_1v_1) \\ \text{for all } (u_1, v_2)(v_1, v_2) \in X^0 \backslash X,$$

where

$$X = \{(u, u_2)(u, v_2) \mid u \in V_1, u_2v_2 \in X_2\} \cup \\ \{(u_1, w)(v_1, w) \mid w \in V_2, u_1u_2 \in X_1\}.$$

We see that $\sigma_1 \circ \sigma_2 = \sigma_1 \times \sigma_2$ and $\mu_1 \circ \mu_2 = \mu_1\mu_2$ on $X$.

**Proposition 2.11.5** *Let $G$ be the composition $G_1[G_2]$ of graph $G_1$ with graph $G_2$. Let $(\sigma_i, \mu_i)$ be a partial subgraph of $G_i$, $i = 1, 2$. Then $(\sigma_1 \circ \sigma_2, \mu_1 \circ \mu_2)$ is a partial fuzzy subgraph of $G_1[G_2]$.*

*Proof* We have already seen in the proof of Proposition 2.11.1 that

$$(\mu_1 \circ \mu_2)((u_1, u_2)(v_1, v_2)) \leq (\sigma_1 \circ \sigma_2)((u_1, u_2)) \wedge (\sigma_1 \circ \sigma_2)((v_1, v_2))$$

for all $(u_1, u_2)(v_1, v_2) \in X$. Suppose that $(u_1, u_2)(v_1, v_2) \in X^0 \backslash X$ and so $u_1v_1 \in X_1$ and $u_2 \neq v_2$. Then

$$\begin{aligned}(\mu_1 \circ \mu_2)((u_1, u_2)(v_1, v_2)) &= \sigma_2(u_2) \wedge \sigma_2(v_2) \wedge \mu_1(u_1v_1) \\ &\leq \sigma_2(u_2) \wedge \sigma_2(v_2) \wedge \sigma_1(u_1) \wedge \sigma_1(v_1) \\ &= \sigma_1(u_1) \wedge \sigma_2(u_2) \wedge \sigma_1(u_1) \wedge \sigma_2(v_2) \\ &= (\sigma_1 \circ \sigma_2)((u_1, u_2)) \wedge (\sigma_1 \circ \sigma_2)((v_1, v_2)).\end{aligned}$$

■

The fuzzy graph $(\sigma_1 \circ \sigma_2, \mu_1 \circ \mu_2)$ of the previous proposition is called the **composition** of $(\sigma_1, \mu_1)$ with $(\sigma_2, \mu_2)$.

**Theorem 2.11.6** *Let $G$ be the composition $G_1[G_2]$ of graph $G_1$ with graph $G_2$. Let $(\sigma, \mu)$ be a partial subgraph of $G$. Consider the following equations:*

*(i)* $x_i \wedge y_j = \sigma(v_{1i}, v_{2j}), i = 1, \ldots, ; j = 1, \ldots, m;$
*(ii)* $x_i \wedge z_{jk} = \mu(v_{1i}, v_{2j})(v_{1i}, v_{2k})), i = 1, \ldots, n;$ *$j, k$ such that $v_{2j}v_{2k} \in X_2$;*
*(iii)* $y_j \wedge w_{ih} = \mu((v_{1i}v_{2j})(v_{1h}, v_{2j})), j = 1, \ldots, m;$ *$i, h$ such that $v_{1i}v_{1h} \in X$;*
*(iv)* $y_j \wedge y_k \wedge w_{ih} = \mu((v_{1i}, v_{2j})(v_{1h}, v_{2k}))$, *where $(v_{1i}, v_{2j})(v_{1h}, v_{2k}) \in X^0 \backslash X$ for $X$ defined as above.*

*Suppose that a solution to equations $(i)$–$(iv)$ exists. If*

$$\widehat{w}_{ih} \geq \mu(((v_{1i}, v_{2j})(v_{1h}, v_{2k})) \textit{ for all } (i, h) \in I$$

*such that* $(v_{1i}, v_{2j})(v_{1h}, v_{2j}) \in X^0 \backslash X$, *then* $(\sigma, \mu)$ *is a composition of partial fuzzy subgraphs of* $G_1$ *and* $G_2$.

*Proof* The necessary part of the theorem is clear. Suppose that a solution to equations $(i)$–$(iv)$ exists. Then there exists a solution to equations $(i)$–$(iv)$ as determined in the proof of Theorem 2.11.2 for equations $(i)$–$(iii)$ because every $w_{ih} \geq \widehat{w}_{ih}$ and by the hypothesis concerning the $\widehat{w}_{ih}$. Thus, if $\mu_i$, $i = 1, 2$, are defined as in the proof of Theorem 2.11.2, we have that $(\sigma_i, \mu_i)$ is a partial fuzzy subgraph of $G_i$, $i = 1, 2$, and $\sigma = \sigma_1 \circ \sigma_2$ and $\mu = \mu_1 \circ \mu_2$. ■

*Example 2.11.7* Let $G_1 = (V_1, X_1)$ and $G_2 = (V_2, X_2)$ be graphs and let $\sigma_1, \sigma_2, \mu_1, \mu_2$ be fuzzy subsets of $V_1, V_2, X_1, X_2$, respectively. Then $(\sigma_1 \times \sigma_2, \mu_1\mu_2)$ is a partial fuzzy subgraph of $G_1 \times G_2$, but $(\sigma_i, \mu_i)$ is not a partial fuzzy subgraph of $G_i$, $i = 1, 2$ : Let $V_1 = \{u_1, v_1\}$, $V_2 = \{u_2, v_2\}$, $X_1 = \{u_1v_1\}$, and $X_2 = \{u_2v_2\}$. Define the fuzzy subsets $\sigma_1, \sigma_2, \mu_1$, and $\mu_2$ as follows: $\sigma_1(u_1) = \sigma_1(v_1) = \sigma_2(u_2) = \sigma_2(v_2) = 1/2$ and $\mu_1(u_1v_1) = \mu_2(u_2v_2) = 3/4$. Then $(\sigma_i, \mu_i)$ is not a partial fuzzy subgraph of $G_i$, $i = 1, 2$. Now, $x \in V_1$ and $y \in V_2$, $\mu_1\mu_2((x, u_2)(x, v_2)) = \sigma_1(x) \wedge \mu_2(u_2v_2) = 1/2 = \sigma_1(x) \wedge \sigma_2(u_2) \wedge \sigma_2(v_2) = (\sigma_1 \times \sigma_2)(x, u_2)) \wedge (\sigma_1 \times \sigma_2)(x, v_2))$ and similarly, $\mu_1\mu_2((u_1, y)(v_1, y)) = (\sigma_1 \times \sigma_2)(u_1, y)) \wedge (\sigma_1 \times \sigma_2)(v_1, y))$. Thus, $(\sigma_1 \times \sigma_2, \mu_1\mu_2)$ is a partial fuzzy subgraph of $G_1 \times G_2$. Note that for $x_1y_1 \in X_1$ and $x_2y_2 \in X_2$, $(\mu_1 \circ \mu_2)((x_1, x_2)(y_1, y_2)) = \sigma_2(x_2) \wedge \sigma_2(y_2) \wedge \mu_1(x_1y_1) = 1/2 = (\sigma_1 \times \sigma_2)((x_1, x_2)) \wedge (\sigma_1 \times \sigma_2)((y_1, y_2))$. Thus, $(\sigma_1 \circ \sigma_2, \mu_1 \circ \mu_2)$ is a partial fuzzy subgraph of $G_1[G_2]$.

In the previous example, $(\sigma_1 \times \sigma_2, \mu_1\mu_2)$ satisfies the conditions in Theorem 2.11.2. Hence, $(\sigma_1 \times \sigma_2, \mu_1\mu_2)$ is the Cartesian product of partial fuzzy subgraphs $(\tau_i, \nu_i)$ of $G_i$, $i = 1, 2$. In fact, $\tau_i$ and $\nu_i$, $i = 1, 2$, are constant functions with value 1/2.

Consider the union $G = G_1 \cup G_2$ of two graphs $G_1 = (V_1, X_1)$ and $G_2 = (V_2, X_2)$. Let $\mu_i$ be a fuzzy subset of $v_i$ and $\rho_i$ a fuzzy subset of $X_i$, $i = 1, 2$. Define the fuzzy subsets $\sigma_1 \cup \sigma_2$ of $V_1 \cup V_2$ and $\mu_1 \cup \mu_2$ of $X_1 \cup X_2$ as follows:

$$(\sigma_1 \cup \sigma_2)(u) = \begin{cases} \sigma_1(u) & \text{if } u \in V_1 \backslash V_2, \\ \sigma_2(u) & \text{if } u \in V_2 \backslash V_1, \\ \sigma_1(u) \vee \sigma_2(u) & \text{if } u \in V_1 \cap V_2, \end{cases}$$

$$(\mu_1 \cup \mu_2)(uv) = \begin{cases} \mu_1(uv) & \text{if } uv \in X_1 \backslash X_2, \\ \mu_2(uv) & \text{if } uv \in X_2 \backslash X_1, \\ \mu_1(uv) \vee \mu_2(uv) & \text{if } uv \in X_1 \cap X_2. \end{cases}$$

**Proposition 2.11.8** *Let* $G$ *be the union of the graphs* $G_1$ *and* $G_2$. *Let* $(\sigma_i, \mu_i)$ *be a partial fuzzy subgraph of* $G_i$, $i = 1, 2$. *Then* $(\sigma_1 \cup \sigma_2, \mu_1 \cup \mu_2)$ *is a partial fuzzy subgraph of* $G$.

*Proof* Suppose that $uv \in X_1 \backslash X_2$. We have three different cases to consider.

(i) Suppose $u, v \in V_1 \backslash V_2$. Then $(\mu_1 \circ \mu_2)(uv) = \mu_1(uv) \leq \sigma_1(u) \wedge \sigma_1(v) = (\sigma_1 \cup \sigma_2)(u) \wedge (\sigma_1 \cup \sigma_2)(v)$.

(ii) Suppose $u \in V_1 \backslash V_2$ and $v \in V_1 \cap V_2$. Then $(\mu_1 \cup \mu_2)(uv) \leq (\sigma_1 \cup \sigma_2)(u) \wedge (\sigma_1(v) \vee \sigma_2(v) = (\sigma_1 \cup \sigma_2)(u) \wedge (\sigma_1 \cup \sigma_2)(v)$.

(iii) Suppose $u, v \in V_1 \cap V_2$. Then

$$\begin{aligned}(\mu_1 \cup \mu_2)(uv) &\leq (\sigma_1(u) \vee \sigma_2(u)) \wedge (\sigma_1(v) \vee \sigma_2(v)) \\ &= (\sigma_1 \cup \sigma_2)(u) \wedge (\sigma_1 \cup \sigma_2)(v).\end{aligned}$$

Similarly, if $uv \in X_2 \backslash X_1$. Then $(\mu_1 \cup \mu_2)(uv) \leq (\sigma_1 \cup \sigma_2)(u) \wedge (\sigma_1 \cup \sigma_2)(v)$. Suppose that $uv \in X_1 \cap X_2$. Then

$$\begin{aligned}(\mu_1 \cup \mu_2)(uv) &= \mu_1(uv) \vee \mu_2(uv) \\ &\leq (\sigma_1(u) \wedge \sigma_1(v)) \vee \sigma_2(u) \wedge \sigma_2(v) \\ &\leq (\sigma_1(u) \vee \sigma_2(u)) \wedge \sigma_1(v) \vee \sigma_2(v) \\ &= (\sigma_1 \cup \sigma_2)(u) \wedge (\sigma_1 \cup \sigma_2)(v).\end{aligned}$$

■

The fuzzy subgraph $(\sigma_1 \cup \sigma_2, \mu_1 \cup \mu_2)$ of Proposition 2.11.8 is called the **union** of $(\sigma_1, \mu_1)$ and $(\sigma_1, \mu_2)$.

**Theorem 2.11.9** *If $G$ is a union of two fuzzy subgraphs $G_1$ and $G_2$, then every partial fuzzy subgraph $(\sigma, \mu)$ is a union of a partial fuzzy subgraph of $G_1$ and a partial fuzzy subgraph of $G_2$.*

*Proof* Define the fuzzy subsets $\sigma_1, \sigma_2, \mu_1$, and $\mu_2$ of $V_1, V_2, X_1$ and $X_2$, respectively, as follows:

$$\sigma_i(u) = \sigma(u) \text{ if } u \in V_i \text{ and } \mu_i(uv) = \mu(uv) \text{ if } uv \in X_i, i = 1, 2.$$

Then $\mu_i(uv) = \mu(uv) \leq \sigma(u) \wedge \sigma(v) = \sigma_i(u) \wedge \sigma_i(v)$ if $uv \in X_i, i = 1, 2$. Thus, $(\sigma_i, \mu_i)$ is a partial fuzzy subgraph of $G_i, i = 1, 2$. Clearly, $\sigma = \sigma_1 \cup \sigma_2$ and $\mu = \mu_1 \cup \mu_2$. ■

Consider the join $G = G_1 + G_2 = (V_1 \cup V_2, X_1 \cup X_2 \cup X')$ of graphs $G_1 = (V_1, X_2)$ and $G_2 = (V_2, X_2)$, where $X'$ is the set of all edges joining the vertices of $V_1$ and $V_2$ and where we assume $V_1 \cap V_2 = \emptyset$. Let $\sigma_i$ be a fuzzy subset of $V_i$ a fuzzy subset of $X_i$, $i = 1, 2$. Define the fuzzy subsets $\sigma_1 + \sigma_2$ of $V_1 \cup V_2$ and $\mu_1 + \mu_2$ of $X_1 \cup X_2 \cup X'$ as follows:

$$(\sigma_1 + \sigma_2)(u) = (\sigma_1 \cup \sigma_2)(u) \text{ for all } u \in V_1 \cup V_2,$$

$$(\mu_1 + \mu_2)(uv) = \begin{cases} (\mu_1 \cup \mu_2)(uv) \text{ if } uv \in X_1 \cup X_2, \\ \sigma_1(u) \wedge \sigma_2(v) \text{ if } uv \in X', u \in V_1, v \in V_2. \end{cases}$$

**Proposition 2.11.10** *Let $G$ be the join of two graphs $G_1$ and $G_2$. Let $(\sigma_1, \mu_i)$ be a partial fuzzy subgraph of $G_i$, $i = 1, 2$. Then $(\sigma_1 + \sigma_2, \mu_1 + \mu_2)$ is a partial fuzzy subgraph of $G$.*

*Proof* Suppose that $uv \in X_1 \cup X_2$. Then the desired result follows from Proposition 2.11.8. Suppose that $uv \in X'$. Then

$$\begin{aligned}(\mu_1 + \mu_2)(uv) &= \sigma_1(u) \wedge \sigma_2(v) \\ &= (\sigma_1 \cup \sigma_2)(u) \wedge (\sigma_1 \cup \sigma_2)(v) \\ &= (\sigma_1 + \sigma_2)(u) \wedge (\sigma_1 + \sigma_2)(v).\end{aligned}$$

■

The fuzzy subgraph $(\sigma_1 + \sigma_2, \mu_1 + \mu_2)$ of Proposition 2.11.10 is called the **join** of $(\sigma_1, \mu_1)$ and $(\sigma_2, \mu_2)$.

**Definition 2.11.11** Let $(\sigma, \mu)$ be a partial fuzzy subgraph of a graph $G = (V, X)$. Then $(\sigma, \mu)$ is called a **strong partial fuzzy subgraph** of $G$ if $\mu(uv) = \sigma(u) \wedge \sigma(v)$ for all $uv \in X$.

**Theorem 2.11.12** *If $G$ is the join of two subgraphs $G_1$ and $G_2$, then every strong partial fuzzy subgraph $(\sigma, \mu)$ of $G$ is a join of a strong partial fuzzy subgraph of $G_1$ and a strong partial fuzzy subgraph of $G_2$.*

*Proof* Define the fuzzy subsets $\sigma_1, \sigma_2, \mu_1$, and $\mu_2$ of $V_1$, $V_2$, $X_1$ and $X_2$, respectively, as follows: $\sigma_i(u) = \sigma(u)$ if $u \in V_i$ and $\mu_i(uv) = \mu(uv)$ if $uv \in X_i, i = 1, 2$. Then $(\sigma_i, \mu_i)$ is a fuzzy partial subgraph of $G_i$, $i = 1, 2$, and $\sigma = \sigma_1 + \sigma_2$ as in the proof of Theorem 2.11.9. If $uv \in X_1 \cup X_2$, then $\mu(uv) = (\mu_1 + \mu_2)(uv)$ as in the proof of Theorem 2.11.9. Suppose that $uv \in X'$, where $u \in V_1$ and $v \in V_2$. Then $(\mu_1 + \mu_2)(uv) = \sigma_1(u) \wedge \sigma_2(v) = \sigma(u) \wedge \sigma(v) = \mu(uv)$, where the latter equality holds because $(\sigma, \mu)$ is strong. ■

*Example 2.11.13* Let $G_1 = (V_1, X_1)$ and $G_2 = (V_2, X_2)$ be graphs and let $\sigma_1, \sigma_2, \mu_1, \mu_2$ be fuzzy subsets of $V_1, V_2, X_1, X_2$, respectively. Then $(\sigma_1 \cup \sigma_2, \mu_1 \cup \mu_2)$ is a partial fuzzy subgraph of $G_1 \cup G_2$, but $(\sigma_i, \mu_i)$ is not a partial fuzzy subgraph of $G_i$, $i = 1, 2$: Let $V_1 = V_2 = \{u, v\}$ and $X_1 = X_2 = \{uv\}$. Define the fuzzy subsets $\sigma_1, \sigma_2, \mu_1, \mu_2$ be fuzzy subsets of $V_1, V_2, X_1, X_2$, respectively, as follows: $\sigma_1(u) = 1 = \sigma_2(v), \sigma_1(v) = 1/4 = \sigma_2(u), \mu_1(uv) = 1/2 = \mu_2(uv)$. Then $(\sigma_i, \mu_i)$ is not a partial fuzzy subgraph of $G_i, i = 1, 2$. Now, $(\mu_1 \cup \mu_2)(uv) = \mu_1(uv) \vee \mu_2(uv) = 1/2 < 1 = (\sigma_1(u) \vee \sigma_2(u)) \wedge (\sigma_1(v) \vee \sigma_2(v)) = (\sigma_1 \cup \sigma_2)(u) \wedge (\sigma_1 \cup \sigma_2)(v)$. Thus, $(\sigma_1 \cup \sigma_2, \mu_1 \cup \mu_2)$ is a partial fuzzy subgraph of $G_1 \cup G_2$.

The above example can be extended to the case where $V_1 \nsubseteq V_2$, $V_2 \nsubseteq V_1$ and $X_1 \nsubseteq X_2$, $X_2 \nsubseteq X_1$ as follows: Let $V_1 = \{u, v, w\}$, $V_2 = \{u, v, z\}$ and $X_1 = \{uv, uw\}$, $X_2 = \{uv, vz\}$ and $\sigma_1(w) = \sigma_2(z) = 1 = \mu_1(uw) = \mu_2(uz)$.

**Theorem 2.11.14** *Let $G_1 = (V_1, X_1)$ and $G_2 = (V_2, X_1)$ be graphs. Suppose that $V_1 \cap V_2 = \emptyset$. Let $\sigma_1, \sigma_2, \mu_1$, and $\mu_2$ be fuzzy subsets of $V_1$, $V_2$, $X_1$ and $X_2$, respectively. Then $(\sigma_1 \cup \sigma_2, \mu_1 \cup \mu_2)$ is a partial fuzzy subgraph of $G_1 \cup G_2$ if and only if $(\sigma_1, \mu_1)$ and $(\sigma_2, \mu_2)$ are partial fuzzy subgraphs of $G_1$ and $G_2$, respectively.*

*Proof* Suppose that $(\sigma_1 \cup \sigma_2, \mu_1 \cup \mu_2)$ is a partial fuzzy subgraph of $G_1 \cup G_2$. Let $uv \in X_1$. Then $uv \notin X_2$ and $u, v \in V_1 \backslash V_2$. Hence, $\mu_1(uv) = (\mu_1 \cup \mu_2)(uv) \leq (\sigma_1 \cup \sigma_2)(u) \wedge (\sigma_1 \cup \sigma_2)(v) = \sigma_1(u) \wedge \sigma_1(v)$. Thus, $(\sigma_1, \mu_1)$ is partial fuzzy subgraph of $G_1$. Similarly, $(\sigma_2, \mu_2)$ is partial fuzzy subgraph of $G_2$. The converse is Proposition 2.11.8. ■

The following result follows from the proof of Theorem 2.11.12 and Proposition 2.11.10.

**Theorem 2.11.15** *Let $G_1 = (V_1, X_1)$ and $G_2 = (V_2, X_1)$ be graphs. Suppose that $V_1 \cap V_2 = \emptyset$. Let $\sigma_1, \sigma_2, \mu_1$, and $\mu_2$ be fuzzy subsets of $V_1$, $V_2$, $X_1$ and $X_2$, respectively. Then $(\sigma_1 + \sigma_2, \mu_1 + \mu_2)$ is a partial fuzzy subgraph of $G_1 + G_2$ if and only if $(\sigma_1, \mu_1)$ and $(\sigma_2, \mu_2)$ are partial fuzzy subgraphs of $G_1$ and $G_2$, respectively.*

**Definition 2.11.16** Let $(\sigma, \mu)$ be a partial fuzzy subgraph of $(V, T)$, where $T \subseteq V$. Define the fuzzy subsets $\sigma'$ of $V$ and $\mu'$ of $T$ as follows: $\sigma' = \sigma$ and for all $uv \in T$, $\mu'(uv) = 0$ if $\mu(uv) > 0$ and $\mu'(uv) = \sigma(u) \wedge \sigma(v)$ if $\mu(uv) = 0$.

Clearly, $G' = (\sigma', \mu')$ is a fuzzy graph.

**Definition 2.11.17** Let $(\sigma, \mu)$ be a partial fuzzy subgraph of $G = (V, X)$, Then $(\sigma, \mu)$ is said to be **complete** if $X = T$ and for all $uv \in X$, $\mu(uv) = \sigma(u) \wedge \sigma(v)$.

We use the notation $C_m(\sigma, \mu)$ for a complete fuzzy graph, where $|V| = m$.

**Definition 2.11.18** $(\sigma, \mu)$ is called a **fuzzy bigraph** if and only if there exist partial fuzzy subgraphs $(\sigma_i, \mu_i), i = 1, 2,$ of $(\sigma, \mu)$ such that $(\sigma, \mu)$ is the join $(\sigma_1, \mu_1) + (\sigma_2, \mu_2)$, where $V_1 \cap V_2 = \emptyset$ and $X_1 \cap X_2 = \emptyset$. A fuzzy bigraph is said to be **complete** if $\mu(uv) > 0$ for all $uv \in X'$.

We use the notation $C_{m,n}(\sigma, \mu)$ for a complete fuzzy bigraph, where $|V| = m$ and $|V_2| = n$.

**Proposition 2.11.19** $C_{m,n}(\sigma, \mu) = C_m(\sigma, \mu)' + C_n(\sigma, \mu)'$.

# Chapter 3
# Connectivity in Fuzzy Graphs

In graph theory, edge analysis is not very necessary because all edges have the same weight one. But in fuzzy graphs, the strength of an edge is a real number in $[0, 1]$ and hence the properties of edges and paths may vary significantly from that of graphs. So it is important to identify and study the nature of edges of fuzzy graphs. In Chap. 2, we have discussed the strength of connectedness between two vertices $x$ and $y$ in a fuzzy graph $G$. In this chapter, a detailed analysis of the structure of fuzzy graphs based on the strength of connectedness will be made.

## 3.1 Strong Edges in Fuzzy Graphs

In 2002, Bhutani and Rosenfeld introduced the concept of strong edges in fuzzy graphs. In a series of papers, they have discussed strong edges, fuzzy end vertices and geodesics in fuzzy graphs. The discussions in this section are from [42–44]. We shall use the notation $CONN_G(x, y)$ for the strength of connectedness $\mu^{\infty}(x, y)$. Note that a fuzzy graph $G = (\sigma, \mu)$ is connected if $CONN_G(x, y) > 0$ for every pair of vertices $x, y \in \sigma^*$. In the following definitions, $G - xy$ is the fuzzy graph obtained from $G$ by replacing $\mu(xy)$ by 0.

**Definition 3.1.1** Let $G = (\sigma, \mu)$ be a fuzzy graph. An edge $xy$ is said to be **strong** in $G$ if $\mu(xy) > 0$ and $\mu(xy) \geq CONN_{G-xy}(x, y)$. A path $\rho : x = x_0, x_1, \ldots, x_n = y$ from $x$ to $y$ is called a **strong path** if $x_{i-1}x_i$ is strong for all $1 \leq i \leq n$.

In any fuzzy graph, an edge of maximum weight is always strong. But the converse is not true. For example, consider a fuzzy graph $G = (\sigma, \mu)$ with $G^*$ is $K_3$. Let $\sigma^* = \{a, b, c\}$ and $\sigma(a) = \sigma(b) = \sigma(c) = 1$, $0 < \mu(ab) = \mu(ac) < \mu(bc) \leq 1$. Then every edge of $G$ is strong, including the weakest ones.



S. Mathew et al., *Fuzzy Graph Theory*, Studies in Fuzziness and Soft Computing 363, https://doi.org/10.1007/978-3-319-71407-3_3

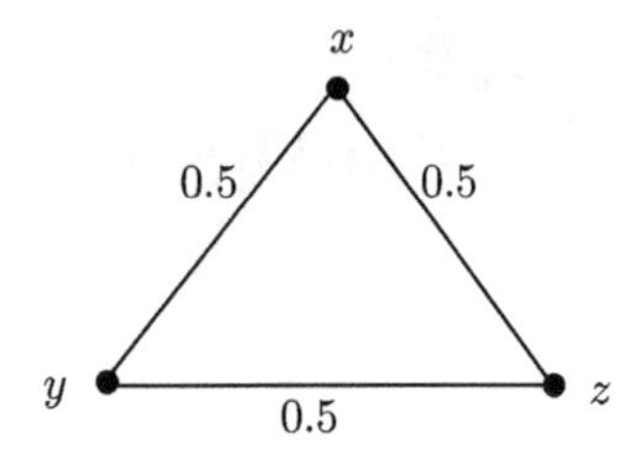


**Fig. 3.1** A strong cycle without fuzzy bridges

**Proposition 3.1.2** *In a fuzzy graph G, every fuzzy bridge is strong.*

*Proof* Let $G = (\sigma, \mu)$ be a fuzzy graph and $xy$ be a fuzzy bridge of $G$. Suppose $xy$ is not strong. Then $\mu(xy) < CONN_{G-xy}(x, y)$. Let $\rho$ be a strongest path from $x$ to $y$ in $G - xy$. The strength of this path is $CONN_{G-xy}(x, y)$. If we adjoin $xy$ to $\rho$ to obtain a cycle, $xy$ becomes the weakest edge of this cycle. Hence, by Theorem 2.2.1, $xy$ is not a fuzzy bridge of $G$. This proves that a fuzzy bridge must be strong. ■

The converse of Proposition 3.1.2 is not generally true, as seen from the example of a triangle in which all the edges have the same weight (Fig. 3.1).

**Proposition 3.1.3** *An edge xy in a fuzzy graph G is strong if and only if* $\mu(xy) = CONN_G(x, y)$.

*Proof* Assume $xy$ is strong. Then $CONN_G(x, y) \geq \mu(xy)$. If a path from $x$ to $y$ contains $xy$, its strength is less than or equal to $\mu(xy)$. If it doesn't contain $xy$, it is in $G - xy$, so its strength is less than or equal to $CONN_{G-xy}(x, y)$, which is less than or equal to $\mu(xy)$ because $xy$ is strong. Hence, in either case, the strength of a path from $x$ to $y$ is at most $\mu(xy)$, so that $CONN_G(x, y) \leq \mu(xy)$.

Conversely, suppose that for $xy \in \mu^*$, $\mu(xy) = CONN_G(x, y)$, then we must have $\mu(xy) \geq CONN_{G-xy}(x, y)$. So $xy$ is strong. ■

**Proposition 3.1.4** *Let* $G = (\sigma, \mu)$ *be connected, and let* $x, y$ *be any two vertices in* $\sigma^*$. *Then there exists a strong path from x to y.*

*Proof* Because $G$ is connected, there exists a path $\rho : x = x_0, x_1, \ldots, x_n = y$ from $x$ to $y$ such that $\mu(x_{i-1}x_i) > 0$ for all $1 \leq i \leq n$. If $x_{j-1}x_j$ is not strong, we must have $\mu(x_{j-1}x_j) < CONN_{G-x_{j-1}x_j}(x_{j-1}, x_j)$. Hence, there exists a path $\rho_j$ from $x_{j-1}$ to $x_j$ whose strength is greater than $\mu(x_{j-1}x_j)$, so that all its edges have weights greater than $\mu(x_{j-1}x_j)$. If some edge on $\rho_j$ is not strong, then this argument can be repeated. Evidently, the argument cannot be repeated arbitrarily often. Hence, eventually we can find a path from $x$ to $y$ on which all the edges are strong. ■

Now, consider a fuzzy tree $G = (\sigma, \mu)$. By definition of a fuzzy tree, there exists a unique maximum spanning tree $F = (\sigma, \nu)$ such that for all edges $xy$ not in $F$, $\mu(xy) < CONN_G(x, y)$. The strong edges of $G$ are precisely the strong edges of $F$ as seen from the following theorem.

**Proposition 3.1.5** *If G is a fuzzy tree, an edge of G is strong if and only if it is an edge of its unique maximum spanning tree.*

*Proof* Let $G = (\sigma, \mu)$ be a fuzzy tree with $F$ its unique maximum spanning tree. If $xy$ is not an edge of $F$, we must have $\mu(xy) < CONN_F(x, y)$. Because $xy$ is strong we have $\mu(xy) \geq CONN_{G-xy}(x, y) \geq CONN_F(x, y)$. Because $xy$ is not an edge of $F$, we get a contradiction.

Conversely, suppose that $xy$ is an edge of $F$, but not a strong edge of $G$. Then $\mu(xy) < CONN_{G-xy}(x, y)$. Let $\rho$ be a path of maximum strength from $x$ to $y$ in $G - xy$. The strength of $\rho$ is $CONN_{G-xy}(x, y)$. So $xy$ is the weakest edge of the cycle formed by adjoining $xy$ to $\rho$. Clearly, $xy$ is a bridge of $G$. So by Theorem 2.2.1, $xy$ cannot be the weakest edge of a cycle; a contradiction. ■

**Corollary 3.1.6** *If G is a fuzzy tree, F is uniquely determined.*

*Proof* The edges of $F$ are just the strong edges of $G$. ■

**Corollary 3.1.7** *An edge of a fuzzy tree is strong if and only if it is a fuzzy bridge.*

*Proof* A strong edge of $G$ must be an edge of $F$ and hence must be a fuzzy bridge of $G$. ■

The converse is true even if $G$ is not a fuzzy tree.

**Proposition 3.1.8** *G is a fuzzy tree if and only if there is a unique strong path in G between any two vertices of G.*

*Proof* By Proposition 3.1.4, there exists a strong path $\rho$ in $G$ between any two vertices $x$ and $y$. By Corollary 3.1.6, $\rho$ lies entirely in $F$, where $F$ is the unique spanning tree associated with $G$. Because $F^*$ is a tree, there is a unique path in $F$ from $x$ to $y$; hence $\rho$ is unique. To prove the converse, note first that a connected fuzzy graph $G$ is a fuzzy tree if and only if, in any cycle of $G$, there is an edge $xy$ such that $\mu(xy) < CONN_{G-xy}(x, y)$ (Theorem 2.3.1). Hence, if $G$ is not a fuzzy tree, there is a cycle $\rho$ in $G$ such that $\mu(xy) \geq CONN_{G-xy}(x, y)$ for every edge $xy$ of $\rho$, such that every edge of $\rho$ is strong. Thus, for any two vertices $u$ and $v$ on $\rho$, there are two strong paths between $u$ and $v$, a contradiction. ■

**Proposition 3.1.9** *In a fuzzy tree G, a strong path between any two vertices u, v is a path of maximum strength between u and v.*

*Proof* Let $\rho$ be the unique strong path between $u$ an $v$. Because every edge of $\rho$ is strong, it is in the unique maximum spanning tree $F$ of $G$. Suppose $\rho$ is not a path of maximum strength from $u$ to $v$. Let $\rho'$ be such a path; then $\rho'$ is not same as $\rho$, so $\rho$ and the reversal of $\rho'$ form a cycle. Because $F^*$ is a tree, it cannot have a cycle; hence some edge $u'v'$ of $\rho'$ must fail to be in $F$. By definition of $F$, we have $\mu(u'v') < CONN_F(u', v')$. Hence, there is a path from $u'$ to $v'$ in $F$, so we can replace every edge $u'v'$ of $\rho'$ that fail to be in $F$ by a path in $F$. This yields a path $\rho^*$ in $F$ from $u$ to $v$. Because it was constructed by replacing edge $u'v'$ of $\rho'$ by paths stronger than these edges, $\rho^*$ is at least as strong as $\rho'$. Thus, $\rho^*$ too can't be the same as $\rho$, so $\rho$ and the reversal of $\rho^*$ form a cycle, and thus we have a cycle in $F$, which is impossible. ■

If $xy$ is a strong edge, we say that $x$ and $y$ are strong neighbors. Hence, we have the following proposition.

**Proposition 3.1.10** *If $G = (\sigma, \mu)$ is a connected fuzzy graph and $\sigma^*$ is not a singleton, then every vertex of $G$ has at least one strong neighbor.*

The proof of Proposition 3.1.10 follows directly from Proposition 3.1.4.

A vertex $x$ in a fuzzy graph $G$ is called a **fuzzy end vertex** if $x$ has exactly one strong neighbor. Evidently an end vertex (a vertex that has only one neighbor in the support) is a fuzzy end vertex.

**Theorem 3.1.11** *A fuzzy cutvertex has at least two strong neighbors.*

*Proof* Let $G = (\sigma, \mu)$ be a fuzzy graph and let $z$ be a fuzzy cutvertex of $G$. Deleting $z$ reduces $CONN_G(x, y)$ for some $x, y \in \sigma^*$. Thus, there was a strongest path $\pi$ from $x$ to $y$ that passed through $z$, say $x, \ldots, u, z, v, \ldots, y$. If $uz$ is not strong, we have $\mu(uz) < CONN_G(u, z)$, when $uz$ is deleted. Thus, there is a path $\rho$ from $u$ to $z$, not involving $uz$, which is stronger than $\mu(uz)$. Let $u'$ be the vertex just preceding $z$ on $\rho$. Because the strength of $\rho$ is at most $\mu(u'z)$, we must have $\mu(u'z) > \mu(uz)$. If $u'z$ is not strong, we can repeat this argument. Because it cannot be repeated infinitely, we eventually find a $u^*$ such that $u^*z$ is strong. Similarly, we eventually find a $v^*$ such that $zv^*$ is strong. If $u^* = v^*$, we would have a path from $x$ to $u^* = v^*$ to $y$ that is stronger than $\pi$, so that deleting $z$ would not reduce $CONN_G(x, y)$, contradiction. Hence, $z$ has at least two strong neighbors. ■

**Corollary 3.1.12** *No vertex can be both a fuzzy cutvertex and a fuzzy end vertex.*

**Definition 3.1.13** A fuzzy graph $G$ such that $G^*$ is a cycle is called **multimin** if $G$ has more than one weakest edge. Recall that $G$ is called **locamin** if every vertex of $G$ lies on a weakest edge.

Because a cycle has at least three vertices, locamin implies multimin. We shall call a fuzzy graph whose support is a cycle as an $F$-cycle.

**Theorem 3.1.14** *A fuzzy cycle is multimin if and only if it is not a fuzzy tree.*

**Theorem 3.1.15** *A fuzzy cycle is multimin if and only if it has no fuzzy end vertices.*

*Proof* In a multimin fuzzy cycle, every edge is strong. Thus, for any vertex $z$, both of the edges on which $z$ lies are strong, so $z$ cannot be a fuzzy end vertex. Conversely, if a fuzzy cycle has only one weakest edge $xy$, it is not hard to see that $y$ is not a strong neighbor of $x$ or vice versa. Hence, only the other neighbors of $x$ and $y$ are strong, so that $x$ and $y$ are fuzzy end vertices. ■

**Theorem 3.1.16** *A multimin fuzzy cycle is locamin if and only if it has no fuzzy cutvertices.*

*Proof* It is not hard to see that if $xy$ is a weakest edge of a fuzzy cycle $G$, then $x$ and $y$ cannot be cutvertices of $G$. Hence, if $G$ is locamin, so that every vertex of $G$ lies on a weakest edge, no vertex of $G$ can be a cutvertex. Conversely, if $G$ is not locamin, then let $x, y, z$ be three consecutive vertices of $G$ such that neither $xy$ nor $yz$ is a weakest edge. It follows easily that deleting $y$ reduces $CONN_G(x, z)$, so that $y$ is a fuzzy cutvertex. ■

**Theorem 3.1.17** *A nontrivial fuzzy tree G has at least two fuzzy end vertices.*

*Proof* Let $F$ be the unique maximum spanning tree of $G$. Because the support of $F$ is a nontrivial tree, it has at least two end vertices. We will prove that these vertices are fuzzy end vertices of $G$. Indeed, suppose $z$ is an end vertex of $F$ and not a fuzzy end vertex of $G$. Then $z$ has at least two strong neighbors $x, y$ so that $\mu(zx) \geq CONN_{G-zx}(z, x)$ and $\mu(zy) \geq CONN_{G-zy}(z, y)$. Because $z$ is an end vertex of $F$, at most one of $zx$ and $zy$ can be edges in $F$. Suppose $zx$ is not an edge of $F$. By definition of a fuzzy tree, this implies that $\mu(zx) < CONN_F(z, x)$. Hence, $CONN_{H-zx}(z, x) \leq \mu(zx) < CONN_F(z, x)$. But $F$ is a fuzzy subgraph of $G$, and because $zx$ is not an edge of $F$, $F$ is also a fuzzy subgraph of $H - zx$. This implies that $CONN_F(z, x) \leq CONN_{H-zx}(z, x)$, contradiction. ■

**Theorem 3.1.18** *A fuzzy cycle G is multimin if and only if it has at least one vertex which is neither a fuzzy cutvertex nor a fuzzy end vertex.*

*Proof* By Theorem 3.1.15, if $G$ is multimin, it has no fuzzy end vertices, and it follows that a vertex that lies on a weakest edge of $G$, cannot be a fuzzy cutvertex. This proves 'only if'. Conversely, if $G$ is not multimin it has a unique weakest edge. We have seen that the vertices that lie on this edge must be fuzzy end vertices. Thus, all the other vertices of $G$ must be fuzzy cutvertices. Hence, every vertex of $G$ is either a fuzzy end vertex or a fuzzy cutvertex, which proves 'if'. ■

**Corollary 3.1.19** *A fuzzy cycle G is a fuzzy tree if and only if every vertex of G is either a fuzzy cutvertex or a fuzzy end vertex.*

*Proof* The proof follows from Theorems 3.1.14 and 3.1.18. ■

In any fuzzy tree $G$, every vertex is either a fuzzy cutvertex or a fuzzy end vertex; indeed, the end vertices of the spanning tree $F$ of $G$ must be fuzzy end vertices, and the other vertices must be fuzzy cutvertices. The converse is not true, i.e., even if every vertex of $G$ is either a fuzzy cutvertex or a fuzzy endvertex, $G$ need not be a fuzzy tree. Consider the following example.

*Example 3.1.20* Let $G = (\sigma, \mu)$ be the triangle $xyz$ with branches $xu$, $yv$ and $zw$ attached to its vertices. Assign unit weight to all edges and vertices (Fig. 3.2). Evidently $x, y, z$ are cutvertices and hence fuzzy cutvertices. Also, $u$, $v$ and $w$ are end vertices, but $G$ is not a fuzzy tree.

**Proposition 3.1.21** *If G is a fuzzy tree, every vertex of G is either a fuzzy cutvertex or a fuzzy end vertex.*

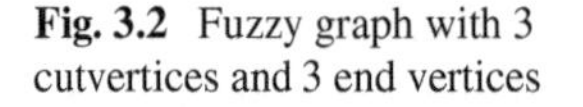

Fig. 3.2 Fuzzy graph with 3 cutvertices and 3 end vertices

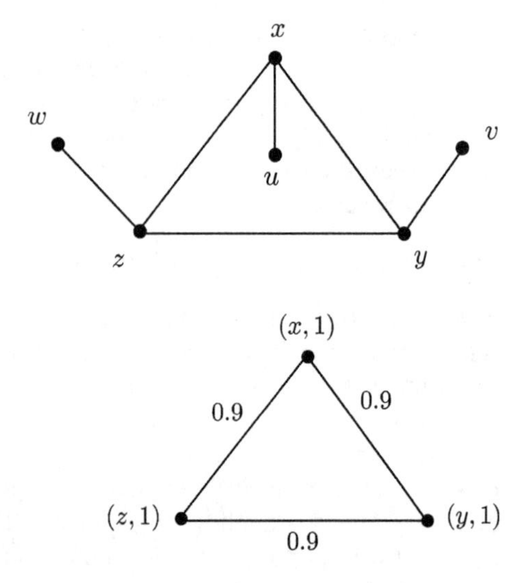


Fig. 3.3 A fuzzy graph with all edges strong

**Proposition 3.1.22** *If* $\mu(uv) = \sigma(u) \wedge \sigma(v)$*, then* $uv$ *is strong.*

*Proof* Any path $\rho$ from $u$ to $v$ in $G - uv$ must contain edges $ux$ and $yv$ for some $x \neq v$ and $y \neq u$. Hence, the strength of $\rho$ is at most $\mu(ux) \wedge \mu(yv) \leq \wedge\{\sigma(u), \sigma(x), \sigma(y), \sigma(v)\} \leq \sigma(u) \wedge \sigma(v) = \mu(u, v)$. ■

The converse of Proposition 3.1.22 is not true. Consider the following example.

*Example 3.1.23* Suppose $G$ has three vertices $x, y, z$ such that $\sigma(x) = \sigma(y) = \sigma(z) = 1$ and $\mu(xy) = \mu(yz) = \mu(zx) = 0.9$. Then all the edges of $G$ are strong, but the $\mu$ values are less than the minimum of corresponding $\sigma$ values (Fig. 3.3).

It was shown in Proposition 3.1.4 that if $G$ is connected, there is a strong path between any two vertices of $G$. Hence, we have the following definition.

**Definition 3.1.24** A strong path $\rho$ from $x$ to $y$ is called a **geodesic** if there is no shorter strong path from $x$ to $y$. The length of a geodesic from $x$ to $y$ will be called the **geodesic distance** from $x$ to $y$, denoted by $d_g(x, y)$.

It is not hard to see that geodesic distance is a metric.

**Definition 3.1.25** Let $G = (\sigma, \mu)$ be a connected fuzzy graph and $S$ be a subset of $\sigma^*$. The **geodesic closure** $(S)$ of $S$ is defined as the set of all vertices that lie on geodesics between vertices of $S$. $S$ is said to be a **geodesic cover** of $G$ if $(S) = \sigma^*$. A minimal cover of $G$ will be called a **geodesic basis** for $G$.

Evidently, $G$ has a basis consisting of a single vertex if and only if $G$ is a singleton. If $G$ has a basis consisting of two vertices $u, v$, then every vertex of $G$ must lie on a geodesic between $u$ and $v$. Hence, for any two vertices $x, y$ of $G$, we have $d_g(x, y) \leq d_g(u, v)$. We have an easy proposition.

**Proposition 3.1.26** *A fuzzy end vertex cannot be on a strong path between any two other vertices.*

**Corollary 3.1.27** *If G is connected, it remains connected when any end vertex is removed from it.*

**Proposition 3.1.28** *A cover of a fuzzy graph G must contain all the fuzzy end vertices of G.*

*Proof* If the end vertex $x$ was not in the cover $S$, it would have to be on a geodesic between two vertices of $S$, contradicting Proposition 3.1.26. ■

**Theorem 3.1.29** *If G is a fuzzy tree, it has a unique basis consisting of its end vertices.*

*Proof* Let $z$ be a non fuzzy end vertex of a fuzzy graph $G$. Then $z$ has at least two strong neighbors $x_0$ and $y_0$. If $x_0$ is not an end vertex, then it has another strong neighbor $x_1$ (different from $z$). If $x_1$ is not an end vertex, then it has another strong neighbor $x_2$ (different from $x_1$), and so on. Because $G$ has no cycles, the $x_i$'s must be distinct and because $G$ is finite, this process must stop, so that some $x_i$ must be an end vertex. Similarly, some $y_j$ must be an end vertex. Thus, $z$ is on a geodesic between the end vertices $x_i$ and $y_j$. Hence, the end vertices form a cover of $G$, so that by Proposition 3.1.28, they must form a unique basis of $G$. ■

The converse of Theorem 3.1.29 is false; $G$ need not be a fuzzy tree even if its end vertices cover it. See the example given below.

*Example 3.1.30* Let $G = (\sigma, \mu)$ be the fuzzy graph with $\sigma^* = \{a, b, c, d, e, f\}$, $\sigma(s) = 1$ for all $s \in \sigma^*$. Let $\mu(ab) = \mu(bc) = \mu(cd) = \mu(da) = 0.9$ and $\mu(ae) = \mu(cf) = 1$ (Fig. 3.4). Clearly, $\{e, f\}$ is a basis for $G$, because all edges are strong. But $G$ is not a fuzzy tree.

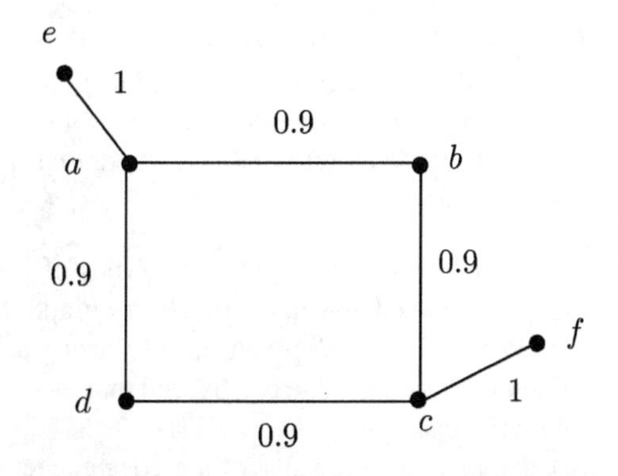


**Fig. 3.4** A non fuzzy tree with a basis of end vertices

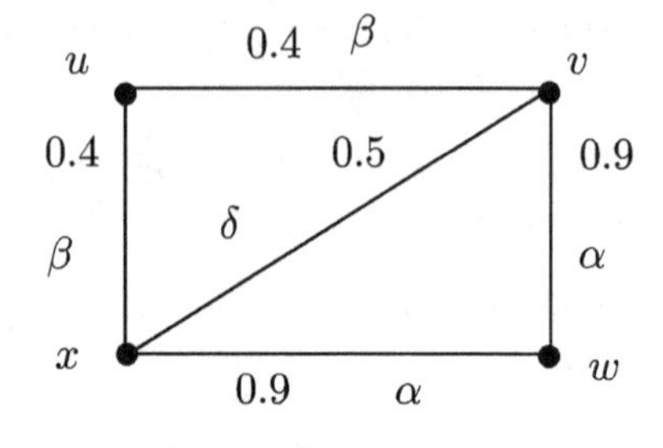


**Fig. 3.5** Fuzzy graph with all type of edges

## 3.2 Types of Edges in Fuzzy Graphs

This section is based on the work by Mathew and Sunitha [116] in 2009. Based on the strength of connectedness between the end vertices of edges in a fuzzy graph $G = (\sigma, \mu)$, the edges of $G$ are divided into three different classes. Note that $CONN_{G-xy}(x, y)$ is the strength of connectedness between $x$ and $y$ in the fuzzy graph obtained from $G$ by deleting the edge $xy$.

**Definition 3.2.1** An edge $xy$ in a fuzzy graph $G = (\sigma, \mu)$ is called $\alpha$-**strong** if $\mu(xy) > CONN_{G-xy}(x, y)$, $\beta$-**strong** if $\mu(xy) = CONN_{G-xy}(x, y)$ and a $\delta$-**edge** if $\mu(xy) < CONN_{G-xy}(x, y)$.

*Remark 3.2.2* The nature of strong edges varies from fuzzy graph to fuzzy graph. For example, a fuzzy tree consists only $\alpha$-strong edges, whereas most of the edges in a complete fuzzy graph are $\beta$-strong. Thus, the division of strong edges into $\alpha$ and $\beta$ will help in understanding the structure of a fuzzy graph properly.

**Definition 3.2.3** A $\delta$-edge $xy$ is called a $\delta^*$-**edge** if $\mu(xy) > \mu(uv)$, where $uv$ is a weakest edge of $G$.

**Definition 3.2.4** A path in a fuzzy graph $G = (\sigma, \mu)$ is called an $\alpha$-**strong path** if all its edges are $\alpha$-strong and is called a $\beta$-**strong path** if all its edges are $\beta$-strong.

*Example 3.2.5* Let $G = (\sigma, \mu)$ with $\sigma^* = \{u, v, w, x\}, \sigma(s) = 1$ for all $s \in \sigma^*$, $\mu(uv) = 0.4 = \mu(xu)$, $\mu(vw) = 0.9 = \mu(wx)$ and $\mu(vx) = 0.5$. Here, $vw$ and $wx$ are $\alpha$-strong edges, $uv$ and $xu$ are $\beta$-strong edges and $vx$ is a $\delta$-edge. Also, $vx$ is a $\delta^*$-edge because $\mu(vx) > \mu(uv)$, where $uv$ is a weakest edge of G. Here $P_1 : x, w, v$ is an $\alpha$-strong $x - v$ path whereas $P_2 : x, u, v$ is a $\beta$-strong $x - v$ path (Fig. 3.5.)

It is interesting to see that, the types of edges cannot be determined by simply examining the edge weights in a fuzzy graph; for, the membership value of a $\delta$-edge can exceed membership values of $\alpha$-strong and $\beta$-strong edges. Also, membership value of a $\beta$-strong edge can exceed that of an $\alpha$-strong edge as can be seen from the following examples.
(*a*) The membership value of a $\delta$-edge exceeds the membership value of a $\beta$-strong edge.

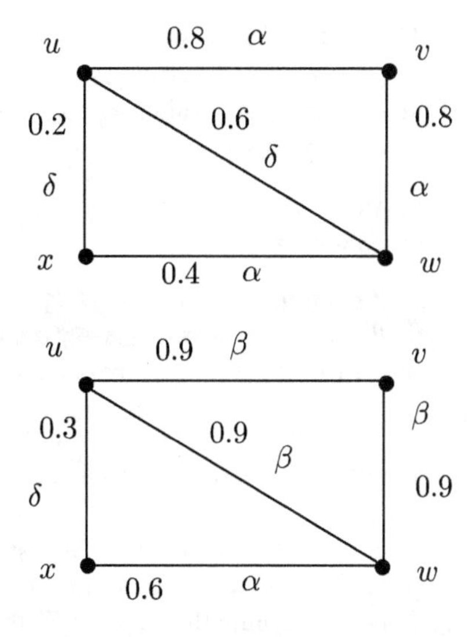


**Fig. 3.6** A fuzzy graph with $\alpha$ and $\delta$ edges

**Fig. 3.7** Membership value-$\beta$-strong exceeds $\alpha$-strong

Consider the fuzzy graph in Example 3.2.5. $\mu(vx) = 0.5 > 0.4 = \mu(uv)$. Here, $vx$ is a $\delta$-edge whereas $uv$ is $\beta$-strong.
($b$) The membership value of a $\delta$-edge exceeds the membership value of an $\alpha$-strong edge. See Example 3.2.6.

*Example 3.2.6* Let $G = (\sigma, \mu)$ with $\sigma^* = \{u, v, w, x\}, \sigma(s) = 1$ for all $s \in \sigma^*$, $\mu(uv) = 0.8 = \mu(vw), \mu(uw) = 0.6$, $\mu(wx) = 0.4$ and $\mu(xu) = 0.2$ (Fig. 3.6). Here, $uv$, $vw$ and $wx$ are $\alpha$-strong edges, whereas $uw$ and $xu$ are $\delta$-edges with $\mu(uw) = 0.6 > 0.4 = \mu(wx)$.

($c$) Membership value of a $\beta$-edge exceeds membership value of an $\alpha$-strong edge.

*Example 3.2.7* Let $G = (\sigma, \mu)$ be such that $\sigma^* = \{u, v, w, x\}$, $\sigma(s) = 1$ for all $s \in \sigma^*$, $\mu(uv) = \mu(uw) = \mu(vw) = 0.9$, $\mu(wx) = 0.6$ and $\mu(xu) = 0.3$ (Fig. 3.7). Here, $uv$,$vw$, $uw$ are $\beta$-strong edges, whereas $wx$ is $\alpha$-strong and $xu$ is a $\delta$-edge with $\mu(uw) = \mu(uv) = \mu(vw) = 0.9 > 0.6 = \mu(wx)$.

In a connected graph $G$, $CONN_G(x, y) = 1$ for all pairs of vertices $x$ and $y$. Also, each $x - y$ path in $G$ is strong as well as strongest. In this section, the type of edges in strongest paths of fuzzy graphs are studied and conditions under which a strong path becomes a strongest path are discussed.

Clearly a strongest path may contain all types of edges. In Example 3.2.5, the strength of the path $P : u, v, x, w$ is 0.4, which is a strongest path from $u$ to $w$ and it contains all types of edges, namely $uv$ is $\beta$-strong, $xw$ is $\alpha$-strong and $vx$ is a $\delta$-edge.

In a graph $G$, each path is strong as well as strongest. But in a fuzzy graph a strongest path need not be a strong path and a strong path need not be a strongest path. In Example 3.2.5, $P_1 : u, v, x, w$ is a strongest $u - w$ path, but not a strong $u - w$ path. Note that $P_2 : u, v, w$ and $P_3 : u, x, w$ are strong $u - w$ paths.

Conversely, $P_4 : v, u, x$ is a strong $v - x$ path which is not a strongest $v - x$ path and $P_5 : v, w, x$ is the strongest $v - x$ path.

A strongest path without $\delta$-edges is a strong path; for, it contains only $\alpha$-strong and $\beta$-strong edges.

**Proposition 3.2.8** ([116]) *A strong path P from x to y is a strongest $x - y$ path in the following cases.*

*(i) P contains only $\alpha$-strong edges.*
*(ii) P is the unique strong $x - y$ path.*
*(iii) All $x - y$ paths in G are of equal strength.*

*Proof* (*i*) Let $G = (\sigma, \mu)$ be a fuzzy graph. Let $P$ be a strong $x - y$ path in $G$ containing only $\alpha$-strong edges. Suppose that $P$ is not a strongest $x - y$ path. Let $Q$ be a strongest $x - y$ path in $G$. Then $P \cup Q$ will contain at least one cycle $C$, in which every edge of $C - P$ will have strength greater than strength of $P$. Thus, a weakest edge of $C$ is an edge of $P$. Let $uv$ be such an edge of $C$. Let $C'$ be the $u - v$ path in $C$, not containing the edge $uv$. Then $\mu(uv) \leq$ strength of $C' \leq CONN_{G-uv}(u, v)$, which implies that $uv$ is not $\alpha$-strong, a contradiction. Thus, $P$ is a strongest $x - y$ path.

(*ii*) Let $G = (\sigma, \mu)$ be a fuzzy graph. Let $P$ be the unique strong $x - y$ path in $G$. Suppose that $P$ is not a strongest $x - y$ path. Let $Q$ be a strongest $x - y$ path in $G$. Then strength of $Q >$ strength of $P$. That is, for every edge $uv$ in $Q$, $\mu(uv) > \mu(x'y')$, where $x'y'$ is a weakest edge of $P$.

**Claim**: $Q$ is a strong $x - y$ path.

For otherwise, if there exists an edge $uv$ in $Q$ which is a $\delta$-edge, then

$$\mu(uv) < CONN_{G-uv}(u, v) \leq CONN_G(u, v)$$

and hence $\mu(uv) < CONN_G(u, v)$. Thus, there exists a path from $u$ to $v$ in $G$ whose strength is greater than $\mu(uv)$. Let it be $P'$. Let $w$ be the last vertex after $u$, common to $Q$ and $P'$ in the $u - w$ sub path of $P'$ and $w'$ be the first vertex before $v$, common to $Q$ and $P'$ in the $w' - v$ sub path of $P'$ (If $P'$ and $Q$ are disjoint $u - v$ paths then $w = u$ and $w' = v$). Then the path $P''$ consisting of the $x - w$ path of $Q$, the $w - w'$ path of $P'$, and the $w' - y$ path of $Q$ is an $x - y$ path in $G$ such that Strength of $P''$ > Strength of $Q$, contradiction to the assumption that $Q$ is a strongest $x - y$ path in $G$. Thus, $uv$ cannot be a $\delta$-edge and hence $Q$ is a strong $x - y$ path in $G$. Thus, we have another strong path from $x$ to $y$, other than $P$, which is a contradiction to the assumption that $P$ is the unique strong $x - y$ path in $G$. Hence, $P$ should be a strongest $x - y$ path in $G$.

(*iii*) If every path from $x$ to $y$ have the same strength, then each such path is a strongest $x - y$ path. In particular, a strong $x - y$ path is a strongest $x - y$ path. ■

We observe that if all edges of a fuzzy graph $G$ are $\beta$-strong, as in graphs without bridges, then each strongest path is a strong path but the converse need not be true. For consider the fuzzy graph $G = (\sigma, \mu)$ with $\sigma^* = \{u, v, w, x, y\}$, $\sigma(s) = 1$ for all

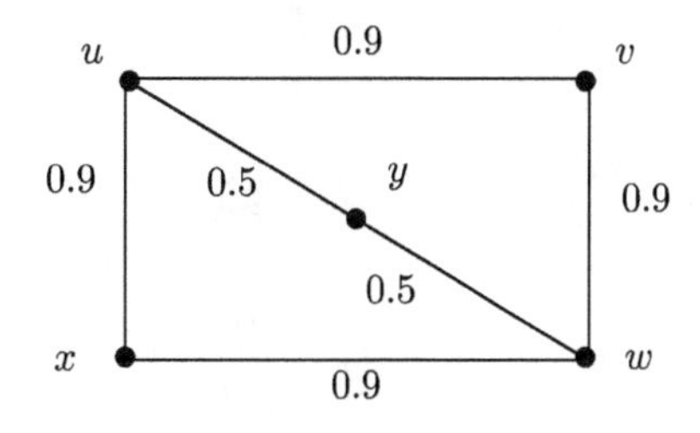


**Fig. 3.8** Fuzzy graph with all strong edges

$s \in \sigma^*$ and $\mu(uv) = \mu(vw) = \mu(wx) = \mu(xu) = 0.9$, $\mu(uy) = \mu(yw) = 0.5$. Here all edges are $\beta$-strong and $P = u, y, w$ is a strong $u - w$ path but it is not a strongest $u - w$ path (Fig. 3.8.)

Next we characterize fuzzy bridges of a fuzzy graph using the concept of $\alpha$-strong edges. Note that in a fuzzy graph, a fuzzy bridge is strong but not conversely (Proposition 3.1.2).

**Theorem 3.2.9** (Characterization of fuzzy bridges in a fuzzy graph) *Let $G = (\sigma, \mu)$ be a fuzzy graph. Then an edge $xy$ of $G$ is a fuzzy bridge if and only if it is $\alpha$-strong.*

*Proof* Let $G = (\sigma, \mu)$ be a fuzzy graph. Let $xy$ be a fuzzy bridge in $G$. Then by Theorem 2.2.1, $\mu(xy) > CONN_{G-xy}(x, y)$, which shows that $xy$ is $\alpha$-strong.

Conversely, suppose that $xy$ is $\alpha$-strong. Then by definition, it follows that $xy$ is the unique strongest path from $x$ to $y$ and the removal of $xy$ will reduce the strength of connectedness between $x$ and $y$. Thus, $xy$ is a fuzzy bridge. ■

Recall that if an edge $xy$ of $G$ is a fuzzy bridge, then $CONN_G(x, y) = \mu(xy)$ (Theorem 2.2.14). The converse need not be true. In Example 3.2.5, edges $uv$ and $xu$ are $\beta$-strong and are not fuzzy bridges.

One of the characterizations for a fuzzy bridge is that it is in every maximum spanning tree (MST) of $G$ (Corollary 2.2.4). So we have the following corollary.

**Corollary 3.2.10** *An edge $xy$ of a connected fuzzy graph $G$ is $\alpha$-strong if and only if $xy$ is in every MST of $G$.*

**Corollary 3.2.11** *Let $G = (\sigma, \mu)$ be a fuzzy graph with $|\sigma^*| = n$, then the number of $\alpha$-strong edges in $G$ is at most $n - 1$.*

If $G$ is a fuzzy tree, then the removal of an $\alpha$-strong edge reduces the strength of connectedness between its end vertices and also between some other pair of vertices. Also, note that the internal vertices of $F$ are the fuzzy cutvertices of $G$ and hence $w$ is a fuzzy cutvertex if and only if $w$ is a common vertex of at least two $\alpha$-strong edges.

Also, if $w$ is a common vertex of at least two fuzzy bridges, then $w$ is a fuzzy cutvertex. Hence, it follows that, if $w$ is a common vertex of at least two $\alpha$-strong edges in a fuzzy graph $G$, then $w$ is a fuzzy cutvertex. But the converse is not true. See Examples 3.2.12 and 3.2.13.

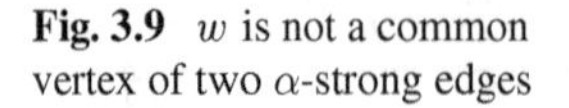

**Fig. 3.9** $w$ is not a common vertex of two $\alpha$-strong edges

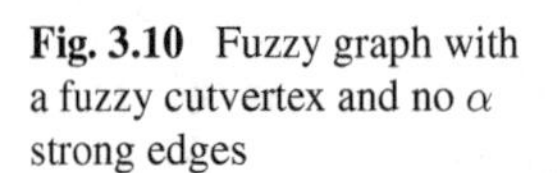

**Fig. 3.10** Fuzzy graph with a fuzzy cutvertex and no $\alpha$ strong edges

*Example 3.2.12* Let $G = (\sigma, \mu)$ with $\sigma^* = \{u, v, w, x\}$, $\sigma(s) = 1$ for all $s \in \sigma^*$ and $\mu(uv) = \mu(vw) = \mu(uw) = 0.5$, $\mu(xu) = 0.4$ and $\mu(wx) = 0.9$. Then $wx$ is the only $\alpha$-strong edge, $xu$ is a $\delta$-edge, $uv$, $vw$ and $uw$ are $\beta$-strong edges. Also, $w$ is a fuzzy cutvertex, but it is not a common vertex of two or more $\alpha$-strong edges (Fig. 3.9).

*Example 3.2.13* Let $G = (\sigma, \mu)$ with $\sigma^* = \{u, v, w, x, y\}$, $\sigma(s) = 1$ for all $s \in \sigma^*$ and $\mu(uv) = 0.5 = \mu(xy)$, $\mu(ux) = \mu(vy) = \mu(vw) = \mu(uw) = \mu(xw) = \mu(yw) = 0.8$. Here $G$ has no $\alpha$-strong edges. $xy$ and $uv$ are $\delta$-edges and all other edges are $\beta$-strong. $w$ is a fuzzy cutvertex and there are no $\alpha$-strong edges incident on $w$ (Fig. 3.10)

Now, we present a condition under which an edge of a fuzzy tree becomes $\alpha$-strong. In a fuzzy tree $G$, an edge of $G$ is strong if and only if it is an edge of $F$, where $F$ is the associated unique maximum spanning tree of $G$ (Proposition 3.1.5). Actually these strong edges are $\alpha$-strong and there are no $\beta$-strong edges in a fuzzy tree. The characterization of fuzzy trees using $\beta$-strong edges is given. We know that a loop less graph is a tree if and only if there exists a unique path between any two vertices in it. Analogues to this, it is shown that $G$ is a fuzzy tree if and only if there exists a unique $\alpha$-strong path between any two vertices in $G$.

**Theorem 3.2.14** *A connected fuzzy graph $G$ is a fuzzy tree if and only if it has no $\beta$-strong edges.*

*Proof* Let $G = (\sigma, \mu)$ be a fuzzy tree and let $F = (\sigma, \nu)$ be its maximum spanning tree. Now, all edges in $F$ are $\alpha$-strong (Proposition 2.3.4 and Theorem 3.2.9). Suppose $xy$ is a $\beta$-strong edge in $G$. Then $xy$ is not in $F$ and by definition of a fuzzy tree,

$$\mu(xy) < CONN_F(x, y) \tag{3.1}$$

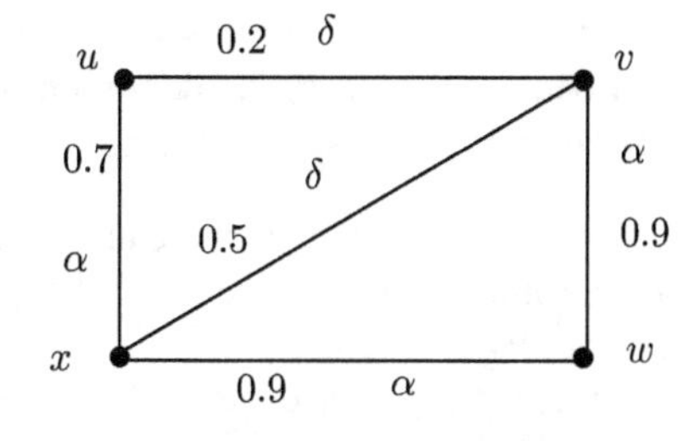


**Fig. 3.11** A fuzzy tree with $\delta^*$ edge

Also, because $F$ is a subgraph of $G$,

$$CONN_F(x, y) \leq CONN_{G-xy}(x, y). \tag{3.2}$$

From (3.1) and (3.2), $\mu(xy) < CONN_{G-xy}(x, y)$, which implies that $xy$ is a $\delta$-edge, which is a contradiction. Thus, $G$ contains no $\beta$-strong edges.

Conversely, suppose that $G$ is connected and has no $\beta$-strong edges. If $G$ has no cycles, then $G$ is a fuzzy tree. Now, assume that $G$ has cycles. Let $C$ be a cycle in $G$. Then $C$ will contain only $\alpha$-strong edges and $\delta$-edges. Also, all edges of $C$ cannot be $\alpha$-strong because otherwise it will contradict the definition of $\alpha$-strong edges. Thus, there exists at least one $\delta$-edge in $C$. Then by Theorem 2.3.1, it follows that $G$ is a fuzzy tree. ■

Thus, all strong edges of a fuzzy tree are $\alpha$-strong and hence Proposition 3.1.5 can be restated as follows.

**Theorem 3.2.15** *An edge xy in a fuzzy tree $G = (\sigma, \mu)$ is $\alpha$-strong if and only if xy is an edge of the spanning tree $F = (\sigma, \nu)$ of G.*

A fuzzy tree can have $\delta^*$-edges as seen from the following example.

*Example 3.2.16* Let $G = (\sigma, \mu)$ with $\sigma^* = \{u, v, w, x\}$, $\sigma(s) = 1$ for all $s \in \sigma^*$ and $\mu(uv) = 0.2$, $\mu(xu) = 0.7$, $\mu(vw) = 0.9 = \mu(wx)$, $\mu(vx) = 0.5$. Then $G$ is a fuzzy tree with $vw$, $wx$ and $xu$ as $\alpha$-strong and $vx$ and $uv$ as $\delta$-edges. Also, $vx$ is a $\delta^*$-edge because $\mu(vx) > \mu(uv)$, where $uv$ is a weakest edge of $G$ (Fig. 3.11.)

**Theorem 3.2.17** *G is a fuzzy tree if and only if there exists a unique $\alpha$-strong path between any two vertices in G.*

*Proof* The proof follows from Proposition 3.1.8 and Theorem 3.2.14. ■

$G$ is a fuzzy tree if and only if $G$ has a unique MST and all edges in the MST are $\alpha$-strong edges. In general, we have the following theorem.

**Theorem 3.2.18** *Let T be any spanning tree of a fuzzy graph G. Then T is an MST of G if and only if T contains no $\delta$-edges. Further, an MST T is unique for G if and only if T contains no $\beta$-strong edges.*

*Proof* The first part follows from the definitions of $\delta$ edge and MST and the second part follows from the definition of $\beta$-strong edge, Theorem 3.2.15 above and Theorem 2.3.19. ■

Note that the strength of the unique $x$-$y$ path in any MST of $G$ gives $CONN_G(x, y)$ and it follows from Theorem 3.2.18 that there exists strong $x$-$y$ path between any two vertices $x$ and $y$ of $G$.

Next the types of edges in fuzzy cycles are discussed. It is observed that there are no $\delta$-edges in a fuzzy cycle $G$. For, if $uv$ is a $\delta$-edge in $G$, then it becomes the unique weakest edge of $G$, which contradicts that $G$ is a fuzzy cycle. Also, a fuzzy cycle cannot have all its edges $\alpha$-strong because the weakest edges in the fuzzy cycle cannot be $\alpha$-strong and note that these weakest edges are $\beta$-strong edges and all other edges are $\alpha$-strong. This leads to the following theorem.

**Theorem 3.2.19** *Let $G$ be a fuzzy graph such that $G^*$ is a cycle. Then $G$ is a fuzzy cycle if and only if $G$ has at least two $\beta$-strong edges.*

Note that in a fuzzy graph $G$ such that $G^*$ is a cycle, $w$ is a fuzzy cutvertex if and only if it is a common vertex of at least two fuzzy bridges and using Theorem 3.2.9 we have the following theorem.

**Theorem 3.2.20** *Let $G$ be a fuzzy graph such that $G^*$ is a cycle. If $G$ contains at most one $\alpha$-strong edge, then $G$ has no fuzzy cutvertices.*

Converse of Theorem 3.2.20 is not true. The condition for the converse to be true is given in the following theorem whose proof is obvious.

**Theorem 3.2.21** *If there exists a unique strongest path between any two vertices $x$, and $y$ in a fuzzy graph $G$, then it is a strong $x - y$ path.*

Now, we discuss types of edges in a complete fuzzy graph (CFG). In the following results, the number of $\beta$-strong edges in a CFG is calculated and the existence of a $\beta$-strong path between any two vertices of a CFG is proved. In a complete graph, there are no bridges, but a CFG may contain fuzzy bridges. Hence, we have the following two lemmas.

**Lemma 3.2.22** *A complete fuzzy graph has no $\delta$-edges.*

*Proof* Let $G$ be a complete fuzzy graph. If possible assume that $G$ contains a $\delta$-edge $uv$ (say). Then

$$\mu(uv) < CONN_{G-uv}(u, v).$$

That is, there exists a stronger path $P$ other than the edge $uv$ from $u$ to $v$ in $G$. Let $\mu(uv) = p$ and the strength of the path $P$ be $q$. Then $p < q$. Let $w$ be the first vertex in $P$ after $u$. Then

$$\mu(uw) > p. \tag{3.3}$$

Similarly, let $x$ be the last vertex in $P$ before $v$. Then

$$\mu(xv) > p. \tag{3.4}$$

Because $\mu(uv) = p$, at least one of $\sigma(u)$ or $\sigma(v)$ should be $p$. Now, $G$ being a CFG, (3.3) gives a contradiction if $\sigma(u) = p$ and (3.4) gives a contradiction if $\sigma(v) = p$; which completes the proof. ■

The following result is obvious.

**Lemma 3.2.23** *There exists at most one α-strong edge in a CFG.*

Using Lemmas 3.2.22 and 3.2.23 we have the following two theorems.

**Theorem 3.2.24** *Let $G = (\sigma, \mu)$ be a CFG with $|\sigma^*| = n$. Then the number of β-strong edges in G is ${}^nC_2$ or ${}^nC_2 - 1$, where ${}^nC_2$ denotes the number of combinations of n things taken two at a time given by the formula ${}^nC_2 = \frac{n!}{2!(n-2)!}$.*

**Theorem 3.2.25** *Let $G = (\sigma, \mu)$ be a CFG. Then there exists a β-strong path between any two vertices of G.*

## 3.3 Vertex Connectivity and Edge Connectivity of Fuzzy Graphs

As mentioned before, along with Rosenfeld [154], Yeh and Bang [186] also introduced fuzzy graphs independently in 1975. They emphasized on the connectivity of fuzzy graphs and provided several results related to it. There are a large number of papers on connectivity of fuzzy graphs, available in the literature. But, this section is based on [114], which generalizes the Yeh and Bang parameters of connectivity.

**Definition 3.3.1** Let $G = (\sigma, \mu)$ be a fuzzy graph. The **strong degree** of a vertex $v \in \sigma^*$ is defined as the sum of membership values of all strong edges incident at $v$. It is denoted by $d_s(v)$. Also, if $N_s(v)$ denotes the set of all strong neighbors of $v$, then $d_s(v) = \sum\limits_{u \in N_s(v)} \mu(uv)$.

*Example 3.3.2* Let $G = (\sigma, \mu)$ be a fuzzy graph with $\sigma^* = \{u, v, w\}$, $\sigma(s) = 1$ for all $s \in \sigma^*$, $\mu(uv) = 0.4$, $\mu(vw) = 0.5$ and $\mu(wu) = 0.6$ (Fig. 3.12). Here, $d_s(u) = 0.6$, $d_s(v) = 0.5$, $d_s(w) = 1.1$.

The existence of a strong path between any two vertices of a fuzzy graph was shown in Proposition 3.1.4. As a consequence, we can find at least one strong edge incident at each vertex of a nontrivial connected fuzzy graph.

**Proposition 3.3.3** *In a non trivial connected fuzzy graph $G = (\sigma, \mu)$, $0 < d_s(v) \leq d(v)$ for all vertices $v \in \sigma^*$.*

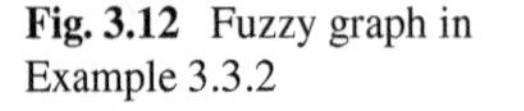
Fig. 3.12 Fuzzy graph in Example 3.3.2

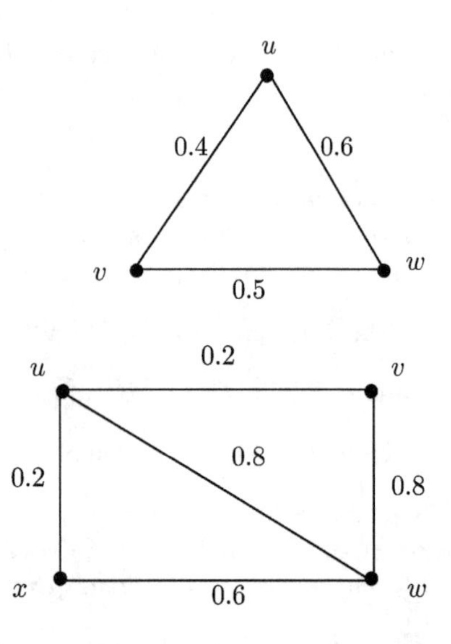


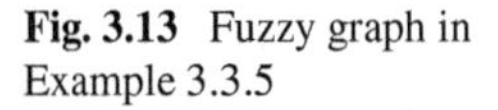
Fig. 3.13 Fuzzy graph in Example 3.3.5

As in graphs, we can define minimum and maximum strong degree of a fuzzy graph as given below.

**Definition 3.3.4** The **minimum strong degree** of $G$ is $\delta_s(G) = \wedge\{d_s(v) \mid v \in \sigma^*\}$ and **maximum strong degree** of $G$ is $\Delta_s(G) = \vee\{d_s(v), v \in \sigma^*\}$.

*Example 3.3.5* Let $G = (\sigma, \mu)$ with $\sigma^* = \{u, v, w, x\}$, $\sigma(s) = 1$ for all $s \in \sigma^*$ and $\mu(uv) = 0.2 = \mu(xu)$, $\mu(vw) = 0.8 = \mu(uw)$, $\mu(wx) = 0.6$. In $G$ (Fig. 3.13), all edges except $uv$ and $ux$ are strong. Thus, $d_s(u) = 0.8 = d_s(v)$, $d_s(w) = 2.2$ and $d_s(x) = 0.6$. Hence, $\delta_s(G) = 0.6$ and $\Delta_s(G) = 2.2$.

As in the case of graphs, $\delta_s(G) \leq d_s(v) \leq \Delta_s(G)$, for all $v \in \sigma^*$. Also, $d(v) = d_s(v)$ for every vertex $v$ in a graph. The next result, similar to the handshaking lemma in graphs is trivial.

**Proposition 3.3.6** *The sum of strong degrees of all vertices in an fuzzy graph is equal to twice the sum of membership values of all strong edges in G.*

In a complete fuzzy graph, all edges are strong and hence $d_s(v) = d(v)$ for all $v \in \sigma^*$. Also, strong degree of a vertex $v$ in a CFG is given by $d_s(v) = \sum_{u \neq v} \wedge\{\sigma(u), \sigma(v)\}$, where $u \in \sigma^*$.

**Proposition 3.3.7** *In a CFG there exists at least one pair of vertices u and v such that* $d_s(u) = d_s(v)$.

The next lemma is related to the minimum and maximum degrees of a CFG. Because $\mu(uv) = \sigma(u) \wedge \sigma(v)$ for every edge $uv$ in a CFG $G$, the minimum and maximum degrees of vertices of $G$ can be evaluated in terms of the membership values of its vertices.

**Lemma 3.3.8** *Let* $G = (\sigma, \mu)$ *be a CFG with* $\sigma^* = \{u_1, u_2, \ldots, u_n\}$ *such that* $\sigma(u_1) \le \sigma(u_2) \le \sigma(u_3) \le \cdots \le \sigma(u_n)$. *Then* $u_1u_j$ *is an edge of minimum weight at* $u_j$ *for* $2 \le j \le n$ *and* $u_iu_n$ *is an edge of maximum weight at* $u_i$ *for* $1 \le i \le n-1$. *Also,*

$$d(u_1) = \delta_s(G) = (n-1)\sigma(u_1)$$

*and*

$$d(u_n) = \Delta_s(G) = \sum_{i=1}^{n-1} \sigma(u_i).$$

*Proof* Throughout the proof, we suppose that $\sigma(u_1) < \sigma(u_2) \le \sigma(u_3) \le \cdots \le \sigma(u_{n-1}) < \sigma(u_n)$. If there are more than one vertex with minimum vertex strength or maximum vertex strength, the proof will be similar. First we prove that for $2 \le j \le n$, $u_1u_j$ is an edge of minimum weight at $u_j$. If possible, suppose that $u_1u_l$, $2 \le l \le n$ is not an edge of minimum weight at $u_l$. Also, let $u_ku_l$, $2 \le k \le n, k \ne l$ be an edge of minimum weight at $u_l$. Being a CFG, $\mu(u_1u_l) = \sigma(u_1) \wedge \sigma(u_l)$ and $\mu(u_ku_l) = \sigma(u_k) \wedge \sigma(u_l)$.

Because $\mu(u_ku_l) < \mu(u_1u_l)$, we have, $\sigma(u_k) \wedge \sigma(u_l) < \sigma(u_1) \wedge \sigma(u_l) = \sigma(u_1)$. That is, either $\sigma(u_k) < \sigma(u_1)$ or $\sigma(u_l) < \sigma(u_1)$. Because $l, k \ne 1$, this is a contradiction to our assumption that $\sigma(u_1)$ is the unique minimum vertex degree. Thus, for $2 \le j \le n$, $u_1u_j$ is an edge of minimum weight at $u_j$.

Next we prove that $u_iu_n$ is an edge of maximum weight at $u_i$ for $1 \le i \le n-1$. On the contrary suppose that $u_ku_n$, $1 \le k \le n-1$ is not an edge of maximum weight at $u_k$ and let $u_ku_r$, $1 \le r \le n-1$, $k \ne r$ be an edge of maximum weight at $u_k$. Then $\mu(u_ku_r) > \mu(u_ku_n)$ and hence $\sigma(u_k) \wedge \sigma(u_r) > \sigma(u_k) \wedge \sigma(u_n) = \sigma(u_k)$, which implies that $\sigma(u_r) > \sigma(u_k)$. Therefore, $\mu(u_ku_r) = \sigma(u_k) = \mu(u_ku_n)$, which is a contradiction to our assumption. Thus, $u_ku_n$ is an edge of maximum weight at $u_k$.

Now, we have,

$$d_s(u_1) = \sum_{i=2}^{n} \mu(u_1u_i) = \sum_{i=2}^{n} (\sigma(u_1) \wedge \sigma(u_i)) = \sum_{i=2}^{n} \sigma(u_1) = (n-1)\sigma(u_1).$$

If possible suppose that $d_s(u_1) \ne \delta_s(G)$ and let $u_k, k \ne 1$ be a vertex in $G$ with minimum strong degree.

Now, $d_s(u_1) > d_s(u_k)$ implies $\sum_{i=2}^{n} \mu(u_1u_i) > \sum_{k \ne 1, j \ne k} \mu(u_ku_j)$.

That is, $\sum_{i=2}^{n} (\sigma(u_1) \wedge \sigma(u_i)) > \sum_{k \ne 1, j \ne k} (\sigma(u_k) \wedge \sigma(u_j))$.

Because $\sigma(u_1) \wedge \sigma(u_i) = \sigma(u_1)$ for $i = 2, 3, \ldots, n$, $\sigma(u_k) \wedge \sigma(u_1) = \sigma(u_1)$ and for all other indices $j$, $\sigma(u_k) \wedge \sigma(u_j) > \sigma(u_1)$, it follows that

$$(n-1)\sigma(u_1) > \sum_{k \ne 1, j \ne k} (\sigma(u_k) \wedge \sigma(u_j)) > (n-1)\sigma(u_1).$$

That is, $d_s(u_1) > d_s(u_1)$, a contradiction. Thus, $d_s(u_1) = \delta_s(G) = (n-1)\sigma(u_1)$.

Finally we show that $d_s(u_n) = \Delta_s(G) = \sum_{i=1}^{n-1} \sigma(u_i)$. Because $\sigma(u_n) > \sigma(u_i)$ for $i = 1, 2, \ldots, n-1$ and $G$ is a CFG, $\mu(u_n u_i) = \sigma(u_n) \wedge \sigma(u_i) = \sigma(u_i)$.

Therefore, $d_s(u_n) = \sum_{i=1}^{n-1} \mu(u_n u_i) = \sum_{i=1}^{n-1} \sigma(u_i)$.

If possible suppose that $d_s(u_n) \neq \Delta_s(G)$. Let $u_l$, $1 \leq l \leq n-1$ be a vertex in $G$ such that $d_s(u_l) = \Delta_s(G)$ and $d_s(u_n) < d_s(u_l)$.

Now,

$$
\begin{aligned}
d_s(u_l) &= \sum_{i=1}^{l-1} \mu(u_i u_l) + \sum_{i=l+1}^{n-1} \mu(u_i u_l) + \mu(u_n u_l) \\
&\leq \sum_{i=1}^{l-1} \sigma(u_i) + (n-l)\sigma(u_l) + \sigma(u_l) \leq \sum_{i=1}^{n-1} \sigma(u_i) = d_s(u_n).
\end{aligned}
$$

That is, $d_s(u_l) \leq d_s(u_n)$, a contradiction to our assumption. Thus, the proposition is proved. ■

In graphs, all vertices are assumed to have the same membership value 1, whereas in fuzzy graphs the membership value of a vertex is always a real number in (0, 1]. So to each fuzzy graph, we can associate a sequence of real numbers namely the vertex strength sequence or node strength sequence abbreviated as *n-s* sequence which is given below.

**Definition 3.3.9** Let $G = (\sigma, \mu)$ be a fuzzy graph with $|\sigma^*| = n$. Then the **vertex-strength sequence** or **node-strength sequence** (***n-s* sequence**) of $G$ is defined to be $(p_1, p_2, \ldots, p_n)$ with $p_1 \leq p_2 \leq \cdots \leq p_n$, where $p_i$, $0 < p_i \leq 1$ is the strength of vertex $i$ when vertices are arranged so that their strengths are non decreasing. In particular, $p_1$ is the smallest vertex strength and $p_n$ is the largest vertex strength.

Example 3.3.10 illustrates Definition 3.3.9.

*Example 3.3.10* Let $G = (\sigma, \mu)$ with $\sigma^* = \{a, b, c, d\}$ and $\sigma(a) = \sigma(c) = \sigma(d) = 0.3$, $\sigma(b) = 0.4$. Then the vertex-strength sequence of $G$ is (0.3, 0.3, 0.3, 0.4) or $(0.3^3, 0.4)$.

By observing the *n-s* sequence, one can determine the number of vertices of minimum strong degree and maximum strong degree in a CFG as in the next proposition.

**Proposition 3.3.11** *Let $G = (\sigma, \mu)$ be a CFG with $|\sigma^*| = n$. Then the following conditions hold.*

*(i) If the n-s sequence of G is of the form $(p_1^{n-1}, p_2)$, then $\delta_s(G) = \Delta_s(G) = (n-1)p_1 = d_s(u_i)$, $i = 1, 2, \ldots, n$.*

*(ii) If the n-s sequence of G is of the form* $(p_1^{r_1}, p_2^{n-r_1})$ *with* $0 < r_1 \leq n-2$, *then there exist exactly* $r_1$ *vertices with degree* $\delta_s(G)$ *and* $n - r_1$ *vertices with degree* $\Delta_s(G)$.

*(iii) If the n-s sequence of G is of the form* $(p_1^{r_1}, p_2^{r_2}, \ldots, p_k^{r_k})$ *with* $r_k > 1$ *and* $k > 2$, *then there exist exactly* $r_1$ *vertices with degree* $\delta_s(G)$ *and exactly* $r_k$ *vertices with degree* $\Delta_s(G)$.

*(iv) If the n-s sequence of G is of the form* $(p_1^{r_1}, p_2^{r_2}, \ldots, p_{k-1}^{r_{k-1}}, p_k)$ *with* $k > 2$, *then there exist exactly* $1 + r_{k-1}$ *vertices with degree* $\Delta_s(G)$.

*Proof* The proofs of $(i)$ and $(ii)$ are obvious. We present proofs for $(iii)$ and $(iv)$. $(iii)$ Let $v_i^{(j)}$, $j = 1, 2, \ldots, r_i$ be the set of vertices in $G$ with $d_s(v_i^{(j)}) = p_i$, $1 \leq i \leq k$. By Lemma 3.3.8, we have, $d_s(v_1^{(j)}) = \delta_s(G) = (n-1)p_1$ for $j = 1, 2, \ldots, r_1$. No vertex with strength more than $p_1$ can have degree $\delta_s(G)$ because $\mu(v_i^{(j)} v_{i+1}^{(l)}) = \sigma(v_i^{(j)}) > p_1$ for $2 \leq i \leq k$, $j = 1, 2, \ldots, r_i$, $l = 1, 2, \ldots, r_{i+1}$. Thus, there exists exactly $r_1$ vertices with strong degree $\delta_s(G)$.

Next we prove that $d_s(v_k^t) = \Delta_s(G), t = 1, 2, \ldots, r_k$.

Because $\sigma(v_k^t)$ is the maximum vertex strength, we have $\mu(v_k^t v_k^j) = p_k$, $t \neq j$; $t, j = 1, 2, \ldots, r_k$ and $\mu(v_k^t v_i^j) = \sigma(v_k^t) \wedge \sigma(v_i^j) = \sigma(v_i^j)$ for $t = 1, 2, \ldots, r_k$; $j = 1, 2, \ldots, r_i$; $i = 1, 2, \ldots, k-1$. Thus, for $t = 1, 2, \ldots, r_k$,

$$
\begin{aligned}
d_s(v_k^t) &= \sum_{i=1}^{k-1} \sum_{j=1}^{r_i} \sigma(v_i^j) + (r_k - 1)p_k \\
&= \sum_{i=1}^{n-1} \sigma(u_i) = \Delta_s(G), \text{ by Lemma 3.3.8}
\end{aligned}
$$

Now, if $u$ is a vertex such that $\sigma(u) = p_{k-1}$, we have,

$$
\begin{aligned}
d_s(u) &= \sum_{i=1}^{k-2} \sum_{j=1}^{r_i} \mu(uv_i^j) + (r_{k-1} - 1 + r_k)p_{k-1} \\
&= \sum_{i=1}^{k-2} \sum_{j=1}^{r_i} \sigma(v_i^j) + \sum_{j=1}^{r_{k-1}} \sigma(v_{k-1}^j) + (r_k - 1)p_{k-1} \\
&< \sum_{i=1}^{k-2} \sum_{j=1}^{r_i} \sigma(v_i^j) + \sum_{j=1}^{r_{k-1}} \sigma(v_{k-1}^j) + (r_k - 1)p_k = \Delta_s(G).
\end{aligned}
$$

Thus, there exists exactly $r_k$ vertices with degree $\Delta_s(G)$.

$(iv)$ Let $v_k^{(1)} = v_k$ be the vertex in $G$ such that $d_s(v_k) = p_k$. Then by Lemma 3.3.8, $d_s(v_k) = \Delta_s(G) = \sum_{i=1}^{n-1} \sigma(u_i)$. Now, let $v_{k-1}^t$, $t = 1, 2, \ldots, r_{k-1}$ be the vertices in $G$ with $d_s(v_{k-1}^t) = p_{k-1}$. Then for $t = 1, 2, \ldots, r_{k-1}$,

$$d_s(v_{k-1}^t) = \sum_{i=1}^{k-2} \sum_{j=1}^{r_i} \mu(v_i^j v_{k-1}^t) + \sum_{l \neq m} \mu(v_{k-1}^l v_{k-1}^m) + \mu(v_{k-1}^l v_k).$$

But, $\mu(v_i^j v_{k-1}^t) = \sigma(v_i^j)$ for $i = 1, 2, \ldots, k-2$ and $j = 1, 2, \ldots, r_i$.

$$\mu(v_{k-1}^l v_{k-1}^m) = p_{k-1} \text{ and } \mu(v_{k-1}^l v_k) = p_{k-1}.$$

Thus,

$$\begin{aligned} d_s(v_{k-1}^t) &= \sum_{i=1}^{k-2} \sum_{j=1}^{r_i} \sigma(v_i^j) + (r_{k-1} - 1)p_{k-1} + p_{k-1} \\ &= \sum_{i=1}^{k-2} \sum_{j=1}^{r_i} \sigma(v_i^j) + r_{k-1}p_{k-1} \\ &= \sum_{i=1}^{n-1} \sigma(u_i) = \Delta_s(G). \end{aligned}$$

Thus, there exist $r_{k-1} + 1$ vertices with strong degree $\Delta_s(G)$.

Now, if $u$ is a vertex such that $\sigma(u) < p_{k-1}$, as in the proof of (iii), we can show that $d_s(u) < \Delta_s(G)$. Thus, there exist exactly $r_{k-1} + 1$ vertices with strong degree $\Delta_s(G)$ and the proof is complete. ■

Yeh and Bang [186] introduced two connectivity parameters for fuzzy graphs namely vertex connectivity and edge connectivity. In this section, we generalize these definitions using the concepts of strong edges. Both vertex connectivity and edge connectivity are related with sets disconnecting the fuzzy graph. But in a fuzzy set up, we need only the reduction of strength of connectedness between some pair of vertices. The definitions of disconnection and vertex connectivity are given below.

**Definition 3.3.12** A **disconnection** of a fuzzy graph $G = (\sigma, \mu)$ is a vertex set $D$ whose removal results in a disconnected or a single vertex fuzzy graph. The **weight** of $D$ is defined to be $\sum_{v \in D} \wedge\{\mu(vu) \mid \mu(vu) \neq 0\}$.

**Definition 3.3.13** The **vertex connectivity** of a fuzzy graph $G$, denoted by $\Omega(G)$, is defined to be the minimum weight of a disconnection in $G$.

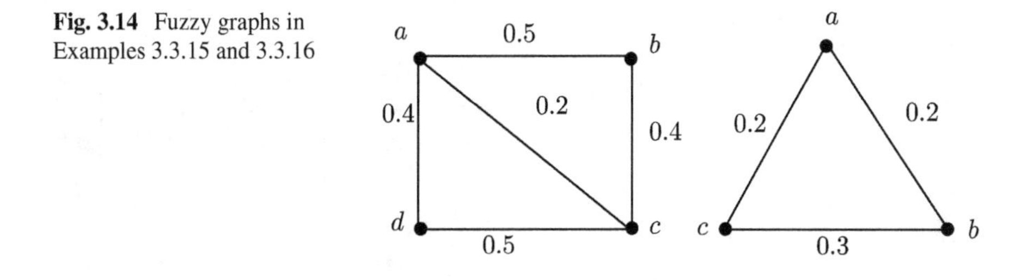


**Fig. 3.14** Fuzzy graphs in Examples 3.3.15 and 3.3.16

The generalized definitions from [114] are given below.

**Definition 3.3.14** Let $G = (\sigma, \mu)$ be a connected fuzzy graph. A set of vertices $X = \{v_1, v_2, \ldots, v_m\} \subset \sigma^*$ is said to be a **fuzzy vertex cut** or **fuzzy node cut** (**FNC**) if either, $CONN_{G-X}(x, y) < CONN_G(x, y)$ for some pair of vertices $x, y \in \sigma^*$ such that both $x, y \neq v_i$ for $i = 1, 2, \ldots, m$ or $G - X$ is trivial.

If there are $n$ vertices in $X$, then $X$ is called an $n$-FNC. Clearly a 1-FNC is a singleton set $X = \{u\}$, where $u$ is a fuzzy cutvertex.

*Example 3.3.15* Let $G = (\sigma, \mu)$ with $\sigma^* = \{a, b, c, d\}$, $\sigma(s) = 1$ for all $s \in \sigma^*$ and $\mu(ab) = \mu(cd) = 0.5$, $\mu(ad) = \mu(bc) = 0.4$, $\mu(ac) = 0.2$. Then $S = \{b, d\}$ is a 2-FNC for, $CONN_{G-S}(a, c) = 0.2 < 0.4 = CONN_G(a, c)$ (Fig. 3.14).

*Example 3.3.16* Let $G = (\sigma, \mu)$ with $\sigma^* = \{a, b, c\}$, $\sigma(s) = 1$ for all $s \in \sigma^*$ with $\mu(a, b) = \mu(ca) = 0.2$, $\mu(bc) = 0.3$. $G$ has no fuzzy cutvertices, but all the three pairs of vertices are fuzzy vertex cuts because the removal of any pair of vertices results in a trivial fuzzy graph.

By Proposition 3.1.4, there exists at least one strong edge incident on every vertex of a nontrivial connected fuzzy graph. The following definition is based on this result.

**Definition 3.3.17** Let $X$ be a fuzzy vertex cut in $G$. The **strong weight** of $X$, denoted by $s(X)$ is defined as $s(X) = \sum_{x \in X} \mu(xy)$, where $\mu(xy)$ is the minimum of the weights of strong edges incident at $x$.

**Definition 3.3.18** The **fuzzy vertex connectivity** of a connected fuzzy graph $G$ is defined as the minimum strong weight of fuzzy vertex cuts of $G$. It is denoted by $\kappa(G)$.

*Example 3.3.19* Let $G = (\sigma, \mu)$ with $\sigma^* = \{a, b, c, d\}$ and $\mu(ab) = 0.2$, $\mu(bc) = 0.5$, $\mu(cd) = 0.4$, $\mu(da) = \mu(ac) = 0.3$ (Fig. 3.15). Then $X_1 = \{c\}$ is the only 1-FNC (i.e., $c$ is a fuzzy cutvertex) with $s(X_1) = 0.3$. The only 2-FNC in $G$ is $X_2 = \{a, c\}$ and $s(X_2) = 0.6$. Also, any three vertices of $G$ form a 3-FNC with $s(\{a, b, c\}) = s(\{a, b, d\}) = s(\{b, c, d\}) = 1.1$ and $s(\{a, c, d\}) = 0.9$. Thus, $\kappa(G) = 0.3$.

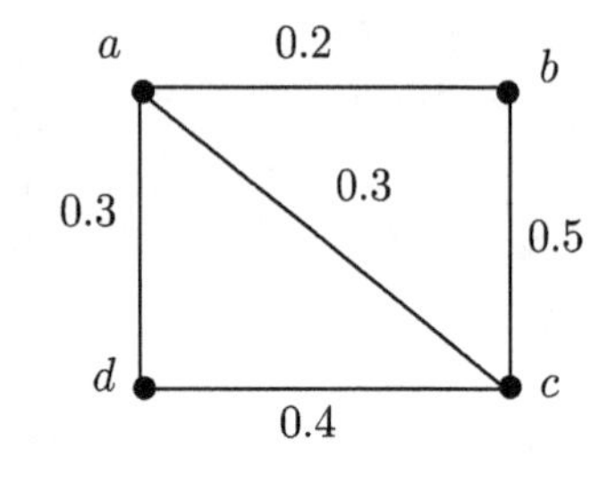


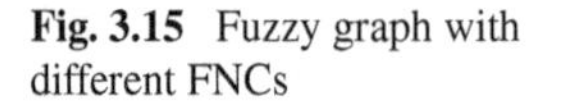
**Fig. 3.15** Fuzzy graph with different FNCs

In [186], the notion of edge connectivity of a fuzzy graph is defined. As mentioned before, this definition is more close to a graph rather than a fuzzy graph. But in a fuzzy graph the reduction of flow is more important than the total disruption of the flow.

The following definitions of a cut-set and edge connectivity are due to Yeh and Bang [186].

**Definition 3.3.20** Let $G$ be a fuzzy graph and $\{V_1, V_2\}$ be a partition of its vertex set. The set of edges joining vertices of $V_1$ and vertices of $V_2$ is called a **cut-set** of $G$, denoted by $(V_1, V_2)$ relative to the partition $\{V_1, V_2\}$. The weight of the cut-set $(V_1, V_2)$ is defined as $\sum\limits_{u \in V_1, v \in V_2} \mu(uv)$.

**Definition 3.3.21** Let $G$ be a fuzzy graph. The **edge connectivity** of $G$ denoted by $\lambda(G)$ is defined to be the minimum weight of cut-sets of $G$.

A generalized definition of fuzzy edge cuts is given below. The edges which are not strong need not be considered because such edges do not contribute towards the strength of connectedness between any pair of vertices.

**Definition 3.3.22** Let $G = (\sigma, \mu)$ be a fuzzy graph. A set of strong edges $E = \{e_1, e_2, \ldots, e_n\}$ with $e_i = u_i v_i$, $i = 1, 2, \ldots, n$ is said to be a **fuzzy edge cut** or **fuzzy arc cut** (**FAC**) if either $CONN_{G-E}(x, y) < CONN_G(x, y)$ for some pair of vertices $x, y \in \sigma^*$ with at least one of $x$ or $y$ different from both $u_i$ and $v_i$, $i = 1, 2, \ldots, n$, or $G - E$ is disconnected.

If there are $n$ edges in $E$ of Definition 3.3.22, then it is called an $n$-FAC. Among all fuzzy edge cuts, an edge cut with one edge (1-FAC) is a special type of fuzzy bridge and we have the following definition.

**Definition 3.3.23** A 1-FAC is called a **fuzzy bond** (**f-bond**).

Note that f-bonds are special type of fuzzy bridges. Not all fuzzy bridges are f-bonds. For example, by Theorem 2.3.19, fuzzy bridges of a fuzzy tree are f-bonds. Also, all bridges in a graph different from $K_2$ are bonds.

*Example 3.3.24* Let $G = (\sigma, \mu)$ with $\sigma^* = \{a, b, c, d, e\}$, $\sigma(x) = 1$ for all $x \in \sigma^*$ with $\mu(ab) = 0.5$, $\mu(ac) = 0.2$, $\mu(bc) = 0.6$, $\mu(cd) = 0.9$, $\mu(da) = 1$,

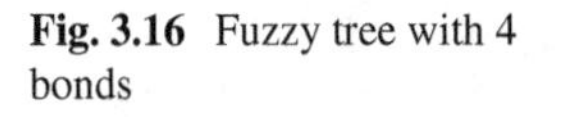

**Fig. 3.16** Fuzzy tree with 4 bonds

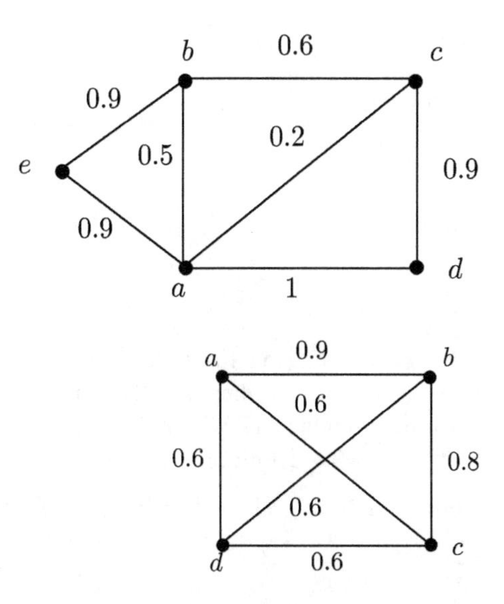


**Fig. 3.17** A non fuzzy tree with all bridges being bonds

$\mu(ae) = \mu(be) = 0.9$ (Fig. 3.16). There are 4 fuzzy bonds (1-FAC) in this fuzzy graph namely edges $ad, ae, dc$ and $eb$. Also, $E = \{ab, dc\}$ is a 2-FAC because $0.6 = CONN_{G-E}(e, c) < CONN_G(e, c) = 0.9$.

As noted, all fuzzy bridges of a fuzzy tree are f-bonds. But there are other examples of non fuzzy trees with this property as seen from the following example.

*Example 3.3.25* Let $G = (\sigma, \mu)$ with $\sigma^* = \{a, b, c, d, \}$, $\sigma(x) = 1$ for all $x \in \sigma^*$ with $\mu(ab) = 0.9$, $\mu(bc) = 0.0.8$, $\mu(cd) = \mu(da) = \mu(ac) = \mu(bd) = 0.6$ (Fig. 3.17). Here $G$ is not a fuzzy tree and there are two fuzzy bridges namely edge $ab$ and edge $bc$ which are f-bonds because deletion of each of these edges from $G$ reduces the strength of connectedness between $a$ and $c$ from 0.8 to 0.6.

In graphs, if $uv$ is a bridge, then at least one of $u$ or $v$ must be a cutvertex. But in fuzzy graphs, if $uv$ is a fuzzy bridge, it is not necessary that at least one of $u$ or $v$ is a fuzzy cutvertex. Note that blocks in fuzzy graphs and CFG can contain fuzzy bridges but no fuzzy cutvertices. But for a fuzzy bond, we have the following proposition.

**Proposition 3.3.26** *At least one of the end vertices of a fuzzy bond is a fuzzy cutvertex.*

*Proof* Let $G = (\sigma, \mu)$ be a fuzzy graph and $e = uv$ be an f-bond in $G$. Being an f-bond, the deletion of $e$ from $G$ reduces the strength of connectedness between $x$ and $y$ with at least one of them different from $u$ and $v$. If both $x$ and $y$ are different from $u$ and $v$, then $u$ as well as $v$ will be fuzzy cutvertices. If one of $x$ or $y$ coincides with $u$ or $v$, then $u$ or $v$ which is neither $x$ nor $y$ will be a fuzzy cutvertex. ■

Because blocks and complete fuzzy graphs contain no fuzzy cutvertices, from Proposition 3.3.26, it follows that no fuzzy bridge in a CFG or in a block can be a

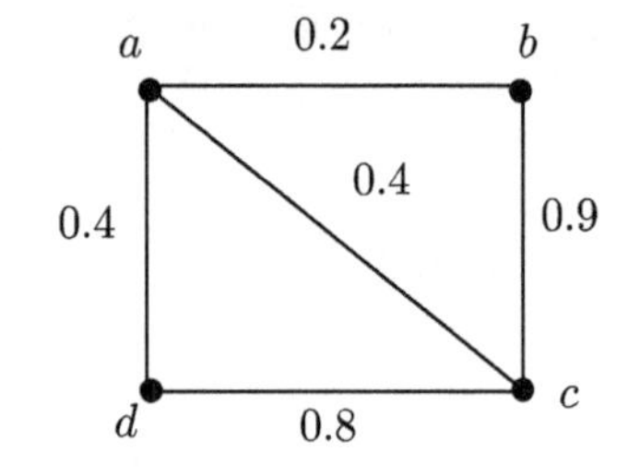


**Fig. 3.18** Fuzzy graph $G$ with $\kappa(G) = 0.4$ and $\kappa'(G) = 0.8$

fuzzy bond. Also, in a fuzzy tree, if a fuzzy bond has exactly one of its end vertices as a fuzzy cutvertex, then the other end vertex must be a fuzzy end vertex.

Strong weight of a FAC and the fuzzy edge connectivity of a fuzzy graph are given in the following definitions.

**Definition 3.3.27** The **strong weight** of a fuzzy edge cut $E$ is defined as $s'(E) = \sum_{e_i \in E} \mu(e_i)$.

**Definition 3.3.28** The **fuzzy edge connectivity** $\kappa'(G)$ of a connected fuzzy graph $G$ is defined to be the minimum strong weight of fuzzy edge cuts of $G$.

*Example 3.3.29* Let $G = (\sigma, \mu)$ with $\sigma^* = \{a, b, c, d\}$, $\sigma(x) = 1$ for all $x \in \sigma^*$ with $\mu(ab) = 0.2$, $\mu(bc) = 0.9$, $\mu(cd) = 0.8$, $\mu(da) = 0.4$, $\mu(ac) = 0.4$ (Fig. 3.18). Then $E_1 = \{bc\}$ and $E_2 = \{cd\}$ are the only 1-FACs (fuzzy bonds)of $G$ with $s'(E_1) = 0.9$ and $s'(E_2) = 0.8$. But $E_3 = \{ad, ac\}$ is a 2-FAC in $G$ with weight $s'(E_3) = 0.8$. Among all fuzzy edge cuts of $G$, $E_3$ has the minimum strong weight and hence $\kappa'(G) = 0.8$. Also, note that $\kappa(G) = 0.4$.

In a tree with at least three vertices, $\kappa(G) = \kappa'(G) = 1$. This is due to the fact that all edges in a tree are strong with strength one and so we have the fuzzy analogue given by the following theorem.

**Theorem 3.3.30** *In a fuzzy tree $G = (\sigma, \mu)$, $\kappa(G) = \kappa'(G) = \wedge\{\mu(xy) \mid xy$ is a strong edge in $G$.*

*Proof* Let $G = (\sigma, \mu)$ be a fuzzy tree. Consider the unique maximum spanning tree $F$ of $G$. An edge $xy$ in $G = (\sigma, \mu)$ is a fuzzy bridge if and only if $xy$ is an edge of the maximum spanning tree $F = (\sigma, \nu)$ of $G$. All these fuzzy bridges are fuzzy bonds. Also, all edges in $F$ are strong. Thus, each strong edge in $F$ is a 1-FAC of $G$. Clearly the strong weight of each such 1-FAC is $\mu(xy)$. Hence, fuzzy edge connectivity $\kappa'(G)$ of $G$ is the minimum weight of all edges in $F$ and hence the minimum weight of all strong edges in $G$.

Now, every internal vertex of $F$ is a fuzzy cutvertex of $G$ and hence are 1-fuzzy vertex cuts of $G$. Hence, fuzzy vertex connectivity $\kappa(G)$ of $G$ is the minimum weight of all edges in $F$ and hence the minimum weight of all strong edges in $G$. ■

The next example shows that, Theorem 3.3.30 does not hold in a fuzzy graph generally.

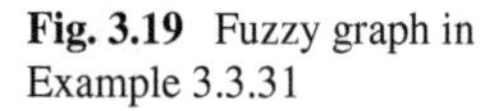
Fig. 3.19 Fuzzy graph in Example 3.3.31

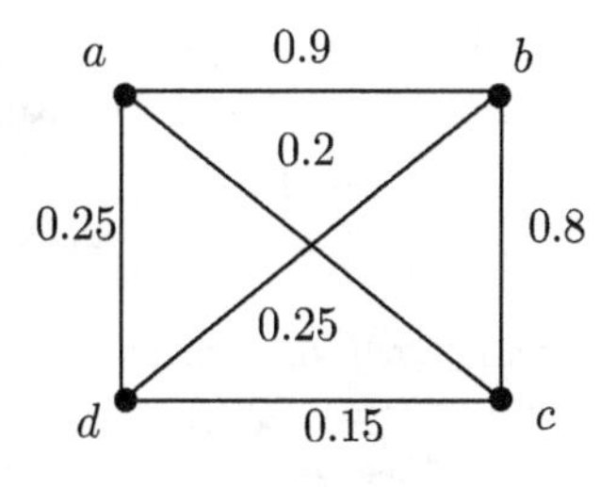


*Example 3.3.31* Let $G = (\sigma, \mu)$ with $\sigma^* = \{a, b, c, d\}$ with $\mu(ab) = 0.9$, $\mu(bc) = 0.8$, $\mu(cd) = 0.15$, $\mu(da) = 0.25$, $\mu(ac) = 0.2$, $\mu(bd) = 0.25$ (Fig. 3.19). Here $G$ is not a fuzzy tree. Edges $ab$ and $bc$ are the only fuzzy bridges in $G$. Clearly these are f-bonds (1-FACs). Now, $s'(\{ab\}) = 1$ and $s'(\{bc\}) = 0.6$. But the fuzzy edge connectivity of $G$ is 0.5 as $E = \{da, db\}$ is a 2-FAC with minimum strong weight. Note that removal of $E$ from $G$ reduces the strength of connectedness between $d$ and $c$ from 0.25 to 0.15.

Next we present the fuzzy analogue of a famous result regarding vertex connectivity, edge connectivity and minimum degree of a graph due to Hassler Whitney.

**Theorem 3.3.32** ([114]) *In a connected fuzzy graph $G = (\sigma, \mu)$, $\kappa(G) \leq \kappa'(G) \leq \delta_s(G)$.*

*Proof* First we prove the second inequality. Let $G = (\sigma, \mu)$ be a connected fuzzy graph. Let $v$ be a vertex in $G$ such that $d_s(v) = \delta_s(G)$. Let $E$ be the set of strong edges incident at $v$. If these are the only edges incident at $v$, then $G - E$ is disconnected. If not, let $vu$ be a edge which is not strong incident at $v$. Then $u$ is a vertex different from the end vertices of edges in $E$. By definition of a strong edge, $\mu(uv) < CONN_G(u, v)$, which implies that there exists a strongest $u - v$ path say $P$ in $G$ which should definitely pass through one of the strong edges at $v$. Thus, the removal of $E$ from $G$ will reduce the strength of connectedness between $v$ and $u$. Thus, in both cases, $E$ is a fuzzy edge cut. The strong weight of this FAC is $\delta_s(G)$. Hence, it follows that $\kappa'(G) \leq \delta_s(G)$.

Next we prove $\kappa(G) \leq \kappa'(G)$. Let $E$ be a FAC with strong weight $\kappa'$. We have the following cases.

**Case 1**: Every edge in $E$ has one vertex in common $v$ (say).

In this case, $E = \{e_i = vv_i, i = 1, 2, \ldots, n\}$. Let $X = \{v_1, v_2, \ldots, v_n\}$. Then clearly $X$ is a fuzzy vertex cut. Now, $\wedge_{u\in\sigma_*}\mu(v_iu) \leq \mu(vv_i)$. Therefore,

$$\sum_i (\wedge_{u\in\sigma_*} \mu(v_iu)) \leq \mu(v_1v) + \mu(v_2v) + \cdots + \mu(v_nv).$$

That is, $\kappa(G) \leq \kappa'(G)$.

**Case 2**: Not all edges in $E$ have a vertex in common.

Let $E = \{e_i = u_i v_i \mid i = 1, 2, \ldots, n\}$ for some $n$. Let $X_1 = \{u_1, u_2, \ldots, u_n\}$ and $X_2 = \{v_1, v_2, \ldots, v_n\}$. By assumption,

$$CONN_{G-E}(x, y) < CONN_G(x, y)$$

for some pair of vertices $x, y \in \sigma^*$ with at least one of $x$ or $y$ different from both $u_i$ and $v_i$ for $i = 1, 2, \ldots, n$.

**Sub Case 1**: $x$ and $y$ are not members of $X_1 \cup X_2$

In this case, take $X = X_1$ or $X = X_2$. Then clearly, $X$ is a fuzzy vertex cut because its deletion from $G$ reduces the strength of connectedness between $x$ and $y$ and,

$$\kappa(G) \leq \text{strong weight of } X \leq \text{strong weight of } E = \kappa'(G).$$

**Sub Case 2**: Either $x$ or $y$ is in $X_1 \cup X_2$

Without loss of generality suppose that $x$ is in $X_1 \cup X_2$. Let $x \in X_1$. Then take $X = X_2$. Clearly $X$ is a fuzzy vertex cut, for; the deletion of $X$ from $G$ will reduce the strength of connectedness between $x$ and $y$. Thus,

$$\kappa(G) \leq \text{strong weight of } X \leq \text{strong weight of } E = \kappa'(G).$$

Thus, in all cases, $\kappa(G) \leq \kappa'(G) \leq \delta_s(G)$. ■

The generalized parameters coincide with Yeh and Bang parameters in a CFG and their values are equal to the minimum strong degree of the fuzzy graph as given in the next corollary.

**Corollary 3.3.33** *In a CFG, $G = (\sigma, \mu)$, $\kappa(G) = \kappa'(G) = \delta_s(G)$.*

*Proof* Let $G = (\sigma, \mu)$ be a CFG such that $|\sigma^*| = n$. Because $G$ is complete, the deletion of any set $E$ of $n - 2$ edges from $G$ will not reduce the strength of connectedness between any pair of vertices in $G$ different from the vertices adjacent to the edges in $E$. Any set of $n - 1$ edges incident at a vertex $u$ in $G$ is a FAC with strong weight $d_s(u) = d(u)$. Let $v$ be a vertex in $G$ such that $d_s(v) = \delta_s(G)$. Clearly the set of edges incident at $v$ is a FAC with minimum strong weight. Therefore, $\kappa'(G) = d_s(v) = \delta_s(G)$.

Now, we prove that $\kappa(G) = \delta_s(G)$.

If possible suppose that $\kappa(G) \neq \delta_s(G)$. By Theorem 3.3.32, $\kappa(G) \leq \kappa'(G) \leq \delta_s(G)$. Hence, $\kappa(G) < \delta_s(G)$.

But because the deletion of $i$ vertices $1 \leq i \leq n - 2$ results again in a nontrivial CFG, any FNC should contain at least $n - 1$ vertices. Among such fuzzy vertex cuts, the one which does not contain a vertex $v$ such that $d_s(v) = \delta_s(G)$ say $S_1$ will have the minimum strong weight because the set of edges adjacent with vertices in $S_1$ with one of its end at $v$ are the edges with minimum weights at vertices of $S_1$. Thus, $\kappa(G) = s(S_1) < \delta_s(G)$. Now, let $E_1$ be the set of all edges incident with the vertex $v$. Then $E_1$ is a FAC such that $s'(E_1) = s(S_1) < \delta_s(G)$, which contradicts the fact that $\kappa'(G) = \delta_s(G)$. Hence, $\kappa(G) = \kappa'(G) = \delta_s(G)$. ■

Even if the values of these parameters coincide on a fuzzy graph, it is not necessary that the given fuzzy graph is a CFG as seen from the following example.

*Example 3.3.34* Let $G = (\sigma, \mu)$ be a fuzzy graph with $\sigma^* = \{a, b, c\}$ with $\sigma(a) = 0.9$, $\sigma(b) = 1$, $\sigma(c) = 0.8$, $\mu(a, b) = 0.2$, $\mu(bc) = 0.1$, $\mu(ac) = 0.1$. Then $\kappa(G) = \kappa'(G) = \delta_s(G) = 0.2$, but $G$ is not a CFG.

Now, we discuss relationships between strong connectivity parameters and Yeh and Bang parameters. The strong parameters always produce smaller values than Yeh and Bang parameters as seen from Theorem 3.3.37.

**Theorem 3.3.35** ([186]) *Let G be a fuzzy graph, then* $\Omega(G) \leq \lambda(G) \leq \delta(G)$.

For given real numbers *a*,*b* and *c*, there exists a fuzzy graph with vertex connectivity *a*, edge connectivity *b* and the minimum degree *c*.

**Theorem 3.3.36** ([186]) *For any real numbers a, b and c such that* $0 < a \leq b \leq c$, *there exists a fuzzy graph G with* $\Omega(G) = a$, $\lambda(G) = b$ *and* $\delta(G) = c$.

**Theorem 3.3.37** *Let* $G = (\sigma, \mu)$ *be a connected fuzzy graph. Then* $\kappa'(G) \leq \lambda(G)$.

*Proof* Let $G = (\sigma, \mu)$ be a connected fuzzy graph with edge connectivity $\lambda(G)$. Let $E = (V_1, V_2)$ be a cut-set in $G$ with minimum weight. That is, the weight of $E = \lambda(G)$. Because $E$ partitions the vertex set into two sets namely $V_1$ and $V_2$, the removal of $E$ from $G$ disconnects $G$. Let $G_1 = (\sigma_1, \mu_1)$ and $G_2 = (\sigma_2, \mu_2)$ be the fuzzy subgraphs of $G$ induced by $V_1$ and $V_2$ respectively. Let $x \in \sigma_1^*$ and $y \in \sigma_2^*$. Then $CONN_{G-E}(x, y) = 0 < CONN_G(x, y)$. Hence, $E$ is a FAC in $G$. Now, $\kappa'$ being the minimum strong weight of all FACs, it follows that $\kappa'(G) \leq weight(E) = \lambda(G)$, which completes the proof. ■

In a fuzzy tree, edge connectivity and minimum degree are upper bounds for both $\kappa(G)$ and $\kappa'(G)$ as seen from the following theorem.

**Theorem 3.3.38** *Let* $G = (\sigma, \mu)$ *be a fuzzy tree. Then* $\kappa(G) = \kappa'(G) \leq \lambda(G) \leq \delta(G)$.

The proof of Theorem 3.3.38 follows from the above theorems. Note that when the fuzzy graph is complete, all these four parameters coincide with value equal to minimum strong degree of the fuzzy graph.

**Theorem 3.3.39** *Let* $G = (\sigma, \mu)$ *be a CFG. Then* $\kappa(G) = \kappa'(G) = \Omega(G) = \lambda(G) = \delta_s(G)$.

*Proof* First we prove that in a CFG, $\lambda(G) = \delta_s(G)$. Let $G = (\sigma, \mu)$ be a CFG with $|\sigma^*| = n$. Because $G$ is complete the deletion of $n - 2$ edges will not disconnect the graph. So any cut-set in $G$ will contain at least $n - 1$ edges. Let $v$ be a vertex in $G$ such that $d_s(v) = \delta_s(G)$. Then because $uv$, $u \in \sigma^*$, $u \neq v$ is an edge with minimum weight at $u$, the cut-set $(V_1, V_2)$ with $V_1 = \{v\}$ and $V_2 = \sigma^* - \{v\}$ will have the minimum weight and by Lemma 3.3.8, it is equal to $d_s(v) = \delta_s(G)$. So

$$\lambda(G) = \delta_s(G). \tag{3.5}$$

Next we prove that $\Omega(G) = \delta_s(G)$. Because $G$ is complete, the deletion of $i$ vertices, $1 \leq i \leq n-2$ results again in a CFG. Therefore, any disconnection $D$ will contain at least $n-1$ vertices and the removal of $D$ will results in a trivial fuzzy graph. Among such disconnections D, the one, not containing the vertex $v$ will have the minimum weight. Thus,

$$\Omega(G) = \delta_s(G). \tag{3.6}$$

By Corollary 3.3.33,

$$\kappa(G) = \kappa'(G) = \delta_s(G). \tag{3.7}$$

Now, from (3.5)–(3.7), $\kappa(G) = \kappa'(G) = \Omega(G) = \lambda(G) = \delta_s(G)$. ■

The condition in Theorem 3.3.39 is not sufficient for a fuzzy graph to be a CFG as seen from Example 3.3.34 above. $G$ is not a CFG even if $\kappa(G) = \kappa'(G) = \Omega(G) = \lambda(G) = \delta_s(G) = 0.2$.

The word clustering means the classification of observations into groups such that the degree of 'association' is high among the members of a group and is less among the members of different groups. Graph theoretically clustering is a partitioning of the graph based on qualitative aspects. Both the introductory articles on fuzzy graph theory by Rosenfeld and Yeh and Bang were intended to present clustering techniques. Rosenfeld introduced distance based clustering while Yeh and Bang introduced connectivity based clustering. Yeh and Bang presented a series of processes like single linkage, $k$-linkage, $k$-edge connectivity, $k$-vertex connectivity and complete linkage to extract fuzzy graph clusters. In [114], $k$-edge connectivity procedure is modified using the newly defined parameters of connectivity.

If $\lambda(G)$ denotes the edge connectivity of a fuzzy graph $G$, then $G$ is called $\tau$-edge connected if $G$ is connected and $\lambda(G) \geq \tau$ and a $\tau$-edge component of $G$ is a maximal $\tau$-edge connected subgraph of $G$. Analogues to this using the concept of fuzzy edge connectivity $\kappa'$, we have the following definitions.

**Definition 3.3.40** A fuzzy graph $G = (\sigma, \mu)$ is called $t$-**fuzzy edge connected** if $G$ is connected and $\kappa'(G) \geq t$ for some $t \in (0, \infty)$.

Thus, if $\kappa'(G) = t'$, then $G$ is $t$-fuzzy edge connected for all $t$ such that $t \leq t'$.

**Definition 3.3.41** A $t$-**fuzzy edge component** of $G = (\sigma, \mu)$ is a maximal $t$-fuzzy edge connected fuzzy subgraph of $G = (\sigma, \mu)$.

Note that by a maximal $t$-fuzzy edge connected subgraph, we mean a fuzzy subgraph $H$ of $G$, induced by a set of vertices in $G$ such that $\kappa'(H) = t$. The above concepts are illustrated in the following example.

*Example 3.3.42* Let $G = (\sigma, \mu)$ be a fuzzy graph with $\sigma^* = \{a, b, c, d\}$ with $\sigma(a) = \sigma(b) = \sigma(c) = \sigma(d) = 1$ and $\mu(ab) = \mu(ac) = 0.2$, $\mu(bc) = 0.3$, $\mu(bd) = 0.1$, $\mu(cd) = 0.4$.

Here $\kappa'(G) = 0.3$. Hence, $G$ is $t$-fuzzy edge connected for all $t$ such that $t \leq 0.3$. Thus, $G$ itself is a $t$-fuzzy edge component for all $t$ such that $0 < t \leq 0.3$. Next let $t = 0.4$. Then the 0.4-fuzzy edge components of $G$ are $H_1 = \langle\{a, b, c\}\rangle$ and $H_2 = \langle\{d\}\rangle$. Here $\kappa'(H_1) = 0.4$.

Now, using the definition of $t$-fuzzy edge components, we have the definition of a fuzzy cluster of level $t$ as follows.

**Definition 3.3.43** Let $G = (\sigma, \mu)$ be a fuzzy graph. A collection $C$ of vertices in $G$ is called a **fuzzy cluster** of level $t$ if the fuzzy subgraph of $G$ induced by $C$ is a $t$-fuzzy edge component of $G$.

We use cohesive matrix $M$ [186] to find the maximal $t$-edge connected components of a fuzzy graph $G$.

**Definition 3.3.44** ([186]) Let $G = (\sigma, \mu)$ be a fuzzy graph. An **element** of $G$ is defined to be either a vertex or an edge. The **cohesiveness** of an element denoted by $h(e)$, is the maximum value of edge connectivity of the subgraphs of $G$ containing $e$.

**Definition 3.3.45** ([186]) Let $G = (\sigma, \mu)$ be a fuzzy graph. The **cohesive matrix** $M$ of $G$ is defined as $M = (m_{i,j})$, where $m_{i,j}$ = the cohesiveness of the edge $v_i v_j$ if $i \neq j$ and the cohesiveness of the vertex $v_i$ if $i = j$.

Note that a vertex $v \in \sigma^*$ is said to be in a cluster of level $t$ if $v$ belongs to a $t$-fuzzy edge component of $G$. Thus, finding the $t$-fuzzy edge components of $G$ is equivalent to the extraction of clusters from $G$. This process of finding $t$-fuzzy edge components and thus finding the fuzzy clusters in $G$ based on fuzzy edge connectivity $\kappa'$ is termed $t$-fuzzy edge connectivity procedure.

$t$-**fuzzy edge connectivity procedure**:

Step-1: Obtain the *Cohesive matrix* $M$ of the fuzzy graph $G = (\sigma, \mu)$.
Step-2: Obtain the *t-threshold graph* $G_t$ of $M$.
Step-3: The maximal complete subgraphs of $G_t$ are the *t-fuzzy edge components*.

**Illustration: Cancer detection problem**

Based on the location of the cells in the low magnification image of a tissue sample, surgically removed from a human patient, it is possible to construct a graph $G$ with vertices as cells, called cell graph [187]. By analyzing the physical features of the cells, for example color and size, we can assign a membership value to the vertices of $G$. This value will range over $(0, 1]$ depending on the nature of the cell; that is healthy, inflammatory or cancerous. Also, edges of $G$ can assign a membership value based on the distance between the cells. Thus, the cell graph can be converted to a fuzzy graph in this manner. By applying the above clustering procedure to such a fuzzy graph, the cancerous cell clusters can be detected at the cellular level in principle. This process classifies cell clusters in a tissue into different phases of cancer depending on the distribution, density and the fuzzy connectivity of the cell clusters within the tissue. Moreover, this process helps in examining the dynamics and progress of cancer qualitatively.

Consider the fuzzy graph $G$ given by the following fuzzy matrix representing a fuzzy cell graph consisting of ten cells. Assume that the vertices with weights more than 0.5 represent cancerous cells, vertices with weights between 0.2 and 0.5 inflammatory cells and between 0 and 0.2, healthy cells. Let the vertices of $G$ be $\{a, b, c, d, e, f, g, h, i, j\}$ and let,

$A =$

| | | | | | | | | | |
|---|---|---|---|---|---|---|---|---|---|
| 0.00 | 0.13 | 0.00 | 0.00 | 0.00 | 0.10 | 0.00 | 0.00 | 0.00 | 0.00 |
| 0.13 | 0.00 | 0.30 | 0.00 | 0.00 | 0.00 | 0.15 | 0.00 | 0.00 | 0.00 |
| 0.00 | 0.30 | 0.00 | 0.50 | 0.00 | 0.00 | 0.00 | 0.40 | 0.00 | 0.00 |
| 0.00 | 0.00 | 0.50 | 0.00 | 0.90 | 0.00 | 0.00 | 0.00 | 0.70 | 0.00 |
| 0.00 | 0.00 | 0.00 | 0.90 | 0.00 | 0.00 | 0.00 | 0.00 | 0.00 | 1.00 |
| 0.10 | 0.00 | 0.00 | 0.00 | 0.00 | 0.00 | 0.14 | 0.00 | 0.00 | 0.00 |
| 0.00 | 0.15 | 0.00 | 0.00 | 0.00 | 0.14 | 0.00 | 0.20 | 0.00 | 0.00 |
| 0.00 | 0.00 | 0.40 | 0.00 | 0.00 | 0.00 | 0.20 | 0.00 | 0.60 | 0.00 |
| 0.00 | 0.00 | 0.00 | 0.70 | 0.00 | 0.00 | 0.00 | 0.60 | 0.00 | 0.80 |
| 0.00 | 0.00 | 0.00 | 0.00 | 1.00 | 0.00 | 0.00 | 0.00 | 0.80 | 0.00 |

The cohesive matrix $M$ of $G$ is given below.

$M =$

| | | | | | | | | | |
|---|---|---|---|---|---|---|---|---|---|
| 0.00 | 0.13 | 0.13 | 0.13 | 0.13 | 0.13 | 0.13 | 0.13 | 0.13 | 0.13 |
| 0.13 | 0.00 | 0.30 | 0.30 | 0.30 | 0.14 | 0.20 | 0.30 | 0.30 | 0.30 |
| 0.13 | 0.30 | 0.00 | 0.50 | 0.50 | 0.14 | 0.20 | 0.50 | 0.50 | 0.50 |
| 0.13 | 0.30 | 0.50 | 0.00 | 0.90 | 0.14 | 0.20 | 0.60 | 0.80 | 0.90 |
| 0.13 | 0.30 | 0.50 | 0.90 | 0.00 | 0.14 | 0.20 | 0.60 | 0.80 | 1.00 |
| 0.13 | 0.14 | 0.14 | 0.14 | 0.14 | 0.00 | 0.14 | 0.14 | 0.14 | 0.14 |
| 0.13 | 0.20 | 0.20 | 0.20 | 0.20 | 0.14 | 0.00 | 0.20 | 0.20 | 0.20 |
| 0.13 | 0.30 | 0.50 | 0.60 | 0.60 | 0.14 | 0.20 | 0.00 | 0.60 | 0.60 |
| 0.13 | 0.30 | 0.50 | 0.80 | 0.80 | 0.14 | 0.20 | 0.60 | 0.00 | 0.80 |
| 0.13 | 0.30 | 0.50 | 0.90 | 1.00 | 0.14 | 0.20 | 0.60 | 0.80 | 0.00 |

For any value $t \in (0, \infty)$, we can find the threshold graph $G_t$ from $M$. The $t$-fuzzy edge components are the maximal complete subgraphs of $G_t$. The corresponding vertices in these components form clusters of level $t$. For example, the threshold graph for $t = 0.5$ is given below.

$G_{0.5} =$

| | | | | | | | | | |
|---|---|---|---|---|---|---|---|---|---|
| 0 | 0 | 0 | 0 | 0 | 0 | 0 | 0 | 0 | 0 |
| 0 | 0 | 0 | 0 | 0 | 0 | 0 | 0 | 0 | 0 |
| 0 | 0 | 0 | 1 | 1 | 0 | 0 | 1 | 1 | 1 |
| 0 | 0 | 1 | 0 | 1 | 0 | 0 | 1 | 1 | 1 |
| 0 | 0 | 1 | 1 | 0 | 0 | 0 | 1 | 1 | 1 |
| 0 | 0 | 0 | 0 | 0 | 0 | 0 | 0 | 0 | 0 |
| 0 | 0 | 0 | 0 | 0 | 0 | 0 | 0 | 0 | 0 |
| 0 | 0 | 1 | 1 | 1 | 0 | 0 | 0 | 1 | 1 |
| 0 | 0 | 1 | 1 | 1 | 0 | 0 | 1 | 0 | 1 |
| 0 | 0 | 1 | 1 | 1 | 0 | 0 | 1 | 1 | 0 |

The different fuzzy clusters of level 0.5 obtained from $G_{0.5}$ are $\{c, d, e, h, i, j\}$, $\{a\}$, $\{b\}$, $\{f\}$ and $\{g\}$.

The fuzzy clusters of all levels are given below.

| Level | Fuzzy Clusters |
|---|---|
| $(1, \infty)$ | $\{a\}, \{b\}, \{c\}, \{d\}, \{e\}, \{f\}, \{g\}, \{h\}, \{i\}, \{j\}$ |
| $(0.9, 1]$ | $\{e, j\}, \{a\}, \{b\}, \{c\}, \{d\}, \{f\}, \{g\}, \{h\}, \{i\}$ |
| $(0.8, 0.9]$ | $\{d, e, j\}, \{a\}, \{b\}, \{c\}, \{f\}, \{g\}, \{h\}, \{i\}$ |
| $(0.6, 0.8]$ | $\{d, e, i, j\}, \{a\}, \{b\}, \{c\}, \{f\}, \{g\}, \{h\}$ |
| $(0.5, 0.6]$ | $\{d, e, h, i, j\}, \{a\}, \{b\}, \{c\}, \{f\}, \{g\}$ |
| $(0.3, 0.5]$ | $\{c, d, e, h, i, j\}, \{a\}, \{b\}, \{f\}, \{g\}$ |
| $(0.2, 0.3]$ | $\{b, c, d, e, h, i, j\}, \{a\}, \{f\}, \{g\}$ |
| $(0.14, 0.2]$ | $\{b, c, d, e, g, h, i, j\}, \{a\}, \{f\}$ |
| $(0.13, 0.14]$ | $\{b, c, d, e, f, g, h, i, j\}, \{a\}$ |
| $(0, 0.13]$ | $\{a, b, c, d, e, f, g, h, i, j\}$ |

From the above fuzzy clusters corresponding to $t = 0.5$ (which is the threshold for cancerous cells), it is observed that $\{d, e, h, i, j\}$ is a cell cluster which is affected seriously by cancer whereas its neighboring area containing the cells $b$, $c$ and $g$ can be found inflammatory. Note that the cells $a$ and $f$ are healthy.

**Comparison Between New and Old Methods**:

As mentioned above, the $t$-fuzzy edge connectivity procedure is more powerful than the $\tau$-edge connectivity procedure. It can be seen from the following example.

Consider the fuzzy graph in Example 3.3.42. The clusters using $\tau$-edge connectivity procedure and fuzzy clusters using $t$-fuzzy edge connectivity procedure are as follows.

**$\tau$-Edge Connectivity Procedure**:

The edge connectivity of the fuzzy graph $G$ in Example 3.3.42 is $\lambda = 0.4$.

Using the Yeh and Bang procedure, we obtain the $\tau$-edge components of $G$ as given below.

| Level | Maximal $\tau$-e.c subgraphs | Clusters |
|---|---|---|
| $(0, 0.4]$ | $\langle\{a, b, c, d\}\rangle$ | $C_1 = \{a, b, c, d\}$ |
| $(0.4, 1]$ | $\langle\{a\}\rangle, \langle\{b\}\rangle, \langle\{c\}\rangle, \langle\{d\}\rangle$ | $C_2 = \{a\}, C_3 = \{b\}, C_4 = \{c\}, C_5 = \{d\}$ |

By this, we get only two types of clusters corresponding to all possible levels namely the full set of vertices and the clusters of singletons. We will find more clusters if we apply the *t-fuzzy edge connectivity procedure* as seen below.

**$t$-Fuzzy Edge Connectivity Procedure**:

The fuzzy edge connectivity $\kappa'$ of the fuzzy graph $G$ in Example 3.3.42 is 0.3. The clusters of level $t$ are obtained as follows.

| Level | Maximal $t$-f.e.c. subgraphs | Fuzzy Clusters |
|---|---|---|
| $(0, 0.3]$ | $\langle\{a, b, c, d\}\rangle$ | $C_1 = \{a, b, c, d\}$ |
| $(0.3, 0.4]$ | $\langle\{a, b, c\}\rangle, \langle\{d\}\rangle$ | $C_2 = \{a, b, c, \}, C_3 = \{d\}$ |
| $(0.4, 1]$ | $\langle\{a\}\rangle, \langle\{b\}\rangle, \langle\{c\}\rangle, \langle\{d\}\rangle$ | $C_4 = \{a\}, C_5 = \{b\}, C_6 = \{c\}, C_7 = \{d\}$ |

Thus, we get three types of clusters corresponding to different levels. When we deal with qualitative data, this will definitely produce more clusters. If the parameter in the above procedure represents the degree of interaction among four research students, Then by the existing method we may observe that there is a minimum interaction between all the students and there are no groups of high interaction. But using the proposed method, we can find a group with more positive interaction and can identify that there is a student who is less active in the whole group.

## 3.4 Menger's Theorem for Fuzzy Graphs

Menger's Theorem is one of the important results in graph theory. It finds the number of internally disjoint paths between a pair of vertices of a graph, in terms of connectivity. The concept of an internally disjoint path in graphs can be replaced by that of a strongest path in fuzzy graphs, as every path in a graph has strength one and is strongest. Mathew and Sunitha [113] generalized Menger's theorem in 2013 and introduced $t$-connected fuzzy graphs. The contents of this section are from [113].

### Strength Reducing Sets

In graph theory, a $u - v$ **separating set** $S$ of vertices is a collection of vertices in $G$ whose removal disconnects the graph $G$ and, $u$ and $v$ belonging to different components of $G - S$ [83]. Similarly, a $u - v$ separating set of edges is defined. Because the reduction in strength is more important and frequent in graph networks, strength reducing sets of vertices and edges are defined as follows.

**Definition 3.4.1** Let $u$ and $v$ be any two vertices in a fuzzy graph $G = (\sigma, \mu)$ such that the edge $uv$ is not strong. A set $S \subseteq \sigma^*$ of vertices is said to be a $u - v$ **strength reducing set (srs) of vertices** if $CONN_{G-S}(u, v) < CONN_G(u, v)$, where $G - S$ is the fuzzy subgraph of $G$ obtained by removing all vertices in $S$.

**Definition 3.4.2** A set of edges $E \subseteq \mu^*$ is said to be a $u - v$ **strength reducing set of edges** if $CONN_{G-E}(u, v) < CONN_G(u, v)$, where $G - E$ is the fuzzy subgraph of $G$ obtained by removing all edges in $E$.

**Definition 3.4.3** A $u - v$ strength reducing set of vertices(edges) with $n$ elements is said to be a **minimum** $u - v$ **strength reducing set** of vertices (edges) if there exist no $u - v$ strength reducing set of vertices(edges) with less than $n$ elements. A minimum $u - v$ strength reducing set of vertices is denoted by $S_G(u, v)$ and a minimum $u - v$ strength reducing set of edges is denoted by $E_G(u, v)$.

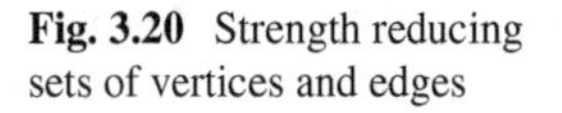

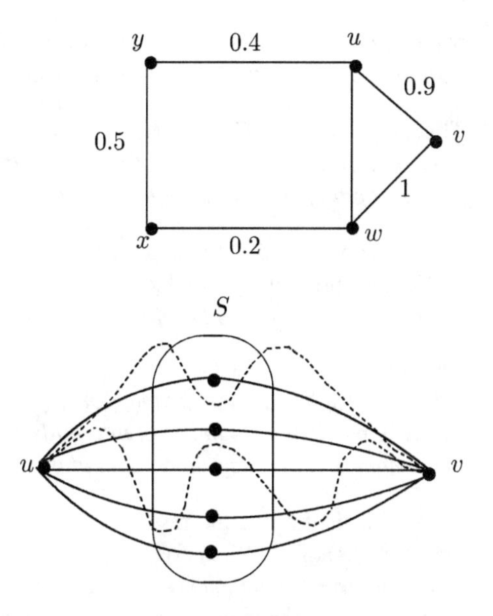


**Fig. 3.20** Strength reducing sets of vertices and edges

**Fig. 3.21** Srs $S$ of vertices. All strongest paths pass through $S$

*Example 3.4.4* (Figure 3.20) Let $G = (\sigma, \mu)$ be a fuzzy graph with $\sigma^* = \{u, v, w, x, y\}$, $\sigma(s) = 1$ for all $s \in \sigma^*$, $\mu(uv) = 0.8$, $\mu(vw) = 0.9$, $\mu(uy) = 0.3$, $\mu(yx) = 0.4$, $\mu(xw) = 0.1$, $\mu(uw) = 1$. Because edge $uw$ is strong, there are no $u - w$ strength reducing set of vertices in $G$. $E = \{uw\}$ is a $u - w$ strength reducing set of edges. $S = \{w\}$ is a $u - v$ strength reducing set of vertices and $E = \{uw\}$ is a $u - v$ strength reducing set of edges.

Note that any $u - v$ separating set of vertices or edges in the underlying graph $G^* = (\sigma^*, \mu^*)$ is a strength reducing set. The following are the characterizations for vertex and edge strength reducing sets.

**Theorem 3.4.5** *Let $G = (\sigma, \mu)$ be a connected fuzzy graph and $u, v$ be any two vertices in $G$ such that $uv$ is not strong. Then a set $S$ of vertices in $G$ is a $u - v$ strength reducing set if and only if every strongest path from $u$ to $v$ contains at least one vertex of $S$.*

*Proof* Suppose that $S$ is a $u - v$ strength reducing set of vertices in $G$ and let $P$ be a strongest $u - v$ path in $G$. If $P$ contains no vertex of $S$, the removal of $S$ keeps $P$ intact and hence $G - S$ contains $P$. Thus, $CONN_{G-S}(u, v) = CONN_G(u, v)$, which contradicts the fact that $S$ is a $u - v$ strength reducing set of vertices. Thus, $P$ must contains at least one member of $S$. It is obvious that this result is not true when edge $uv$ is strong. Any strong edge $uv$ is a strongest $u - v$ path containing no vertex from $S$.

Conversely, suppose that every strongest path from $u$ to $v$ contains at least one vertex of $S$, where $S \subseteq \sigma^*$ and $u$, $v$ not in $S$ (see Fig. 3.21). The dashed lines represent paths which are not strongest). Then the removal of $S$ destroys all strongest $u - v$

paths in $G$ and hence $CONN_{G-S}(u, v) < CONN_G(u, v)$. Hence, it follows that $S$ is a $u - v$ strength reducing set of vertices in $G$. ■

**Theorem 3.4.6** *Let $G = (\sigma, \mu)$ be a connected fuzzy graph and u,v any two vertices in G. Then a set E of edges in G is a u-v strength reducing set if and only if every strongest path from u to v contains at least one edge of E.*

The proof is similar to that of Theorem 3.4.5.

Next we present a generalization of one of the celebrated results in Graph theory due to Karl. Menger (1927).

**Theorem 3.4.7** [113] (Generalization of the vertex version of Menger's Theorem) *Let $G = (\sigma, \mu)$ be a fuzzy graph. For any two vertices $u, v \in \sigma^*$ such that uv is not strong, the maximum number of internally disjoint strongest u-v paths in G is equal to the number of vertices in a minimal u-v strength reducing set.*

*Proof* We shall prove the result by induction on the strong size $ss(G)$ (number of strong edges) of $G$. When $ss(G) = 0$, $G = (\sigma, \mu)$ is such that $\mu^* = \phi$ and the result is trivially true for any pair of vertices $u, v \in \sigma^*$.

Assume that the theorem is true for all fuzzy graphs $G = (\sigma, \mu)$ with strong size less than $m$, where $m \geq 1$. Let $G$ be a fuzzy graph of strong size $m$. Let $u, v \in \sigma^*$ such that $uv$ is not strong. If $u$ and $v$ are in different components of $G = (\sigma, \mu)$, the theorem is obviously true. So assume that $u$ and $v$ belongs to the same component of $G = (\sigma, \mu)$. Then either $uv$ is not in $\mu^*$ or $uv$ is a $\delta$-edge. In both cases, $u - v$ strength reducing sets of vertices exist in $G$. (If $uv$ is strong, then reduction of any number of vertices will not reduce the strength of connectivity between $u$ and $v$ and hence no strength reducing set of vertices exists.)

Now, suppose that $S_G(u, v)$ is a minimum strength reducing set of vertices in $G$ with $|S_G(u, v)| = k \geq 1$. By Theorem 3.4.5, each strongest $u - v$ path must contain at least one member from $S_G(u, v)$. Hence, any $u - v$ strength reducing set must contain at least as many vertices as the number of internally disjoint strongest $u - v$ paths. In other words, there exists at most $k$ internally disjoint strongest $u - v$ paths. We show that $G$ contains exactly $k$ internally disjoint strongest $u - v$ paths.

If $k = 1$, then $|S_G(u, v)| = 1$. Let $S_G(u, v) = \{w\}$. Then $CONN_{G-\{w\}}(u, v) < CONN_G(u, v)$. That is, $w$ is a fuzzy cutvertex of $G$. So every strongest $u - v$ path must pass through $w$. Hence, the number of internally disjoint $u - v$ paths is one and the result is true. So assume that $k \geq 2$.

**Case 1**: (Figure 3.22) $G$ has a minimum $u - v$ strength reducing set of vertices containing a vertex $x$ such that both $ux$ and $xv$ are $\alpha$-strong edges.

Let $S_G(u, v)$ be the minimum $u - v$ strength reducing set of vertices with the above mentioned property. Then $S_G(u, v) - \{x\}$ is a minimum $u - v$ strength reducing set in $G - \{x\}$ having $k - 1$ vertices. Because both $ux$ and $xv$ are $\alpha$-strong edges, they are clearly strong and hence $ss(G - \{x\}) < ss(G)$. By induction, it follows that $G - \{x\}$ contains $k - 1$ internally disjoint strongest $u - v$ paths. Because $ux$ and $xv$ are $\alpha$-strong, $P$ is a strongest $u - v$ path. Thus, we have $k$ internally disjoint strongest $u - v$ paths in $G$.

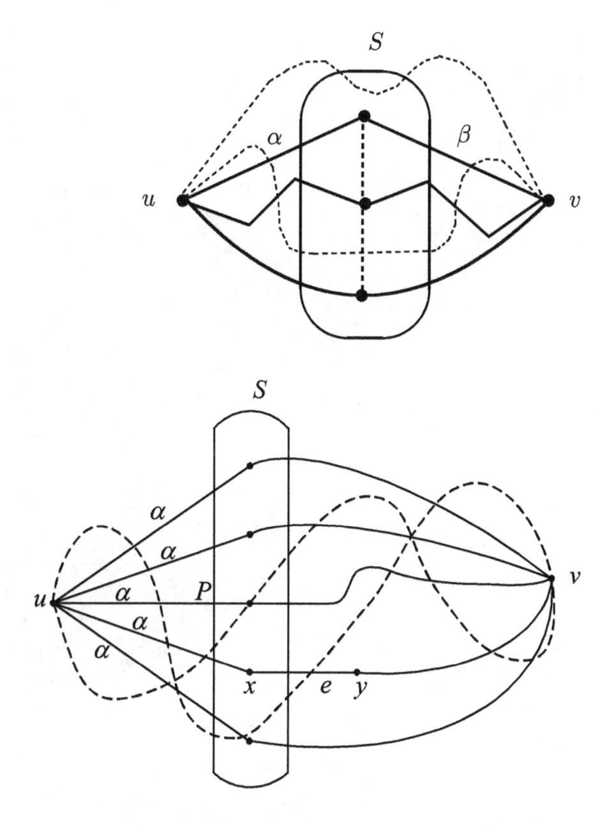


**Fig. 3.22** Case I of the proof

**Fig. 3.23** Case 2 of the proof

**Case 2**: (Figure 3.23) For every minimum $u - v$ strength reducing set $S_G(u, v)$ in $G$, either every vertex in $S_G(u, v)$ is an $\alpha$-strong neighbor of $u$ (i.e., if $w$ is a vertex in $S_G(u, v)$, then $uw$ is an edge which is the unique strongest $u - w$ path.) but not of $v$ or every vertex in $S_G(u, v)$ is an $\alpha$-strong neighbor of $v$ but not of $u$.

Suppose that every vertex in $S_G(u, v)$ is an $\alpha$-strong neighbor of $u$, but not of $v$. Consider a strongest $u - v$ path $P$ in $G$. Let $x$ be the first vertex of $P$ which is in $S_G(u, v)$. Then $ux$ is $\alpha$-strong and because $xv$ is not $\alpha$-strong, there exist at least one vertex say $y$ other than $u$ and $v$ such that $xy$ is $\beta$-strong. Denote the edge $xy$ by $e$.

**Claim**: Every $u - v$ strength reducing set in $G - \{e\}$ has exactly $k$ vertices.

On the contrary assume that there exist a minimum $u - v$ strength reducing set in $G - \{e\}$ with $k - 1$ vertices say $Z = \{z_1, z_2, \ldots, z_{k-1}\}$. Then $Z \cup \{x\}$ is a minimum $u - v$ strength reducing set in $G$. Note that every $z_i$, $i = 1, 2, \ldots, k - 1$ and $x$ are $\alpha$-strong neighbors of $u$. Because $Z \cup \{y\}$ also is a minimum $u - v$ strength reducing set in $G$, it follows that $y$ is an $\alpha$-strong neighbor of $u$ contradicting the fact that edge $xy$ is $\beta$-strong (The edges $ux$, $uy$ and $xy$ forms a triangle with edge $xy$ as the weakest edge. The unique weakest edge of a cycle is a $\delta$-edge). Thus, $k$ is the minimum number of vertices in a $u - v$ strength reducing set in $G - \{e\}$. Because $ss(G - \{e\}) < ss(G)$, it follows by induction that there are $k$ internally disjoint $u - v$ paths in $G - \{e\}$ and hence in $G$.

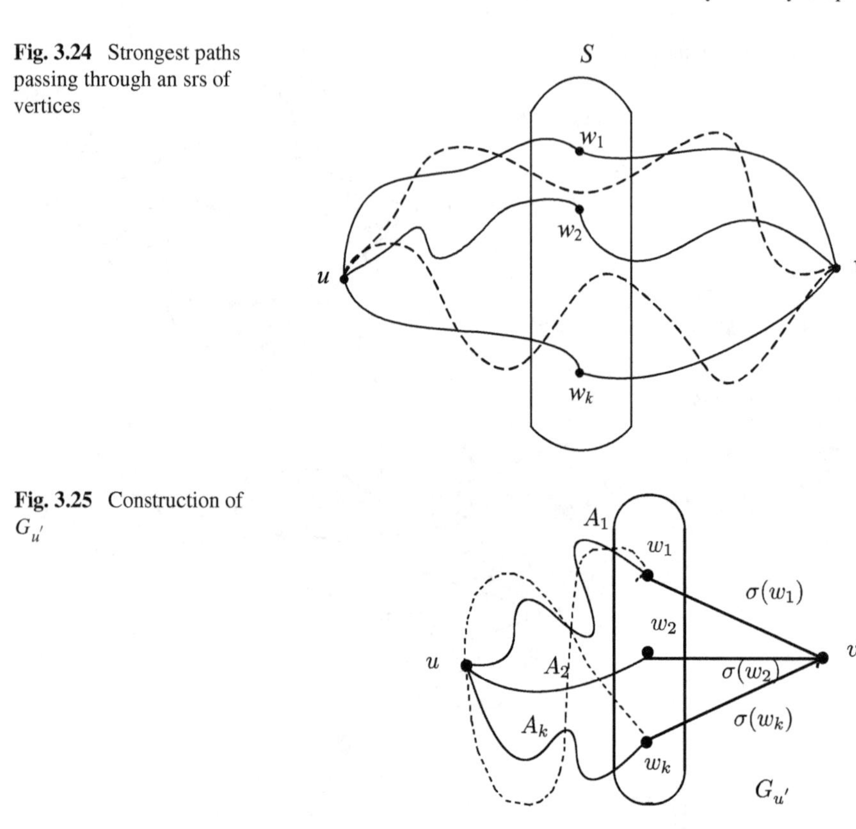


**Fig. 3.24** Strongest paths passing through an srs of vertices

**Fig. 3.25** Construction of $G_{u'}$

**Case 3**: (Figure 3.24) There exist a $u - v$ strength reducing set $W$ in $G$ such that no member of $W$ is an $\alpha$-strong neighbor of both $u$ and $v$ and $W$ contains at least one vertex which is not an $\alpha$-strong neighbor of $u$ and at least one vertex which is not an $\alpha$-strong neighbor of $v$.

Let $W$ be a minimum $u - v$ strength reducing set with $k$ elements having the above properties. Let $W = \{w_1, w_2, \ldots, w_k\}$. Consider all strongest paths from $u$ to $v$. Then because $W$ is minimum, $w_i, i = 1, 2, \ldots, k$ must belong to at least one such path. Let $G_u$ be the fuzzy subgraph of $G$ consisting of all $u - w_i$ sub paths of all strongest $u - v$ paths in which $w_i \in W$ is the only vertex of the path belonging to $W$. Note that if $CONN_G(u, v) = t$, then $\mu(xy) \geq t$ for all edge $xy$ in these paths.

Let $G_u'$ be the fuzzy graph constructed from $G_u$ by adding a new vertex $v'$ and joining $v'$ to each vertex $w_i$ for $i = 1, 2, \ldots, k$ (see Fig. 3.25). Let $\sigma(v') = 1$ and $\mu(w_i v') = \sigma(w_i)$ for all $i = 1, 2, \ldots, k$. The fuzzy graphs $G_v$ and $G_v'$ are defined similarly (see Fig. 3.26).

Because $W$ contains a vertex that is not an $\alpha$-strong neighbor of $u$ and a vertex that is not an $\alpha$-strong neighbor of $v$ (Note that all newly introduced edges are strong), we have $ss(G_u') < ss(G)$ and $ss(G_v') < ss(G)$.

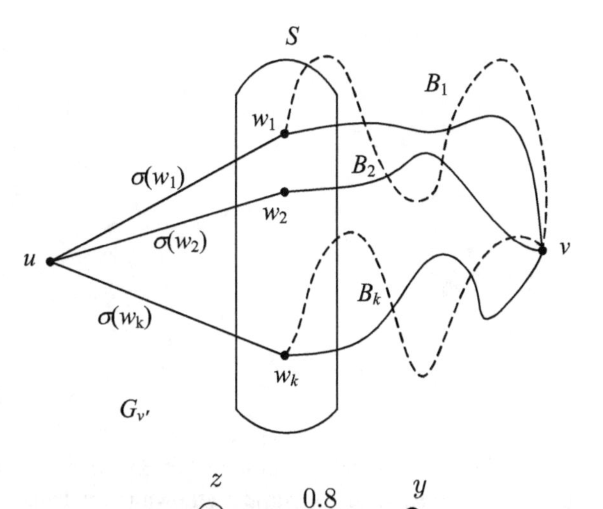

**Fig. 3.26** Construction of $G_{v'}$

**Fig. 3.27** Illustration to generalization of Menger's theorem

Clearly $S_{G'_u}(u, v') = k$ and $S_{G'_v}(u', v) = k$. So by induction $G'_u$ contains $k$ internally disjoint $u - v'$ paths say $A_i, i = 1, 2, \ldots, k$, where $A_i$ contains $w_i$. Also, $G'_v$ contains $k$ internally disjoint $u' - v$ paths say $B_i, i = 1, 2, \ldots, k$, where $B_i$ contains $w_i$. Let $A'_i$ be the $u - w_i$ sub paths of $A_i$ and $B'_i$ be the $w_i - v$ sub path of $B_i$ for $1 \leq i \leq k$. Now, $k$ internally disjoint strongest $u - v$ paths can be formed by joining $A_i$ and $B_i$ for $i = 1, 2, \ldots, k$ and the theorem is proved by induction. ∎

Next we state the edge version of Theorem 3.4.7. The proof is very similar.

**Theorem 3.4.8** [113] (Generalization of the edge version of Menger's Theorem) *Let $G = (\sigma, \mu)$be a connected fuzzy graph and let $u, v \in \sigma^*$. Then the maximum number of edge disjoint strongest u–v paths in G is equal to the number of edges in a minimum (with respect to the number of edges) u–v strength reducing set.*

**Illustration to Theorems** 3.4.7 **and** 3.4.8 Consider the following fuzzy graph $G$ on 7 vertices (Fig. 3.27).

Consider the vertices $u$ and $x$ in the fuzzy graph $G = (\sigma, \mu)$. The edge $ux$ is not strong because $CONN_G(u, x) > \mu(ux)$. Hence, the statement of the theorem can be checked for $u$ and $x$. Note that for a pair $\{x, y\}$ such that edge $xy$ is strong, there do not exist an $x - y$ srs of vertices. In $G$, there are 3 $u - x$ strongest paths with strength equal to 0.6 (which are shown in thick lines). According to Theorem 3.4.7,

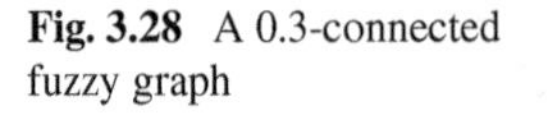
**Fig. 3.28** A 0.3-connected fuzzy graph

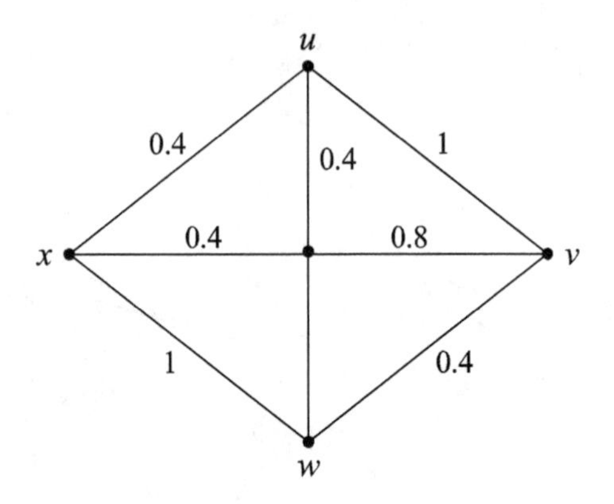


any minimum srs of vertices must contain 3 vertices. It can be easily verified. $\{w, t, z\}$ is a minimum $u - x$ srs of vertices (shown in circles).

Also, the number of edges in any minimum $u - x$ srs of edges is 3, which is same as the number of edge disjoint strongest $u - x$ paths.

**Definition 3.4.9** Let $G$ be a connected fuzzy graph and $t \in (0, \infty)$. $G$ is called $t$-**connected** if $\kappa(G) \geq t$ and $G$ is called $t$-**edge connected** if $\kappa'(G) \geq t$.

In other words a fuzzy graph $G$ is $t$-connected if there exist no fuzzy vertex cut with strong weight less than $t$ and is $t$-edge connected if there exist no fuzzy edge cut with strong weight less than $t$.

*Example 3.4.10* (Figure 3.28) Let $G = (\sigma, \mu)$ be a fuzzy graph with $\sigma^* = \{u, v, w, x, y\}$, $\mu(uv) = \mu(xw) = 1$, $\mu(yw) = 0.8$, $\mu(xu) = \mu(xy) = \mu(uy) = \mu(yv) = \mu(vw) = 0.4$. There are many fuzzy vertex cuts in $G$. Vertex $w$ is a fuzzy vertex cut with strong weight 0.4. $\{w, x\}$ and $\{w, v\}$ are 2-FNCs with strong weight 0.8 each. $\{u, v, w\}$ is a 3-FNC with strong weight 1.2. $\{u, v, w, y\}$ is a 4-FNC with strong weight 1.6. The minimum strong weight of all fuzzy vertex cuts in $G$ is 0.4 and hence $\kappa(G) = 0.4$. Thus, $G$ is $t$-connected for all $t$ such that $t \in (0, 0.4]$. Also, $xw$, $xy$ and $uv$ are fuzzy bridges and hence are fuzzy edge cuts with strong weights 1, 0.8, and 1 respectively. There are many fuzzy edge cuts in $G$ with strong weight more than 1. Thus, $\kappa'(G) = 0.8$ and $G$ is $t$-edge connected for all $t$ such that $t \in (0, 0.8]$.

The concept of 2-connected graphs have been generalized in fuzzy graph theory in a different form. A graph is 2-connected if and only if it has no cutvertices. That is, if and only if it is a block. According to Whitney's theorem, a graph is 2-connected if and only if any two vertices of $G$ are connected by at least two internally disjoint paths. This is true in fuzzy graphs also. As a consequence of Theorem 2.7.4, we have the following result.

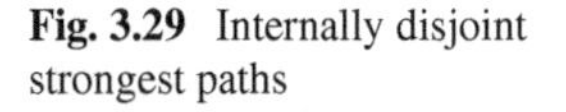

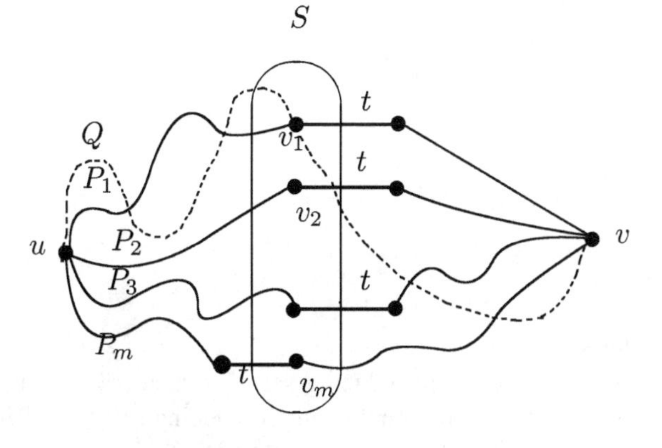


**Fig. 3.29** Internally disjoint strongest paths

**Theorem 3.4.11** *Let G be a connected fuzzy graph. G is a block if and only if sum of strengths of all internally disjoint strongest paths is at least* $2CONN_G(u, v)$ *for every pair of vertices* $u, v \in G$.

Now, we give characterizations of $t$-connected fuzzy graphs and $t$-edge connected fuzzy graphs as follows.

**Theorem 3.4.12** *Let G be a connected fuzzy graph. Then G is t-connected if and only if* $mCONN_G(u, v) \geq t$ *for every pair of vertices u and v in G, where m is the number of internally disjoint strongest* $u - v$ *paths in G.*

*Proof* First assume that $G$ is $t$-connected. Then $\kappa(G) \geq t$. We prove that for every pair of vertices $u, v \in \sigma^*$, the sum of strengths of all internally disjoint strongest paths is at least $t$. On the contrary assume that there exists a pair of vertices $u, v \in \sigma^*$ such that $mCONN_G(u, v) < t$, where $m$ is the number of internally disjoint strongest $u - v$ paths.

Let $S$ be a minimal $u - v$ strength reducing set of vertices in $G$ with minimum strong weight (See Fig. 3.29). By Menger's theorem, $|S| = m$. If $P_1, P_2, \ldots, P_m$ denote the internally disjoint $u - v$ paths, each $P_i$ must contain at least one of the vertices from $S$ (by Theorem 3.4.5) and no vertex can appear in more than one $P_i$, $i = 1, 2, \ldots, m$. Thus, each path $P_i$ contains exactly one vertex $v_i$(say) from $S$. Also, if there exists a strongest $u - v$ path $Q$ other than $P_i$, $i = 1, 2, \ldots, m$, $Q$ has to share a vertex of $S$ with $P_i$ for some $i$. Because $S$ is a $u - v$ strength reducing set, it is a fuzzy vertex cut. Also, because $S$ is a strength reducing set of minimum strong weight, one of the edges incident at $v_i$ in $P_i$ must have strength equal to $CONN_G(u, v)$. Therefore, strong weight of $S = |S|\, CONN_G(u, v) = mCONN_G(u, v) < t$. Thus, there exist a fuzzy vertex cut with strong weight less than $t$ and hence $\kappa(G) < t$, a contradiction to our assumption. Thus, the sum of strengths of all internally disjoint strongest $u - v$ paths is at least $t$.

Conversely, suppose that the sum of strengths of all internally disjoint strongest paths is at least $t$. To show that $\kappa(G) \geq t$. If possible suppose that $\kappa(G) < t$. Then there exist a fuzzy vertex cut $S$ such that strong weight of $S$ is less than $t$. Also,

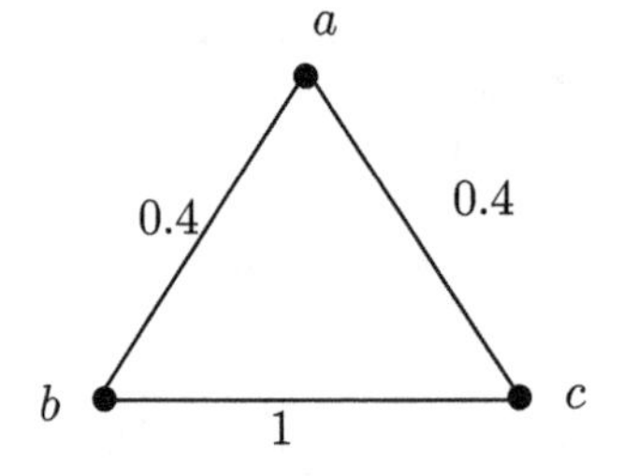


**Fig. 3.30** Illustration to Theorem 3.4.12

for some pair of vertices $u, v \in G$, $CONN_{G-S}(u, v) < CONN_G(u, v)$ and hence $S$ is a $u - v$ strength reducing set of vertices. By Menger's theorem, $m =$ number of vertices in a minimal strength reducing set $\leq |S|$. Therefore, $mCONN_G(u, v) \leq |S|\, CONN_G(u, v) \leq$ strong weight of $S < t$. That is, the sum of strengths of all internally disjoint strongest $u - v$ paths is less than $t$, which is a contradiction. ■

**Illustration to Theorem** 3.4.12. Consider the following fuzzy graph $G = (\sigma, \mu)$ on 3 vertices (Fig. 3.30).

$G$ has no fuzzy cutvertices. All pairs of vertices in $G$ are 2-FNCs with strong weight 0.8. Thus, $\kappa(G) = 0.8$. There are two internally disjoint strongest paths each, between $a$ and $b$ and between $a$ and $c$. Thus, by Theorem 3.4.11, $2CONN_G(a, b) = 2(0.4) = 0.8 = \kappa(G)$ and the result holds. Also, we have a unique strongest path between $b$ and $c$ with strength 1. Hence, $CONN_G(b, c) = 1 > t$, where $t \in (0, 0.8]$ and the result holds.

**Theorem 3.4.13** *Let $G$ be a connected fuzzy graph. Then $G$ is t-edge connected if and only if $mCONN_G(u, v) \geq t$ for every pair of vertices $u$ and $v$ in $G$, where $m$ is the number of edge disjoint strongest $u - v$ paths in $G$.*

*Proof* First assume that $G$ is $t$-edge connected. Then $\kappa'(G) \geq t$. To prove that for every pair of vertices $u, v \in \sigma^*$, the sum of strengths of all edge disjoint strongest paths is at least $t$. On the contrary assume that there exists a pair of vertices $u, v \in \sigma^*$ such that $mCONN_G(u, v) < t$, where $m$ is the number of edge disjoint strongest $u - v$ paths.

Let $E$ be a minimum $u - v$ strength reducing set of edges in $G$ with minimum strong weight (This is when any edge $xy$ in $E$ has strength $CONN_G(x, y)$). By Menger's theorem, $|E| = m$. If $P_1, P_2, \ldots, P_m$ denote the edge disjoint $u - v$ paths, each $P_i$ must contain at least one of the edges from $E$ and no edge can appear in more than one path. Thus, each path $P_i$ contains exactly one edge from $E$. Also, if there exists a strongest $u - v$ path $Q$ other than $P_i$, $i = 1, 2, \ldots, m$, it has to share an edge of $E$ with $P_i$ for some $i$. Because $E$ is a $u - v$ strength reducing set of edges, it is fuzzy edge cut of $G$ and hence strong weight of $E = mCONN_G(u, v) < t$ and thus there exist a FAC with strong weight less than $t$ and hence $\kappa'(G) < t$, a contradiction to our assumption (Fig. 3.31).

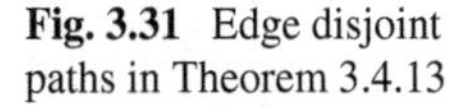

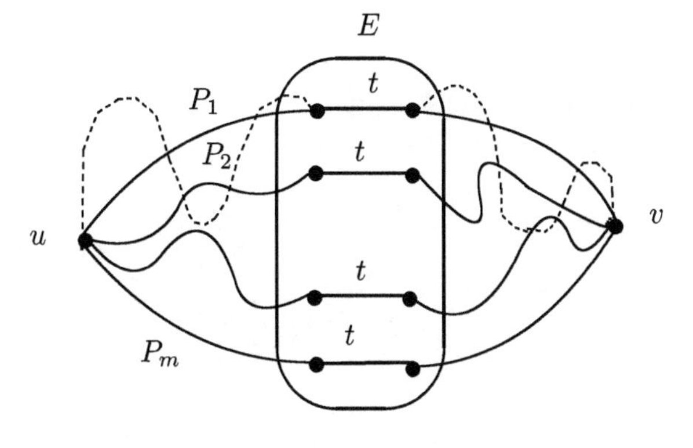


**Fig. 3.31** Edge disjoint paths in Theorem 3.4.13

Conversely, suppose that the sum of strengths of all edge disjoint strongest paths is at least $t$ for every pair of vertices $u$ and $v$ in $G$. To show that $\kappa'(G) \geq t$. If $\kappa'(G) < t$, then there exist a fuzzy edge cut say $E$ with strong weight less than $t$. Also, by definition of a FAC, $CONN_{G-E}(u, v) < CONN_G(u, v)$ for some pair of vertices $u$ and $v$ in $G$. Thus, $E$ is a $u - v$ strength reducing set of edges. By Menger's theorem, $m =$ number of edges in a minimal $u - v$ strength reducing set of edges. Hence, $m \leq |E|$. Therefore, $mCONN_G(u, v) \leq |E|\ CONN_G(u, v) \leq$ strong weight of $E < t$, which is a contradiction to our assumption. ■

# Chapter 4
# More on Blocks in Fuzzy Graphs

As defined in Chap. 2, a fuzzy graph without fuzzy cutvertices is called a block (nonseparable). Rosenfeld introduced this concept in 1975. In contrast to the classical concept of blocks in graphs, the study of blocks in fuzzy graphs is challenging due to the complexity of cutvertices. Note that cutvertices of a fuzzy graph are those vertices which reduce the strength of connectedness between some pair of vertices on its removal from the fuzzy graph rather than the total disconnection of the fuzzy graph. In this chapter, we concentrate on blocks of fuzzy graphs. This work is from [28–30].

## 4.1 Blocks of a Fuzzy Graph

Blocks of a graph are maximal connected induced subgraphs without cutvertices. Hence, a maximal connected fuzzy subgraph without fuzzy cutvertices induced by a subset of vertices is the natural definition for a block of a fuzzy graph. The properties of blocks of a fuzzy graph are very much different from that of blocks of a graph. Two blocks of a graph share at most one vertex, and it is a cutvertex of the graph. But, two blocks of a fuzzy graph may share one or more vertices and they need not be fuzzy cutvertices of the fuzzy graph. For the fuzzy graph in Fig. 4.1, blocks $B_1$ and $B_2$ (Fig. 4.2a, b) share two vertices, $v$ and $w$. $v$ is a fuzzy cutvertex of $G$ while $w$ is not. In the examples, $\sigma$ may be chosen in any way satisfying the definition of a fuzzy graph.

**Definition 4.1.1** A maximal connected fuzzy subgraph of $G = (\sigma, \mu)$, which is a block and induced by a subset of $V$ is called a **block** of $G$. If $G$ is a block, then $G$ itself is a block of $G$.

*Example 4.1.2* Consider the fuzzy graph $G = (\sigma, \mu)$ with $\sigma^* = \{u, v, w, x, y\}$, $\sigma(s) = 1$ for all $s \in \sigma^*$, $\mu(uv) = 0.6$, $\mu(vy) = 1$, $\mu(vx) = 0.7$, $\mu(xw) = \mu(uw) = 0.5$ and $\mu(vw) = 0.5$ (Fig. 4.1). $G$ is not a block because $v$ is a fuzzy cutvertex. The blocks of $G$ are $B_1$, $B_2$ and $B_3$ (Fig. 4.2a–c).



S. Mathew et al., *Fuzzy Graph Theory*, Studies in Fuzziness and Soft Computing 363, https://doi.org/10.1007/978-3-319-71407-3_4

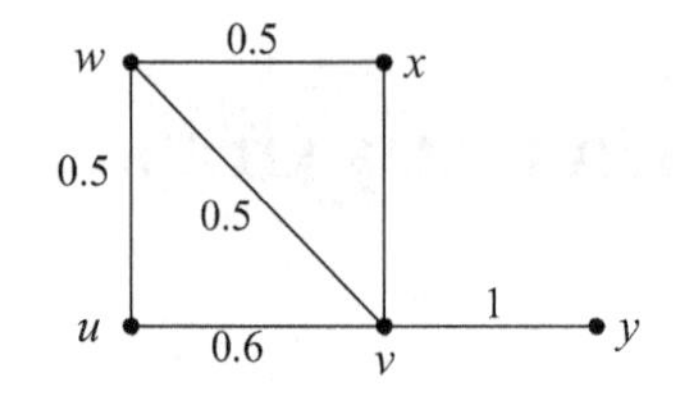


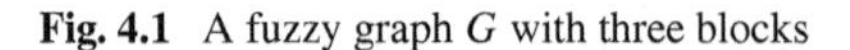
**Fig. 4.1** A fuzzy graph $G$ with three blocks

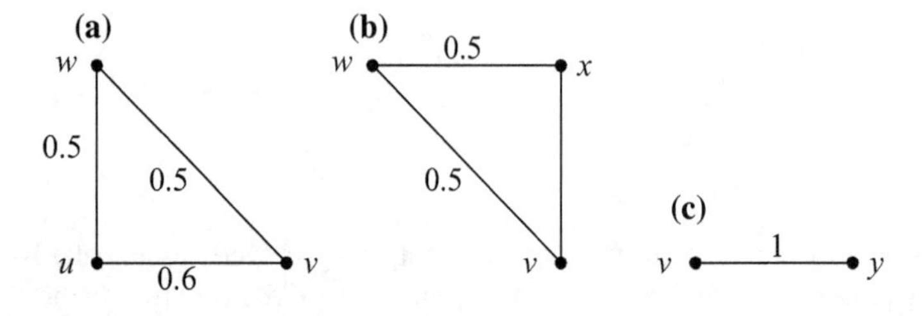


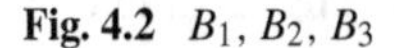
**Fig. 4.2** $B_1, B_2, B_3$

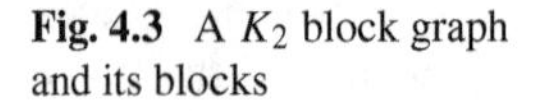
**Fig. 4.3** A $K_2$ block graph and its blocks

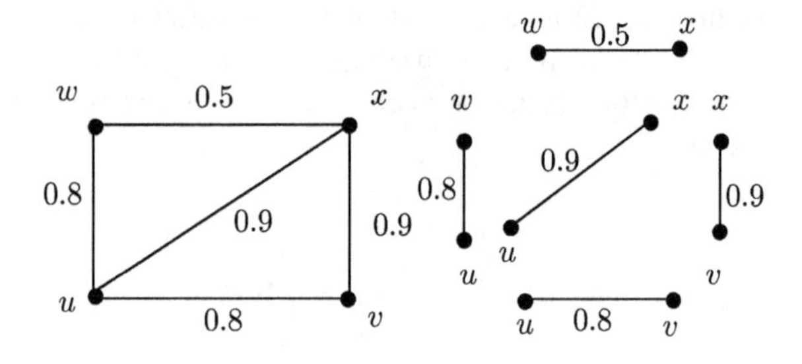


The support of a block $G = (\sigma, \mu)$ is always a block because a cutvertex in $G^*$ is always a fuzzy cutvertex of $G$, independent of $\sigma$ and $\mu$. In certain fuzzy graphs, support of every block will be $K_2$. Such fuzzy graphs have some common properties and we discuss them separately.

**Definition 4.1.3** A fuzzy graph $G = (\sigma, \mu)$ is a $K_2$ **block graph** if the support of every block of $G$ is $K_2$.

A fuzzy graph whose support is a tree is always a $K_2$ block graph.

*Example 4.1.4* Let $G = (\sigma, \mu)$ be the fuzzy graph given in Fig. 4.3, with $\sigma^* = \{u, v, w, x\}, \sigma(s) = 1$ for all $s \in \sigma^*$, $\mu(wx) = 0.5, \mu(uv) = \mu(uw) = 0.8, \mu(ux) = \mu(vx) = 0.9$. $G$ is a $K_2$ block graph.

Stars are a special type of trees in graphs. A star has exactly one internal vertex. We denote a star by $S_n$, where $n$ is the number of leaves. Analogous to star graphs, a fuzzy star is defined as follows.

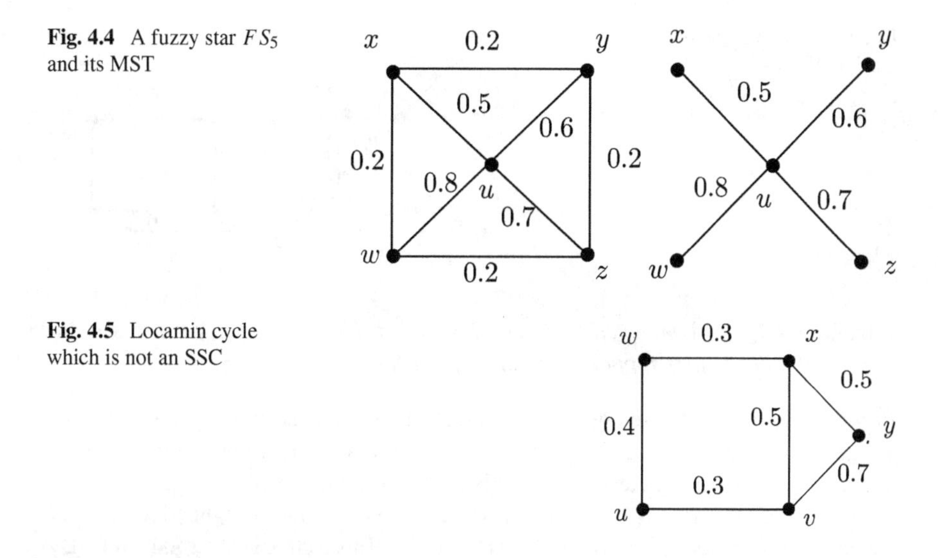


**Fig. 4.4** A fuzzy star $FS_5$ and its MST

**Fig. 4.5** Locamin cycle which is not an SSC

**Definition 4.1.5** A **fuzzy star** is a fuzzy tree whose unique maximum spanning tree is a star. A fuzzy star $G = (\sigma, \mu)$ with $|\sigma^*| = n + 1$ is denoted by $FS_n$.

*Example 4.1.6* A fuzzy star $FS_5$ and its maximum spanning tree are given in Fig. 4.4.

The following are some results on strongest strong cycles, blocks and fuzzy stars.

**Lemma 4.1.7** *A strongest strong cycle $C$ in $G = (\sigma, \mu)$ is locamin.*

*Proof* Suppose $C$ is not locamin. Then there exists a vertex $u$ in $C$ such that weights of both the edges (say, $w_1u$ and $w_2u$) incident with it are greater than the weight of a weakest edge in $C$. Thus, the strength of $w_1 - u - w_2$ path is greater than the other $w_1 - w_2$ path in $C$, which is not possible. ■

But, the converse of Lemma 4.1.7 is not true. A locamin cycle need not be a strongest strong cycle as seen from Fig. 4.5.

In Fig. 4.5, the cycle $C : w, u, v, x, w$ is locamin but not strongest strong.

**Lemma 4.1.8** *Let $G = (\sigma, \mu)$ be a fuzzy graph. If $C$ is a strongest strong cycle in $G$, then there exists a block of $G$ containing $C$.*

*Proof* Let $C$ be a strongest strong cycle in $G = (\sigma, \mu)$. Let $H = \langle P \rangle$, where $P$ is the set of vertices constituting $C$. Every pair of vertices in $H$ except those joined by $\alpha$-strong edges are joined by two internally disjoint strongest paths. Hence, $H$ is a block according to Theorem 2.7.4. Therefore, either $H$ itself is a block of $G$ or it is properly contained in a block of $G$. ■

In Fig. 4.1, two cycles $C_1 : w, u, v, w$ and $C_2 : w, x, v, w$ are both strongest strong cycles in $G$. Each of them constitutes a block of $G$.

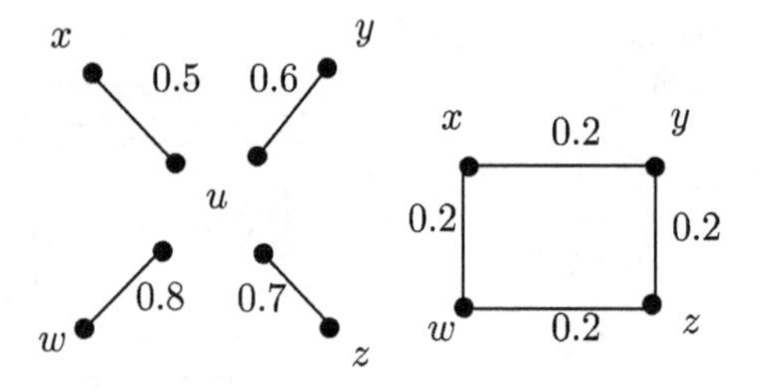


**Fig. 4.6** Blocks of the fuzzy graph in Fig. 4.4

**Corollary 4.1.9** *A connected fuzzy graph $G = (\sigma, \mu)$ without fuzzy bridges has a nontrivial block whose support is different from $K_2$.*

*Proof* $G$ is not a fuzzy tree because all the edges in the maximum spanning tree of a fuzzy tree are fuzzy bridges by Proposition 2.3.4. The maximum weight edge in $G$ (say, $uv$) is $\beta$-strong. Hence, there exists a strongest $u - v$ path $P$ different from $uv$ in $G$ whose strength is $\mu(uv)$. Because $uv$ is the maximum weight edge in $G$, all the edges in $P$ have weight $\mu(uv)$. Thus, $uv \cup P$ forms a strongest strong cycle. By Lemma 4.1.8, $G$ has a nontrivial block whose support is different from $K_2$. ■

**Proposition 4.1.10** *A fuzzy star $FS_{n-1}$ has at least $n - 1$ blocks.*

*Proof* $FS_{n-1}$ is a fuzzy star with $n$ vertices. Its maximum spanning tree is a star, say $F$. Let $u$ be the vertex of degree $n - 1$ in $F^*$ and $u_i, i = 1, 2, \ldots, n - 1$ be the remaining $n-1$ vertices. Let $B$ be a block containing $uu_i$ for some $i = 1, 2, \ldots, n-1$. Suppose $B$ contains a vertex $u_j$, $j \neq i$. Because $u$ is incident with two $\alpha$-strong edges $uu_i$ and $uu_j$, it is a fuzzy cutvertex of $B$ according to Theorem 2.2.11. This is not possible because $B$ is a block. Therefore, $uu_i$ is a block of $FS_{n-1}$. Hence, $FS_{n-1}$ has at least $n - 1$ blocks. ■

The blocks of the fuzzy graph in Fig. 4.4 are given in Fig. 4.6. $G$ is a fuzzy star with five vertices and has five blocks.

**Proposition 4.1.11** *Let $G = (\sigma, \mu)$ be a fuzzy graph. If all the edges incident with a vertex $u$ in $G$ are $\alpha$-strong, then each of these edges forms a block of $G$.*

*Proof* Let $deg_{G^*}(u) = k$. Suppose $u_i$, $i = 1, 2, \ldots, k$ are the neighbors of $u$ and $B$ is a block to which $uu_i$ belongs. $deg_{B^*}(u) = 1$, otherwise $u$ becomes a common vertex of at least two $\alpha$-strong edges and by Theorems 2.2.11 and 3.2.9, becomes a fuzzy cutvertex of $B$, which is not possible. We prove $deg_{B^*}(u_i) = 1$. Assume $u_i$ is adjacent to a vertex $v$ different from $u$ in $B^*$. Then $u - u_i - v$ is the only $u - v$ path in $B$ making $u_i$, a fuzzy cutvertex of $B$ which is not possible. That is, $B$ does not contain any vertex other than $u$ and $u_i$. Hence, $uu_i$, $i = 1, 2, \ldots, k$ form $k$ blocks of $G$. ■

In Fig. 4.4, all the edges $xu$, $yu$, $zu$ and $wu$ incident with $u$ are $\alpha$-strong. Each of them forms a block as seen in Fig. 4.6.

According to Lemma 4.1.8, associated with a strongest strong cycle, there is a nontrivial block whose support is different from $K_2$. The following lemma proves the existence of a strongest strong cycle in every nontrivial block with at least three vertices. This lemma is then used to characterize a $K_2$ block graph.

**Lemma 4.1.12** ([29]) *A nontrivial block* $G = (\sigma, \mu)$ *with* $|\sigma^*| \geq 3$ *contains a strongest strong cycle.*

*Proof* Let $G = (\sigma, \mu)$ be a nontrivial block with $|\sigma^*| \geq 3$. Let $a, b$ be two vertices in $G$ not joined by an $\alpha$-strong edge such that $CONN_G(a, b) = \vee\{CONN_G(x, y) \mid x, y \in \sigma^*$ not joined by an $\alpha$-strong edge$\}$. In other words, if $x$ and $y$ are two vertices in $\sigma^*$ not joined by an $\alpha$-strong edge and $P$ is an $x - y$ path, then

$$s(P) \leq CONN_G(x, y) \leq CONN_G(a, b). \tag{4.1}$$

Because $G$ is a block, there are two internally-disjoint strongest $a - b$ paths say, $P_1$ and $P_2$. Let $C = P_1 \cup P_2$. A weakest edge in $C$ has weight, $CONN_G(a, b)$.

Claim: $C$ is a strongest strong cycle.

Let $x$ and $y$ be two vertices in $C$ not joined by an $\alpha$-strong edge. Because strength of an $x - y$ path cannot exceed $CONN_G(a, b)$, an $x - y$ path whose strength is $CONN_G(a, b)$ is a strongest $x - y$ path. We prove the strength of the two $x - y$ paths in $C$ is $CONN_G(a, b)$.

Case 1: $x$ and $y$ are adjacent in $C$.

This implies, $\mu(xy) = CONN_G(a, b)$. The path $C - xy$ also has strength, $CONN_G(a, b)$. Thus, it follows that $xy$ is $\beta$-strong.

Case 2: $x$ and $y$ are not adjacent in $C$.

Let $P$ and $P'$ be the two $x - y$ paths in $C$. Suppose the weights of all the edges in one path, say $P$ is greater than $CONN_G(a, b)$. Then the strength of $P$ is greater than $CONN_G(a, b)$ which is not possible according to Eq. (4.1). Therefore, both $P_1$ and $P_2$ contain an edge whose weight is $CONN_G(a, b)$. It follows that both $P_1$ and $P_2$ are strongest $x - y$ paths.

The weakest edges in $C$ are $\beta$-strong as proved in Case 1 and the remaining edges are $\alpha$-strong. Therefore, $C$ is a strongest strong cycle in $G$. ■

In Fig. 4.7, $G$ is a block. $C : w, u, v, w$ is a strongest strong cycle in $G$.

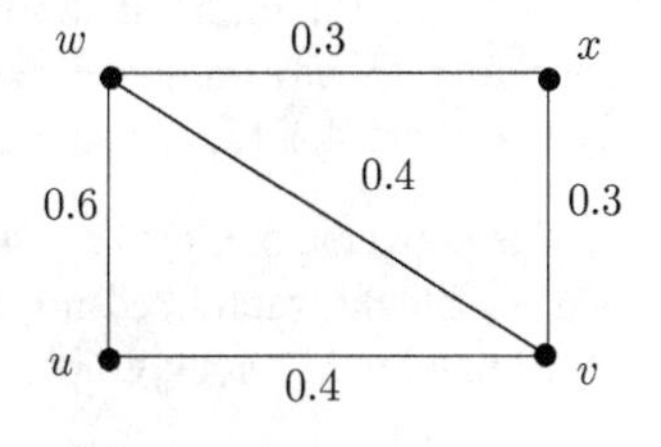


**Fig. 4.7** A block and its strongest strong cycle

**Theorem 4.1.13** *Let $G = (\sigma, \mu)$ be a fuzzy graph on $n$ vertices without isolated vertices. Then $G$ is a $K_2$ block graph if and only if no fuzzy subgraph of $G$ induced by a subset of $V$ has a strongest strong cycle.*

*Proof* Suppose a fuzzy subgraph, say $H$ of $G$ induced by a subset of $V$, has a strongest strong cycle $C : u_1, u_2, \ldots, u_k, u_1, k \leq n$. Then $H' = \langle\{u_1, u_2, \ldots, u_k\}\rangle$ is a block. Because $H'$ is a fuzzy subgraph of $G$ induced by $\{u_1, u_2, \ldots, u_k\}$, either $H'$ itself is a block of $G$ or there is a block of $G$ properly containing it. It follows that $G$ has a nontrivial block whose support is different from $K_2$.

Assume no fuzzy subgraph of $G$ induced by a subset of $V$ has a strongest strong cycle. By Lemma 4.1.12, the support of every block of $G$ is $K_2$. This implies that $G$ is a $K_2$ block graph. ■

In Example 4.1.4 (Fig. 4.3), $G$ is a $K_2$ block graph. Its fuzzy subgraphs induced by three vertices whose supports are not trees are namely $H_1 = < \{w, u, x\} >$ and $H_2 = < \{u, x, v\} >$. $H_1$ and $H_2$ are not strongest strong because $wx$ is a $\delta$-edge in $H_1$ and $uv$ is a $\delta$-edge in $H_2$.

**Corollary 4.1.14** *If $G$ is a fuzzy graph without isolated vertices and has no locamin cycles, then $G$ is a $K_2$ block graph.*

**Corollary 4.1.15** *Let $G = (\sigma, \mu)$ be an edge-disjoint fuzzy graph without isolated vertices. $G$ is a $K_2$ block graph if and only if no cycle is strongest strong in $G$.*

*Proof* Cycles in $G$ are fuzzy subgraphs induced by the vertices constituting the cycle because $G$ is edge-disjoint. Suppose $G = (\sigma, \mu)$ is a $K_2$ block graph. From Theorem 4.1.13, no cycle in $G$ is strongest strong as a fuzzy subgraph induced by the vertices constituting the cycle and the same holds in $G$. Out of the fuzzy subgraphs induced by subsets of $V$ different from $K_2$, the supports of cycles alone do not contain cutvertices. If the cycles are not strongest strong in $G$, then they are not strongest strong in any fuzzy subgraph induced by a subset of $V$. It follows from Theorem 4.1.13 that $G$ is a $K_2$ block graph. Therefore, the number of nontrivial blocks whose supports are different from $K_2$ for an edge-disjoint fuzzy graph is equal to the number of strongest strong cycles. ■

**Corollary 4.1.16** *Let $G = (\sigma, \mu)$ be an edge-disjoint connected fuzzy graph. If $G$ is a fuzzy tree, then $G$ is a $K_2$ block graph.*

*Proof* For a tree $G$, the corollary is true. Suppose $G = (\sigma, \mu)$ is a fuzzy tree which is not a tree. Let $C$ be a cycle in $G$. For a pair of vertices $x, y$ in $C$ the two $x - y$ paths in $C$ are the only $x - y$ paths in $G$. If $C$ has more than one weakest edge, then $G$ has at least two maximum spanning trees which is a contradiction to Theorem 2.3.19. Therefore, $C$ has a unique weakest edge and is not strongest strong from Lemma 4.1.7. By Corollary 4.1.15, $G$ is a $K_2$ block graph. ■

The converse of Corollary 4.1.15 is not true. An edge-disjoint fuzzy graph which is a $K_2$ block graph need not be a fuzzy tree. The fuzzy graph $G$ in Fig. 4.8 is an edge-disjoint $K_2$ block graph, but it is not a fuzzy tree.

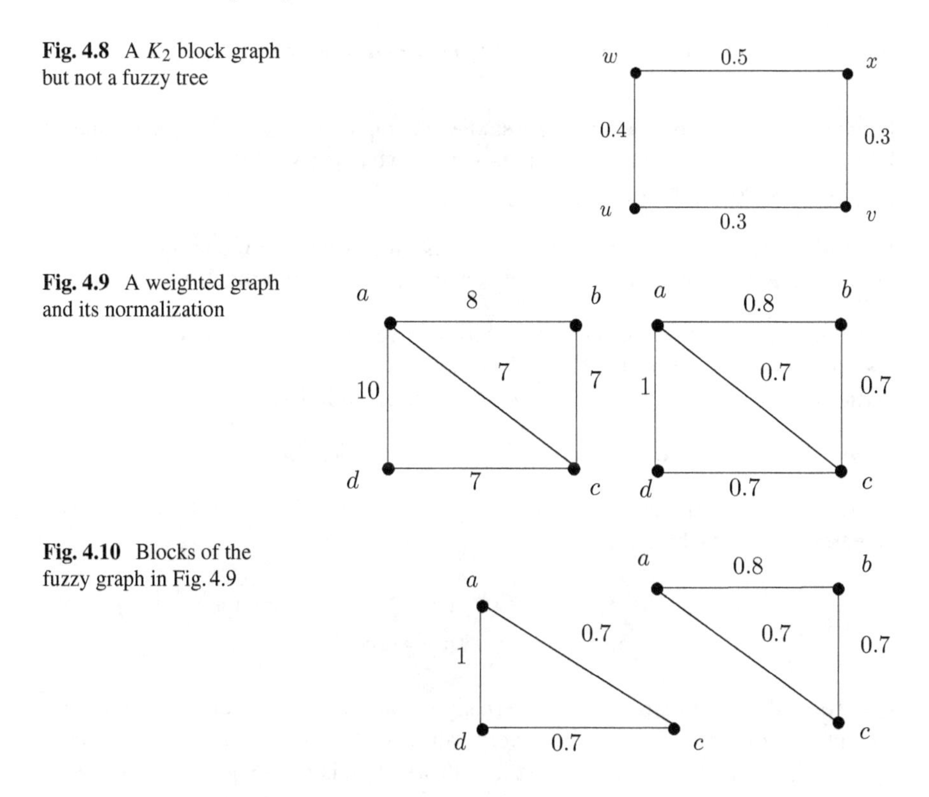


**Fig. 4.8** A $K_2$ block graph but not a fuzzy tree

**Fig. 4.9** A weighted graph and its normalization

**Fig. 4.10** Blocks of the fuzzy graph in Fig. 4.9

**An Application**: Consider an undirected network of roads connecting certain cities. It can be represented by a weighted graph, where vertices, edges and weights of edges denote the cities, roads and capacities respectively. By the capacity of a road, we refer to the maximum number of vehicles that can pass through it per hour. On normalization, we obtain a fuzzy graph. For simplicity, consider a network of four cities given in Fig. 4.9.

Assume that the flow through an edge is equal to its capacity. Consider traffic flow from $b$ to $d$ in $G$. The maximum flow is 0.8 and it occurs through the path $bad$. During certain occasions like festivals, traffic flow may be restricted through $a$. This reduces the traffic flow from $b$ to $d$ to a great extent. A similar problem cannot occur in case of the flow between $a$ and $c$ because not all of the maximum flow paths pass through a common city. So, it is always desirable to have two independent maximum flow paths between every pair of cities. In graph-theoretic terms, $G$ should be a block. The model of a large road network need not be a block. There may be instances when no flow or two independent paths of same flow (even if, low) is better than a higher flow through two intersecting paths. Blocks of a fuzzy graph are small networks-satisfying the above mentioned property-contained in a large network modelled using a fuzzy graph. For example, $G$ is not a block and the blocks of $G$ are given in Fig. 4.10. In the blocks of $G$, there is no flow between $b$ and $d$ in comparison, with the single maximum flow path in $G$.

## 4.2 Critical Blocks and Block Graph of a Fuzzy Graph

Critical graphs have been extensively studied in graph literature in different contexts like coloring, connectivity, etc. In this section, we discuss critical blocks in fuzzy graphs. This work is based on [28]

**Definition 4.2.1** A fuzzy graph $G = (\sigma, \mu)$ is called a **critical block** if $G$ is a block and $G - v$ is not a block for all $v \in \sigma^*$. Otherwise, $G$ is noncritical.

In this section, a block refers to a block in fuzzy graph. In the examples, we assume $\sigma(x) = 1$, for all $x \in V$ for convenience. Clearly, blocks on two and three vertices are not critical. Some critical blocks are given in Example 4.2.2.

*Example 4.2.2* Both fuzzy graphs in Fig. 4.11 are critical blocks.

According to Lemma 4.1.12, every block $G = (\sigma, \mu)$ on at least three vertices contains a strongest strong cycle, $C$. Let

$$c = \vee\{CONN_G(x, y) \mid x, y \in \sigma^*,\ x \text{ and } y \text{ are not joined by an } \alpha\text{-strong edge.}\}$$

$C$ is formed by the union of two internally-disjoint paths of strength $c$. Also, the strength of connectedness between every pair of vertices lying on $C$ and not joined by an $\alpha$-strong edge is $c$. In $C$, an edge with weight $c$ is $\beta$-strong and an edge with weight greater than $c$ is $\alpha$-strong.

**Proposition 4.2.3** *Let $G = (\sigma, \mu)$ be a block on at least four vertices. Let $C$ be a strongest strong cycle in $G$. If $C$ is spanning and the only strongest strong cycle in $G$, then $G$ is critical.*

*Proof* Let $x, y$ be two vertices in $G$, not joined by an $\alpha$-strong edge. Let $Q_1$ and $Q_2$ be the two $x - y$ strongest paths in $C$. Because $C$ is unique, there is no other pair of internally-disjoint $x - y$ strongest paths. Hence, $x$ and $y$ are joined by exactly two internally disjoint strongest paths in $G$.

We next prove that $G - v$ is not a block, $v \in \sigma^*$.

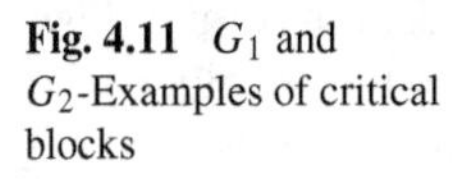
**Fig. 4.11** $G_1$ and $G_2$-Examples of critical blocks

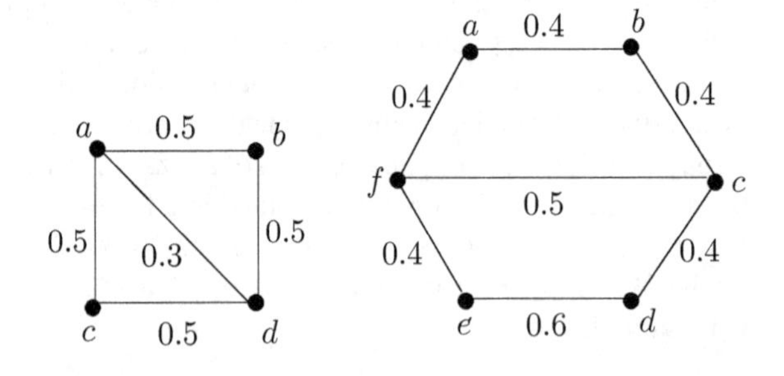

Suppose $v$ is different from both $x$ and $y$. Removal of $v$ results in exactly one internally disjoint strongest path between $x$ and $y$. According to Theorem 2.7.4, $G - v$ is not a block. Suppose, $v = x$. Let $v_1$ and $v_2$ be the neighbors of $x$ in $C$.

**Claim**: $v_1$ and $v_2$ are not joined by an $\alpha$-strong edge.

Assume on the contrary that $v_1$ and $v_2$ are joined by an $\alpha$-strong edge.

**Case** 1: Both $v_1x$ and $xv_2$ are $\beta$ -strong

The triangle $x, v_1, v_2, x$ is a strongest strong cycle, which is a contradiction to the uniqueness of $C$.

**Case** 2: Exactly one of $v_1x$ and $xv_2$ is $\alpha$-strong

Suppose $xv_1$ is $\alpha$-strong. If so, $v_1$ is incident with two $\alpha$-strong edges which is not possible according to Theorems 2.2.11 and 3.2.9. Thus, it follows that $v_1$ and $v_2$ are not joined by an $\alpha$-strong edge. So, removal of $x$ results in exactly one internally disjoint strongest $v_1 - v_2$ path. According to Theorem 2.7.4, $G - x$ is not a block. The same argument holds for $v = y$. Therefore, $G$ is a critical block. ■

In graphs, all cycles are critical blocks. Proposition 4.2.3 is a generalization of this result. $G_1$ in Fig. 4.11 has a unique and spanning strongest strong cycle. It is a critical block. But, the converse is not true as seen from $G_2$. $G_2$ is a critical block, but all the three cycles are strongest strong. The uniqueness as well as the spanning property of $C$ are required for a block to be critical. Consider the fuzzy graphs $G_1$ and $G_2$ in Fig. 4.12. $G_1$ has a spanning strongest strong cycle satisfying conditions in Lemma 4.1.12. It is spanning but not unique while $G_2$ has a unique strongest strong cycle satisfying the conditions in Lemma 4.1.12 but it is not spanning. Neither $G_1$ nor $G_2$ is critical.

**Lemma 4.2.4** *Let $G = (\sigma, \mu)$ be a connected fuzzy graph. If $uv$ is an $\alpha$-strong edge in $G$ which is not a fuzzy bond, then $CONN_G(u, x) = CONN_G(v, x)$ for all $x \neq u, v$.*

*Proof* Assume, there exists a vertex $y \neq u, v$ such that $CONN_G(u, y) \neq CONN_G(v, y)$. Without loss of generality, let $CONN_G(u, y) > CONN_G(v, y)$. Because $uv$ is not a fuzzy bond, removal of $uv$ does not reduce the strength of connectedness between any pair of vertices other than $u$ and $v$. That is, for every pair of vertices except $u, v$, there is a strongest path not containing $uv$. Let $P_1$ and $P_2$ be strongest $u-y$ and $v-y$ paths, respectively which do not contain $uv$. By assumption, $s(P_1) > s(P_2)$. So, $P_1$ cannot contain $v$. The following argument assumes that $P_2$ does not contain $u$.

**Fig. 4.12** Non critical fuzzy blocks

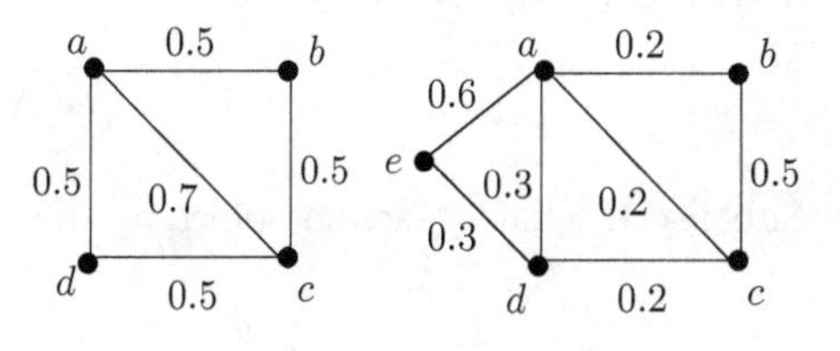

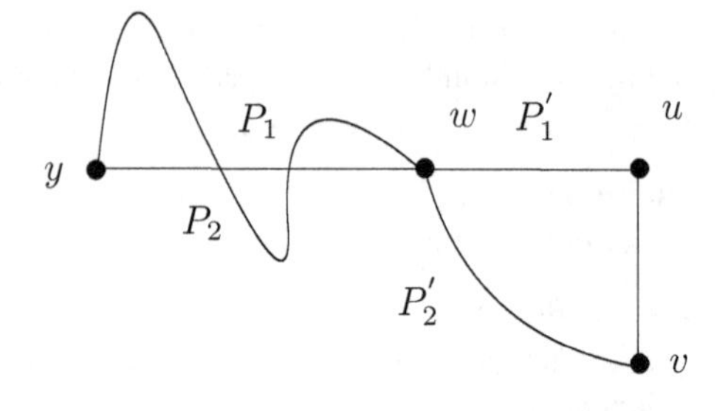


**Fig. 4.13** Graph in Lemma 4.2.4

Let $w$ be the first common vertex of $P_1$ and $P_2$, as one moves along $P_1$ starting from $u$. Therefore, $u - w$ subpath of $P_1$ ($P_1'$), $w - v$ subpath of $P_2$ ($P_2'$) and $uv$ constitute a cycle (Fig. 4.13). Because $uv$ is not a weakest edge in this cycle,

$$\mu(uv) > s(P_1') \wedge s(P_2') \geq s(P_1) \wedge s(P_2) = s(P_2).$$

If $P_2$ contains $u$, then $w$ coincides with $u$ and $P_1'$ vanishes. $uv$ together with $P_2'$ form a cycle. Hence, $\mu(uv) > s(P_2')$. Therefore, the strength of the $v - y$ path $P_1 \cup uv$ is greater than $s(P_2)$. This is not possible because $P_2$ is a strongest $v - y$ path. Therefore, $CONN_G(u, x) = CONN_G(v, x)$ for all $x \neq u, v$. ■

**Proposition 4.2.5** *Let $G = (\sigma, \mu)$ be a block with at least four vertices. If $C = (\tau, \upsilon)$ is a strongest strong cycle in $G$ as mentioned in Lemma 4.1.12 and $x \in \sigma^*$, then $CONN_G(x, y_1) = CONN_G(x, y_2)$, where $y_1, y_2 \in \tau^*$ not joined to $x$ by an $\alpha$-strong edge and are different from $x$.*

*Proof* Let $x \in \sigma^*$ and $y_1, y_2 \in \tau^*$ not joined to $x$ by an $\alpha$-strong edge and different from $x$. For $x \in \tau^*$, clearly $CONN_G(x, y_1) = CONN_G(x, y_2)$. Suppose $x \notin \tau^*$.

**Case 1**: $x$ is joined to $C$ by an $\alpha$-strong edge, $xy$ (say).

Because a block does not contain fuzzy bonds, $xy$ is not a fuzzy bond. According to Lemma 4.2.4, $CONN_G(x, y_1) = CONN_G(y, y_1)$ and $CONN_G(x, y_2) = CONN_G(y, y_2)$. Because $xy$ is $\alpha$-strong, neither $y, y_1$ nor $y, y_2$ is joined by an $\alpha$-strong edge. Hence, $CONN_G(y, y_1) = CONN_G(y, y_2)$. So, $CONN_G(x, y_1) = CONN_G(x, y_2)$.

**Case 2**: $x$ is not joined to $C$ by an $\alpha$-strong edge.

Assume for some $y_1, y_2$ not joined to $x$ by $\alpha$-strong edge, $CONN_G(x, y_1) < CONN_G(x, y_2)$. Let $P$ be an $x - y_2$ strongest path and $Q$ be a $y_2 - y_1$ strongest path in $C$. $P \cup Q$ gives an $x - y_1$ walk.

$$s(P) > CONN_G(x, y_1). \tag{4.2}$$

**Subcase** 1: $y_1$ and $y_2$ are not joined by an $\alpha$-strong edge.

$$s(Q) = c \geq CONN_G(x, y_2) > CONN_G(x, y_1). \tag{4.3}$$

**Subcase** 2: $y_1$ and $y_2$ are joined by an $\alpha$-strong edge. Then $Q = y_1y_2$.

$$s(Q) > c \geq CONN_G(x, y_2) > CONN_G(x, y_1). \tag{4.4}$$

From (4.2)–(4.4), it can be seen that the strength of every edge in $P \cup Q$ is strictly greater than $CONN_G(x, y_1)$, which is not possible.

Therefore, $CONN_G(x, y_1) = CONN(x, y_2)$, for all $y_1, y_2 \in \tau^*$ not joined to $x$ by an $\alpha$-strong edge for all $x \neq y_1, y_2 \in \sigma^*$. ■

**Lemma 4.2.6** *In a critical block,* $G = (\sigma, \mu)$ *the following conditions hold.*

$(i)$ *Every vertex* $z \in \sigma^*$ *is an internal vertex of an* $x - y$ *strongest path.*

$(ii)$ *There exists a pair of vertices, not joined by an* $\alpha$*-strong edge which is joined by at most two internally-disjoint strongest paths.*

*Proof* $(i)$ Suppose there exists a vertex $z \in \sigma^*$ which is not an internal vertex of any $x - y$ strongest path. So, the removal of $z$ does not remove any $x - y$ strongest path. In $G - z$, every pair of vertices except those joined by $\alpha$-strong edges are still joined by two internally- disjoint strongest paths and hence, $G - z$ is also a block, which is not possible. Therefore, every vertex $z \in \sigma^*$ is an internal vertex of an $x - y$ strongest path.
$(ii)$ Suppose every pair of vertices in $G$ except those joined by $\alpha$-strong edges are joined by at least three internally-disjoint strongest paths. Let $z \in \sigma^*$. Removal of $z$ removes at most one internally-disjoint strongest path between $x$ and $y$, $x, y \neq z$. This implies, $G - z$ is still a block which is not possible. Hence, there exists a pair of vertices, not joined by an $\alpha$-strong edge which is joined by at most two internally-disjoint strongest paths. ■

The converse is not true as seen from $G_2$ in Fig. 4.12. $a$ and $d$ are joined by only two internally-disjoint strongest paths. But, $G_2$ is noncritical. Also, every vertex in $G_2$ is an internal vertex of some strongest path.

The block-graph of a graph $G = (V, E)$ is a graph whose vertex set is the set of blocks of $G$ and two vertices are adjacent if the corresponding blocks of $G$ share a vertex [83]. The block-graph of a fuzzy graph is defined below.

**Definition 4.2.7** Let $G = (\sigma, \mu)$ be a fuzzy graph. The **block-graph** of $G$ is a fuzzy graph $B_f(G) = (V', \sigma', \mu')$, where $V'$ is the set of blocks of $G$. $\sigma'$ and $\mu'$ are defined as follows. Let $b, b_1, b_2 \in V'$ :

$$\sigma'(b) = \begin{cases} \sigma(u) & \text{if } b \text{ is a trivial block, } \{u\} \\ \vee\{CONN_b(x, y) \mid x, y \in V(b)\} & \text{if } b \text{ is a nontrivial block} \end{cases}$$

$$\mu'(b_1b_2) = \begin{cases} 0 \quad \text{if } b_1 \text{ and } b_2 \text{ do not share a vertex of } G. \\ \vee\{CONN_{b_1}(a_1, x) \wedge CONN_{b_2}(x, a_2) \mid a_1 \in V(b_1^*)\backslash V(b_2^*), \\ \quad a_2 \in V(b_2^*)\backslash V(b_1^*),\ x \in V(b_1^*) \cap V(b_2^*)\} \\ \quad \text{if } b_1 \text{ and } b_2 \text{ share a vertex of } G. \end{cases}$$

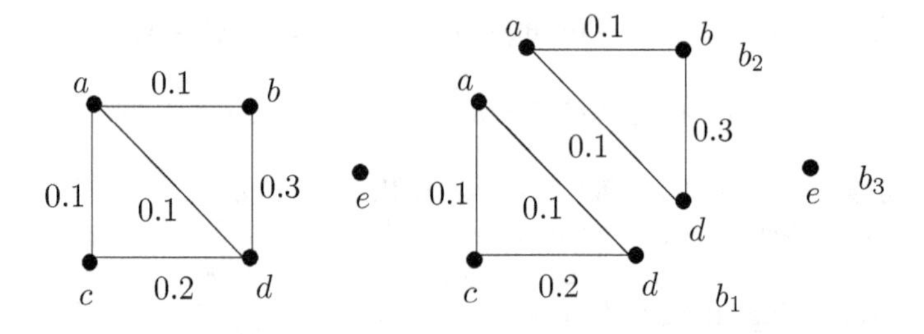


**Fig. 4.14** Fuzzy graph $G$ and its blocks

**Fig. 4.15** Block graph of the fuzzy graph in Example 4.2.8



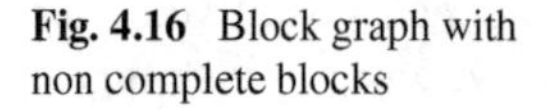

**Fig. 4.16** Block graph with non complete blocks

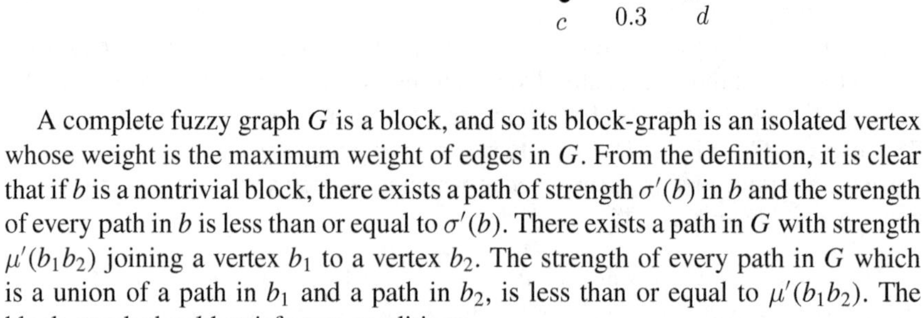


A complete fuzzy graph $G$ is a block, and so its block-graph is an isolated vertex whose weight is the maximum weight of edges in $G$. From the definition, it is clear that if $b$ is a nontrivial block, there exists a path of strength $\sigma'(b)$ in $b$ and the strength of every path in $b$ is less than or equal to $\sigma'(b)$. There exists a path in $G$ with strength $\mu'(b_1b_2)$ joining a vertex $b_1$ to a vertex $b_2$. The strength of every path in $G$ which is a union of a path in $b_1$ and a path in $b_2$, is less than or equal to $\mu'(b_1b_2)$. The block-graph should satisfy two conditions.

$(i)$ $B_f(G)$ should give maximum information about $G$.

$(ii)$ $B_f(G)$ must be a fuzzy graph.

*Example 4.2.8* Let the fuzzy graph $G = (\sigma, \mu)$ with $\sigma^* = \{a, b, c, d, e\}$ be given by $\sigma(a) = \sigma(b) = \sigma(c) = \sigma(d) = \sigma(e) = 1, \mu(ab) = \mu(ad) = \mu(ac) = 0.1, \mu(cd) = 0.2, \mu(bd) = 0.3$ and $\mu(xy) = 0$ for all other $x, y \in \sigma^*$. $G$ and its blocks are given in Fig. 4.14 and its block graph is given in Fig. 4.15.

For graphs, if $H$ is a block-graph, then the blocks of $H$ are complete [83]. But, this is not true for fuzzy graphs. Consider the fuzzy graph in Fig. 4.16.

The blocks and block graph of the fuzzy graph in Fig. 4.16 are given in Fig. 4.17. The cycle $b_1, b_2, b_3, b_4, b_1$ is a block of $B_f(G)$ and is not complete.

But, the result holds for trees as seen from the following lemma.

**Lemma 4.2.9** *If $G = (\sigma, \mu)$ is a tree, then the blocks of $B_f(G)$ are complete.*

*Proof* Let $B$ be a block of $B_f(G)$. $G$ is $K_2$ block-graph. By definition of $B_f(G)$, the weight of an edge in $B_f(G)$ is the minimum of the weights of the vertices incident

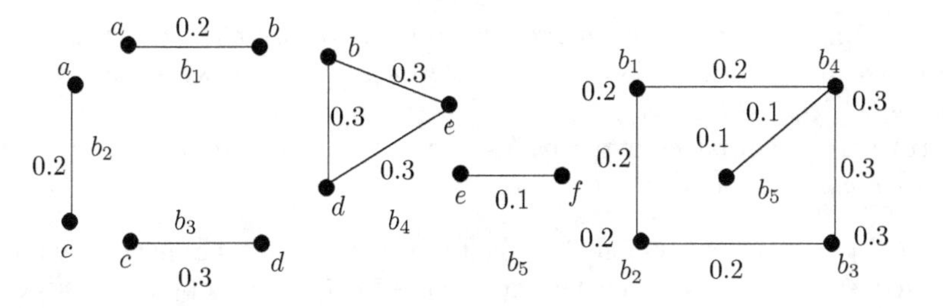


Fig. 4.17 Blocks and block graph of fuzzy graph in Fig. 4.16

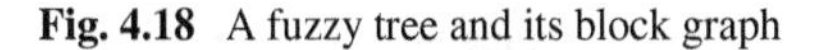


Fig. 4.18 A fuzzy tree and its block graph

with it. This is true for every edge in $B$. In order to prove $B$ is complete, it is enough to prove $B^*$ is complete. It is true for blocks on two and three vertices. Suppose $B$ has at least four vertices and assume, $B^*$ is not complete. Then there exists a pair of vertices $u$ and $v$ which are not adjacent. Because $B^*$ is a block, $u$ and $v$ lie on a cycle of length at least four [83]. $u$ and $v$ correspond to edges in $G$. It follows that there are two paths joining a vertex incident with $u$ and a vertex incident with $v$, in $G$. This is not possible because $G$ is a tree. Therefore, our assumption is wrong and $B^*$ is complete. ■

A fuzzy tree $G$ and its block graph $B_f(G)$ are shown in Fig. 4.18.

## 4.3 More on Blocks in Fuzzy Graphs

In [110], four conditions which are true and equivalent in blocks on at least three vertices without fuzzy bridges, has been given. But, it is observed that the result is correct for all fuzzy blocks on at least three vertices, leading to the following characterization.

**Theorem 4.3.1** ([109]) *In a fuzzy graph $G = (\sigma, \mu)$ with $|\sigma^*| \geq 3$, the following conditions are equivalent.*

*(i) $G$ is a block.*

*(ii) Every pair of vertices except those joined by $\alpha$-strong edges are joined by two internally-disjoint strongest strong paths.*

*(iii) There is a cycle containing a vertex $u$ and a strong edge $xy$ formed by two strongest strong $u - x$ or $u - y$ paths for every vertex-strong edge pair except for the pairs where the strong edge is $\alpha$-strong and is incident with the vertex.*

*(iv) There is a cycle containing two strong edges $uv$ and $xy$ formed by two strongest strong $u - x$ paths or $u - y$ paths for every pair of strong edges $uv$ and $xy$.*

*(v) There is a strongest strong path joining two vertices, not containing the third for every three distinct vertices $x$, $y$ and $z$.*

*Proof* $(i) \Rightarrow (ii)$: Outline- Let $G'$ be the fuzzy graph obtained from $G$ by removing all $\delta$-edges. $G'$ consists of only the strong edges in $G$. Instead of proving that every pair of vertices except those joined by an $\alpha$-strong edge is joined by two internally-disjoint strongest strong paths in $G$, it is enough to prove the following for $G'$.

(1) $CONN_{G'}(x, y) = CONN_G(x, y)$ for all distinct $x, y$ in $G$. (2) $G'$ is a block.

Because every pair of vertices in a fuzzy graph is joined by a strongest strong path, $CONN_{G'}(x, y) = CONN_G(x, y)$ for all distinct $x, y$ in $G$. Let $w$ be a vertex in $G'$. Next we prove that $w$ is not a fuzzy cutvertex of $G'$. Suppose that $x$ and $y$ are two distinct vertices in $G'$ different from $w$. If $xy$ is strong, then the removal of $w$ cannot reduce the strength of connectedness between $x$ and $y$. If $x$ and $y$ are not joined by an $\alpha$-strong edge, then there is a strongest strong $x - y$ path in $G'$ not containing $w$. Suppose not. Then every strongest strong $x - y$ path in $G'$ contains $w$. Therefore, every strongest strong $x - y$ path in $G$ contains $w$. So, $w$ is an internal vertex of every maximum spanning tree of $G$ and hence an f-cutvertex of $G$. This is not possible. Therefore, there is an $x - y$ strongest path in $G'$ which does not contain $w$. Because $w$ is arbitrary, $G'$ is a block. So, every pair of vertices in $G'$ not joined by an $\alpha$-strong edge is joined by two internally disjoint strongest paths in $G'$ and hence joined by two internally disjoint strongest strong paths in $G$.

$(ii) \Rightarrow (iii)$: A fuzzy graph satisfying condition $(ii)$ is a block. It does not contain $\alpha$-strong paths of length greater than or equal to two because a common vertex of two $\alpha$-strong edges is a fuzzy cutvertex (Theorem 2.2.11).

**Case** 1: $xy$ is a $\beta$-strong edge.

In any fuzzy graph, for a $\beta$-strong edge $xy$ there exists a strongest strong path $P$ joining $x$ and $y$, different from $xy$. So, $xy$ together with $P$ constitute the required cycle.

**Case** 2: $u$ is not incident with $xy$.

**Subcase** 1: Exactly one of $u, x$ and $u, y$ is joined by an $\alpha$-strong edge.

Let $ux$ be an $\alpha$-strong edge. Because $uy$ is not $\alpha$-strong, there exists two internally-disjoint strongest strong $u - y$ paths and at least one of them, say $P$ does not intersect with the $u - x - y$ path. $\mu(ux) > \mu(xy) \wedge CONN_G(u, y)$ because an $\alpha$-strong edge cannot be a weakest edge in any cycle (Fig. 4.19).

**Claim**: $\mu(xy) = CONN_G(u, y)$

Suppose that $\mu(xy) \neq CONN_G(u, y)$. If $\mu(xy) < CONN_G(u, y)$, then the $x - y$ path given by $xu \cup P$ has strength greater than $\mu(xy)$, which is not possible because $xy$ is strong. If $\mu(xy) > CONN_G(u, y)$, then the path $u - x - y$ has strength greater than $CONN_G(u, y)$, which is not possible. So, $\mu(xy) = CONN_G(u, y)$.

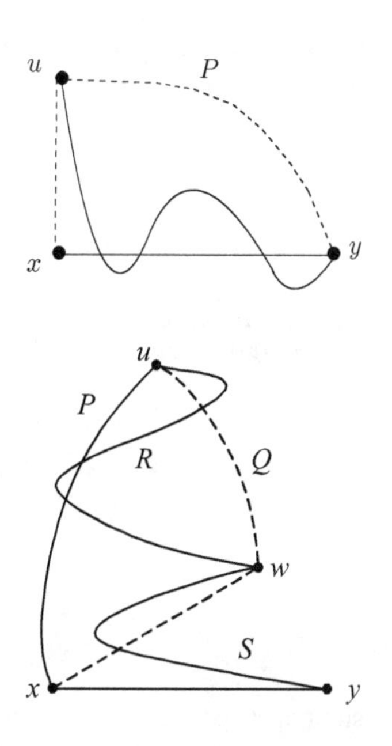


**Fig. 4.19** Theorem 4.3.1 $((ii) \Rightarrow (iii))$ - Case 2, Subcase 1

**Fig. 4.20** Theorem 4.3.1 $((ii) \Rightarrow (iii))$ - Case 2, Subcase 2

Because the strength of the path $u - x - y$ is $\mu(xy)$, path $u - x - y$ is a strongest $u - y$ path. Hence, $P$ together with $u - x - y$ path, gives the required cycle.

**Subcase** 2: Neither $ux$ nor $uy$ is $\alpha$-strong.

Both $u, x$ and $u, y$ are joined by two internally disjoint strongest strong paths. Suppose that $CONN_G(u, x) \leq CONN_G(u, y)$. Let $P$ and $Q$ be two internally disjoint strongest strong $u - x$ paths and $R$ be a strongest strong $u - y$ path. Let $w$ be the last vertex in $R$ which either lies on $P$ or $Q$, moving from $u$ to $y$. Let $w$ lies on $Q$. $u - w$ subpath of $Q$, $w - y$ subpath of $R$ together with edge $yx$ gives a strong $u - x$ path, say $S$ internally-disjoint with $P$ (Fig. 4.20).

**Claim**: $S$ is a strongest $u - x$ path.

The strengths of $u - w$ subpath of $Q$ and $w - y$ subpath of $R$ are greater than or equal to $CONN_G(u, x)$. The strength of an $x - y$ path, in particular, $x - w$ subpath of $Q$ together with $w - y$ subpath of $R$ is greater than or equal to $CONN_G(u, x)$. Because $xy$ is strong, $\mu(xy) \geq CONN_G(u, x)$. The strength of each of the three subpaths, constituting $S$ is greater than or equal to $CONN_G(u, x)$. Hence, $S$ is a strongest $u - x$ path. $P \cup S$ is the required cycle.

$(iii) \Rightarrow (iv)$

**Case** 1: $uv$ and $xy$ are adjacent, Fig. 4.21.

Let $u$ be different from both $x$ and $y$. There exists a cycle $C$ containing $u$ and $xy$, formed by two strongest strong $u - x$ paths or $u - y$ paths. If $u$ is adjacent to $x$ in $C$, $C$ is the required cycle. If not, remove the $u - x$ path in $C$ not containing $y$ and add $ux$ to obtain a new cycle, $C'$. Replacing subpath of a path by another path whose

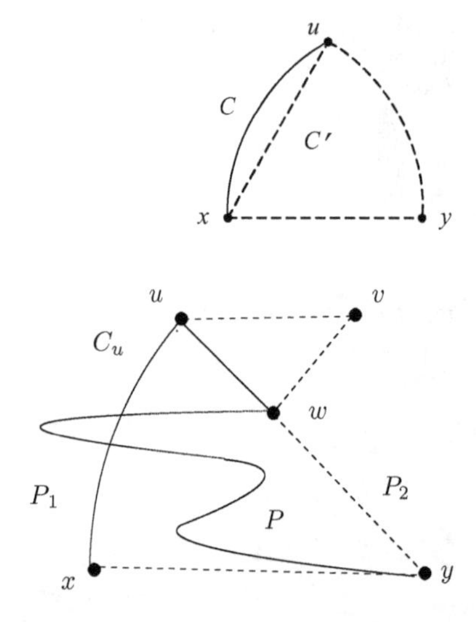


**Fig. 4.21** Theorem 4.3.1 $((iii) \Longrightarrow (iv))$ - Case 1

**Fig. 4.22** Theorem 4.3.1 $((iii) \Rightarrow (v))$ - Case 2

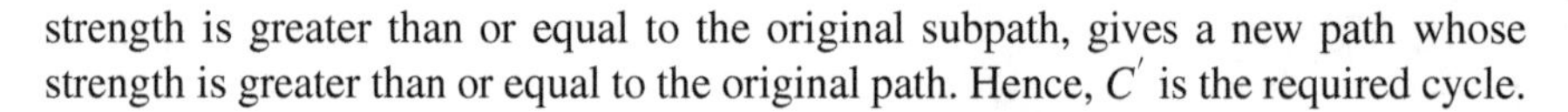

strength is greater than or equal to the original subpath, gives a new path whose strength is greater than or equal to the original path. Hence, $C'$ is the required cycle.

**Case** 2: $uv$ and $xy$ are not adjacent.

Let $C_u$ be a cycle mentioned in $(iii)$, containing $u$ and $xy$ and $C_v$ be one containing $v$ and $xy$. If either $C_u$ contains $v$ or $C_v$ contains $u$, there is nothing to prove. Suppose not. Let $c_u$ and $c_v$ be the strengths of the strongest strong paths in $C_u$ and $C_v$ respectively. Assume that $c_u \leq c_v$. Let $C_u$ be formed by two strongest strong $u - x$ paths, $P_1$ and $P_2$. Let $P$ be the $v - y$ path in $C_v$, not containing $xy$. Assume that $w$ is the last vertex at which $P$ intersects $C_u$, moving from $y$ to $v$ and let it lie on $P_2$. Strength of every path in $C_u$ and every subpath of $P$ is greater than or equal to $CONN_G(u, x)$. A cycle containing $uv$ and $xy$ is constructed using paths in $C_u$ and subpaths of $P$. Strengths of both the $u - x$ paths in this cycle are greater than or equal to $CONN_G(u, x)$. Hence, the cycle is formed by two strongest strong $u - x$ paths (Fig. 4.22).

The paths $P_1$, edge $xy$, $y - w$ subpath of $P_2$, $w - v$ subpath of $P$ and edge $vu$ together give the required cycle. $uv$ does not lie on $C_u$ or $C_v$. But, there is a $u - v$ path with strength greater than or equal to $CONN_G(u, x)$, namely, the union of $u - w$ subpath of $P_2$ and $w - v$ subpath of $P$. Hence, $\mu(uv) \geq CONN_G(u, x)$.

$(iv) \Rightarrow (v)$

Suppose that $(iv)$ is true and $(v)$ is not true.

In this case, there exists three distinct vertices $w_1$, $w_2$ and $x$ such that $x$ lies on every $w_1 - w_2$ strongest strong path. Let $P$ be a strongest strong $w_1 - w_2$ path and $u$ and $y$ be the two neighbors of $x$ in $P$. Let $C$ be the cycle mentioned in $(iv)$, containing $ux$ and $xy$ and $Q$ be the $u - y$ path in it, which is different from $u - x - y$. Let $a$ be the last vertex at which $Q$ intersects $y - w_1$ subpath of $P$ and $b$, the first vertex at which $Q$ intersects $u - w_2$ subpath of $P$ moving from $y$ to $u$, along $Q$.

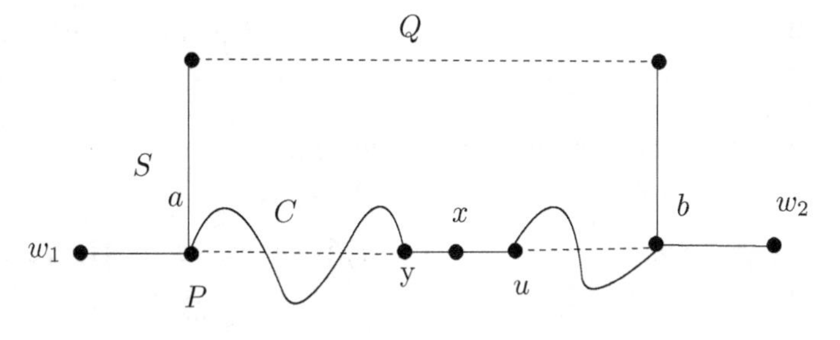


**Fig. 4.23** Theorem 4.3.1 ($(iv) \Rightarrow (v)$)

Consider the $w_1 - w_2$ path $S$ given by the union of $w_1 - a$ subpath of $P$, $a - b$ subpath of $Q$ and $b - w_2$ subpath of $P$. The strengths of $w_1 - a$ subpath and $(b - w_2)$ subpath of $P$ are clearly greater than or equal to $CONN_G(w_1, w_2)$ because $P$ is a strongest strong path. The strengths of the two strongest strong paths constituting $C$ is $\mu(ux) \wedge \mu(xy)$. Because the membership values of both the edges are greater than or equal to $CONN_G(w_1, w_2)$, $s(Q) \geq CONN_G(w_1, w_2)$ and so is the strength of the $a - b$ subpath of $Q$. Hence, $s(S) \geq CONN_G(w_1, w_2)$. So, $S$ is a strongest strong $w_1 - w_2$ path, a contradiction as it does not contain $x$. It follows that our assumption is wrong. Therefore, for any three distinct vertices $w_1$, $w_2$ and $x$ there is a strongest strong $w_1 - w_2$ path which does not contain $x$ (Fig. 4.23).

$(v) \Rightarrow (i)$

Let $x$ be a vertex in $G$. Because there is a strongest $w_1 - w_2$ path which does not contain $x$ for all $w_1, w_2 \in G$, $x$ is not a fuzzy cutvertex. It follows that $G$ is a block. ■

In graph theory, there are six equivalent conditions for blocks [83] which are given in the following theorem.

**Theorem 4.3.2** *Let G be a connected graph of order at least three. The following conditions are equivalent.*

$(i)$ *G is a block.*

$(ii)$ *Every two vertices lie on a common cycle.*

$(iii)$ *Every vertex and edge lie on a common cycle.*

$(iv)$ *Every two edges lie on a common cycle.*

$(v)$ *For any two vertices and an edge, there is a path joining the two vertices containing the edge.*

$(vi)$ *For any three distinct vertices, there is a path joining any two vertices containing the third.*

$(vii)$ *For three distinct vertices, there is a path joining any two vertices not containing the third.*

Of these six characterizations for blocks, four of them namely, $(ii)$, $(iii)$, $(iv)$, and $(vii)$ are generalized by Theorem 4.3.1. The remaining two does not hold for all blocks in fuzzy graphs. All paths in graphs are strongest strong. Now, consider the following theorem.

**Theorem 4.3.3** *Let $G = (\sigma, \mu)$ be a fuzzy graph on at least three vertices. $G$ is a block if and only if for any three distinct vertices $x$, $y$ and $z$ in $\sigma^*$, the following properties are equivalent.*
*$(i)$ There is an $x - y$ strongest path containing $z$.*
*$(ii)$ $CONN_G(x, y) = CONN_G(x, z) \wedge CONN_G(z, y)$.*

*Proof* First suppose that $G = (\sigma, \mu)$ is a block. We prove that the given conditions are equivalent.
$(i) \Rightarrow (ii)$
For three distinct vertices $x$, $y$ and $z$ in $\sigma^*$,

$$CONN_G(x, y) \geq CONN_G(x, z) \wedge CONN_G(z, y).$$

If there exists a strongest $x - y$ path containing $z$ (clearly, $x$ and $y$ are not joined by an $\alpha$-strong edge), then

$$CONN_G(x, y) \leq CONN_G(x, z) \wedge CONN_G(z, y).$$

It follows that,

$$CONN_G(x, y) = CONN_G(x, z) \wedge CONN_G(z, y).$$

Thus, if $(i)$ is true, then $(ii)$ also is true in $G$.
$(ii) \Rightarrow (i)$
In a block, $(ii)$ does not hold for any $\alpha$-strong edge. If $x$ and $y$ are joined by an $\alpha$-strong edge in a block, then $CONN_G(x, y) > CONN_G(x, z) \wedge CONN_G(z, y)$ for all $z \neq x, y$. Hence, if $(ii)$ is true for a pair of vertices $x$ and $y$, then $x$ and $y$ are not joined by an $\alpha$-strong edge.

**Case** 1: One of $x, z$ or $z, y$ is joined by an $\alpha$-strong edge.
Suppose that $xz$ is $\alpha$-strong. So,

$$CONN_G(x, z) > CONN_G(x, y) \wedge CONN_G(y, z),$$

which is equal to $CONN_G(x, y)$. It follows that $CONN_G(x, y) = CONN_G(z, y)$. Because $z$ and $y$ are not joined by an $\alpha$-strong edge, $z$ and $y$ are joined by two internally disjoint strongest paths, say $P_1$ and $P_2$. At least one of $P_1$ and $P_2$ (say, $P_1$) does not contain $xz$. $xz \cup P_1$ constitute an $x - y$ strongest path containing $z$.

**Case** 2: Neither $x, z$ nor $z, y$ is joined by an $\alpha$-strong edge.
Let $CONN_G(x, y) = CONN_G(x, z)$. Let $P$ be a strongest $x - z$ path. If at least one of $P_1$ and $P_2$ (say, $P_1$) does not intersect $P$, then $P \cup P_1$ form an $x - y$ path with strength, $CONN_G(x, z)$ and hence, a strongest $x - y$ path containing $z$.
Suppose that both $P_1$ and $P_2$ intersect $P$. Let $w$ be the first vertex in $P$ which is common to either $P_1$ or $P_2$ (say, $P_1$). The $x - w$ subpath of $P$ (vertices except $w$ are

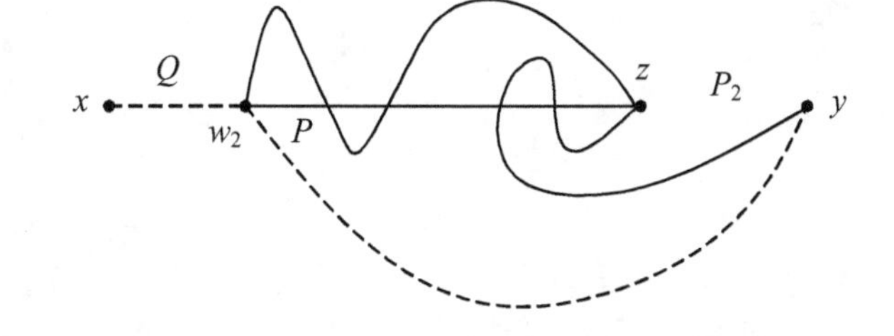


**Fig. 4.24** Theorem 4.3.3 - Case 2 - $P_1$ and $P_2$ intersecting $P$

neither in $P_1$ nor in $P_2$)–$(w - z)$ subpath of $P_1$–$P_2$ constitute an $x - y$ path, $Q$ containing $z$ and all of whose edges have weight, greater than or equal to $CONN_G(x, z)$ and hence, is a strongest $x - y$ path containing $z$. Figure 4.24.

To prove the converse, assume that $G$ is not a block. Let $w$ be a fuzzy cutvertex in $G$ and let it lie on every strongest path between vertices $u$ and $v$. Clearly,

$$CONN_G(u, v) = CONN_G(u, w) \wedge CONN_G(w, v). \tag{4.5}$$

Assume,

$$CONN_G(w, v) = CONN_G(w, u) \wedge CONN_G(u, v). \tag{4.6}$$

We prove that $(i)$ is not true, i.e., no $w - v$ strongest path contains $u$.

From (4.5) and (4.6), it follows that

$$CONN_G(w, v) = CONN_G(u, v). \tag{4.7}$$

If $w$ and $v$ are joined by an $\alpha$-strong edge, there is nothing to prove. Suppose that $w$ and $v$ are not joined by an $\alpha$-strong edge. If there is no $w - v$ path containing $u$, there is nothing to prove. Suppose that $Q$ is a $w - v$ path containing $u$. We prove that $Q$ is not a strongest $w - v$ path. The $u - v$ subpath of $Q$, say $Q'$ does not contain $w$. Hence, $Q'$ is not a strongest $u - v$ path. So, the strength of $Q'$ is less than $CONN_G(u, v)$ and so, is the strength of $Q$. From (4.7), $Q$ is not a strongest $w - v$ path. Because $Q$ is an arbitrary $w - v$ path containing $u$, no $w - v$ strongest path contains $u$. Hence, the proof. ■

'Strongest strong' path, in place of 'strongest' path in Theorem 4.3.3 gives a stronger characterization.

**Theorem 4.3.4** *Let $G = (\sigma, \mu)$ be a fuzzy graph on at least three vertices. $G$ is a block if and only if for any three distinct vertices $x$, $y$ and $z$ in $G$, the following properties are equivalent.*

$(i)$ *There is an $x - y$ strongest strong path containing $z$.*

$(ii)$ $CONN_G(x, y) = CONN_G(x, z) \wedge CONN_G(z, y)$.

Let $G$ be a block on at least three vertices. Then $G$ contains a strongest strong cycle $C$. For every three distinct vertices $x$, $y$ and $z$ in $C$, where $x$ and $y$ are not

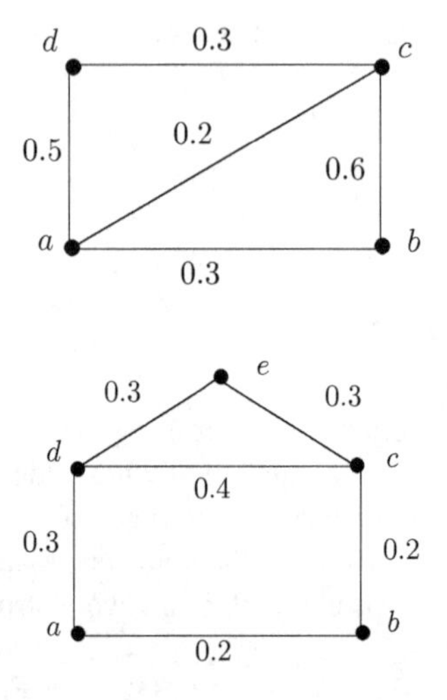


**Fig. 4.25** Connectivity transitive

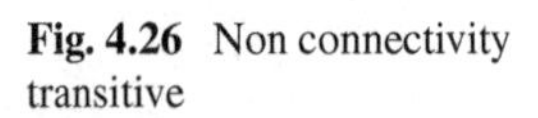

**Fig. 4.26** Non connectivity transitive

joined by an $\alpha$-strong edge, either $(i)$ or $(ii)$ is true. So, in every block on at least three vertices, there exists at least three distinct vertices $x$, $y$ and $z$, for which $(i)$ or $(ii)$ is true. Hence, one can think of blocks in which for all $x$, $y$ and $z$, where $x$ and $y$ are not joined by an $\alpha$-strong edge, condition $(i)$ or $(ii)$ is true. Such blocks are termed as connectivity-transitive blocks.

*Example 4.3.5* The block in Fig. 4.25 is a connectivity-transitive block while the block in Fig. 4.26 is not connectivity-transitive because $CONN_G(c, e) = 0.3$ while $CONN_G(c, b) = 0.2$.

Similarly, cyclically-transitive blocks can be defined using cycle-connectivity. To have a better understanding of connectivity (cyclically)-transitive blocks, we first discuss connectivity (cyclically)-transitive graphs and study some of their properties.

## 4.4 Connectivity-Transitive and Cyclically-Transitive Fuzzy Graphs

In this section, two important subcategories of blocks in fuzzy graphs are discussed. We start with a lemma.

**Lemma 4.4.1** *Let $G = (\sigma, \mu)$ be a fuzzy graph with $|\sigma^*| \geq 3$. The following conditions are equivalent.*

*$(i)$ $CONN_G(x, y) = CONN_G(x, z) \wedge CONN_G(z, y)$ for all $x, y, z \in \sigma^*$ such that $xy$ is not $\alpha$-strong.*

*$(ii)$ $CONN_G(x, y)$ is equal for every distinct pair $x, y \in \sigma^*$ such that $x$ and $y$ are not joined by an $\alpha$-strong edge.*

*Proof* Suppose that $(i)$ is true. Let $CONN_G(x, y) = a$ for some $x$ and $y$, not joined by an $\alpha$-strong edge. We prove that the strength of connectedness between every pair of vertices except those joined by an $\alpha$-strong edge is $a$. If $z$ is a vertex not joined by an $\alpha$-strong edge to $x$, then

$$CONN_G(x, z) = CONN_G(x, y) \wedge CONN_G(y, z).$$

Therefore, $CONN_G(x, z) \leq a$.

If $CONN_G(x, z) < a$, then $CONN_G(x, y) < a$ from the relation

$$CONN_G(x, y) = CONN_G(x, z) \wedge CONN_G(z, y),$$

which is not possible. Hence,

$$CONN_G(x, z) = a = CONN_G(x, y). \tag{4.8}$$

Similarly, $CONN_G(y, z') = a$, where $z'$ is a vertex not joined to $y$ by an $\alpha$-strong edge.

Let $u$ and $v$ be two distinct vertices, not joined by an $\alpha$-strong edge and both different from $x$ and $y$.

Scenario in Fig. 4.27 is not possible because the weakest edge of a cycle cannot be $\alpha$-strong. So, at least one of the pairs of vertices joined by an $\alpha$-strong edge in Fig. 4.27 is not $\alpha$-strong. Without loss of generality, let $x$ and $u$ be not joined by an $\alpha$-strong edge.

Therefore, $CONN_G(u, v) = CONN_G(u, x) = a$ from (4.8).

Now, assume that $(ii)$ is true. That is, $CONN_G(x, y) = a$ for every distinct pair of vertices $x$ and $y$, not joined by an $\alpha$-strong edge.

To prove: $CONN_G(u, v) = CONN_G(u, w) \wedge CONN_G(w, v)$ for all distinct $u, v$ not joined by an $\alpha$-strong edge and $w$, is any vertex different from $u$ and $v$.

**Case** 1: Neither $uw$ nor $vw$ is $\alpha$-strong.

$CONN_G(u, v) = CONN_G(u, w) = CONN_G(w, v) = a$. Clearly,

$$CONN_G(u, v) = CONN_G(u, w) \wedge CONN_G(w, v).$$

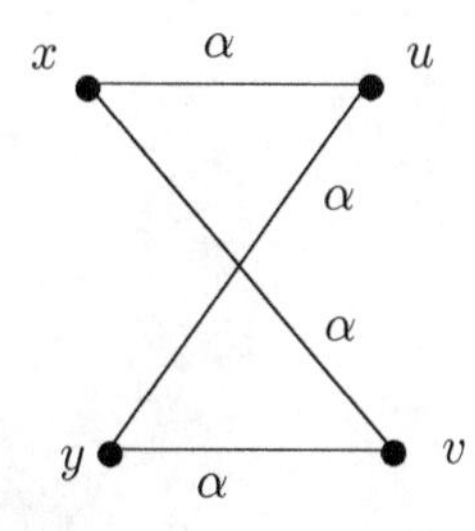


**Fig. 4.27** An impossible situation

**Case** 2: Both $uw$ and $vw$ are $\alpha$-strong.

$u - w - v$ is strongest $u - v$ path and hence,

$$CONN_G(u, v) = CONN_G(u, w) \wedge CONN_G(w, v).$$

**Case** 3: Exactly one of $u, w$ and $v, w$ is joined by an $\alpha$-strong edge.

Let $uw$ be an $\alpha$-strong edge. Because $CONN_G(u, v) = CONN_G(v, w) = a$, it is enough to prove $CONN_G(u, w) \geq a$. For any three distinct vertices $x, y, z$ in a fuzzy graph $G$,

$$CONN_G(x, y) \geq CONN_G(x, z) \wedge CONN_G(z, y).$$

So, $CONN_G(u, w) \geq CONN_G(u, v) \wedge CONN_G(v, w) = a$. Hence,

$$CONN_G(u, v) = CONN_G(u, w) \wedge CONN_G(w, v).$$

■

**Definition 4.4.2** A fuzzy graph $G$ in which

$$CONN_G(x, y) = CONN_G(x, z) \wedge CONN_G(z, y)$$

for every three distinct vertices $x$, $y$ and $z$, where $x$ and $y$ are not joined by an $\alpha$-strong edge is called a **connectivity-transitive fuzzy graph**. Equivalently, a fuzzy graph $G$ in which $CONN_G(x, y)$ is same for every pair of vertices except for those joined by an $\alpha$-strong edge is called a connectivity-transitive fuzzy graph (Fig. 4.28).

If a connectivity-transitive fuzzy graph is disconnected, then it is totally disconnected. Also, in a connectivity-transitive fuzzy graph, all the $\beta$-strong edges have the same weight. The converse is not true as seen from Fig. 4.31.

**Lemma 4.4.3** *A connected fuzzy graph $G$ is connectivity-transitive if and only if all the strong paths, except $\alpha$-strong edges have the same strength.*

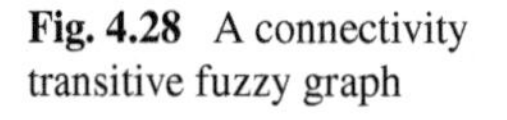

**Fig. 4.28** A connectivity transitive fuzzy graph

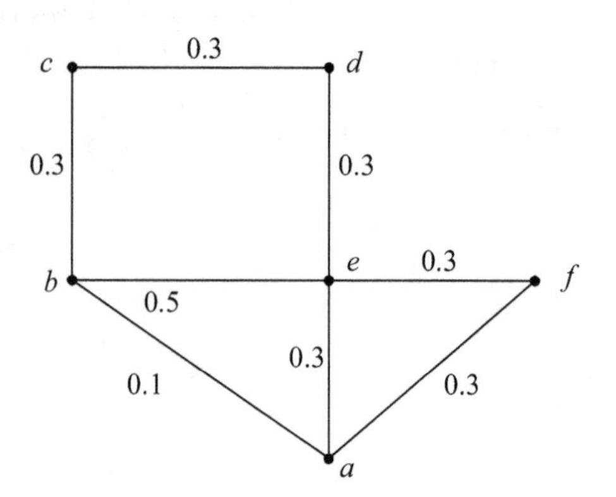

*Proof* In any fuzzy graph $G$, every pair of vertices is joined by a strongest strong path. If all strong paths except $\alpha$-strong edges in $G$ have the same strength, then clearly $G$ is connectivity-transitive.

Conversely, assume that $G$ is a connectivity-transitive fuzzy graph with $CONN_G(x, y) = a$ for all distinct $x, y$, not joined by an $\alpha$-strong edge. Suppose that $P$ is a strong path in $G$. If $G$ has only two vertices, then there is nothing to prove. So, assume it has at least three vertices. If $P$ is $\beta$-strong, $s(P) = a$. If $P$ is an $\alpha$-strong path of length at least two, then $P$ is a strongest path and hence, $s(P) = a$. Let $xy$ be an $\alpha$-strong edge in $G$. Then there exists a vertex $u$ which is not joined by an $\alpha$-strong edge to at least one of $x$ and $y$. Let $u$ and $y$ be not joined by an $\alpha$-strong edge. Then

$$CONN_G(u, y) = CONN_G(u, x) \wedge CONN_G(x, y).$$

Because $CONN_G(u, y) = a$, $CONN_G(x, y) \geq a$. If $P$ of length greater than or equal to two is not an $\alpha$-strong path, then $\beta$-strong edges are weakest edges in $P$. So, $s(P) = a$. ■

**Corollary 4.4.4** *A connectivity-transitive fuzzy graph is a $\theta$-fuzzy graph.*

**Lemma 4.4.5** *In any fuzzy graph on at least three vertices, the following conditions are equivalent.*

*(i)* $C^G_{x,y} = C^G_{x,z} \wedge C^G_{y,z}$.

*(ii)* $C^G_{x,y}$ *is equal for every three distinct vertices $x$, $y$ and $z$, where $x$ and $y$ are not joined by an $\alpha$-strong edge and $C^G_{x,y}$ is the cycle connectivity between $x$ and $y$ in $G$.*

*Proof* $(i) \Rightarrow (ii)$ can be proved using the technique used in Lemma 4.4.1.

To prove: $(ii) \Rightarrow (i)$

Let $C^G_{x,y} = c$ for all $x \neq y$ not joined by an $\alpha$-strong edge. If $uv$ is an $\alpha$-strong edge, then

$$C^G_{u,v} \leq c. \tag{4.9}$$

**Case** 1: $c = 0$.

$C^G_{x,y} = 0$ for every pair of distinct vertices in $G$ and hence $(i)$ is true.

**Case** 2: $c > 0$.

**Subcase** 1: Neither $x, z$ nor $z, y$ is joined by an $\alpha$-strong edge. $C^G_{x,y} = c = C^G_{x,z} = C^G_{z,y}$.

**Subcase** 2: Both $xz$ and $zy$ are $\alpha$-strong.

Clearly, $x$ and $y$ are not joined by an $\alpha$-strong edge. Hence, $x$ and $y$ lie on a common strong cycle, $C$ of strength $c$. At least one of the $x - y$ paths, say $P$ in $C$, does not intersect the $\alpha$-strong $x - y$ path, $x - z - y$. $P \cup (x - z - y)$ constitute a strong cycle containing $xz$ whose strength is greater than or equal to $c$. Therefore, $C^G_{x,z} \geq c$. From (4.9), $C^G_{x,z} = c$. Similarly, $C^G_{z,y} = c$ (Figs. 4.29 and 4.30).

Hence, $C^G_{x,y} = c = C^G_{x,z} = C^G_{z,y}$.

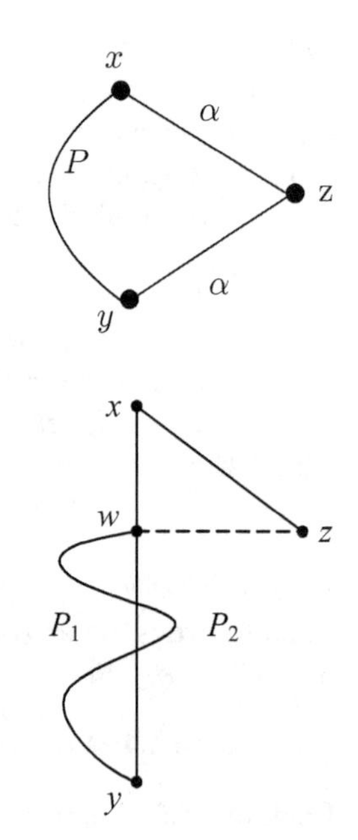


**Fig. 4.29** Lemma 4.4.5 – Case 2, Subcase 2

**Fig. 4.30** Lemma 4.4.5-Case 2, Subcase 3

**Subcase** 3: Exactly one of $x$, $z$ or $z$, $y$ is joined by an $\alpha$-strong edge.

Let $xz$ be an $\alpha$-strong edge. $xy$ lies on a strong cycle $C$ of strength $c$. At most one of the two $x - y$ paths in $C$ contains $xz$. Let $P_1$ be the $x - y$ path which does not contain $xz$. Similarly, let $P_2$ be a $y - z$ path which does not contain $xz$. Let $w$ be the first common vertex of $P_1$ and $P_2$, as one moves from $z$ to $y$ along $P_2$. Therefore, $z - w$ subpath of $P_2$, $w - x$ subpath of $P_1$ together with $xz$ constitute an $x - z$ strong cycle with strength greater than or equal to $c$. Therefore, $C^G_{x,z} \geq c$. From (4.9), $C^G_{x,z} = c$. Hence, $C^G_{x,y} = c = C^G_{x,z} = C^G_{z,y}$. Therefore, $C^G_{x,y} = C^G_{x,z} \wedge C^G_{z,y}$ for all distinct $x$, $y$ not joined by an $\alpha$-strong edge and any vertex $z$ different from both $x$ and $y$. ■

**Lemma 4.4.6** *In any fuzzy graph with at least three vertices, the following conditions are equivalent.*

*(i)* $C^G_{x,y} = C^G_{x,z} \wedge C^G_{y,z}$.

*(ii)* $C^G_{x,y}$ *is equal for every three distinct vertices* $x$, $y$ *and* $z$.

The proof of Lemma 4.4.6 is trivial.

Condition $(ii)$ in Lemma 4.4.5 and condition $(ii)$ in Lemma 4.4.6 are equivalent. An outline of the proof is as follows.

Let $C^G_{x,y} = c$ for all distinct $x$ and $y$ not joined by an $\alpha$-strong edge. In the proof of Lemma 4.4.5, $C^G_{u,v} = c$, where $uv$ is an $\alpha$-strong edge incident with at least one of $x$ and $y$. Now, it is enough to prove, $C^G_{u,v} = c$, where $uv$ is an $\alpha$-strong edge with both $u$ and $v$ different from $x$ and $y$. Clearly, both the pairs $u, x$ and $v, x$ cannot be joined by an $\alpha$-strong edge. Suppose that $u$ and $x$ are not joined by an $\alpha$-strong edge.

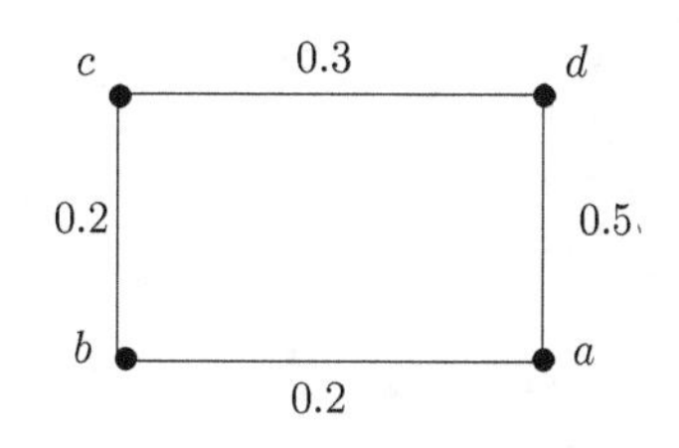


**Fig. 4.31** A cyclically transitive fuzzy graph

Then

$$C^G_{u,x} = C^G_{u,v} \wedge C^G_{v,x}.$$

Because $C^G_{u,x} = c$, $C^G_{u,v} \geq c$. Using (4.9), $C^G_{u,v} = c$.

So, $C^G_{x,y} = c$, for all distinct $x$ and $y$. Hence, the collections of fuzzy graphs satisfying the conditions in Lemmas 4.4.5 and 4.4.6, respectively, are the same.

**Definition 4.4.7** A fuzzy graph $G$ in which $C^G_{x,y} = C^G_{x,z} \wedge C^G_{y,z}$ for every three distinct vertices $x$, $y$ and $z$ is called a **cyclically-transitive** fuzzy graph.

A connectivity-transitive fuzzy graph need not be cyclically-transitive (Fig. 4.28) and vice versa (Fig. 4.31).

**Lemma 4.4.8** *A fuzzy graph $G$ is cyclically-transitive if and only if all the strong cycles in $G$ have the same strength.*

*Proof* If all strong cycles in $G$ have the same strength, then $G$ is cyclically-transitive.

Conversely, to prove that all the strong cycles will have same strength in a cyclically-transitive graph, consider a $\beta$-strong edge $e$ in $G$. By definition, $\mu(e) = CONN_G(x, y) = C^G_{x,y}$. So, in a cyclically-transitive fuzzy graph $G$, all $\beta$-strong edges have the same weight. Because the strength of a strong cycle is the weight of some $\beta$-strong edge, all strong cycles will have the same strength. ■

**Corollary 4.4.9** *All cyclically-transitive fuzzy graphs are $\theta$-fuzzy graphs.*

Even though, both connectivity-transitive fuzzy graphs and cyclically-transitive graphs are $\theta$-fuzzy graphs, a $\theta$-fuzzy graph need not be connectivity-transitive or cyclically-transitive.

**Theorem 4.4.10** *In blocks, connectivity-transitive, cyclically-transitive and $\theta$-fuzzy graphs are equivalent.*

*Proof* In blocks with less than three vertices, there are either no edges or exactly one edge and hence the theorem is trivial. Let $G$ be a connectivity-transitive block on at least three vertices. In a connectivity-transitive fuzzy graph, all strong paths have the same strength. Because every two vertices in a block on at least three vertices lie on a common strong cycle, all the strong cycles have the same strength. Hence, $G$ is cyclically-transitive.

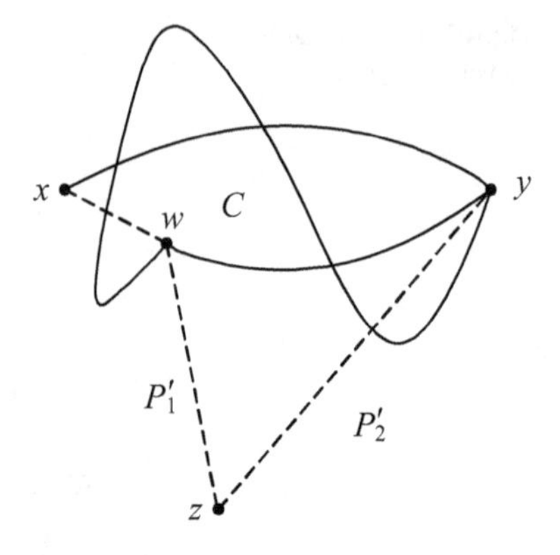


**Fig. 4.32** Theorem 4.4.10-Case I

If $G$ is cyclically-transitive, then all the strong cycles have the same strength and so $G$ is a $\theta$-fuzzy graph. Assume that $G$ is a $\theta$-fuzzy block. We prove, $G$ is cyclically-transitive. If $x$, $y$ and $z$ are three distinct vertices in $G$, we prove they either lie on a common cycle or there are at least two vertices which lie on a common $x-y$, $y-z$ and $z-x$ strong cycle.

Let $C$ be an $x-y$ strong cycle and $C'$ be a $y-z$ strong cycle constituted by two $y-z$ paths, $P_1'$ and $P_2'$. $C'$ intersects $C$ at $y$. $C'$ may or may not intersect $C$ at any other vertex. Cases 1 and 2 give all the different cases when $C'$ intersects $C$, at a point different from $y$ while case 3 is the case where $C'$ intersects $C$ only at $y$.

**Case** 1: Exactly one of $P_1'$ and $P_2'$ intersects $C$.

Suppose that $P_1'$ intersects $C$ and $w$ is the first vertex common to both $P_1'$ and $C$, moving from $z$ to $y$ along $P_1'$ (Fig. 4.32). The $z-w$ subpath of $P_1'$–$(w-y)$ path in $C$ containing $x$–$P_2'$, form a strong cycle $C''$ containing $x$, $y$ and $z$. Hence,

$$C^G_{x,y} = C^G_{y,z} = C^G_{x,z} = s(C'').$$

**Case** 2: Both $P_1'$ and $P_2'$ intersects $C$.

Let $w_i$ be the first vertex common to $P_i'$ and $C$, moving from $z$ to $y$ along $P_i'$, $i = 1, 2$.

**Subcase** 1: $w_1$ and $w_2$ lie on the same $x-y$ path in $C$ (Fig. 4.33).

Suppose that $w_1$ is closer to $x$ than $w_2$. The $z-w_1$ subpath of $P_1'$, $w_1-y$ path of $C$ containing $x$, $y-w_2$ path of $C$ not containing $x$ and $w_2-z$ subpath of $P_2'$ constitute a strong cycle $C''$ containing $x$, $y$ and $z$. So, $C^G_{x,y} = C^G_{y,z} = C^G_{x,z} = s(C'')$.

**Subcase** 2: $w_1$ and $w_2$ lie on different $x-y$ paths in $C$.

Clearly, $w_1$ and $w_2$ lie on $C$ and $C'$. $z-w_1$ subpath of $P_1'$, $w_1-w_2$ path of $C$ containing $x$ and $w_2-z$ subpath of $P_2'$ form a common strong cycle containing $w_1$ and $w_2$. Therefore,

$$C^G_{x,y} = C^G_{y,z} = C^G_{x,z} = C^G_{w_1,w_2}.$$

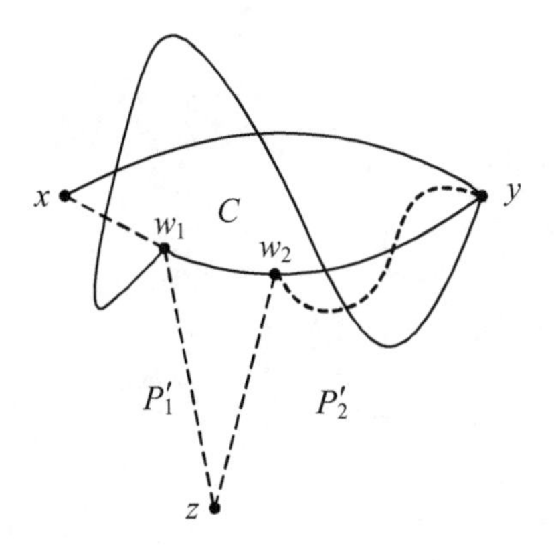


**Fig. 4.33** Theorem 4.4.10-Case2, Subcase1

**Case** 3: $C'$ intersects $C$ only at $y$.

There exists an $x - z$ strong path $P$ which does not contain $y$. Let $w_1$ be the last vertex in $P$ intersecting $C$ and $w_2$ be the first vertex in $P$, intersecting $C'$, moving from $x$ to $z$ along $P$. Assume that $w_1$ precedes $w_2$ moving from $x$ to $z$ along $P$. $x - w_1$ path of $C$ not containing $y$, $w_1 - w_2$ subpath of $P$, $w_2 - z$ path of $C'$ not containing $y$, $z - y$ path of $C'$ not containing $w_2$ and $y - x$ path of $C$ not containing $w_1$, together constitute a strong cycle say $C''$ containing $x$, $y$ and $z$. It follows that

$$C^G_{x,y} = C^G_{y,z} = C^G_{x,z} = s(C'').$$

Because $x$, $y$ and $z$ are distinct arbitrary vertices in $G$, $C^G_{x,y}$ is same for all distinct vertices $x$, $y$ in $G$. Hence, $G$ is cyclically-transitive.

In a block with at least three vertices, $C^G_{x,y} = CONN_G(x, y)$ for all $x$, $y$ not joined by an $\alpha$-strong edge. Therefore, a cyclically-transitive block is connectivity-transitive. ■

# Chapter 5
# More on Connectivity and Distances

In this chapter, we discuss more connectivity concepts and distances in fuzzy graphs. The first three sections deal with connectivity and the rest distances. Starting with the first paper of Rosenfeld [154], connectivity was an intense area of research in fuzzy graph theory. Several other authors including Bhattacharya [37], Mordeson [125–128], Bhutani [42–44], Sunitha and Vijayakumar [166–169], Mathew and Sunitha [110–117] also have contributed much to the study of connectivity in fuzzy graphs. A variety of distances also has been considered in fuzzy graphs in recent years [101, 102, 106, 171].

## 5.1 Connectedness and Acyclic Level of Fuzzy Graphs

The concepts of connectivity and acyclicity are considered in this section. This work is from [62]. Connectedness of different fuzzy graph structures by levels were discussed in Sect. 2.6. This is a continuation of Sect. 2.6.

**Definition 5.1.1** Let $G = (\sigma, \mu)$ be a fuzzy graph. Then the **connectedness level** of $G$ is the value $C(G) = \wedge\{\mu^{\infty}(x, y) \mid x, y \in V,\ x \neq y\}$

Obviously, $G$ is connected if and only if $C(G) > 0$. Moreover, if $C(G) > 0$, then for all $t \in (0, 1]$ such that $t \leq C(G)$, $G^t$ is connected.

*Remark 5.1.2* From Definition 2.6.24, a fuzzy graph $G$ is weakly connected if there is some $t$-cut of $G$ which is connected. Also, we see that a fuzzy graph is weakly connected if and only if $\exists t \in [0, 1]$ such that $\wedge\{\mu^{\infty}(x, y) \mid x, y \in \sigma^t\} \geq t$.

By Proposition 2.6.27, connectedness implies weak connectedness, but not conversely. We see that weak connectedness is meaningful only if $\sigma(x) < 1$ for some $x \in V$ because in this situation the set of vertices changes with the variations of



S. Mathew et al., *Fuzzy Graph Theory*, Studies in Fuzziness
and Soft Computing 363, https://doi.org/10.1007/978-3-319-71407-3_5

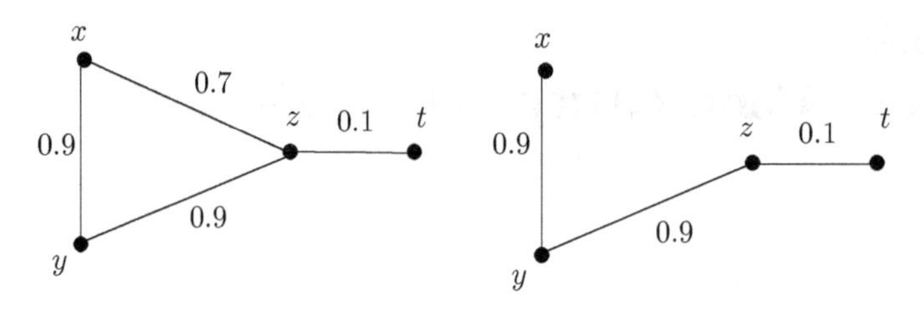


**Fig. 5.1** Fuzzy tree given in Example 5.1.3

membership degree. A fuzzy graph $G = (\sigma, \mu)$ is called an **acyclic fuzzy graph** if there is a fuzzy subgraph $F = (\sigma, \tau)$ of $G$ such that $F$ is a forest and for all $x, y \in \sigma^*$, $\mu(xy) > 0$ and $\tau(xy) = 0$ implies $\tau^\infty(x, y) > \mu(xy)$.

The concept of an acyclic fuzzy graph plays a vital role in general fuzzy graph theory, as it is related to connectedness problems. Further, it is closely related to the concept of a tree. For example, one may define a fuzzy tree as an acyclic and connected fuzzy graph. The following example will lead us to a new notion of an acyclic graph.

*Example 5.1.3* Let $G = (\sigma, \mu)$ be the fuzzy graph given by $\sigma^* = \{x, y, z, t\}$, $\sigma(s) = 1$ for all $s \in \sigma^*$, $\mu(xy) = \mu(yz) = 0.9$, $\mu(xz) = 0.7$ and $\mu(zt) = 0.1$. Then $G$ is a fuzzy tree. Both $G$ and its maximum spanning tree are given in Fig. 5.1 The importance of this fuzzy tree is that there is no $t$-cut of $G$, which is connected in the classical sense.

Given a graph $G$, the **cyclomatic number** of $G$ is defined as $m - n + p$, where $n$, $m$ and $p$ denote the number of vertices, edges and connected components of $G$, respectively.

**Definition 5.1.4** Let $G = (\sigma, \mu)$ be a fuzzy graph with $|\sigma^*| = n$. We call $h(G, .) : [0, 1] \longrightarrow \mathbb{N} \cup \{0\}$ defined by $h(G, t) =$ the cyclomatic number of $G^t$, the **cyclomatic function** of $G$.

If $G$ is a fuzzy graph and $t \in [0, 1]$, we let $n^t$, $m^t$ and $p^t$ denote the number of vertices, edges and connected components of $G^t$, respectively. Then $h(G, t) = m^t - n^t + p^t$.

The following properties of $h(G, .)$ are important for further definitions and developments.

**Proposition 5.1.5** *For every $t \in [0, 1]$, $h(G, t) \geq 0$.*

**Proposition 5.1.6** *$h(G, .)$ is a piecewise constant function with finite jumps.*

These two propositions follow directly from the definition of $h(G, .)$.

Let $G = (\sigma, \mu)$ be a fuzzy graph. If we remove an edge from $G$ to obtain a fuzzy graph $G'$, then $m' = m - 1$, $n' = n$ and $p' \leq p + 1$, where $n'$, $m'$ and $p'$ are the number of vertices, edges and connected components of $G'$, respectively. Hence, $h(G', .) = m' - n' + p' \leq (m-1) - n + (p+1) = h(G, .)$. Now, suppose a vertex $v$ and edges connected to $v$ are removed from $G$. Suppose the number of such edges is $k$. If we remove the edges one at a time, but not $v$, then the resulting graph has less than or equal to $p+k$ connected components. Finally, removing $v$, the resulting graph $G'$ has less than or equal to $p+k-1$ connected components. Hence, $h(G', .) = m' - n' + p' \leq (m-k) - (n-1) + (p+k-1)$. This reasoning leads to the following.

**Proposition 5.1.7** ([62]) *$h(G, .)$ is non increasing in $t$, i.e., for all $t$, $t' \in [0, 1]$, $t \leq t' \Longrightarrow h(G, t) \leq h(G, t')$.*

*Proof* By the properties of $t$-cuts, $m^t \leq n^t \Longrightarrow m^{t'} - k_1$, $k_1 \in \mathbb{Z}$, $k_1 \geq 0$ and $n^t \leq n^{t'} \Longrightarrow n^t = n^{t'} - k_2$, $k_2 \in \mathbb{Z}$, $k_2 \geq 0$.

We cannot conclude anything about the variation of $p^t$ with $t$, because the number of connected components of $G^t$ may increase, decrease or remain the same as $t$ ranges in $[0, 1]$. Thus, for $t \geq t'$, $p^t = pt' - k_3$ for some $k_3 \in \mathbb{Z}$. In this situation,

$$\begin{aligned} h(G, t) &= (m^{t'} - k_1) - (n^{t'} - k_2) + (p^{t'} - k_3) \\ &= h(G, t') + (k_2 - k_1 - k_3) \\ &= h(G, t'). \end{aligned}$$

We will prove that $k > 0$ is impossible. Two possibilities must be considered:

(*i*) $k_2 = 0$. This assumption implies that the vertex set does not change from $G^{t'}$ to $G^t$. Thus, $k_3 \leq 0$, because connected components cannot decrease by a possible edge suppression. Moreover, because a new connected component appears only by edge suppression, $-k_3 \leq k_1 \Longrightarrow k \leq 0$.

(*ii*) $k_2 > 0$. If we denote $h$ to be the number of eliminated connected components of $G^{t'}$, obviously $k_3 \leq h$ and $k_2 \geq h$. Let us write $h = k_2 - s$; $s \in \{0, \ldots, k_2\}$. From the definition of $s$, we can derive $k_1 \geq s$ and thus $k = k_2 - k_3 - k_1 \leq k_2 - k_2 + s - k_1 \Longrightarrow k \leq s - k_1 \leq 0$. ■

Let $H = \{t \in [0, 1] \mid h(G, t) = 0\}$. By Propositions 5.1.6 and 5.1.7, we can assure only two possibilities for $H$: (*i*) $H = \emptyset$ (*ii*) $H = (0, 1]$. So we have the following definition.

**Definition 5.1.8** The **acyclic level** of a fuzzy graph $G$ is $S(G) = \wedge\{t \mid t \in H\}$ and $S(G) = \infty$ if $H = \emptyset$.

The following result can be easily proved from the definition and properties of cyclomatic function.

**Proposition 5.1.9** *There are no cycles in $G^t$ if and only if $S(G) < \infty$ and $t > S(G)$.*

Two variants of acyclic fuzzy graphs can be formulated by means of $S(G)$.

**Definition 5.1.10** The fuzzy graph $G = (\sigma, \mu)$ is said to be **fully acyclic** if $S(G) = 0$.

A fully acyclic fuzzy graph is a forest and conversely. Because $S(G) = 0$, it is equivalent to say the graph formed by the edges with nonzero membership degree must be acyclic. However, this nomenclature emphasizes the acyclic situation underlying in such a fuzzy graph.

**Definition 5.1.11** A fuzzy graph $G = (\sigma, \mu)$ is said to be **acyclic by $t$-cuts** if there exists a $t \in [0, 1]$ such that $G^t$ has no cycles.

Obviously, $G$ will be acyclic by $t$-cuts if and only if $S(G) \neq \infty$.

**Proposition 5.1.12** *Every acyclic fuzzy graph is acyclic by t-cuts.*

*Proof* Let us assume that $G$ is an acyclic fuzzy graph such that $S(G) = \infty$. This implies there is a cycle $L$ in $G$ such that $\mu(xy) = 1$ for every edge $xy$ belonging to $L$. Let $\bar{x}\bar{y}$ be an edge of $L$. Let $\zeta$ denotes the membership function of the fuzzy set of edges in the fuzzy subgraph of $G$ which appears when $\bar{x}\bar{y}$ is suppressed. Then $\zeta^{\infty}(\bar{x}\bar{y}) = 1 \Longrightarrow \zeta^{\infty}(\bar{x}\bar{y}) = \mu(\bar{x}\bar{y})$. Hence, $G$ cannot be an acyclic fuzzy graph and thus we conclude that $S(G) \neq \infty$. ■

The converse of Proposition 5.1.12 does not hold. Consider the fuzzy graph $G$ in Example 5.1.13.

*Example 5.1.13* Let $G = (\sigma, \mu)$ be the fuzzy graph defined by $\sigma^* = \{a, b, c, d\}$, $\sigma(x) = 1$ for all $x \in \sigma^*$, $\mu(ab) = \mu(cd) = 0.9$, $\mu(ad) = \mu(bc) = 0.6$. Obviously, $G^{0.75}$ is an acyclic graph, but the definition of an acyclic fuzzy graph never holds for the edges $ad$ and $bc$.

We now consider several definitions for a fuzzy tree. By using the concepts of connectedness and acyclicity, some fuzzy tree definitions can be introduced. Recall the definitions of Sect. 2.6, we shall make a comparison of results in these sections very soon (Fig. 5.2).

**Definition 5.1.14** The fuzzy graph $G = (\sigma, \mu)$ is called a **full fuzzy tree** if it satisfies the conditions $C(G) > 0$ and $S(G) = 0$.

**Definition 5.1.15** The fuzzy graph $G = (\sigma, \mu)$ is a **complete fuzzy tree** if there exits $t \in [0, 1]$ such that $G^t$ is a tree and $\sigma^t = V$.

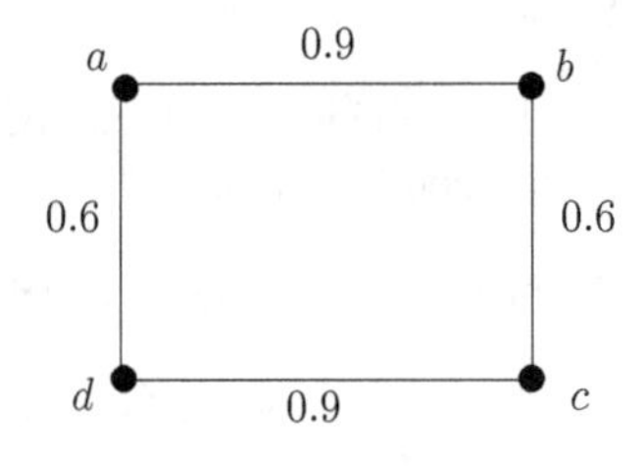


**Fig. 5.2** A fuzzy graph which is acyclic by $t$-cuts

**Lemma 5.1.16** *$G = (\sigma, \mu)$ is a complete fuzzy tree if and only if it satisfies the conditions $C(G) > 0$ and $S(G) < C(G)$.*

*Proof* Let $G^{\bar{t}}$ be the tree which appears in Definition 5.1.15. It is a connected acyclic graph so that $C(G) > \bar{t}$ and $S(G) < \bar{t}$. Therefore, $S(G) < \bar{t} < C(G)$. Let us assume the above conditions hold for $G$. For every $\bar{t} \in (S(G), C(G)]$, $G^{\bar{t}}$ is a tree. Therefore, to prove it is a complete fuzzy tree, it suffices to prove $\sigma^{\bar{t}} = V$, that is, $\sigma(x) \geq \bar{t}$, for all $x \in V$. Let $x \in V$. By the definition of $C(G)$, we have for all $y \in V$, $x \neq y \Longrightarrow \mu^{\infty}(x, y) \geq C(G)$. Because $\mu^{\infty}(x, y)$ is the strength of the strongest chain joining $x$ and $y$, we can assure that $\exists z \in V$ such that $\mu(xz) \geq \mu^{\infty}(x, y)$. Moreover, $\mu(xz) \leq \sigma(x) \wedge \sigma(z)$ and thus $\sigma(x) \geq \mu(xz) \geq \mu^{\infty}(x, y) \geq C(G) \geq \bar{t}$. ■

**Proposition 5.1.17** *If $G$ is a complete fuzzy tree, then for all $t, t' \in (S(G), C(G)]$, $G^t = G^{t'}$.*

*Proof* By Lemma 5.1.16, $\sigma^t = \sigma^{t'} = V$. Therefore, $n = n^{t'} = |\sigma^*|$. Moreover, $G^t$ and $G^{t'}$ are trees and so $|\mu^t| = |\mu^{t'}| = |\sigma^*| - 1$, i.e., both trees have the same number of edges. Now, if $t \leq t'$, then $\sigma^t \supseteq \sigma^{t'}$ and $\mu^t \supseteq \mu^{t'}$. Hence, $\sigma^t = \sigma^{t'}$ and $\mu^t \supseteq \mu^{t'}$. Thus, $G^t = G^{t'}$. ■

By Definition 2.6.8(iii), $G = (\sigma, \mu)$ is called a **weak fuzzy tree** if there is $t' \in (0, 1]$ such that $G^{t'}$ is a tree.

**Theorem 5.1.18** *$G$ is a week fuzzy tree if and only (i) $G$ is weekly connected and (ii) $S(G) < \bar{t}$, $\bar{t}$ being some level such that $G^{\bar{t}}$ is connected.*

The proof of this theorem is similar to that of Lemma 5.1.16. Obviously all these definitions are related as seen in the next proposition.

**Proposition 5.1.19** *The following implications hold.*

*(i) If $G$ is a full fuzzy tree, then $G$ is a complete fuzzy tree.*

*(ii) If $G$ is a complete fuzzy tree, then $G$ is a fuzzy tree and in fact, $G$ is a weak fuzzy tree.*

*Proof* (i) This statement follows from Definition 5.1.14 and Lemma 5.1.16.

(ii) Let $G$ be a complete fuzzy tree. From Lemma 5.1.16, $C(G) > 0$ and $S(G) < C(G)$. For every $t \in (S(G), C(G)]$ we can define the fuzzy subgraph of $G$, $F = (\sigma, \nu)$, where

$$\nu(xy) = \begin{cases} t & \text{if } xy \in \mu^t \\ 0 & \text{otherwise} \end{cases}$$

Because $G^t$ is a tree, obviously $F$ is a full fuzzy tree. Moreover, $\nu^{\infty}(x, y) = t$ if $x \neq y$. Thus, $F$ is acyclic and $G$ is connected. Therefore, $G$ is a fuzzy tree. Thus, $G$ is a complete fuzzy tree implies $G$ is a weak fuzzy tree follows from Lemma 5.1.16 and Definition 2.6.8 with $\bar{t} = C(G)$. ■

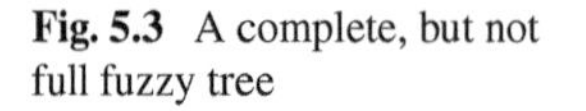
**Fig. 5.3** A complete, but not full fuzzy tree

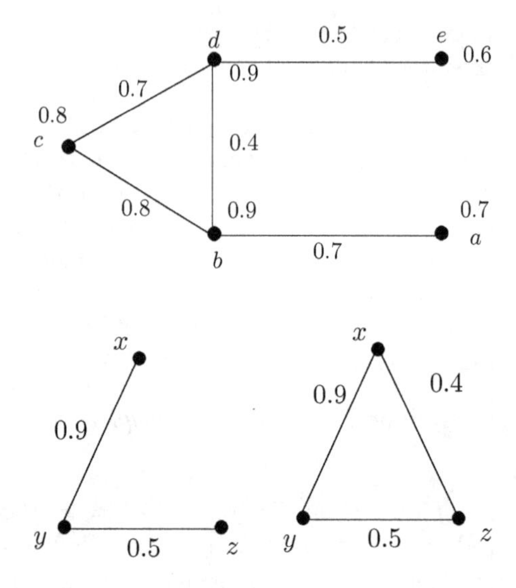


**Fig. 5.4** Fuzzy graphs in Example 5.1.20

The converse of $(i)$ in Proposition 5.1.19 is not true as seen from the fuzzy graph given in Fig. 5.3.

In Fig. 5.3, $S(G) = 0.4$, $C(G) = 0.5$ and $V^{0.5} = V$. Hence, $G$ is a complete fuzzy tree, but not a full fuzzy tree. Also, the fuzzy graph in Example 5.1.3 shows that a fuzzy tree need not be complete.

A crisp graph is acyclic if and only if it's cyclomatic number is 0. Hence, it follows easily that the definitions of a full fuzzy forest given in this section and in Sect. 2.6 agree. However, the following example shows that this is not the case for full fuzzy trees.

*Example 5.1.20* Let $V = \{x, y, z\}$. Let $\sigma$ be the fuzzy subset of $V$ and $\mu$ be the fuzzy subset of $E = \{xy, yz\}$ defined as follows: $\sigma(x) = \sigma(y) = \sigma(z) = 1$ and $\mu(xy) = 0.9$ and $\mu(yz) = 0.5$. Then $C(G) > 0$ and $S(G) = 0$. So $G$ is a full fuzzy tree as per the definition in this section. However, $G$ is not a full fuzzy tree in the sense of Sect. 2.6 because $G^t$ is not a tree for $0.5 < t \leq 0.9$, and in fact $G$ is not a partial fuzzy tree; it is a full fuzzy forest.

If a fuzzy graph $G$ is a full fuzzy tree as per this section, then $G$ is a weak fuzzy tree in the sense of Definition 2.6.8, but the converse does not hold. Let $G$ be the second fuzzy graph given in Fig. 5.4. Then $G$ is not a full fuzzy tree as per this section because $S(G) \neq 0$. However, $G$ is a weak fuzzy tree in the sense of Definition 2.6.8 because $G^t$ is a tree for $0.4 < t \leq 0.5$. In fact, $G$ is a fuzzy tree in the sense of Definition 2.6.8.

## 5.2 Cycle Connectivity of Fuzzy Graphs

The contents of this section are from [112]. Recall the definition of cycle connectivity (Definition 2.8.12) between a pair of vertices in a fuzzy graph discussed in Chap. 2. Cycle connectivity is a measure of connectedness of a fuzzy graph and it is always less than or equal to the strength of connectedness between any two vertices $u$ and $v$. In a graph, the cycle connectivity of any two vertices $u$ and $v$ is 1 if $u$ and $v$ belongs to a common cycle and 0, otherwise. We define cycle connectivity of a fuzzy graph as follows.

**Definition 5.2.1** Let $G = (\sigma, \mu)$ be a fuzzy graph. The **cycle connectivity** of $G$ is defined as $CC(G) = \vee\{C^G_{u,v} \mid u, v \in \sigma^*\}$. That is, the cycle connectivity of a fuzzy graph $G$ is defined as the maximum cycle connectivity of different pairs of vertices in $G$.

Note that for a graph $G$, $CC(G) = 1$ if $G$ is cyclic and $CC(G) = 0$ if $G$ is a tree.

*Example 5.2.2* Let $G = (\sigma, \mu)$ be such that $\sigma^* = \{a, b, c, d\}$, $\sigma(x) = 1$ for all $x \in \sigma^*$ and $\mu(ab) = \mu(bc) = 0.1$, $\mu(db) = 0.3$, $\mu(ac) = 1$, $\mu(cd) = 0.3$, $\mu(ad) = 0.4$. Then $C^G_{a,b} = 0.1$, $C^G_{a,c} = 0.3$, $C^G_{a,d} = 0.3$, $C^G_{b,c} = 0.1 = C^G_{b,d}$. Thus, $CC(G) = 0.3$.

In fact, the cycle connectivity of a fuzzy graph is the maximum of strengths of the cycles in $G$. We have two results whose proofs are obvious.

**Theorem 5.2.3** *A fuzzy graph $G$ is a fuzzy tree if and only if $CC(G) = 0$.*

**Proposition 5.2.4** *The cycle connectivity of a fuzzy cycle $G$ is the strength of $G$.*

Thus, all locamine cycles and multimin cycles have cycle connectivity equal to their strength. Now, we shall find the cyclic connectivity of a complete fuzzy graph.

**Theorem 5.2.5** *Let $G$ be a complete fuzzy graph with vertices $v_1, v_2, \ldots, v_n$ such that $\sigma(v_i) = t_i$ and $t_1 \leq t_2 \leq \cdots \leq t_{n-2} \leq t_{n-1} \leq t_n$. Then $CC(G) = t_{n-2}$.*

*Proof* Assume the conditions of the Theorem. Because any three vertices of $G$ are adjacent, any three vertices are on a 3-cycle. Also, all edges in a CFG are strong. Thus, to find the strength of all cycles in $G$, it is sufficient to find the strength of all 3-cycles of $G$. Among all such 3-cycles, the 3-cycle formed by the vertices of largest vertex strengths will have the maximum strength. Clearly, the cycle $C = v_{n-2}v_{n-1}v_nv_{n-2}$ is of maximum strength in $G$ with strength $t_{n-2} \wedge t_{n-1} \wedge t_n = t_{n-2}$. Thus, $CC(G) = t_{n-2}$. ■

Now, we discuss the concepts of cyclic cutvertices and cyclic bridges. As cutvertices and bridges affect the connectivity of a fuzzy graph on their removal from the graph, cyclic cutvertices and cyclic bridges affect the cycle connectivity of a fuzzy graph on their removal from the fuzzy graph.

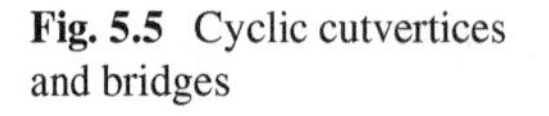

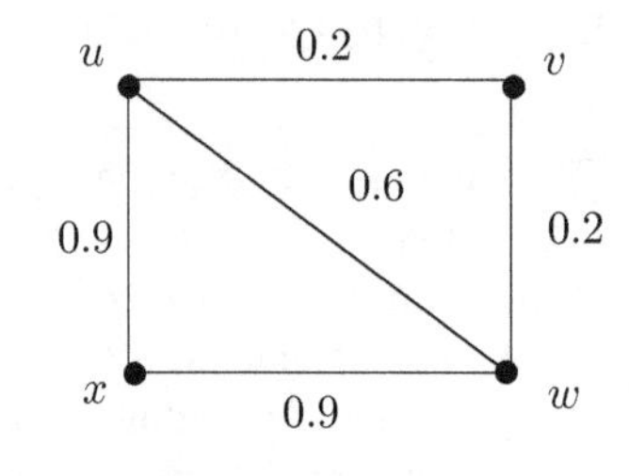


**Fig. 5.5** Cyclic cutvertices and bridges

**Definition 5.2.6** A vertex $w$ in a fuzzy graph is called a **cyclic cutvertex** if $CC(G - w) < CC(G)$ and an edge $uv$ of a fuzzy graph is called a **cyclic bridge** if $CC[G - uv] < CC(G)$.

*Example 5.2.7* Let $G = (\sigma, \mu)$ with $\sigma^* = \{u, v, w, x\}$ and $\mu(uv) = \mu(vw) = 0.2$, $\mu(wx) = \mu(xu) = 0.9$, $\mu(uw) = 0.6$ (Fig. 5.5). In $G$, $w$, $x$ and $u$ are cyclic cutvertices and edges $uw$, $wx$ and $ux$ are cyclic bridges.

**Definition 5.2.8** A fuzzy graph $G$ is said to be **cyclically balanced** if $G$ has no cyclic cutvertices and cyclic bridges.

For example, a fuzzy graph containing two disjoint cycles with the same strength is cyclically balanced. Even edge disjoint graphs are not cyclically balanced as it always contains a cyclic cutvertex (vertex common to the cycles). We have an obvious proposition.

**Proposition 5.2.9** *In a fuzzy graph $G$, if an edge $uv$ is a cyclic bridge, then both $u$ and $v$ are cyclic cutvertices.*

**Proposition 5.2.10** *Let $G$ be a fuzzy graph such that $G^*$ is a cycle. Then*

*(i) $G$ has neither cyclic cutvertices nor cyclic bridges if $G$ is a fuzzy tree.*

*(ii) All edges in $G$ are cyclic bridges and all vertices in $G$ are cyclic cutvertices if $G$ is a strong cycle.*

*Proof* (i) follows from the fact that a fuzzy tree has no strong cycles.

(ii) If $G$ is a strong cycle, then $CC(G) =$ strength of $G$. The removal of any edge or vertex will reduce its cycle connectivity to 0. ■

From Proposition 5.2.10(ii), it follows that every fuzzy bridge of a fuzzy cycle is a cyclic bridge and every fuzzy cutvertex of a fuzzy cycle is a cyclic cutvertex. This result can be generalized as follows.

**Proposition 5.2.11** *Let $G$ be a fuzzy graph containing at most one fuzzy cycle. Then every fuzzy bridge of $G$ is a cyclic bridge and every fuzzy cutvertex of $G$ is a cyclic cutvertex.*

**Proposition 5.2.12** *Let $G$ be a fuzzy graph such that $G$ has a unique fuzzy cycle $C$ such that strength of $C = CC(G)$. Then every fuzzy bridge in $C$ is a cyclic bridge and every fuzzy cutvertex in $C$ is a cyclic cutvertex.*

From Proposition 5.2.10(ii) it follows that every edge of a CFG with exactly three vertices are cyclic cutvertices and all its edges are cyclic bridges. But when a CFG contains more number of vertices it is not true as seen from the following theorem.

**Theorem 5.2.13** *Let $G = (\sigma, \mu)$ be a CFG with $|\sigma^*| \geq 4$. Let $v_1, v_2, \ldots, v_n \in \sigma^*, \sigma(v_i) = c_i$ for $i = 1, 2, \ldots, n$ and $c_1 \leq c_2 \leq \cdots \leq c_n$. Then $G$ has a cyclic cutvertex (or cyclic bridge) if and only if $c_{n-3} < c_{n-2}$. Further, there will be exactly three cyclic cutvertices (or cyclic bridges) in a CFG (if they exist).*

*Proof* Let $v_1, v_2, \ldots, v_n \in \sigma^*, \sigma(v_i) = c_i$ for $i = 1, 2, \ldots, n$ and $c_1 \leq c_2 \leq \cdots \leq c_n$. Suppose that $G$ has a cyclic cutvertex $u$ (say). Then $CC(G - u) < CC(G)$. That is, $u$ belongs to a unique cycle $C$ with $\alpha =$ strength of $C >$ strength of $C'$ for any other cycle $C'$ in $G$. Because $c_1 \leq c_2 \leq \cdots \leq c_n$, it follows that the strength of the cycle $v_{n-2}v_{n-1}v_n$ is $\alpha$. Hence,

$$u \in \{v_{n-2}, v_{n-1}, v_n\}. \tag{5.1}$$

To prove $c_{n-3} < c_{n-2}$. Suppose not, i.e., $c_{n-3} = c_{n-2}$. Then $C_1 = v_n v_{n-1} v_{n-2}$ and $C_2 = v_n v_{n-1} v_{n-3}$ have the same strength and hence the removal of $v_{n-2}$, $v_{n-1}$ or $v_n$ will not reduce $CC(G)$, which is a contradiction to (5.1). Hence, $c_{n-3} < c_{n-2}$.

Conversely, suppose that $c_{n-3} < c_{n-2}$. To prove $G$ has a cyclic cutvertex. Because $c_n \geq c_{n-1} \geq c_{n-2}$ and $c_{n-2} > c_{n-3}$, all cycles of $G$ have strength less than the strength of $v_n v_{n-1} v_{n-2}$. Thus, the deletion of any vertex in $\{v_n, v_{n-1}, v_{n-2}\}$ will reduce the cycle connectivity of $G$. Hence, $v_n$, $v_{n-1}$ and $v_{n-2}$ are cyclic cutvertices of $G$. Also, from the proof, it follows that there is three cyclic cutvertices if they exist.

The case of cyclic bridges is similar. ■

Now, we have a theorem whose proof is obvious.

**Theorem 5.2.14** *Let $G$ be an edge disjoint fuzzy graph which is not a tree. A cutvertex of $G$ is a cyclic cutvertex if it is the common vertex of all cycles of $G$ or it is a vertex of a unique cycle having maximum strength.*

**Definition 5.2.15** Let $G = (\sigma, \mu)$ be a fuzzy graph. A **cyclic vertex cut** or a **cyclic node cut** ($CNC$) of $G$ is a set of vertices $X \subseteq \sigma^*$ such that $CC(G - X) < CC(G)$, provided $CC(G) > 0$, where $CC(G)$ is the cycle connectivity of $G$.

*Example 5.2.16* Let $G = (\sigma, \mu)$ be a fuzzy graph (Fig. 5.6) with $\sigma^* = \{a, b, c, d\}$, $\sigma(s) = 1$ for all $s \in \sigma^*$ and $\mu(ab) = \mu(ad) = 0.9$ and $\mu(bc) = \mu(cd) = \mu(da) = \mu(ca) = \mu(bd) = 0.5$. Here, $abcda$ and $acda$ are cycles of strength 0.5. Also, $CC(G) = 0.5$. Set $X = \{\{a, c\}\}$ is a $2 - CNC$ of $G$.

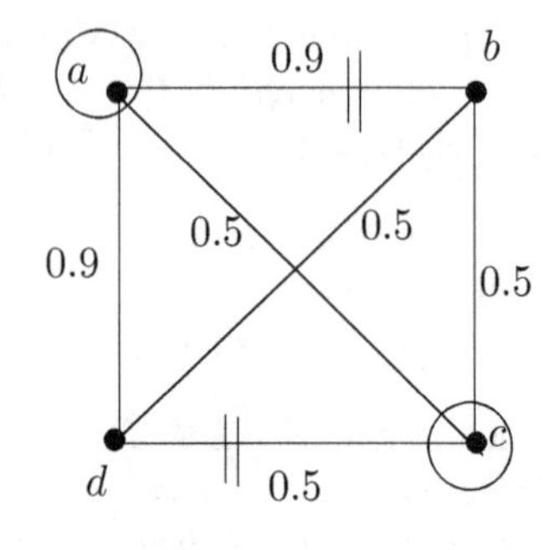


**Fig. 5.6** CNC and CAC in a fuzzy graph

**Definition 5.2.17** Let $X$ be a cyclic vertex cut of $G$. The **strong weight** of $X$ is defined as $S_c(X) = \sum_{x \in X} \mu(xy)$, where $\mu(xy)$ is the minimum of weights of strong edges incident on $x$. **Cyclic vertex connectivity** of a fuzzy graph $G$, denoted by $\kappa_c(G)$, is the minimum of strong weights of cyclic vertex cuts in $G$.

In Example 5.2.16, $C_1 = \{\{a, c\}\}$, $C_2 = \{\{a, d\}\}$, $C_3 = \{\{b, c\}\}$, $C_4 = \{\{b, d\}\}$, $C_5 = \{\{a, b\}\}$ and $C_6 = \{\{c, d\}\}$ are 2-*CNC*s with $S(C_1) = 1, S(C_2) = 1, S(C_3) = 1$, $S(C_4) = 1$, $S(C_5) = 1$, $S(C_6) = 1$. Thus, the cyclic vertex connectivity is 1.

**Definition 5.2.18** Let $G = (\sigma, \mu)$ be a fuzzy graph. A **cyclic edge cut** or a **cyclic arc cut** (**CAC**) of $G$ is a set of edges $Y \subseteq \mu^*$ such that $CC(G - Y) < CC(G)$, provided $CC(G) > 0$, where $CC(G)$ is the cyclic connectivity of $G$.

*Example 5.2.19* In Fig. 5.6, all cycles are of strength 0.5. Also, $CC(G) = 0.5$. Thus, $Y = \{ab, cd, ad\}$ is a 3-*CAC* of $G$.

**Definition 5.2.20** Let $G = (\sigma, \mu)$ be a fuzzy graph. The **strong weight** of a cyclic edge cut $Y$ of $G$ is defined as $S_c^{'}(Y) = \sum_{e_i \in \mu^*} \mu(e_i)$, where $e_i$ is a strong edge of $Y$. The **cyclic edge connectivity** of $G$, denoted by $\kappa_c^{'}(G)$, is the minimum of strong weights of cyclic edge cuts in $G$.

For the fuzzy graph in Fig. 5.6, cyclic edge connectivity is $\mu(ab) + \mu(cd) + \mu(ad) = 0.6 + 0.6 + 0.5 = 1.7$.

Next we discuss relationship between cyclic vertex connectivity and vertex connectivity of a complete fuzzy graph.

**Theorem 5.2.21** *For a complete fuzzy graph $G$, $\kappa_c(G) \leq \kappa(G)$.*

*Proof* Let $G$ be a complete fuzzy graph and let the vertices of $G$ be labeled as $v_1, v_2, \ldots, v_n$ such that $d(v_1) \leq d(v_2) \leq \cdots \leq d(v_n)$. Let $v_1$ be a vertex such that $d(v_1) = \delta_s(G)$ (Fig. 5.7).

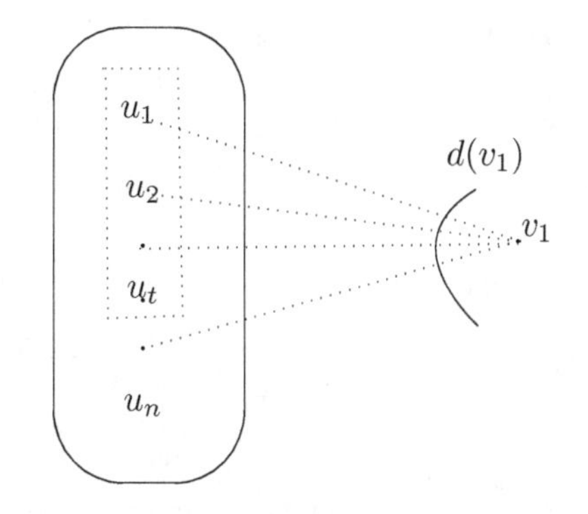


**Fig. 5.7** Minimum strong degree of a vertex

**Case 1**: $v_1$ is a cyclic cutvertex.

In this case, $\{v_1\}$ is a cyclic cutset and therefore,

$$S_c\{v_1\} = \underset{v_i \in \mu^*}{\wedge} \mu(v_1 v_i), \text{ for } i = \{2, \ldots, n\} \leq \sum_{v_i \in \mu^*} \mu(v_1 v_i) = \delta_s(G).$$

Now, because $\kappa_c(G) = \vee\{S_c(V)\}$, where $V$ is a cyclic cutset of $G$, we have, $\kappa_c(G) \leq S_c\{v_1\} = \delta_s(G) = \kappa(G)$.

**Case 2**: $\{v_1\}$ is not a cyclic cutvertex.

Let $F = \{u_1, u_2, \ldots, u_t\}$ be a cyclic cutset such that $S_c(F) = \kappa_c(G)$. Now, $\kappa_c(G) = S_c(F) = \sum_{i=1}^{t} \wedge\{\mu(u_i u_j) \mid u_i u_j \in \mu^*, \quad j \neq i, \quad j = 1, 2, \ldots, n\}$. Clearly, $\kappa_c(G) = \sum_{i=1}^{t} \mu(u_i v_1) \leq d(v_1) = \delta_s(G) = \kappa(G)$. ∎

**Corollary 5.2.22** *For a fuzzy tree $G = (\sigma, \mu)$, $\kappa_c(G) \leq \kappa(G)$.*

Proof follows from the fact that there are no strong cycles in a fuzzy tree and hence $\kappa_c(G) = 0$.

**Theorem 5.2.23** *A vertex in a fuzzy graph is a cyclic cutvertex if and only if it is a common vertex of all strong cycles with maximum strength.*

*Proof* Let $G$ be a fuzzy graph. Let $w$ be a cyclic cutvertex of $G$. Then $CC(G - w) < CC(G)$. That is, $\vee\{s(C)$, where $C$ is a strong cycle in $G - w\} < \vee\{S_c(C')$, where $C'$ is a strong cycle in $G\}$. Therefore, all strong cycles in $G$ having maximum strength will be removed by the deletion of $w$. Hence, $w$ is a common vertex of all strong cycles with maximum strength.

Conversely, let $w$ be a common vertex of all strong cycles with maximum strength. Then the removal of $w$ results in the deletion of all strong cycles with maximum strength. Hence, it will results in the reduction of cycle connectivity of $G$. Thus, $w$ is a cyclic cutvertex of $G$. ∎

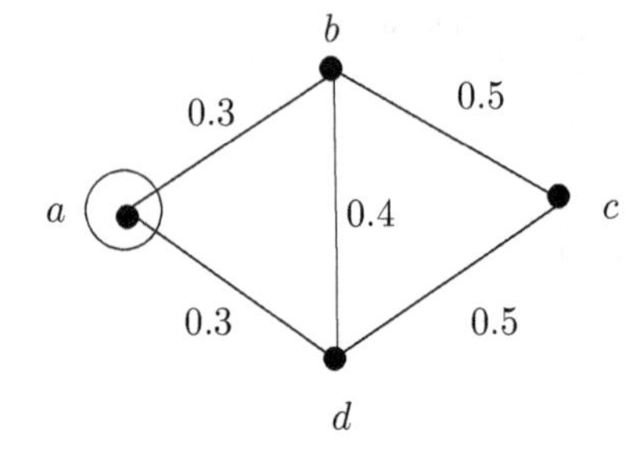


**Fig. 5.8** Cyclic endvertex of a graph

**Definition 5.2.24** A vertex $w \in \sigma^*$ of a fuzzy graph $G$ is said to be a **cyclic fuzzy endvertex**, if it lies on a strong cycle, but which is not a cyclic cutvertex.

*Example 5.2.25* Let $G = (\sigma, \mu)$ with $\sigma^* = \{a, b, c, d\}$, $\sigma(s) = 1$ for all $s \in \sigma^*$ and $\mu(ab) = \mu(ad) = 0.3$, $\mu(bc) = \mu(cd) = 0.5$ and $\mu(bd) = 0.4$ (Fig. 5.8). Here, $abcda$ is a strong cycle. Also, $CC(G) = 0.4$. $a$ is a cyclic fuzzy endvertex of $G$.

**Theorem 5.2.26** *Let $G = (\sigma, \mu)$ be a fuzzy graph. Then no cyclic cutvertex is a fuzzy endvertex of $G$.*

*Proof* Let $G = (\sigma, \mu)$ be a fuzzy graph. Let $w$ be a cyclic cutvertex of $G$, then $w$ lies on a strong cycle with maximum strength in $G$. Clearly, $w$ has at least two strong neighbors in $G$. Hence, $w$ cannot be a fuzzy endvertex of $G$.

Conversely, if $w$ is a fuzzy endvertex of $G$ with $|N_s(w)| = 1$, then $w$ cannot lie on a strong cycle in $G$, which implies that $w$ is not a cyclic cutvertex. ∎

**Corollary 5.2.27** *No cyclic cutvertex is a cyclic fuzzy endvertex of $G$.*

**Theorem 5.2.28** *Let $G = (\sigma, \mu)$ be a complete fuzzy graph with $|\sigma^*| \geq 4$. Let $v_1, v_2, \ldots, v_n \in \sigma^*$ and $\sigma(v_i) = c_i$ for $i = 1, 2, \ldots, n$ and $c_1 \leq c_2 \leq \cdots \leq c_n$. Then $G$ is cyclically balanced if and only if $c_{n-3} = c_{n-2}$.*

*Proof* Let $v_1, v_2, \ldots, v_n \in \sigma^*$ and $\sigma(v_i) = c_i$ for $i = 1, 2, \ldots, n$ and $c_1 \leq c_2 \leq \cdots \leq c_n$. If possible suppose that $G$ is cyclically balanced. To prove that $c_{n-3} = c_{n-2}$. Suppose not, that is $c_{n-3} < c_{n-2}$. Because $c_{n-2} \leq c_{n-1} \leq c_n$ and $c_{n-3} < c_{n-2}$, all cycles of $G$ have strengths less than that of strength of $v_n v_{n-1} v_{n-2} v_n$. Hence, the deletion of any of the vertices in $\{v_n, v_{n-1}, v_{n-2}\}$ reduce the cycle connectivity of $G$. Hence, $v_n$, $v_{n-1}$ and $v_{n-2}$ are cyclic cutvertices of $G$, which is a contradiction to the fact that $G$ is cyclically balanced.

Conversely, suppose that $c_{n-3} = c_{n-2}$. Then $C_1 = v_n v_{n-1} v_{n-2}$ and $C_2 = v_n v_{n-1} v_{n-3}$ have the same strength and hence the removal of $v_{n-2}$, $v_{n-1}$ or $v_n$ will not reduce the cyclic connectivity of $G$. That is, there does not exist any cyclic cutvertex in $G$. Hence, the fuzzy graph $G = (\sigma, \mu)$ is cyclically balanced. ∎

**Theorem 5.2.29** *Let $G$ be a complete fuzzy graph. $G$ is cyclically balanced if there exists a $K_4$ as a sub graph of $G$ in which every cycle is of equal maximum strength.*

*Proof* Let $G$ be a complete fuzzy graph with $|\sigma^*| \geq 4$. Let $v_1, v_2, \ldots, v_n \in \sigma^*$, $\sigma(v_i) = c_i$ for $i = 1, 2, \ldots, n$ and $c_1 \leq c_2 \leq \cdots \leq c_n$. Let $K_4$ be a fuzzy subgraph of $G^*$ with vertex set $\{v_{n-3}, v_{n-2}, v_{n-1}, v_n\}$ such that $c_{n-3} \leq c_{n-2} \leq c_{n-1} \leq c_n$. Suppose all the strong cycles in the fuzzy graph induced by the vertices of this $K_4$ are of equal maximum strength. This happens only when $c_{n-3} = c_{n-2}$. Thus, by Theorem 5.2.28, $G$ is cyclically balanced. ■

**Theorem 5.2.30** *Let $G = (\sigma, \mu)$ be a complete fuzzy graph and $v \in \sigma^*$ such that $d_s(v) = \Delta_s(G)$. Then $v$ lies on a strong cycle $C$ such that $CC(G)$ is the strength of $C$.*

*Proof* Let $G = (\sigma, \mu)$ be a complete fuzzy graph and $v \in \sigma^*$ be a vertex such that $d_s(v) = \Delta_s(G)$. Let $v_1, v_2, \ldots, v_n \in \sigma^*$, $\sigma(v_i) = c_i$ for $i = 1, 2, \ldots, n$ and $c_1 \leq c_2 \leq \cdots \leq c_n$. Because $c_{n-2} \leq c_{n-1} \leq c_n$ and $c_{n-3} < c_{n-2}$, all cycles of $G$ have strength less than that of the strength of $v_n v_{n-1} v_{n-2} v_n$. First to prove that for all $v_i$, $d(v_i) < d(v_n)$.

$$d(v_i) = \sum_{j=1, j\neq i}^{n-1} (c_i \wedge c_j) = c_1 + c_2 + \cdots + (n-i)c_n < \sum c_i = d(v_n).$$

Also,

$$\begin{aligned} d(v_n) &= (c_n \wedge c_n) + \sum_{j=1}^{n-2} (c_n \wedge c_i) = (c_{n-1} \wedge c_n) + \sum_{j=1}^{n-2} (c_{n-1} \wedge c_i) \\ &= d(v_{n-1}) = \sum_{i=1}^{n-1} c_i = \Delta_s(G). \end{aligned}$$

Therefore, $v_n$ belongs to the strong cycle $C : c_{n-2}c_{n-1}c_n c_{n-2}$, where strength of this cycle $C$ is equal to the cycle connectivity of the fuzzy graph $G$. ■

Next we give a result analogous to the Whitney's theorem in fuzzy graphs.

**Theorem 5.2.31** *For a complete fuzzy graph, $\kappa_c(G) \leq \kappa_c'(G) \leq \Delta_s(G)$.*

*Proof* Consider all cycles, $C_1, C_2, \ldots, C_n$ having strengths equal to cycle connectivity of $G = (\sigma, \mu)$. Let $X = \{e_1, e_2, \ldots, e_n\}$, where $e_i = u_i v_i$ be one of the edges from each $C_i$. Then $X$ form a cyclic edge cut of $G$. Let $S_c(X)$ be the strong weight of $X$. Then by the definition of cyclic edge connectivity,

$$\kappa_c'(G) \leq S_c(X). \tag{5.2}$$

Let $Y = \{v_1, v_2, \ldots, v_n\}$ be the collection of one of the endvertices of each edge in the cyclic edge cut $X$ of $G$. Then $Y$ form a cyclic vertex cut of $G$. Let $S_c(Y)$ be the strong weight of $Y$. Then

$$S_c(Y) \leq \kappa_c^{'}(G), \tag{5.3}$$

by the definition of cyclic vertex connectivity of $G$. Hence,

$$\kappa_c(G) \leq S_c(Y). \tag{5.4}$$

From Eqs. (5.3) and (5.4), $\kappa_c(G) \leq S_c(Y) \leq \kappa_c^{'}(G) \leq \Delta_s(G)$. ■

**Theorem 5.2.32** *A fuzzy graph $G = (\sigma, \mu)$ with $n \geq 6$ is cyclically balanced if there exist two disjoint cycles $C_1$ and $C_2$ such that Strength of $C_1$ = Strength of $C_2$ = $CC(G)$.*

*Proof* Let $G = (\sigma, \mu)$ be a fuzzy graph with $n \geq 6$ and cycle connectivity of $G$ is equal to $CC(G)$. Let $C_1$ and $C_2$ be two disjoint cycles in $G$ such that Strength of $C_1$ = Strength of $C_2 = CC(G)$.

Suppose $u$ is a vertex not in $(C_1 \cup C_2)$. Then the cycle connectivity of $G - \{u\}$ remains the same. If the vertex $u$ is in $C_1$ and if $u$ is deleted, then the cycle connectivity remains the same, because there exists another cycle $C_2$, with strength equal to the cycle connectivity of the fuzzy graph. (The proof is similar if $u$ is in $C_2$). Thus, $u$ is not a cyclic cutvertex.

Suppose $uv \in E$, but $uv$ is not in $C_1 \cup C_2$. Then the deletion of $uv$ will not reduce the cycle connectivity of the fuzzy graph. If $uv$ is an edge either on $C_1$ or on $C_2$, then also, the removal of $uv$ from any one of these cycles will not affect the cycle connectivity of $G$. Hence, $uv$ is not a cyclic bridge. ■

A fuzzy graph with $n = 4, 5$ is cyclically balanced if and only if there exist at least 4 cycles having equal maximum strength as seen from Fig. 5.9.

Next we will construct cyclically balanced fuzzy graphs on more than 5 vertices.

**Theorem 5.2.33** *For any $n \geq 4$, there is a connected cyclically balanced fuzzy graph $G = (\sigma, \mu)$ with $|\sigma^*| = n$.*

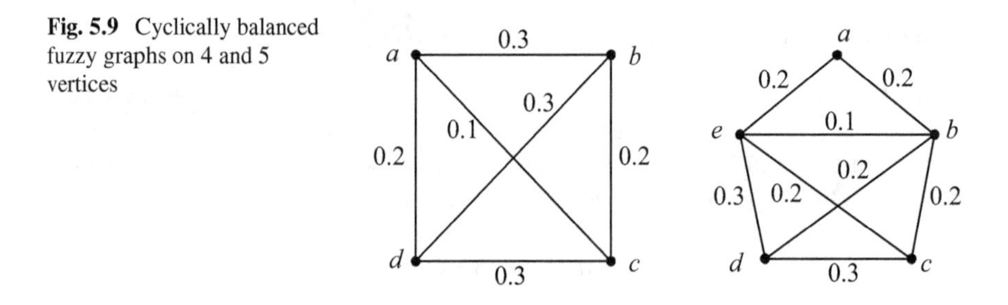


**Fig. 5.9** Cyclically balanced fuzzy graphs on 4 and 5 vertices

*Proof* For $|\sigma^*| = 4$ and 5, we have the examples in Fig. 5.9. For $n \geq 6$, we obtain the fuzzy graphs by induction. For $|\sigma^*| = 6$, let $v_1, v_2, \ldots, v_6$ be the 6 vertices. Construct two disjoint cycles (say) $C_1 = v_1v_2v_3v_1$ and $C_2 = v_4v_5v_6v_4$ with maximum strength.

Join each pair of vertices from the two cycles and make the graph complete. Then the removal of an edge or a vertex will not reduce the cycle connectivity of $G$. So, the newly obtained fuzzy graph is cyclically balanced.

Assume that the result is true for $|\sigma^*| = k$. Let $G_k$ be a cyclically balanced fuzzy graph with $k$ vertices. Then there exist two disjoint cycles of maximum strength in $G_k$.

Let $G_{k+1}$ be the fuzzy graph obtained from $G_k$ by adding one more vertex $u$. Make the fuzzy graph complete by connecting all vertices of $G_k$ with $u$. Also, assign a membership value to all newly joined edges, which is less than or equal to the cycle connectivity of $G_k$.

In this case, if we remove the vertex $u$, then the cycle connectivity of $G_k$ remains the same. In a similar way the removal of any edge incident on $k+1$th vertex $u$ will not change the cycle connectivity of $G$. Therefore, the cycle connectivity of $G_{k+1}$ remains the same. Hence, $G_{k+1}$ is cyclically balanced. ■

## 5.3 Bonds and Cutbonds in Fuzzy Graphs

In this section, two different types of bridges, namely bonds and cutbonds in fuzzy graphs are discussed. Bonds in fuzzy graphs were introduced in [114]. As noted before, the behavior of fuzzy bridges is different in various fuzzy graph structures. Thus, a detailed discussion on special bridges is made in this section. The concept of critical cutvertices in fuzzy graphs are also discussed, and their relations with fuzzy bonds and fuzzy cutbonds are provided. This section is based on [111].

According to Definition 3.3.23, an edge $xy$ in a fuzzy graph $G$ is said to be a fuzzy bond if $CONN_{G-xy}(u, v) < CONN_G(u, v)$ for some pair of vertices $u$ and $v$ with at least one of them different from $x$ and $y$.

*Example 5.3.1* Let $G = (\sigma, \mu)$ with $\sigma^* = \{a, b, c, d, e\}$, $\sigma(s) = 1$ for all $s \in \sigma^*$ and $\mu(ab) = 0.3$, $\mu(ac) = 0.1$, $\mu(bc) = 0.4$, $\mu(cd) = 0.5$, $\mu(da) = 0.6$, $\mu(ae) = \mu(be) = 0.5$. There are 4 fuzzy bonds (1-FAC) in this fuzzy graph namely edges $az$, $ae$, $dc$ and $eb$. Also, $E = \{ab, dc\}$ is a 2-FAC because $0.4 = CONN_{G-E}(e, c) < CONN_G(e, c) = 0.5$.

In graph theory, a minimal cut is a bond. Hence, all bridges are bonds. But in fuzzy graphs this is not true. For example, a complete fuzzy graph can contain a fuzzy bridge and because it has no fuzzy cutvertices, this fuzzy bridge, cannot be a fuzzy bond.

As stated in Proposition 3.3.26, at least one of the end vertices of a fuzzy bond is a fuzzy cutvertex.

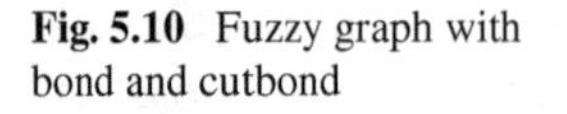

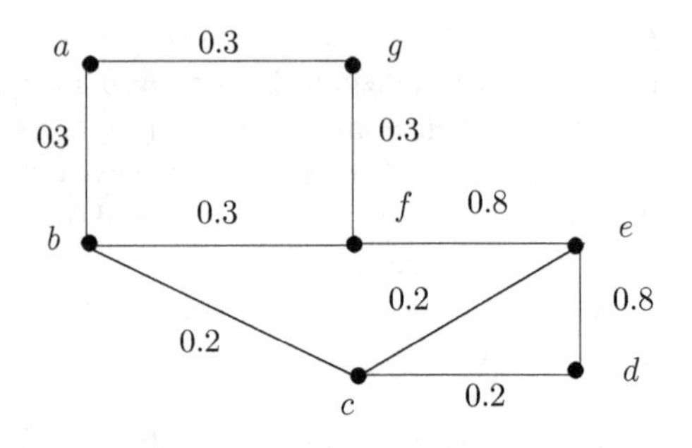


**Fig. 5.10** Fuzzy graph with bond and cutbond

**Definition 5.3.2** An edge $uv$ in a fuzzy graph $G$ is called a **fuzzy cutbond (f-cutbond)** if $CONN_{G-uv}(x, y) < CONN_G(x, y)$ for some pair of vertices $x, y$ in $G$ such that $x \neq u \neq v \neq y$.

Consequently, we can define cutbonds in a graph. Let $G$ be a graph. Then an edge $uv$ is said to be a cutbond if $G - uv$ has at least two nontrivial components.

All fuzzy cutbonds are fuzzy bonds and hence are fuzzy bridges. An edge is a fuzzy bridge if and only if it is $\alpha$-strong. Thus, both fuzzy bonds and fuzzy cutbonds are special type of $\alpha$-strong edges and we some times refer them by $\alpha^*$-edges and $\alpha^{**}$-edges, respectively.

*Example 5.3.3* Let $G = (\sigma, \mu)$ be such that $\sigma^* = \{a, b, c, d, e, f, g\}$, $\sigma(s) = 1$ for all $s \in \sigma^*$ and $\mu(ab) = \mu(ag) = \mu(bf) = \mu(gf) = 0.3$, $\mu(bc) = \mu(cd) = \mu(ce) = 0.2$, $\mu(de) = \mu(ef) = 0.8$ (Fig. 5.10). In this fuzzy graph, $de$ is a fuzzy bond as $0.2 = CONN_{G-de}(g, d) < CONN_G(g, d) = 0.3$. Also, the bridge $ef$ is a fuzzy cutbond because $0.2 = CONN_{G-ef}(g, d) < CONN_G(g, d) = 0.3$. $g$ and $d$ are different from both $e$ and $f$.

Now, we shall discuss about bonds and cutbonds in fuzzy trees. The bonds of a fuzzy tree can be identified from the following theorem.

**Theorem 5.3.4** ([111]) *Let $G = (\sigma, \mu)$ be a fuzzy tree with $|\sigma^*| \geq 3$. An edge $xy$ in $G$ is a fuzzy bond if and only if $xy$ is an edge of the unique maximum spanning tree $F = (\sigma, \nu)$ of $G$.*

*Proof* Let $G = (\sigma, \mu)$ be a fuzzy tree with $|\sigma^*| \geq 3$ and having its unique maximum spanning tree $F = (\sigma, \nu)$. First suppose that $xy$ is a fuzzy bond in $G$. Then $CONN_{G-xy}(u, v) < CONN_G(u, v)$, for two vertices $u$ and $v$ such that at least one of $u$ or $v$ is different from both $x$ and $y$. Thus, all strongest $u - v$ paths in $G$ passes through $xy$. Because $G$ is a fuzzy tree, there exists a unique strongest path between each pair of vertices in $G$. Precisely it is the unique path in $F$. Hence, all strongest $u - v$ paths in $G$ are in $F$. Because all such paths contain $xy$, it follows that $xy$ is an edge in $F$.

Conversely suppose that $xy \in F$. Because $|\sigma^*| \geq 3$, both $x$ and $y$ cannot be fuzzy end vertices of $G$. Thus, one of them say $x$ is a fuzzy cutvertex. Being a fuzzy tree, there exists a unique strongest path between any two vertices of $G$ and it belongs to $F$. Let $w$ be a strong neighbor of $x$ other than $y$. Then there exists a unique strongest path between $y$ and $w$ in $F$ and hence in $G$. Clearly, this path passes through the edge $xy$ and hence it follows that $CONN_{G-xy}(y, w) < CONN_G(y, w)$. Thus, $xy$ is a fuzzy bond of $G$. ∎

From the last theorem, it follows that every strong edge of a fuzzy tree are fuzzy bonds as all strong edges of $G$ belongs to its unique maximum spanning tree (MST) $F$. Also, an edge in a fuzzy graph is a fuzzy bridge if and only if it is an edge in every MST of $G$. But this result is not true in the case of fuzzy bonds. That is, an edge which is in every maximum spanning tree need not be a fuzzy bond. For, if a complete fuzzy graph contains a fuzzy bridge, then it belongs to all MST's, but a CFG has no fuzzy cutvertices. Thus, it follows that a CFG has no fuzzy bonds.

In view of the above, we have the following result.

**Theorem 5.3.5** *Let $G$ be a fuzzy graph. Then $G$ is a fuzzy tree if and only if every strong edge of $G$ is a fuzzy bond of $G$.*

Next we characterize cutbonds in a fuzzy tree.

**Theorem 5.3.6** (Characterization of cutbonds in a fuzzy tree) *Let $G = (\sigma, \mu)$ be a fuzzy tree. An edge $uv$ of $G$ is a fuzzy cutbond if and only if $u$ and $v$ are fuzzy cutvertices of $G$.*

*Proof* Let $G = (\sigma, \mu)$ be a fuzzy tree and $uv$ be a fuzzy cutbond of $G$. Then $CONN_{G-uv}(x, y) < CONN_G(x, y)$, for some $x, y \in \sigma^*$ such that $x \neq u \neq v \neq y$. Thus, every strongest $x - y$ path in $G$ passes through $uv$. Because the deletion of $u$ or $v$ annihilates all such paths, it follows that both $u$ and $v$ are fuzzy cutvertices of $G$.

Conversely suppose that $uv$ is an edge of $G$ such that $u$ and $v$ are fuzzy cutvertices of $G$. We have to show that $uv$ is a fuzzy cutbond. Clearly, $uv$ is an internal edge. (An edge such that $d(u) > 1$ and $d(v) > 1$) of $F$, the unique maximum spanning tree of $G$. Also, $uv$ is the unique strongest $u - v$ path in $F$ and hence in $G$ with strength $\mu(uv)$. Now, let $x_1, x_2, y_1, y_2 \in \sigma^*$ be vertices of $G$ such that $CONN_{G-u}(x_1, x_2) < CONN_G(x_1, x_2)$ and $CONN_{G-v}(y_1, y_2) < CONN_G(y_1, y_2)$. Let $P_1$ be the unique strongest $x_1 - x_2$ path passing through $u$ and $P_2$ be the unique strongest $y_1 - y_2$ path passing through $v$. Let $u'$ be a vertex in $P_1$ adjacent to $u$ and $v'$ be a vertex in $P_2$ adjacent to $v$. Then $CONN_{G-uv}(x', y') < CONN_G(x', y')$, where $u \neq x' \neq y' \neq v$. Thus, it follows that $uv$ is a fuzzy cutbond of $G$. ∎

**Corollary 5.3.7** *A CFG has no fuzzy cutbonds.*

The proof follows from the fact that a CFG has no fuzzy bonds.

**Corollary 5.3.8** *If $G$ be a block, then no fuzzy bridge of $G$ is a fuzzy bond of $G$.*

The converse of Corollary 5.3.8 is not true.

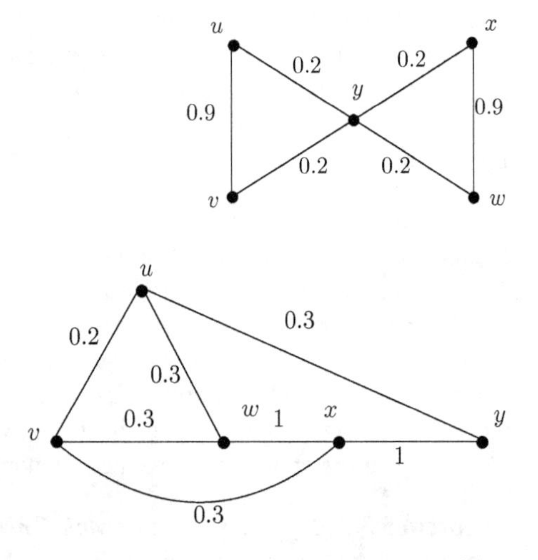


**Fig. 5.11** Fuzzy Graph in Example 5.3.9

**Fig. 5.12** Counter example for the general case of Theorem 5.3.10

*Example 5.3.9* Let $G = (\sigma, \mu)$ be a fuzzy graph with $\sigma^* = \{u, v, w, x, y\}$, $\sigma(s) = 1$ for all $s \in \sigma^*$ with $\mu(uv) = 0.9 = \mu(xw)$, $\mu(uy) = \mu(vy) = \mu(wy) = \mu(xy) = 0.2$ (Fig. 5.11). In this fuzzy graph, edges $uv$ and $wx$ are fuzzy bridges, which are not fuzzy bonds. But vertex $y$ is a fuzzy cutvertex of $G$.

Using the property of fuzzy cutvertices in a fuzzy tree, we have another easy result.

**Theorem 5.3.10** *Let $G$ be a fuzzy tree. Then an edge $xy$ of $G$ is a fuzzy cutbond if and only if $d_s(u) > 1$ and $d_s(v) > 1$.*

The above theorem is not true in general as seen from the following example.

*Example 5.3.11* Let $G = (\sigma, \mu)$ be a fuzzy graph with $\sigma^* = \{u, v, w, x, y\}$, $\sigma(s) = 1$ for all $s \in \sigma^*$ with $\mu(uv) = 0.2$, $\mu(uy) = \mu(uw) = \mu(vw) = \mu(vx) = 0.3$, $\mu(wx) = \mu(xy) = 1.0$ (Fig. 5.12). Clearly, $u$ and $v$ are vertices with the property, $d_s(u) > 1$ and $d_s(v) > 1$. But the edge $uv$ is not even a strong edge.

**Corollary 5.3.12** *Let $G$ be a fuzzy tree and $F$ its unique MST. An edge $uv$ is a fuzzy cutbond of $G$ if and only if $xy$ is a cutbond of $F$.*

*Proof* Let $G$ be a fuzzy tree and let $uv$ be a fuzzy cutbond of $G$. Then by Theorem 5.3.10, $u$ and $v$ are vertices in $G$ such that $d_s(u) > 1$ and $d_s(v) > 1$. Thus, it follows that $uv$ is an internal edge of $F$ and that it is a cutbond of $F$. ■

Let $G = (V, E)$ be a graph and $uv$ be an edge in $G$. Then $uv$ is said to be a **pendant edge** if either $\deg_G(u) = 1$ or $\deg_G(v) = 1$.

**Corollary 5.3.13** *Let $G = (\sigma, \mu)$ be a fuzzy tree with $|\sigma^*| = n$ and $F$ its unique MST. Then the number of fuzzy cutbonds of $G$ is $(n - 1) - l$, where $l$ is the number of pendent edges of $F$.*

When the given fuzzy graph $G$ is a fuzzy cycle, we can determine the upper bounds for the number of fuzzy bonds and fuzzy cutbonds in $G$. Note that any fuzzy cycle contain at least two $\beta$-strong edges. Thus, a 3-fuzzy cycle will not have fuzzy bonds or cutbonds. The maximum number of fuzzy bonds occurs when all $\beta$-strong edges of $G$ form a path. If there are $s$, $\beta$-strong edges which form a $\beta$-strong path of length $s$, the remaining $n - s$ edges are $\alpha$-strong and clearly they are all fuzzy bonds. Excluding the pendent edges from the $\alpha$-strong path formed by these $n - s$ fuzzy bonds, we get $n - s - 2$ fuzzy cutbonds. These idea can be summarized in the following theorem.

**Theorem 5.3.14** *Let $G = (\sigma, \mu)$ be a fuzzy cycle with $|\sigma^*| = n$, where $n \geq 4$. Then $G$ has at most $n - s$ fuzzy bonds and $n - s - 2$ fuzzy cutbonds, where $s$ is the number of $\beta$-strong edges of $G$.*

It is known that the common vertex of two $\alpha$-strong edges is a fuzzy cutvertex. Thus, the common vertex of two bonds also is a fuzzy cutvertex. Further when these bonds are cutbonds, we will have fuzzy cutvertices with a nice property.

**Definition 5.3.15** In a fuzzy graph, a vertex common to two or more fuzzy cutbonds is defined to a **critical fuzzy cutvertex** (**$c$-fuzzy cutvertex**).

Clearly any $c$-fuzzy cutvertex is a fuzzy cutvertex.

**Definition 5.3.16** Let $G$ be a fuzzy graph. Let $x$ be a vertex in $G$. The **strong components** of $G - x$ are the maximal strong edge (in $G$) induced subgraphs of $G - x$.

*Example 5.3.17* Let $G = (\sigma, \mu)$ be a fuzzy graph such that $\sigma^* = \{u, v, w, x, y, z\}$ with $\mu(uv) = 1$, $\mu(vw) = 0.5$, $\mu(wx) = 0.3$, $\mu(xy) = 0.2$, $\mu(yz) = 0.1$, $\mu(zu) = 0.6$, $\mu(wz) = 0.4$. The strong components of $G - w$ are the fuzzy subgraph $\langle\{u, v, z\}\rangle$ and edge $xy$ (Fig. 5.13).

**Theorem 5.3.18** *Let $G = (\sigma, \mu)$ be a fuzzy tree and let $w$ be a critical fuzzy cutvertex of $G$. Then $G - w$ will have at least two strong components.*

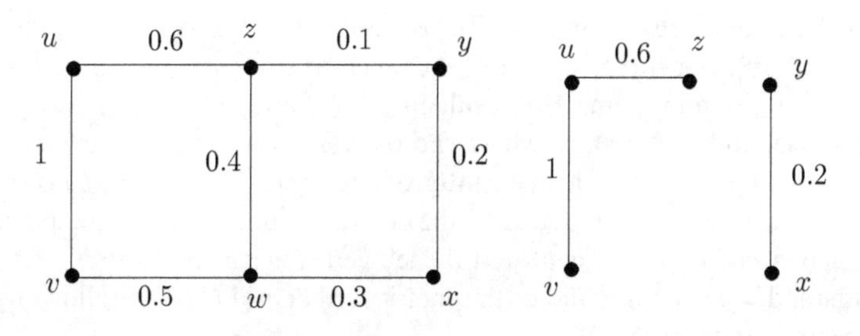


**Fig. 5.13** A fuzzy graph $G$ and strong components of $G - w$

*Proof* Because $w$ is a $c$-fuzzy cutvertex, it must be a common vertex of at least two fuzzy cutbonds of $G$. By Corollary 5.3.12, $c$ must be the common vertex of at least two cutbonds of $F$, the unique MST of $G$. Hence, $w$ is a $c$-cutvertex of $F$. Thus, $F - w$ will have at least two non trivial components. Note that all edges in $F$ are strong. Also, edges which are not in $F$ are not strong. Hence, the non trivial components of $F - w$ are precisely the strong components of $G - w$ and the conclusion follows. ■

## 5.4 Metrics in Fuzzy Graphs

Rosenfeld [154] introduced the first metric in fuzzy graphs as given in the following definition.

**Definition 5.4.1** The $\mu$-**distance** $\delta(u, v)$ is the smallest $\mu$ length of any $u - v$ path, where the $\mu$ length of a path $\rho : u_0, u_1, \ldots, u_n$ is

$$l(\rho) = \sum_{i=1}^{n} \frac{1}{\mu(u_{i-1}u_i)}.$$

If $n = 0$, then define $l(\rho) = 0$.

If $n \geq 1$, then clearly $l(\rho) \geq 1$.

**Theorem 5.4.2** *In a connected fuzzy graph $G$, $\delta(u, v)$ is a metric.*

*Proof* (*i*) $\delta(x, y) = 0$ if and only if $x = y$ because $l(\rho) = 0$ if and only if $\rho$ has length 0.

(*ii*) $\delta(x, y) = \delta(y, x)$ because the reversal of a path is a path and $\mu$ is symmetric.

(*iii*) $\delta(x, z) \leq \delta(x, y) + \delta(y, z)$ because the concatenation of a path from $x$ to $y$ and a path from $y$ to $z$ is a path from $x$ to $z$ and, $l$ is additive for concatenation of paths. ■

Based on this metric, Bhattacharya [37] introduced the concepts of eccentricity and center in fuzzy graphs. The **eccentricity** of a vertex $v$ is defined as $e(v) = \vee_u \delta(u, v)$. A **central vertex** of a connected fuzzy graph is a vertex whose eccentricity is the minimum. The result in graphs $r(G) \leq d(G) \leq 2r(G)$ can be easily extended to fuzzy graphs, where **radius** $r(G) = \wedge\{e(v) \mid v \in \sigma^*\}$ and **diameter** $d(G) = \vee\{e(v) \mid v \in \sigma^*\}$. The **center** of a fuzzy graph $G = (\sigma, \mu)$ is the fuzzy subgraph $\langle C(G)\rangle = (\tau, \nu)$, induced by the central vertices of $G$. Also, a connected fuzzy graph is said to be **self centered** if each vertex is a central vertex.

Sunitha and Vijayakumar studied this metric further in [171]. The following results in this section are from [171].

**Theorem 5.4.3** *A connected fuzzy graph $G = (\sigma, \mu)$ is self centered if*

$$CONN_G(u, v) = \mu(uv)$$

*for all $u, v \in \sigma^*$ and $r(G) = \frac{1}{\mu(uv)}$, where $\mu(uv)$ is the least.*

*Proof* $CONN_G(u, v) = \mu(uv)$ for all $u, v \in \sigma^*$ implies that $G^* = (\sigma^*, \mu^*)$ is complete. Also, each edge $uv$ is a strongest $u - v$ path. It follows that weight of the weakest edge in any other strongest $u - v$ path is $\mu(uv)$ and hence $\mu$-length of a strongest $u - v$ path is at least $\frac{1}{\mu(uv)}$. Now, let $\rho : u = u_0, u_1, \ldots, u_n = v$ be any $u - v$ path which is not strongest. Then strength of $\rho$ is strictly less than $\mu(uv)$. Thus, $\mu$-length of $\rho$ is strictly greater than $\frac{1}{\mu(uv)}$ and hence $\delta(u, v) = \frac{1}{\mu(uv)}$. Now, $e(u) = \vee_v \delta(u, v) = \vee_v \frac{1}{\mu(uv)} = \frac{1}{\wedge_v \mu(uv)}$.

**Claim**: $e(v_i) = e(v_j)$ for all $v_i \neq v_j$

If not, let

$$e(v_i) < e(v_j) \tag{5.5}$$

and let $u_i$ and $u_j$ be such that $e(v_i) = \frac{1}{\mu(v_i u_i)}$ and $e(v_j) = \frac{1}{\mu(v_j u_j)}$. (Note that $u_i$ may or may not be equal to $u_j$.) Consider the path $\rho : v_j, v_i, u_j$. Then $\mu(v_j v_i) \geq \mu(v_i u_i)$ and $\mu(v_i u_j) \geq \mu(v_i u_i)$. Thus, $\mu(v_j v_i) \wedge \mu(v_i u_j) > \mu(v_j u_j)$ by (5.5). That is, strength of $\rho > \mu(v_j u_j)$. Thus, the strength of a $v_j - u_j$ path exceeds $\mu(v_j u_j)$, which contradicts our assumption that every edge is a strongest path. Interchanging $i$ and $j$, a similar argument holds and thus $e(v_i) = e(v_j)$ for all $v_i \neq v_j$. Hence, $G$ is self centered. ■

**Corollary 5.4.4** *A complete fuzzy graph is self centered and $r(G) = \frac{1}{\sigma(u)}$, where $\sigma(u)$ is least.*

The condition in Theorem 5.4.3 is not necessary for a fuzzy graph to be self centered as seen from the following example.

*Example 5.4.5* Let $G = (\sigma, \mu)$ be the fuzzy graph given by $\sigma^* = \{a, b, c, d\}$, $\sigma(s) = 1$ for all $s \in \sigma^*$, $\mu(ab) = 1$, $\mu(bc) = 0.2$, $\mu(cd) = 1$ and $\mu(ad) = 0.2$ (Fig. 5.14). The eccentricity of each vertex is 6. Hence, it is self centered. But, $CONN_G(a, c) = CONN_G(b, d) = 0.2$ and $\mu(ac) = \mu(bd) = 0$.

For any real number $c > 0$, there exist self centered fuzzy graphs of diameter $c$. Also, for any two real numbers $a, b$ such that $a \leq b \leq 2a$, there exists a fuzzy graph

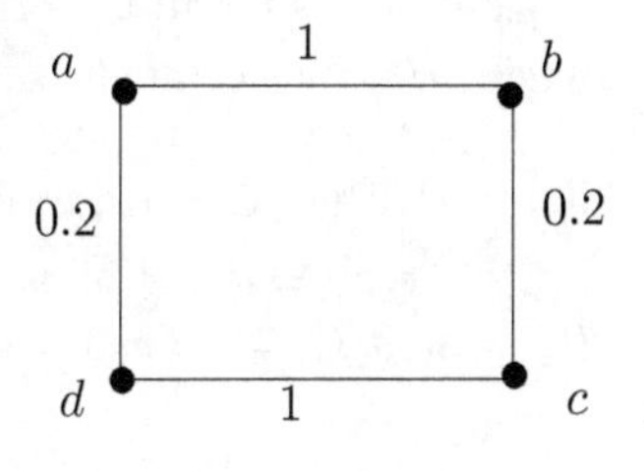


**Fig. 5.14** A self centered fuzzy graph

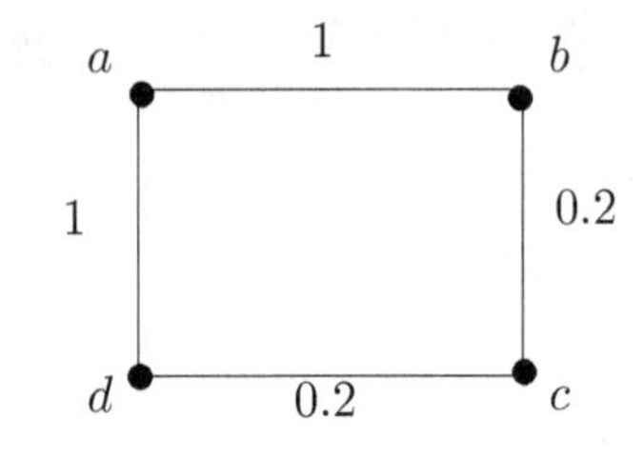


**Fig. 5.15** Fuzzy graph in Example 5.4.7

$G$ such that $r(G) = 1$ and $d(G) = b$. Also, an obvious necessary condition for a fuzzy graph to be self centered can be obtained in terms of eccentric vertices. Note that an eccentric vertex of a vertex $v$ is a vertex $v^*$ such that $e(v) = \delta(v, v^*)$.

**Theorem 5.4.6** *If $G = (\sigma, \mu)$ is a self centered fuzzy graph, then each vertex of $G$ is eccentric.*

The condition in Theorem 5.4.6 is not sufficient as seen from the following example.

*Example 5.4.7* Let $G = (\sigma, \mu)$ be the fuzzy graph given by $\sigma^* = \{a, b, c, d\}, \sigma(s) = 1$ for all $s \in \sigma^*$, $\mu(ab) = \mu(ad) = 1$ and $\mu(bc) = \mu(cd) = 0.2$ (Fig. 5.15). Then clearly, $e(a) = e(c) = 6,\ e(b) = e(d) = 2,\ a^* = c,\ b^* = d, c,\ \ c^* = a,\ d^* = b, c$. Each vertex is eccentric, but $G$ is not self centered.

It is well known that for a given connected graph $H$, there exists a connected super subgraph of $G$ such that $\langle C(G)\rangle \cong H$ (Embedding Theorem). A generalization of this result to a fuzzy set up is given in the following theorem.

**Theorem 5.4.8** [171] (Embedding Theorem) *Let $H = (\sigma', \mu')$ be a connected fuzzy graph with diameter $d$. Then there exists a connected fuzzy graph $G = (\sigma, \mu)$ such that $\langle C(G)\rangle \cong H$. Also, $r(G) = d$ and $d(G) = 2d$.*

*Proof* Construct $G = (\sigma, \mu)$ from $H$ as follows. Take two vertices $u$ and $v$ such that $\sigma(u) = \sigma(v) = \frac{1}{d}$ and join all vertices of $H$ to both $u$ and $v$ with $\mu(uw) = \mu(vw) = \frac{1}{d}$ for all $w$ in $H$. Put $\sigma = \sigma'$ for all vertices in $H$ and $\mu = \mu'$ for all edges in $H$.

**Claim**: $G = (\sigma, \mu)$ is a fuzzy graph.

First note that $\sigma(u) \leq \sigma(w)$ for all $w \in H$. If possible let $\sigma(u) > \sigma(w)$ for at least one vertex $w$ in $H$. Then $\frac{1}{d} > \sigma(w)$, That is, $d < \frac{1}{\sigma(w)} \leq \frac{1}{\mu(ww')}$, where the last inequality holds for all $w' \in H$ because $H$ is a fuzzy graph. That is, $\frac{1}{\mu(ww')} > d$ for all $w' \in H$, which contradicts that $d(H) = d$. Therefore, $\sigma(u) \leq \sigma(w)$ for all $w \in H$ and $\mu(uw) = \sigma(u) \wedge \sigma(w) = \sigma(u) = \frac{1}{d}$. Similarly, $\mu(vw) = \sigma(v) \wedge \sigma(w) = \frac{1}{d}$ for all $w \in H$. Thus, $G = (\sigma, \mu)$ is a fuzzy graph.

Also, $e(w) = d$ for al $w \in H$ and $e(u) = e(v) = \frac{1}{\mu(uw)} + \frac{1}{\mu(wv)} = 2d$. Thus, $r(G) = d, d(G) = 2d$ and $\langle C(G)\rangle \cong H$. ■

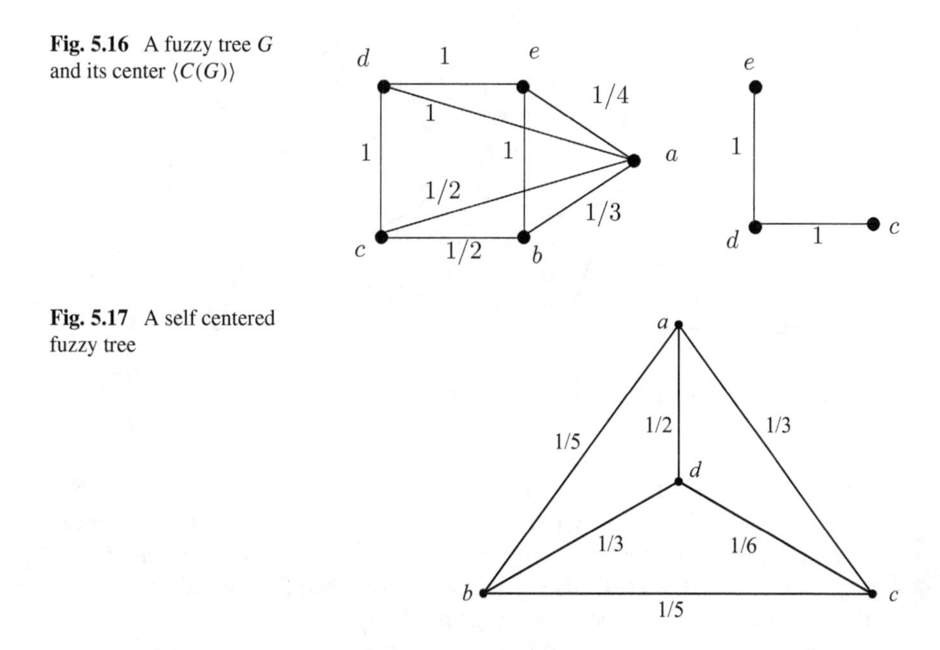


**Fig. 5.16** A fuzzy tree $G$ and its center $\langle C(G) \rangle$

**Fig. 5.17** A self centered fuzzy tree

We know that the center of a tree is either $K_1$ or $K_2$. But this is not true for fuzzy trees as seen from the following example.

*Example 5.4.9* Let $G = (\sigma, \mu)$ be the fuzzy graph given by $\sigma^* = \{a, b, c, d, e\}$, $\sigma(s) = 1$ for all $s \in \sigma^*$, $\mu(ab) = 1/3$, $\mu(ac) = 1/2$, $\mu(ad) = 1$, $\mu(ae) = 1/4$, $\mu(bc) = 1/2$, $\mu(bd) = 1/2$, $\mu(be) = 1$, $\mu(cd) = 1$, $\mu(de) = 1$. Then $G$ is a fuzzy tree, but its center is different from $K_1$ and $K_2$ (See Fig. 5.16).

Next examples shows a self centered fuzzy tree.

*Example 5.4.10* Let $G = (\sigma, \mu)$ be the fuzzy graph with $\sigma^* = \{a, b, c, d\}$, $\sigma(s) = 1$ for all $s \in \sigma^*$, $\mu(ab) = \mu(bc) = 1/5$, $\mu(ac) = \mu(bd) = 1/3$, $\mu(cd) = 1/6$, $\mu(ad) = 1/2$ (Fig. 5.17). This fuzzy graph is a fuzzy tree and is self centered. Note that $e(a) = e(b) = e(c) = e(d) = 5$.

Next we have a result concerning the center of fuzzy trees.

**Theorem 5.4.11** *Let $H = (\sigma', \mu')$ be a fuzzy tree with diameter $d$. Then there exists a fuzzy tree $G = (\sigma, \mu)$ such that $\langle C(G) \rangle$ is isomorphic to $H$.*

*Proof* Put $t = \vee\{\sigma'(w) \mid w \in \sigma'^*\}$. Construct $G = (\sigma, \mu)$ from $H = (\sigma', \mu')$ as follows.

Take two vertices $u$ and $v$ with $\sigma(u) = \sigma(v) = \frac{1}{d}$ and join all vertices in $H$ to both $u$ and $v$. Let $w$ and $w'$ be any two vertices in $H$. Put $\mu(uw) = \frac{1}{d}$; $\mu(wv) = 1/d + \frac{1}{t}$; $\mu(uw') = 1/d + \frac{1}{t}$; $\mu(w'v) = \frac{1}{d}$ and put $1/d + \frac{1}{t}$ as the strength of all the other new edges. Also, put $\sigma = \sigma'$ for all new vertices in $H$ and $\mu = \mu'$ for all edges in $H$.

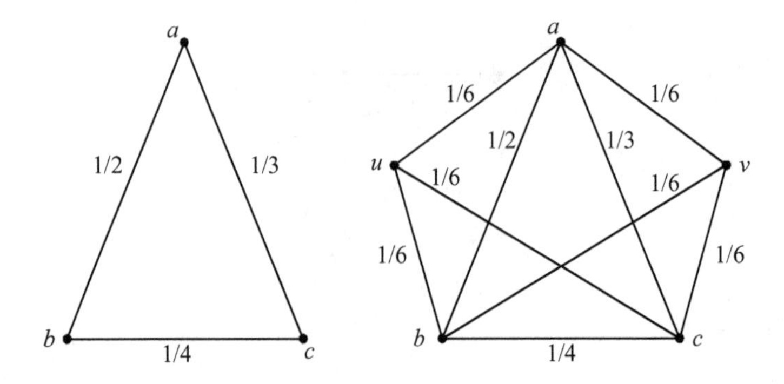


**Fig. 5.18** Fuzzy trees H and G with H as the center of G

**Claim 1**. $G = (\sigma, \mu)$ is a fuzzy graph.

As in the proof of Theorem 5.4.8, $\sigma(u) \leq \sigma(w)$ and $\sigma(v) \leq \sigma(w)$ for all vertices $w \in H$. So $\mu(uw) = \sigma(u) \wedge \sigma(w)$ and $\mu(w'v) = \sigma(w') \wedge \sigma(v)$. Also, because $1/d + \frac{1}{t} < \frac{1}{d}\mu(wv) < \sigma(w) \wedge \sigma(v)$, $\mu(uw') < \sigma(u) \wedge \sigma(w')$ and the inequality holds for all the other new edges. Hence, $G = (\sigma, \mu)$ is a fuzzy graph.

**Claim 2**. $C(G)$ is isomorphic to $H$.

Note that $\mu(w_i w_j) \leq t$ for every edge $w_i w_j$ in $H$. That is,

$$\frac{1}{t} \leq \frac{1}{\mu(w_i w_j)}. \tag{5.6}$$

Now, $u^* = v$, $v^* = u$ and $e(v) = e(u) = 1/\frac{1}{d} + 1/1/d + \frac{1}{t} = 2d + \frac{1}{t}$.

Also, $w^* = v$ and $w'^* = u$, $e(w) = e(w') = 1/1/d + \frac{1}{t} = d + \frac{1}{t}$ and all other vertices in $H$ have eccentricity equal to $d + \frac{1}{t}$ by Eq. (5.6), with $u$ and $v$ as their eccentric vertices. Thus, each vertex in $H$ is a central vertex of $G$ with $r(G) = d + \frac{1}{t}$, $d(G) = 2d + \frac{1}{t}$ and $C(G) \cong H$.

Finally, we claim that $G = (\sigma, \mu)$ is a fuzzy tree; for $H$ being a fuzzy tree, it has a spanning subgraph $F_H$, which is a tree, satisfying the requirements. Now, $F_H$ together with the edges $uw$ and $w'v$ is the required spanning subgraph of $G$. ■

A fuzzy graph $H$ on three vertices and the graph $G$ whose center is isomorphic to $H$ are given in Fig. 5.18.

In Fig. 5.18, $d(H) = 4$, $t = 1/2$, $e(u) = e(v) = 10$, $e(a) = e(b) = e(c) = 6$. It is not hard to see that the center of a fuzzy tree need not be a fuzzy tree.

Other than the $\mu$ distance of Rosenfeld, several other distances in fuzzy graphs were proposed by different authors in the past. They include geodesic distance of Bhutani and Rosenfeld (Definition 3.1.24), strong sum distance [175], $\delta$-distance, detour distance [101, 102], detour $\mu$-distance [136], and so on. The strong geodesic distance, known as $g$-distance is a direct generalization of distance in graphs. Mathew and Mathew introduced $\alpha$-strong distance as follows.

**Definition 5.4.12** Let $G = (\sigma, \mu)$ be a fuzzy graph. Let $u$ and $v$ be any two vertices in $\sigma^*$. Then the **$\alpha$-distance** between $u$ and $v$ is defined and denoted by $d_\alpha(u, v) = \wedge_P \sum_{e \in P} \mu(e)$, where $P$ is an $\alpha$-strong path between $u$ and $v$, 0 if $u = v$, $\infty$ if there exists no $\alpha$-strong $u - v$ path.

$d_\alpha$ satisfies the properties of a metric and hence $(\sigma^*, d_\alpha)$ qualifies as metric space. Using this distance in fuzzy graphs, $\alpha$-eccentricity, $\alpha$-center, $\alpha$-diametral vertices, $\alpha$-periphery, and so on, can be similarly defined as in crisp graphs. A vertex $u \in \sigma^*$ is called **$\alpha$-isolated** if there is no $\alpha$-strong edge incident on $u$. An $\alpha$ strong edge $e = uv$ is called $\alpha$-isolated, if there exists no other $\alpha$ strong edge adjacent with $e$. Note that this function is not a metric if $\alpha$-strong edges are replaced by $\beta$ strong or simply strong edges.

**Proposition 5.4.13** *Let $(G = \sigma, \mu)$ be a fuzzy graph. If $e = uv$ is an $\alpha$-isolated edge of $G$, then $e_\alpha(u) - e_\alpha(v) = 0$*

*Proof* Because $e = uv$ is $\alpha$-isolated, $e$ is adjacent with no $\alpha$-strong edges. That is, $e$ is the only $\alpha$-strong edge, which is incident at $u$ and $v$. By calculating the $\alpha$-eccentricity of $u$, we see that the farthest vertex from $u$ is $v$ and vice-versa. Thus, $e_\alpha(u) = e_\alpha(v) = \mu(e)$ and hence $e_\alpha(u) - e_\alpha(v) = 0$. ■

If $G = (\sigma, \mu)$ is a connected fuzzy graph, then the result $r_\alpha(G) \leq d_\alpha(G) \leq 2r_\alpha(G)$ is trivial as in case of any distance. Recall that a fuzzy graph $G$ is called $\alpha$-self centered if $G$ is isomorphic with its $\alpha$-center. Thus, we have the following result.

**Proposition 5.4.14** *Let $G = (\sigma, \mu)$ be a connected fuzzy graph such that there exists exactly one $\alpha$-strong edge incident at every vertex and that all the $\alpha$-strong edges are of equal weight. Then $G$ is $\alpha$-self centered.*

*Proof* Given that all vertices of $G$ are incident with exactly one $\alpha$-strong edge each, and all the $\alpha$-strong edges are of equal weight. This means if $e = uv$ is $\alpha$-strong, then there will be no other $\alpha$-strong edges incident at $u$ or $v$. Hence, $e_\alpha(u) = \mu(e) = e_\alpha(v)$. Thus, $e_\alpha(u) = \mu(e)$ for every vertex $u$ in $\sigma^*$. This proves that $G$ is $\alpha$-self centered. ■

**Proposition 5.4.15** *A connected fuzzy graph $G = (\sigma, \mu)$ is $\alpha$-self centered if for any two vertices $u$ and $v$ such that $u$ is an $\alpha$-eccentric vertex of $v$, $v$ should be one of the $\alpha$-eccentric vertices of $u$.*

*Proof* First assume that $G$ is $\alpha$-self centered. Also, assume that $u$ is an $\alpha$-eccentric vertex of $v$. That is, $e_\alpha(v) = d_\alpha(v, u)$. Because $G$ is $\alpha$-self centered, all vertices will have the same $\alpha$-eccentricity. Therefore, $e_\alpha(v) = e_\alpha(u)$. Thus, we get, $e_\alpha(u) = d_\alpha(v, u) = d_\alpha(u, v)$. Thus, $e_\alpha(u) = d_\alpha(u, v)$. That is, $v$ is an $\alpha$-eccentric vertex of $u$. Next assume that $u$ is an $\alpha$-eccentric vertex of $v$. Then $v$ is an $\alpha$-eccentric vertex of $u$. Thus, $e_\alpha(u) = d_\alpha(u, v)$, and $e_\alpha(v) = d_\alpha(v, u)$. But $d_\alpha(u, v) = d_\alpha(v, u)$. Therefore, $e_\alpha(v) = e_\alpha(u)$, where $u$ and $v$ are two arbitrary vertices of $G$. Thus, all vertices of $G$ have the same $\alpha$-eccentricity, and hence $G$ is $\alpha$-self centered. ■

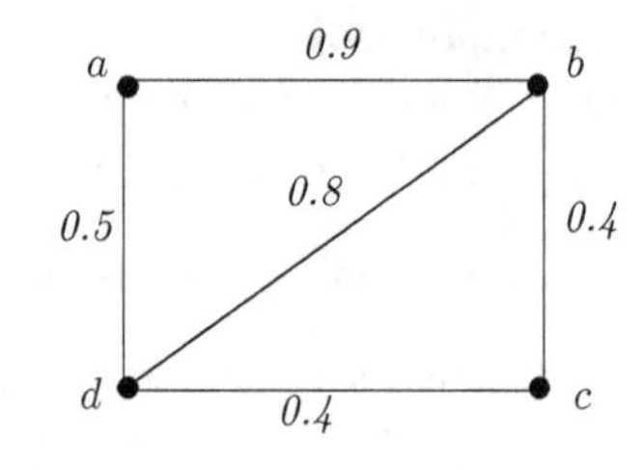


**Fig. 5.19** $\alpha$-distance matrix-example

We have an easy check for a fuzzy graph $G$ to find whether it is $\alpha$-self centered or not. Consider the following definition.

**Definition 5.4.16** Let $G = (\sigma, \mu)$ be a connected fuzzy graph with $|\sigma^*| = n$. The $\alpha$-**distance matrix** $D_\alpha = (d_{i,j})$ is a square matrix of order $n$ and is defined by $(d_{i,j}) = d_\alpha(v_i, v_j)$. Note that the $\alpha$-distance matrix is a symmetric matrix.

Consider the following example.

*Example 5.4.17* Let $G = (\sigma, \mu)$ be the fuzzy graph given in Fig. 5.19, with $\sigma^* = \{a, b, c, d\}$, $\sigma(s) = 1$ for all $s \in \sigma^*$ and $\mu(ab) = 0.9$, $\mu(bc) = \mu(cd) = 0.4$, $\mu(da) = 0.5$, $\mu(bd) = 0.8$.

The $\alpha$-distance matrix of the fuzzy graph in Fig. 5.19, is given by,

$$D_\alpha = \begin{bmatrix} 0 & 0.9 & \infty & 1.7 \\ 0.9 & 0 & \infty & 0.8 \\ \infty & \infty & 0 & \infty \\ 1.7 & 0.8 & \infty & 0 \end{bmatrix}.$$

Next we have a theorem for finding the eccentricities of vertices using the max-max composition of the distance matrix.

**Theorem 5.4.18** *Let $G = (\sigma, \mu)$ be a connected fuzzy graph. The diagonal elements of the max-max composition of the $\alpha$-distance matrix of $G$ with itself are the $\alpha$-eccentricities of the vertices.*

*Proof* Let $D_\alpha = (d_{i,j})$ be the $\alpha$-distance matrix of $G$. Then $(d_{i,j}) = d_\alpha(v_i, v_j)$. In the max-max composition, $D_\alpha \circ D_\alpha$, the $i$th diagonal entry,

$$\begin{aligned} d_{i,i} &= \vee\{d_{i,1} \vee d_{1,i},\ d_{i,2} \vee d_{2,i},\ d_{i,3} \vee d_{3,i},\ \ldots,\ d_{i,n} \vee d_{n,i}\} \\ &= \vee\{d_{i,1},\ d_{i,2},\ d_{i,3},\ \ldots,\ d_{i,n}\} \\ &= \vee\{d_\alpha(v_i, v_1),\ d_\alpha(v_i, v_2),\ d_\alpha(v_i, v_3),\ \ldots,\ d_\alpha(v_i, v_n)\} \\ &= e_\alpha(v_i). \end{aligned}$$

■

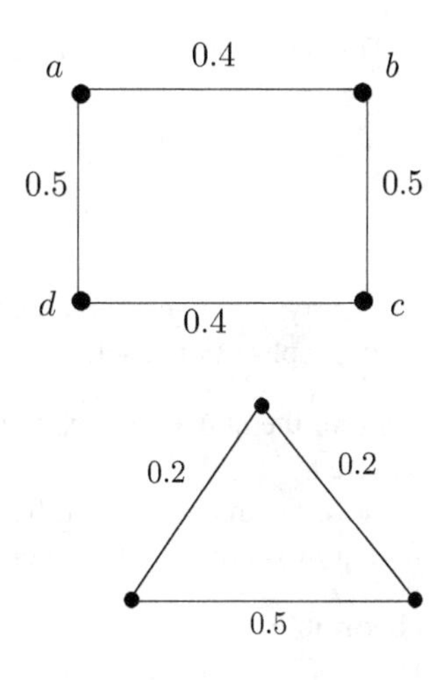


**Fig. 5.20** Max–Max composition

**Fig. 5.21** Max–Max composition 2

**Theorem 5.4.19** *A connected fuzzy graph $G = (\sigma, \mu)$ is $\alpha$-self centered if and only if all the entries in the principal diagonal of the max-max composition of the $\alpha$-distance matrix with itself are the same.*

*Proof* As proved in Theorem 5.4.18, the principal diagonal entries in the max-max composition of the $\alpha$-distance matrix with itself are the $\alpha$-eccentricities of the vertices. If they are the same, then $e_\alpha(u)$ is the same for all $u$ in $G$. Then $G$ is $\alpha$-self centered. ■

We illustrate the above theorem in the following examples.

*Example 5.4.20* Consider the fuzzy graph $G = (\sigma, \mu)$ given in Fig. 5.20.
The $\alpha$-distance matrix and the max-max composition are given below.

$$D_\alpha = \begin{bmatrix} 0 & \infty & \infty & 0.5 \\ \infty & 0 & 0.5 & \infty \\ \infty & 0.5 & 0 & \infty \\ 0.5 & \infty & \infty & 0 \end{bmatrix},$$

$$D_\alpha \circ D_\alpha = \begin{bmatrix} 0.5 & \infty & \infty & 0.5 \\ \infty & 0.5 & 0.5 & \infty \\ \infty & 0.5 & 0.5 & \infty \\ 0.5 & \infty & \infty & 0.5 \end{bmatrix}.$$

The fuzzy graph is $\alpha$-self centered.

Now, consider another example.

*Example 5.4.21* Consider the fuzzy graph shown in Fig. 5.21.

The $\alpha$-distance matrix and the max-max composition are given below.

$$D_\alpha = \begin{bmatrix} 0 & \infty & \infty \\ \infty & 0 & 0.5 \\ \infty & 0.5 & 0 \end{bmatrix}, \ D_\alpha \circ D_\alpha = \begin{bmatrix} \infty & \infty & \infty \\ \infty & 0.5 & 0.5 \\ \infty & 0.5 & 0.5 \end{bmatrix}.$$

Clearly, all diagonal elements in the composition are not the same and hence the fuzzy graph is not $\alpha$-self centered.

From the above two examples, it is clear that a block may or may not be $\alpha$-self centered.

Now, we discuss about the central vertices of fuzzy trees and blocks. In the following theorem, $\alpha$-central vertices of fuzzy trees are characterized.

**Theorem 5.4.22** ([106]) *If a vertex of a fuzzy tree is $\alpha$-central, then it is a common vertex of at least two $\alpha$-strong edges.*

*Proof* Let $G = (\sigma, \mu)$ be a fuzzy tree. Then $G$ has no $\beta$-strong edges. We know that between any two vertices of a connected fuzzy graph $G$, there exists a strong path. As $G$ is independent of $\beta$-strong edges, there exists an $\alpha$-strong path between any two vertices of $G$. Let $u$ be an $\alpha$-central vertex of $G$. We want to prove that two or more $\alpha$-strong edges are incident at $u$. If possible suppose the contrary. Let there be exactly one $\alpha$-strong edge, namely $e$ incident at $u$. Therefore, any $\alpha$-strong path between $u$ and any other vertex of $G$ will contain the edge $e$. This proves that $e_\alpha(u) > r_\alpha(G)$, which is a contradiction to the fact that $u$ is $\alpha$-central. Therefore, our assumption is wrong. Thus, the proof of the theorem is completed. ■

If $u$ is a common vertex of at least two $\alpha$-strong edges, then $u$ is a fuzzy cutvertex of $G$. So from the above theorem it is clear that, if a vertex $u$ of a fuzzy tree $G$ is $\alpha$-central, then it is a fuzzy cutvertex of $G$.

**Theorem 5.4.23** *The $\alpha$-center of a block $G$ contains all $\alpha$-strong edges with minimum weight.*

*Proof* Suppose that $G = (\sigma, \mu)$ is a block. Then $G$ has no fuzzy cutvertices. We know that, if a vertex $u$ in a connected fuzzy graph is common to more than one $\alpha$-strong edge, then it is a fuzzy cutvertex. As $G$ is free from fuzzy cutvertices, at most one $\alpha$-strong edge can be incident at every vertex of $G$. Thus, the $\alpha$-eccentricity, $e_\alpha$ of a vertex $u$ is the weight of the $\alpha$-strong edge incident at $u$.

So the $\alpha$-radius of $G$, $r_\alpha(G)$ is the weight of the smallest $\alpha$-strong edge. Hence, the $\alpha$-center of $G$, $\langle C_\alpha(G) \rangle$ contains all $\alpha$-strong edges of $G$ with minimum weight. This completes the proof of the theorem. ■

The next theorem helps us to find the number of connected components in the $\alpha$-center of a block.

**Theorem 5.4.24** *Let $G = (\sigma, \mu)$ be a block. If there exists a path containing all $\alpha$-strong edges of $G$ with minimum weight alternatively, then $\langle C_\alpha(G) \rangle$ is connected.*

*Proof* By the previous theorem, $\langle C_\alpha(G)\rangle$ consists of all $\alpha$-strong edges of $G$ with minimum weight. Also, in a block, not more than one $\alpha$-strong edge can be incident at any vertex. So if there are $k$ number of $\alpha$-strong edges present in $G$ with minimum weight, then all these edges will be in $\langle C_\alpha(G)\rangle$ . Moreover, they are not adjacent also. Hence, if we can find a path containing all $\alpha$-strong edges with minimum weight alternatively, then it follows that $\langle C_\alpha(G)\rangle$ is connected. ■

**Theorem 5.4.25** *If a connected fuzzy graph $G = (\sigma, \mu)$ is a block with $k$ $\alpha$-strong edges, then $k \leq (|\sigma^*|)/2$.*

*Proof* Suppose that $G = (\sigma, \mu)$ is a block. Then $G$ has no fuzzy cutvertices. Let $k$ be the number of $\alpha$-strong edges in $G$. We have to prove that $k \leq (|\sigma^*|)/2$. If possible suppose the contrary. Let $k > (|\sigma^*|)/2$. Then there will be at least $\lfloor k - (|\sigma^*|)/2 \rfloor$ vertices with more than one $\alpha$-strong edge incident on them. Clearly, these vertices are fuzzy cutvertices of $G$, a contradiction to the fact that $G$ is free from fuzzy cutvertices. So our assumption is wrong and the proof is complete. ■

## 5.5 Detour Distance in Fuzzy Graphs

The **detour distance** between two vertices $u$ and $v$ in a connected graph $G$ is the length of the longest $u - v$ path in $G$ [83]. The **detour $g$-distance** in fuzzy graphs was studied by Linda and Sunitha [101]. The detour $\mu$-distance in fuzzy graphs is discussed in [136]. We discuss some detour $g$-distance parameters in this section.

**Definition 5.5.1** The length of the longest strong $u - v$ path between two vertices $u$ and $v$ in a connected fuzzy graph $G$ is called the **fuzzy detour $g$-distance** from $u$ to $v$, denoted by $D_g(u, v)$. Any $u - v$ strong path of length $D_g(u, v)$ is called a $u - v$ fuzzy $g$-detour. A fuzzy graph $G = (\sigma, \mu)$ is called a **fuzzy $g$-detour graph** if $D_g(u, v) = d_g(u, v)$ for every pair $u$ and $v$ of vertices of $G$.

A fuzzy graph is said to be a fuzzy **$g$-detour graph** if the standard $g$-distance and fuzzy detour $g$-distance coincide.

*Example 5.5.2* Let $G = (\sigma, \mu)$ be the fuzzy graph with $\sigma^* = \{a, b, c\}$, $\sigma(a) = \sigma(b) = \sigma(c) = 1$, $\mu(ab) = 0.6$, $\mu(bc) = 0.7$, $\mu(ac) = 0.8$ (Fig. 5.22). In $G$,

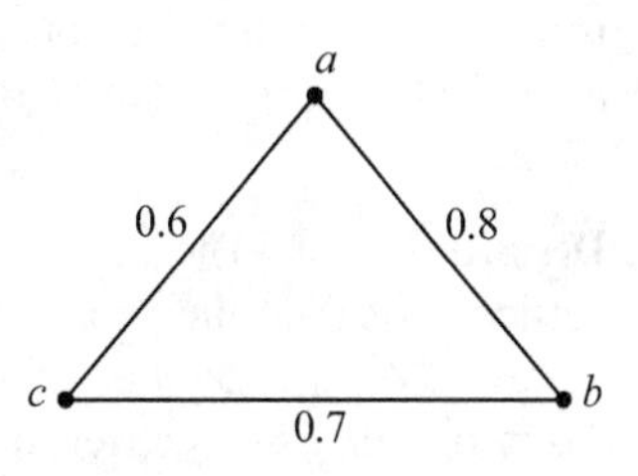


**Fig. 5.22** A fuzzy g-detour graph

$d_g(u_1, u_2) = 2 = D_g(u_1, u_2)$, $d_g(u_1, u_3) = 1 = D_g(u_1, u_3)$, $d_g(u_2, u_3) = 1 = D_g(u_2, u_3)$. Here, $D_g(u, v) = d_g(u, v)$ for every pair $u$ and $v$ of vertices of $G$. Hence, $G$ is a fuzzy $g$-detour graph.

We have two easy propositions.

**Proposition 5.5.3** *If $u$ and $v$ are any two vertices in a connected fuzzy graph $G = (\sigma, \mu)$, then $0 \leq d_g(x, y) \leq D_g(u, v) < \infty$.*

**Proposition 5.5.4** *If $u$ and $v$ are any two vertices in a connected fuzzy graph $G = (\sigma, \mu)$, then $D_g(u, v) = 0$ if and only if $d_g(u, v) = 0$.*

**Theorem 5.5.5** *The fuzzy detour $g$-distance is a metric on the vertex set of every connected fuzzy graph.*

*Proof* Let $G = (\sigma, \mu)$ be a connected fuzzy graph. Note that (i) $D_g(u, v) \geq 0$, (ii) $D_g(u, v) = 0$ if and only if $u = v$ and (iii) $D_g(u, v) = D_g(v, u)$ for every pair $u$, $v$ of vertices of $G$. It remains only to show that the fuzzy detour $g$-distance satisfies the triangle inequality. Let $u$, $v$ and $w$ be any three vertices of $G$. Because the inequality $D_g(u, w) \leq D_g(u, v) + D_g(v, w)$ holds if any two of the three vertices are the same, we assume that $u$, $v$ and $w$ are distinct. Let $P$ be a $u - w$ fuzzy $g$-detour in $G$ of length $D_g(u, w) = k$. Then there exists two cases.

**Case 1**. $v$ lies on $P$.

Let $P_1$ be the $u - v$ sub path of $P$ and $P_2$ be the $v - w$ sub path of $P$. Suppose that the length of $P_1$ is $s$ and the length of $P_2$ is $t$. Then $s + t = k$. Therefore, $D_g(u, w) = k = s + t \leq D_g(u, v) + D_g(v, w)$.

**Case 2**. $v$ does not lie on $P$.

Because there exists a strong path between every pair of vertices, there is a shortest strong path $Q$ from $v$ to a vertex of $P$. Let $x$ be any vertex on $P$ and $Q$ be the $v - x$ geodesic such that no other vertex of $Q$ lies on $P$. Let $r$ be the length of $Q$. Then $r > 0$. Let the $u - x$ sub path $P'$ of $P$ has length $a$ and the $x - w$ sub path $P''$ of $P$ has length $b$. Then $a \geq 0$ and $b \geq 0$. Therefore, $D_g(u, v) \geq a + r$ and $D_g(v, w) \geq b + r$. Thus, $D_g(u, w) = k = a + b < (a + r) + (b + r) \leq D_g(u, v) + D_g(v, w)$. So the triangle inequality holds. ∎

The fuzzy detour $g$-eccentricity, $e_{D_g}(u)$ of a vertex $u$ is the fuzzy detour $g$-distance from $u$ to a vertex farthest from $u$. Let $u^*_{D_g}$ denote set of all fuzzy detour $g$-eccentric vertices of $u$. The fuzzy detour $g$-radius of $G$, $rad_{D_g}(G)$ is the minimum fuzzy detour $g$-eccentricity among the vertices of $G$. A vertex $u$ in $G$ is a fuzzy detour $g$-central vertex if, $e_{D_g}(u) = rad_{D_g}(G)$. The fuzzy detour $g$-diameter of $G$, $diam_{D_g}(G)$ is the maximum fuzzy detour $g$-eccentricity among the vertices of $G$. A vertex $u$ in a connected fuzzy graph $G$ is called fuzzy detour $g$-peripheral vertex if $e_{D_g}(u) = diam_{D_g}(G)$.

**Definition 5.5.6** The fuzzy subgraph of $G$ induced by the fuzzy detour $g$-central vertices is called the **fuzzy detour $g$-centre** of $G$, denoted by $C_{D_g}(G)$. If every vertex of $G$ is fuzzy detour $g$-central vertex, then $C_{D_g}(G) = G$, and $G$ is called **fuzzy detour $g$-self centered**.

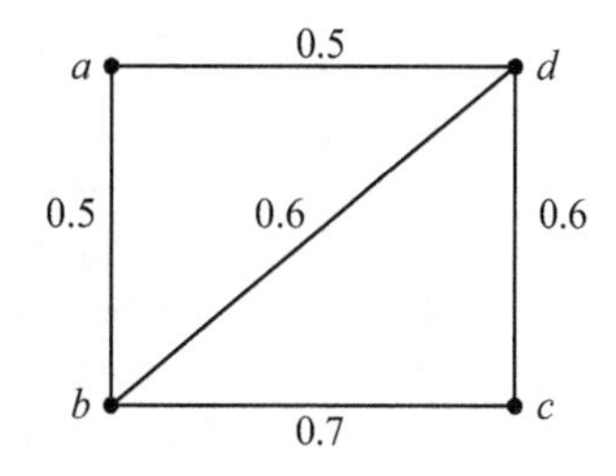


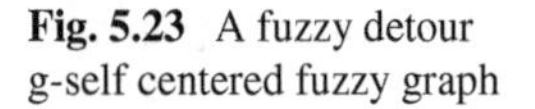
**Fig. 5.23** A fuzzy detour g-self centered fuzzy graph

Note that if $G$ is a fuzzy detour $g$-self centered graph, then $rad_{D_g}(G) = diam_{D_g}(G)$.

*Example 5.5.7* Let $G = (\sigma, \mu)$ be the fuzzy graph with $\sigma^* = \{a, b, c, d\}$, $\sigma(a) = \sigma(b) = \sigma(c) = 1$, $\mu(ab) = 0.5$, $\mu(bc) = 0.6$, $\mu(cd) = 0.7$, $\mu(ad) = 0.5$, and $\mu(bd) = 0.6$ (Fig. 5.23).

Note that all edges in $G$ are strong and the fuzzy detour $g$-distance between two vertices are as follows.

$D_g(u_1, u_2) = 3, D_g(u_1, u_3) = 3, D_g(u_1, u_4) = 3, D_g(u_2, u_3) = 3, D_g(u_2, u_4) = 2, D_g(u_3, u_4) = 3$. $e_{D_g}(u_1) = 3$, $e_{D_g}(u_2) = 3$, $e_{D_g}(u_3) = 3$, $e_{D_g}(u_4) = 3$.

Thus, $rad_{D_g}(G) = diam_{D_g}(G) = 3$. Therefore, $C_{D_g}(G) = G$, and $G$ is a fuzzy detour $g$-self centered fuzzy graph.

**Theorem 5.5.8** *For every non-trivial connected fuzzy graph $G = (\sigma, \mu)$, $rad_{D_g}(G) \leq diam_{D_g}(G) \leq 2rad_{D_g}(G)$.*

*Proof* The inequality $rad_{D_g}(G) \leq diam_{D_g}(G)$ follows from definition. Let $u, v$ be two vertices such that $D_g(u, v) = diam_{D_g}(G)$. Let $w$ be a fuzzy detour $g$-central vertex of $G$. Then the fuzzy detour $g$-distance between $w$ and any other vertex of $G$ is at most fuzzy detour $g$-radius of $G$. By the triangle inequality, $diam_{D_g}(G) = D_g(u, v) \leq D_g(u, w) + D_g(w, v) \leq rad_{D_g}(G) + rad_{D_g}(G) = 2rad_{D_g}(G)$. ■

**Theorem 5.5.9** *The fuzzy detour $g$-centre, $C_{D_g}(G)$ of every connected fuzzy graph $G = (\sigma, \mu)$ lies in a single block of $G^*$.*

*Proof* Assume to the contrary, that $G = (\sigma, \mu)$ is a connected fuzzy graph whose fuzzy detour $g$-centre $C_{D_g}(G)$ is not a subgraph of a single block of $G$. Then there is a cutvertex $v$ of $G^*$ such that $G^* - v$ contains two components $G_1$ and $G_2$ each of which contains vertices of $C_{D_g}(G)$. Let $u$ be a vertex of $G$ such that $D_g(u, v) = e_{D_g}(v)$ and let $P_1$ be a $u - v$ fuzzy $g$-detour in $G$. At least one of $G_1$ or $G_2$ contains no vertex of $P_1$, say $G_2$ contains no vertex of $P_1$. Let $w$ be a fuzzy detour $g$-central vertex of $G$ that belong to $G_2$ and let $P_2$ be a $v - w$ fuzzy $g$-detour. Then $P_1$ followed by $P_2$ produces a $u - w$ fuzzy $g$-detour whose length is greater than that of $P_1$. Hence, $e_{D_g}(w) > e_{D_g}(v)$, which contradicts the fact that $w$ is a fuzzy detour $g$-central vertex of $G$. ■

**Theorem 5.5.10** *Every fuzzy graph is the fuzzy detour g-centre of some fuzzy graph.*

*Proof* Let $G = (V, \sigma, \mu)$ be a fuzzy graph with $n$ vertices, where $\sigma^* = \{u_1, u_2, \ldots, u_n\}$, and let $H = (V', \sigma', \mu')$ be the fuzzy graph obtained by adding $n + 1$ vertices $\{w_1, \ldots, w_n, w_{n+1}\}$ to $G$ as follows. $V' = V \cup \{w_1, \ldots, w_n, w_{n+1}\}$. $\sigma' = \sigma$ for all $u_i \in G, i = 1, \ldots, n$. $\mu' = \mu$ for all $u_iu_j \in G$. Let $c = \wedge\sigma(u_i), i = 1, \ldots, n$. $\sigma'(w_j) = t, 0 < t \leq c, j = 1, \ldots, n + 1$. Then $\mu'(w_ju_i) = t$ for all $u_i$ and $w_j$, $i = 1, \ldots, n$, $j = 1, \ldots, n + 1$. Thus, all edges $w_ju_i$ are strong. Here, for all $u_i \in G$, $e_{D_g}(u_i) = 2n - 1$ and for all $w_j$, $e_{D_g}(w_j) = 2n$. Therefore, $rad_{D_g}(G) = 2n - 1$. Hence, $H = (V', \sigma', \mu')$ is a fuzzy graph with $G = (V, \sigma, \mu)$ as its fuzzy detour $g$-centre. ■

**Theorem 5.5.11** *For each pair a, b of positive real numbers with $a \leq b \leq 2a$, there exists a connected fuzzy graph G with $rad_{D_g}(G) = a$ and $diam_{D_g}(G) = b$.*

*Proof* For $a = b = k \geq 1$ the complete fuzzy graph on $k + 1$ vertices has the desired property. For $a < b \leq 2a$, let $H_1$ and $H_2$ be any two fuzzy graphs such that $H_1$ is of order $a + 1$ and $H_2$ is of order $b - a + 1$ and also such that $H_1^*$ and $H_2^*$ are complete and all edges in $H_1$ and $H_2$ are strong. Now, $G$ be a fuzzy graph of order $b + 1$ obtained by identifying a vertex $v$ of $H_1$ and a vertex of $H_2$. Because $b \leq 2a$, it follows that $b - a + 1 \leq a + 1$. Thus, $e_{D_g}(v) = a$. Because there is a strong path in $G$ which passes through every other vertices of $G$ with initial vertex $x$, where $x \in G - v$, it follows that $e_{D_g}(x) = b$. Hence, $rad_{D_g}(G) = a$ and $diam_{D_g}(G) = b$. ■

With respect to standard $g$-distance note that a necessary condition for a $g$-self centered fuzzy graph is that each vertex is $g$-eccentric, but it is not sufficient. But with respect to fuzzy detour $g$-distance it is sufficient also, as discussed in the next theorem.

**Theorem 5.5.12** ([101]) *A fuzzy graph $G = (\sigma, \mu)$ is fuzzy detour g-self centered if and only if each vertex of G is fuzzy detour g-eccentric.*

*Proof* Assume $G = (\sigma, \mu)$ is a fuzzy detour $g$-self centered fuzzy graph and let $v$ be any vertex of $G$. Let $u \in v^*_{D_g}$. Then $e_{D_g}(v) = D_g(u, v)$ and $G$ being fuzzy detour $g$-self centered fuzzy graph $e_{D_g}(u) = e_{D_g}(v) = D_g(u, v)$, which shows that $v \in u^*_{D_g}$, and $v$ is fuzzy detour $g$-eccentric.

Conversely, assume that each vertex of $G$ is fuzzy detour $g$-eccentric. To prove $G$ is fuzzy detour $g$-self centered. Assume to the contrary, that $G$ is not fuzzy detour $g$-self centered, i.e., $rad_{D_g}(G) \neq diam_{D_g}(G)$. Let $y$ be a vertex in $G$ such that $e_{D_g}(y) = diam_{D_g}(G)$ and let $z \in y^*_{D_g}$. Let $P$ be a $y - z$ fuzzy $g$-detour in $G$. Then there must exists a vertex $w$ on $P$ such that $w$ is not fuzzy detour $g$-eccentric vertex of any vertex of $P$. Also, $w$ is not a fuzzy detour $g$-eccentric vertex of any other vertex. Otherwise if $w$ is a fuzzy detour $g$-eccentric vertex of a vertex $u$ (say), i.e., $w \in u^*_{D_g}$, then we can extend $u - w$ fuzzy $g$-detour to a longer path (to $y$ or to $z$ or to both), which is a contradiction to $w \in u^*_{D_g}$. Therefore, $rad_{D_g}(G) = diam_{D_g}(G)$ and $G$ is fuzzy detour $g$-self centered. ■

**Proposition 5.5.13** *For a fuzzy detour g-self centered fuzzy graph $G = (\sigma, \mu)$, $rad_D(G) = diam_D(G) = n - 1$.*

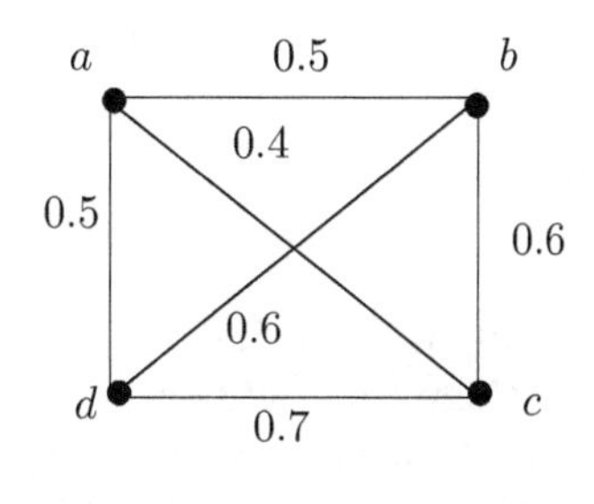


**Fig. 5.24** Counter example for the converse of Theorem 5.5.14

*Proof* Assume $G = (\sigma, \mu)$ is fuzzy detour $g$-self centered. To prove $diam_D$ $(G) = n - 1$. Assume to the contrary that $diam_D(G) = k < n - 1$.

**Claim**: There exists a vertex $x$ in $G$ which is common to all fuzzy detour peripheral paths.

If not let $P_1$ and $P_2$ be two fuzzy detour peripheral paths such that $P_1$ and $P_2$ share no common vertices. Let $y \in P_1$ and $z \in P_2$. Because $G$ is connected, there exists a strong path from $z$ to $y$. Thus, there exist vertices on $P_1$ and $P_2$ with eccentricity greater than $k$, which is not possible. Hence, the claim.

Because $x$ is on every fuzzy detour peripheral path, $e_{D_g}(x) < k$, which is a contradiction to our assumption that $G$ is fuzzy detour $g$-self centered. ■

**Theorem 5.5.14** *A connected fuzzy graph $G = (\sigma, \mu)$ with $|\sigma^*| = n$ such that $G^*$ is complete is fuzzy detour $g$-self centered if each edge is strong. Furthermore, $rad_{D_g}(G) = n - 1$.*

*Proof* Let $\sigma^* = \{v_1, v_2, v_3, \ldots, v_n\}$. Because $G^*$ is complete, each vertex $v_i$ is incident with exactly $n - 1$ edges and all edges are strong. Hence, $e_{D_g}(v_i) = n - 1$, for all $i = 1, 2, \ldots, n$ and $G$ is a fuzzy detour $g$-self centered fuzzy graph with $rad_{D_g}(G) = n - 1$. ■

The condition in Theorem 5.5.14 is not necessary as seen from next example.

*Example 5.5.15* Let $G = (\sigma, \mu)$ be the fuzzy graph with $\sigma^* = \{a, b, c, d\}$, $\sigma(x) = 1$ for all $x \in \sigma^*$, $\mu(ab) = 0.5$, $\mu(bc) = 0.6$, $\mu(cd) = 0.7$, $\mu(ad) = 0.5$, $\mu(ac) = 0.4$, $\mu$ $(bd) = 0.6$, (Fig. 5.24). Here, $e_{D_g}(u_1) = 3$, $e_{D_g}(u_2) = 3$, $e_{D_g}(u_3) = 3$, $e_{D_g}$ $(u_4) = 3$ and $G$ is a fuzzy detour $g$-self centered fuzzy graph with $G^*$ complete, but the edge $ac$ is not strong.

**Corollary 5.5.16** *A complete fuzzy graph on $n$ vertices is fuzzy detour $g$-self centered and $rad_{D_g}(G) = n - 1$.*

**Theorem 5.5.17** *Let $G = (\sigma, \mu)$ be a fuzzy graph and let $u \in v^*_{D_g}$. Then $G$ is fuzzy detour $g$-self centered if and only if $v \in u^*_{D_g}$.*

*Proof* Let $G = (\sigma, \mu)$ be a fuzzy graph and $u$ and $v$ be any two vertices of $G$. Let $u \in v^*_{D_g}$. Assume $G$ is fuzzy detour $g$-self centered. We need to prove that $v \in u^*_{D_g}$. Now, $e_{D_g}(u) = e_{D_g}(v)$,

$$u \neq v \tag{5.7}$$

and

$$D_g(v, u) = e_{D_g}(v). \tag{5.8}$$

From (5.7) and (5.8), $e_{D_g}(u) = D_g(v, u)$ and thus $v \in u^*_{D_g}$.

Conversely, let $G$ be a fuzzy graph and $u$, $v$ be any two vertices of $G$ such that $u \in v^*_{D_g}$ and by assumption $v \in u^*_{D_g}$. Then $e_{D_g}(u) = e_{D_g}(v)$, $u \neq v$. Therefore, $G$ is fuzzy detour $g$-self centered. ■

Because there exists a unique strong path between every pair of vertices in a fuzzy tree, we can see that $g$-distance and fuzzy detour $g$-distance coincide in a fuzzy tree as seen from the following theorem.

**Theorem 5.5.18** *A connected fuzzy graph $G = (\sigma, \mu)$ is a fuzzy $g$-detour graph if and only if $G$ is a fuzzy tree.*

*Proof* Assume $G$ is a fuzzy tree. Then there exists unique strong path between every pair of vertices in $G$. Hence, $D_g(u, v) = d_g(u, v)$ for every pair $u$ and $v$ of vertices of $G$. Therefore, $G$ is a fuzzy $g$-detour graph.

Conversely, assume $G$ is a fuzzy $g$-detour graph on $n$ vertices. That is, $D_g(u, v) = d_g(u, v)$ for every pair of vertices $u$ and $v$ of $G$. When $n = 2$, the result is trivial and $G$ is a fuzzy tree. So let $n \geq 3$. Assume on the contrary that $G$ is not a fuzzy tree. Then there exists at least one pair of vertices $u_1$, $v_1$ having more than one strong path from $u_1$ to $v_1$. Let $P_1$ and $P_2$ be two $u_1 - v_1$ strong paths. Then union of $P_1$ and $P_2$ contains at least one cycle (say) $C$ in $G$. Let $u$ and $v$ be two adjacent vertices in $C$. Then $d_g(u, v) = 1$ and $D_g(u, v) > 1$, which is a contradiction to the assumption that $D_g(u, v) = d_g(u, v)$, and hence, $G$ is a fuzzy tree. ■

**Definition 5.5.19** The fuzzy subgraph of $G = (\sigma, \mu)$ induced by the fuzzy detour $g$-peripheral vertices is called the **fuzzy detour $g$-periphery** of $G$, denoted by $Per_{D_g}(G)$.

**Definition 5.5.20** A connected fuzzy graph $G = (\sigma, \mu)$ is called a **fuzzy detour $g$-eccentric** if every vertex of $G$ is fuzzy detour $g$-eccentric. The fuzzy subgraph of $G$ induced by the set of fuzzy detour $g$-eccentric vertices is called the **fuzzy detour $g$-eccentric fuzzy graph** of $G$, denoted by $Ecc_{D_g}(G)$.

Let $G$ be a non trivial fuzzy graph on $n$ vertices. Then $Per_{D_g}(G) = G$ if and only if every vertex of $G$ has fuzzy detour $g$-eccentricity $n - 1$. With respect to standard geodesic distance every peripheral vertex is an eccentric vertex, but not conversely. But in fuzzy detour $g$-distance, we observe that the converse is also true.

**Theorem 5.5.21** *Let $G = (\sigma, \mu)$ be a connected fuzzy graph. Then $v$ is a fuzzy detour $g$-eccentric vertex if and only if $v$ is a fuzzy detour $g$-peripheral vertex.*

*Proof* Let $v$ be a fuzzy detour $g$-eccentric vertex of $G = (\sigma, \mu)$ and let $v \in u^*_{D_g}$. Let $y$ and $z$ be two fuzzy detour $g$-peripheral vertices. That is, $D_g(y, z) = k = diam_{D_g}(G)$. Let $P$ be a $y - z$ fuzzy $g$-detour and $Q$ be a $u - v$ fuzzy $g$-detour in $G$. We prove that $e_{D_g}(u) = diam_{D_g}(G)$. Assume on the contrary that $e_{D_g}(u) = l < diam_{D_g}(G)$. Then we have two cases.

**Case 1**. $v$ is an internal vertex ($\deg(v) > 1$) of $G$.

Because $G$ is connected, there exists at least one path between every pair of vertices. Therefore, there exists connection between $v$ to $y$ and $v$ to $z$ also. So we can extend $u - v$ fuzzy $g$-detour to $y$ or $z$, which contradicts that $v$ is a fuzzy detour $g$-eccentric vertex of $u$. Hence, $e_{D_g}(u) = diam_{D_g}(G)$. Thus, $v$ is a fuzzy detour $g$-peripheral vertex.

**Case 2**. $v$ is not an internal vertex of $G$.

Let $w$ be the only vertex adjacent to $v$. Then $w$ should belong to $Q$. Now, $G$ being connected, let $w$ be connected to some vertex, say $w'$ of $P$. Then either $w'$ is not in $Q$ or $w'$ is a common vertex of both $P$ and $Q$. In both cases the path from $u$ to $z$ or to $y$ through $w$ and $w'$ is longer than $Q$, which contradicts that $v$ is a fuzzy detour $g$-eccentric vertex of $u$. Hence, $e_{D_g}(u) = diam_{D_g}(G)$. Thus, $v$ is a fuzzy detour $g$-peripheral vertex.

Conversely, assume $v$ is a fuzzy detour $g$-peripheral vertex of $G$. Then there exists at least one more fuzzy detour $g$-peripheral vertex say $u$. Thus, $v$ is a fuzzy detour $g$-eccentric vertex of $u$. ■

**Definition 5.5.22** A vertex $v$ of a connected fuzzy graph $G = (\sigma, \mu)$ is called a **fuzzy detour $g$-boundary vertex** of a vertex $u$ if $D_g(u, v) \geq D_g(u, w)$ for each neighbor $w$ of $v$, while a vertex is a fuzzy detour $g$-boundary vertex of a fuzzy graph $G$ if $v$ is a fuzzy detour $g$-boundary vertex of some vertex of $G$. We denote by $u^b_{D_g}$ the set of all fuzzy detour $g$-boundary vertices of $u$.

*Example 5.5.23* Consider the fuzzy graph $G$ in Fig. 5.25. Here, $u^b_{D_g} = \{t, z, v, x\}$, $y^b_{D_g} = \{t, v, z\}$, $w^b_{D_g} = \{t, z, v\}$, $x^b_{D_g} = \{t, v, z\}$. Hence, the fuzzy detour $g$-boundary vertices of $G$ are $t, z, x$ and $v$.

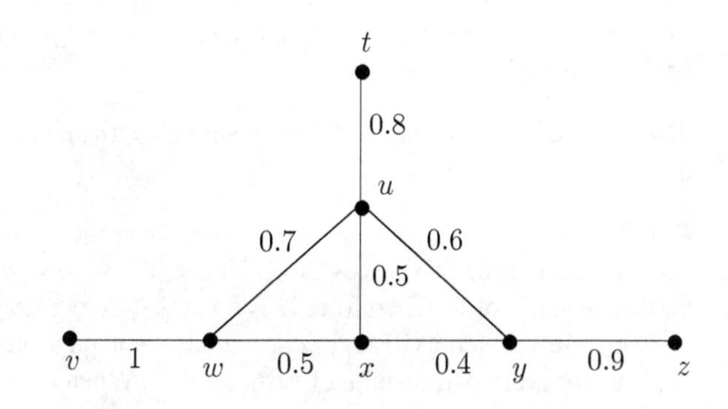


**Fig. 5.25** Fuzzy detour $g$-boundary vertices of $G$

**Definition 5.5.24** A vertex $v$ in a fuzzy graph $G = (\sigma, \mu)$ is called a **complete vertex** if the fuzzy subgraph induced by its strong neighbors form a complete fuzzy graph [172].

**Theorem 5.5.25** *Let $G = (V, \sigma, \mu)$ be a complete fuzzy graph. A vertex $v$ of $G$ is a fuzzy detour $g$-boundary vertex of a vertex distinct from $v$ if and only if $v$ is a complete vertex of $G$.*

*Proof* First, let $v$ be a complete vertex of a connected fuzzy graph $G = (V, \sigma, \mu)$. Then by Definition 5.5.24, the fuzzy subgraph induced by its strong neighbors form a complete fuzzy graph. Let $u$ be a vertex different from $v$. Because $G$ is complete, all edges in $G$ are strong and hence $D_g(u, v) = n - 1 = D_g(u, w)$ for all $w \in N(v)$. Hence, $v$ is a fuzzy detour $g$-boundary vertex of $u$.

Conversely, assume that $v$ is a fuzzy detour $g$-boundary vertex of each vertex distinct from $v$. Because $G$ is complete, all edges in $G$ are strong and $D_g(u, v) = n - 1$ for all $u, v \in \sigma^*$. Hence, the fuzzy subgraph induced by its strong neighbors form a complete fuzzy graph. Therefore, $v$ is a complete vertex. ■

In general, only one implication holds as seen from the result below.

**Theorem 5.5.26** *Let $G = (V, \sigma, \mu)$ be a connected fuzzy graph on $n$ vertices and $v$ be a complete vertex of $G$. Then $v$ is a fuzzy detour $g$-boundary vertex of each vertex distinct from $v$.*

*Proof* First let $v$ be a complete vertex of a connected fuzzy graph $G = (\sigma, \mu)$ and let $u$ be a vertex distinct from $v$. Also, let $u = v_0, v_1, \ldots, v_k = v$ be a $u - v$ fuzzy $g$-detour and let $w$ be a strong neighbor of $v$. Then we have two cases. If $w = v_{k-1}$, then $D_g(u, w) \leq D_g(u, v)$. Therefore, $v$ is a fuzzy detour $g$-boundary vertex of $u$. If $w \neq v_{k-1}$, then because $u = v_0, v_1, \ldots, v_k = v$ is a $u$-$v$ fuzzy $g$-detour, $w$ is a strong neighbor of $v$, the edge $wv_{k-1}$ is strong and $w \neq v_{k-1}$. So, $u = v_0, v_1, \ldots, v_{k-1}, w, v_k = v$ is a longer strong path than $u = v_0, v_1, \ldots, v_k = v$. Therefore, $D_g(u, w) \leq D_g(u, v)$. Hence, $v$ is a fuzzy detour $g$-boundary vertex of $u$. ■

The converse of above theorem need not be true. That is, if $v$ is a fuzzy detour $g$-boundary vertex of all other vertices in a fuzzy graph $G$, then $v$ need not be complete. Consider fuzzy graph in Fig. 5.26. $v$ is a fuzzy detour $g$-boundary vertex of all other vertices. But $v$ is not complete.

**Theorem 5.5.27** *A connected fuzzy graph $G = (\sigma, \mu)$ is a fuzzy $g$-detour graph if and only if $G$ is a fuzzy tree.*

*Proof* Assume $G = (\sigma, \mu)$ is a fuzzy tree. Then there exists a unique strong path between every pair of vertices in $G$. Hence, $D_g(u, v) = d_g(u, v)$ for every pair of vertices $u$ and $v$ of $G$. Therefore, $G$ is a fuzzy $g$-detour graph.

Conversely, assume $G$ is a fuzzy $g$-detour graph on $n$ vertices. That is, $D_g(u, v) = d_g(u, v)$ for every pair $u$ and $v$ of vertices of $G$. When $n = 2$, the result is trivial and

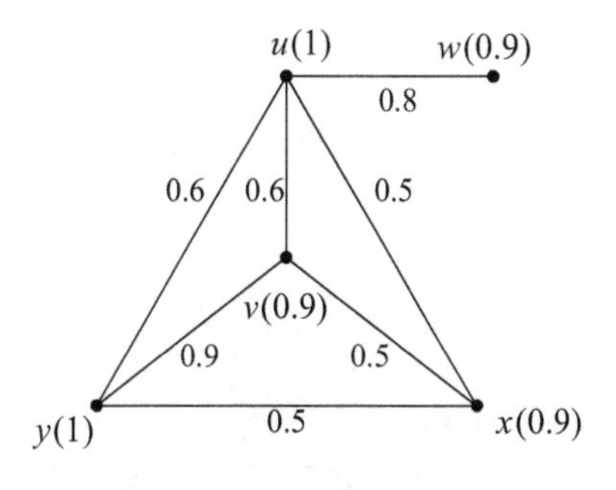


**Fig. 5.26** Counter example for the converse of Theorem 5.5.26

$G$ is a fuzzy tree. So let $n \geq 3$. Assume on the contrary that $G$ is not a fuzzy tree. Then there exists at least one pair of vertices $u_1$ and $v_1$ such that there exists more than one strong path from $u_1$ to $v_1$. Let $P_1$ and $P_2$ be two $u_1$-$v_1$ strong paths. Then union of $P_1$ and $P_2$ contains at least one cycle (say) $C$ in $G$. Let $u$ and $v$ be two adjacent vertices in $C$. Then $d_g(u, v) = 1$ and $D_g(u, v) > 1$, which is a contradiction to the assumption that $D_g(u, v) = d_g(u, v)$, and hence $G$ is a fuzzy tree. ■

**Theorem 5.5.28** *A vertex $v$ in a fuzzy tree $G = (\sigma, \mu)$ is a fuzzy detour $g$-boundary vertex if and only if $v$ is not a fuzzy cutvertex of $G$.*

*Proof* Assume, to the contrary, that there exists a fuzzy tree $G = (\sigma, \mu)$ and a fuzzy cutvertex $v$ of $G$ such that $v$ is a fuzzy detour $g$-boundary vertex of some vertex $u$ in $G$. Let $F$ be the unique maximum spanning tree of $G$. Because $v$ is a fuzzy cutvertex, it is an internal vertex of $F$. Then let $w \in N_S(v)$ be such that it is not on the $u - v$ fuzzy detour in $F$. Because $G$ is a fuzzy tree, $G$ is a fuzzy $g$-detour graph and hence fuzzy detour $g$-distance between any two vertices in $G$ is same as the fuzzy detour $g$-distance between any two vertices in $F$. Hence, we have $D_g(u, w) = D_g(u, v) + D_g(v, w) > D_g(u, v)$, which is a contradiction to our assumption that $v$ is a fuzzy detour $g$-boundary vertex of $u$. Hence, $v$ is not a fuzzy cutvertex.

Conversely, let $v$ be any vertex which is not a fuzzy cutvertex. Then $v$ is an end vertex of unique maximum spanning tree. This means $v$ is fuzzy endvertex and has unique strong neighbor. So any fuzzy $g$-detour from a vertex to $v$ cannot be extended beyond $v$. Hence, $v$ is a fuzzy detour $g$-boundary vertex. ■

**Theorem 5.5.29** *A vertex is a fuzzy detour $g$-boundary vertex of a fuzzy tree if and only if it is a fuzzy endvertex.*

*Proof* Let $G = (\sigma, \mu)$ be a fuzzy tree and $v$ is a fuzzy detour $g$-boundary vertex of some vertex $u$ in $G$. Let $F$ be the unique maximum spanning tree of $G$. Because in fuzzy trees every vertex is either a fuzzy cutvertex or a fuzzy endvertex and no fuzzy cutvertex is a fuzzy detour $g$-boundary vertex, by Theorem 5.5.28, $v$ is a fuzzy endvertex of $G$.

Conversely, assume that $v$ is a fuzzy end vertex of $G$. We prove that $v$ is a fuzzy detour $g$-boundary vertex of $G$. Let $F$ be the unique maximum spanning tree of $G$. Then $v$ is an endvertex of $F$ and $v$ cannot be a fuzzy cutvertex of $G$. Therefore, by Theorem 5.5.28, $v$ is a fuzzy detour $g$-boundary vertex. ■

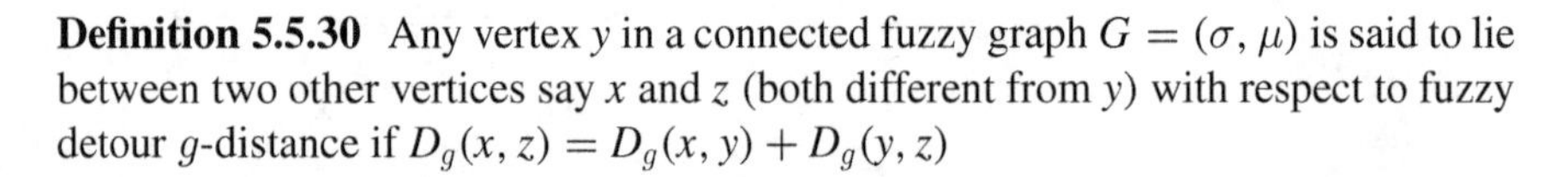


**Fig. 5.27** $Int_{D_g}(G)$ of Example 5.5.23

**Definition 5.5.30** Any vertex $y$ in a connected fuzzy graph $G = (\sigma, \mu)$ is said to lie between two other vertices say $x$ and $z$ (both different from $y$) with respect to fuzzy detour $g$-distance if $D_g(x, z) = D_g(x, y) + D_g(y, z)$

**Definition 5.5.31** A vertex $v$ is a **fuzzy detour $g$-interior vertex** of a connected fuzzy graph $G = (\sigma, \mu)$ if for every vertex $u$ distinct from $v$, there exists a vertex $w$ such that $v$ lies between $u$ and $w$.

**Definition 5.5.32** The **fuzzy detour $g$-interior** of $G = (\sigma, \mu)$, $Int_{D_g}(G)$ is the fuzzy subgraph of $G$ induced by its fuzzy detour $g$-interior vertices.

For the fuzzy graph in Example 5.5.23, the $g$-interior vertices are $u$,$w$ and $y$. The $Int_{D_g}(G)$ is given in Fig. 5.27.

**Theorem 5.5.33** *Let $G = (\sigma, \mu)$ be a connected fuzzy graph. A vertex $v$ is a fuzzy detour $g$-boundary vertex of $G$ if and only if $v$ is not a fuzzy detour $g$-interior vertex of $G$.*

*Proof* Let $v$ be a fuzzy detour $g$-boundary vertex of a connected fuzzy graph $G$, and let $v$ is a fuzzy detour $g$-boundary vertex of some vertex $u$. Assume to the contrary that $v$ is a fuzzy detour $g$-interior vertex of $G$. Because $v$ is a fuzzy detour $g$-interior vertex of $G$, there exists a vertex $w$ distinct from $u$ and $v$ such that $v$ lies between $u$ and $w$. Let $P : u = v_1, v_2, v_3, \ldots, v = v_j, v_{j+1}, \ldots, v_k = w$ be a $u - w$ fuzzy $g$-detour, where $1 < j < k$. However, $v_{j+1} \in N_S(v)$ and $D_g(u, v_{j+1}) > D_g(u, v)$, which contradicts that $v$ is a fuzzy detour $g$-boundary vertex of $u$.

For the converse, let $v$ be a vertex that is not a fuzzy detour $g$-interior vertex of $G$. Hence, there exists at least one vertex $u$ such that for every vertex $w$ distinct from $u$ and $v$, the vertex $v$ does not lie between $u$ and $w$. Let $x \in N(v)$. Then $D_g(u, x) \leq D_g(u, v)$. That is, $v$ is a fuzzy detour $g$-boundary vertex of $u$. ■

**Theorem 5.5.34** *Let $G = (\sigma, \mu)$ be a connected fuzzy graph. If $v$ is a fuzzy end vertex of $G$, then $v$ is not a fuzzy detour $g$-interior vertex.*

*Proof* Suppose $v$ is fuzzy end vertex of a fuzzy graph $G = (\sigma, \mu)$. Then $v$ has exactly one strong neighbor. Then there does not exist a strong fuzzy $g$-detour such that $v$ lies between two other vertices. Hence, $v$ cannot be fuzzy detour $g$-interior vertex. ■

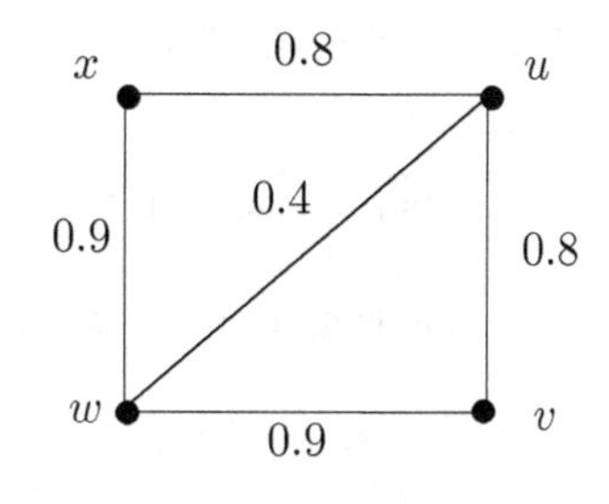


**Fig. 5.28** Fuzzy graph in Example 5.5.35

The converse of Theorem 5.5.34 need not be true as seen from Example 5.5.35.

*Example 5.5.35* Let $G = (\sigma, \mu)$ be the fuzzy graph with $\sigma^* = \{x, u, v, w\}$, $\sigma(s) = 1$ for all $s \in \sigma^*$ and $\mu(xu) = \mu(uv) = 0.8$, $\mu(vw) = \mu(wx) = 0.9$, $\mu(uw) = 0.4$ (Fig. 5.28). Then $w$ is neither fuzzy detour $g$-interior vertex nor fuzzy endvertex. Also, note that $w$ is a fuzzy detour $g$-boundary vertex of $v$ and $x$. Also, $w$ is a fuzzy cutvertex.

**Theorem 5.5.36** *If $u$ is a fuzzy detour $g$-interior vertex of a connected fuzzy graph $G = (\sigma, \mu)$, then $u$ is an internal vertex of every maximum spanning tree of $G$.*

*Proof* Assume that $v$ is fuzzy detour $g$-interior vertex of a connected fuzzy graph $G = (\sigma, \mu)$. Then for every vertex $u$ distinct from $v$, there exists a vertex $w$ such that $v$ lies between $u$ and $w$. Thus, the $u - w$ fuzzy $g$-detour contains $v$ and this strong path will be in every maximum spanning tree of $G$. ■

The converse of Theorem 5.5.36 need not be true. For the fuzzy graph in Example 5.5.35, $w$ is an internal vertex of every maximum spanning tree of $G$. But $w$ is not a fuzzy detour $g$-interior vertex of $G$.

**Theorem 5.5.37** *A vertex $u$ is a fuzzy detour $g$-interior vertex of a fuzzy tree $G = (\sigma, \mu)$ if and only if $u$ is an internal vertex of the unique maximum spanning tree of $G$.*

*Proof* First part follows from Theorem 5.5.36. Conversely assume $u$ is an internal vertex of the unique maximum spanning tree of $G = (\sigma, \mu)$. Then $u$ is fuzzy cutvertex of $G$. Hence, $u$ is not fuzzy detour $g$-boundary vertex of $G$ (Theorem 5.5.28). Therefore, $u$ is fuzzy detour $g$-interior vertex of $G$. ■

**Theorem 5.5.38** *If $u$ is an endvertex of at least one maximum spanning tree of a connected fuzzy graph $G = (\sigma, \mu)$, then $u$ is fuzzy detour $g$-boundary vertex of $G$.*

*Proof* Suppose $u$ is an endvertex of at least one maximum spanning tree of a connected fuzzy graph $G = (\sigma, \mu)$. Then $u$ cannot be an internal vertex of every maximum spanning tree of $G$. Therefore, $u$ is not fuzzy detour $g$-interior vertex of $G$

(Theorem 5.5.36). Thus, by Theorem 5.5.33, $u$ is a fuzzy detour $g$-boundary vertex of $G$. ■

Generally, the converse of Theorem 5.5.38 need not be true. But it is true for fuzzy trees. For the fuzzy graph in Example 5.5.35, $w$ is a fuzzy detour $g$-boundary vertex of $G$. But there does not exist a maximum spanning tree with $w$ as an endvertex.

# Chapter 6
# Sequences, Saturation, Intervals and Gates in Fuzzy Graphs

In this chapter, we discuss three different concepts in fuzzy graphs. The first concept is that of sequences of fuzzy graphs, which allow us to connect a fuzzy graph to a sequence space. Most of the fuzzy graph structures are characterized using different types of sequences. In the second part of this chapter, we discuss saturation in fuzzy graphs and the third part deals with strong intervals and strong gates in fuzzy graphs.

## 6.1 Special Sequences in Fuzzy Graphs

The problem of determining the structure of a fuzzy graph is a challenging one. In [116], an algorithm for the identification of different types of edges of a fuzzy graph is provided. This algorithm is very useful even for fuzzy graphs with a large number of vertices. It divides the edges to the three different types discussed in Chap. 3, namely $\alpha$, $\beta$ and $\delta$ edges. The contents of this section are from [107], in which Mathew and Mathew introduced three different sequences based on the categorization of edges, which will very effectively determine the nature and structure of certain types of fuzzy graphs. We only consider undirected fuzzy graphs without loops and multiple edges in this section.

**Definition 6.1.1** Let $G = (\sigma, \mu)$ be a fuzzy graph with $\sigma^* = \{v_1, v_2, \ldots, v_p\}$ in some order. Then a finite sequence $\alpha_s(G) = (n_1, n_2, n_3, \ldots, n_p)$ is called the **$\alpha$-sequence** of $G$ if $n_i$ represents the number of $\alpha$-strong edges incident at $v_i$. $n_i$ is equal to zero if there are no $\alpha$-strong edges incident at $v_i$.

*Remark 6.1.2* Similar to Definition 6.1.1, we can have the definition of a $\beta$-sequence. If $G = (\sigma, \mu)$ is a fuzzy graph with vertex set $\sigma^* = \{v_1, v_2, \ldots, v_p\}$, a finite sequence $\beta_s(G) = (n_1, n_2, n_3, \ldots, n_p)$ is called the **$\beta$-sequence** of $G$ if $n_i$ is the number of $\beta$-strong edges incident at $v_i$ and equals to zero, if there are no $\beta$-strong edges incident at $v_i$.



S. Mathew et al., *Fuzzy Graph Theory*, Studies in Fuzziness and Soft Computing 363, https://doi.org/10.1007/978-3-319-71407-3_6

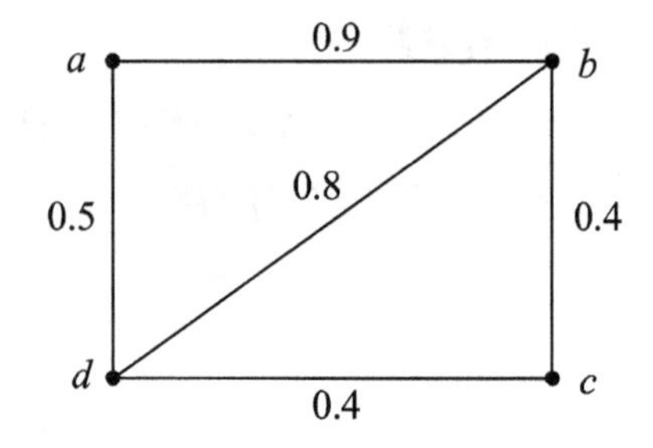


**Fig. 6.1** Fuzzy graph $G$ with all types of edges

**Definition 6.1.3** Let $G = (\sigma, \mu)$ be a fuzzy graph with $\sigma^* = \{v_1, v_2, \ldots, v_p\}$. Then a finite sequence $S_s = (n_1, n_2, n_3, \ldots, n_p)$ is called the **strong sequence** of $G$ if $n_i$ is the number of $\alpha$ and $\beta$ strong edges incident at $v_i$ and equals to zero, if there are no $\alpha$ or $\beta$ strong edges incident at $v_i$.

If there is no confusion regarding the fuzzy graph $G$, we use the notation $\alpha_s$, $\beta_s$ and $S_s$ instead of $\alpha_s(G)$, $\beta_s(G)$ and $S_s(G)$. In the following example (Fig. 6.1), we find these sequences.

*Example 6.1.4* Let $G = (\sigma, \mu)$ with $\sigma^* = \{a, b, c, d\}$, $\sigma(a) = \sigma(b) = \sigma(c) = \sigma(d) = 1$, $\mu(ab) = 0.9$, $\mu(bc) = \mu(cd) = 0.4$, $\mu(da) = 0.5$ and $\mu(bd) = 0.8$ (Fig. 6.1). Clearly, $G$ contains all types of edges. In $G$, $\alpha$, $\beta$ and strong sequences are $\alpha_s = (1, 2, 0, 1)$, $\beta_s = (0, 1, 2, 1)$ and $s_s = (1, 3, 2, 2)$, respectively.

As discussed before, blocks of graphs and fuzzy graphs play important roles in several structural problems in chemistry [182]. Characterization of blocks of fuzzy graphs is a challenging problem. In this section, we present a necessary condition, which must be satisfied by a block and two necessary and sufficient conditions for a block. In the second characterization, we use the notion of strong sequences. These characterizations are comparatively less time consuming than other methods in the literature.

**Definition 6.1.5** A sequence of integers is called a **binary sequence** if it contains only 0's and 1's.

**Theorem 6.1.6** *Let $G = (\sigma, \mu)$ be a connected fuzzy graph. If $G$ is a block, then $\alpha_s(G)$ is a binary sequence.*

*Proof* Suppose that $G = (\sigma, \mu)$ is a block. We have to prove that $\alpha_s(G)$ is binary. That is, we need to prove that $\alpha_s(G)$ contains only 0's and 1's. If possible, suppose the contrary. Suppose that there exists an entry which is at least 2 in $\alpha_s(G)$. Let $n_i = 2$. That is, there are 2 different $\alpha$-strong edges incident at the vertex $v_i$. Now, an edge in $\mu^*$ of $G$ is a fuzzy bridge if and only if it is an $\alpha$-strong edge (Theorem 3.2.9). Also, if a vertex is common to more than one fuzzy bridge, then it is a fuzzy cutvertex (Theorem 2.2.11). Therefore, we see that $v_i$ is a fuzzy cutvertex of $G$, which is a contradiction to our assumption that the fuzzy graph $G$ is free from fuzzy cutvertices because it is a block. So our assumption is wrong. Hence, $n_i < 2$. That is, $n_i = 0$ or 1. Thus, $\alpha_s(G)$ is binary. ■

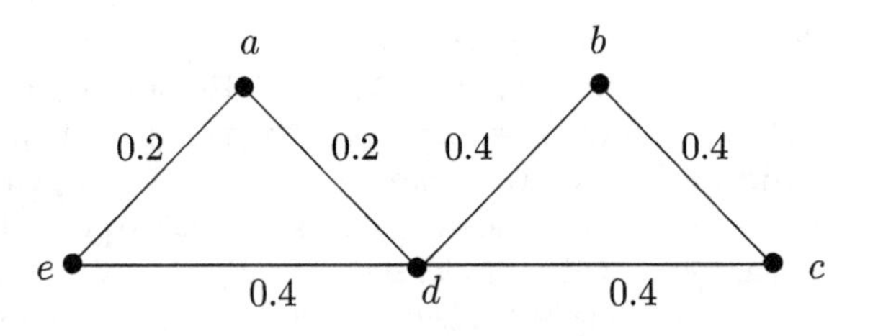


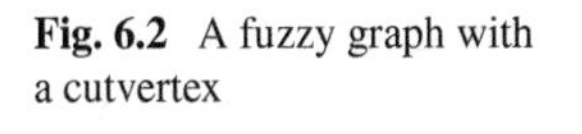
**Fig. 6.2** A fuzzy graph with a cutvertex

Note that the above condition is only necessary. It is not sufficient as seen from the following example.

*Example 6.1.7* Let $G = (\sigma, \mu)$ be the fuzzy graph given in Fig. 6.2. Here, $\sigma^* = \{a, b, c, d, e\}$, $\sigma(s) = 1$ for all $s \in \sigma^*$ and $\mu(ae) = \mu(ad) = 0.2$ and $\mu(bd) = \mu(bc) = \mu(ed) = \mu(dc) = 0.4$. Note that the $\alpha$-sequence of $G$ is, $\alpha_s(G) = (1, 0, 0, 0, 1)$. It is a binary sequence. But the fuzzy graph is not a block because the vertex $d$ is a cutvertex and hence is a fuzzy cutvertex.

The converse of Theorem 6.1.6 is true for a subcategory of fuzzy graphs only. In the next characterization, the underlying graph $G^*$ of $G$ is restricted to be a block. This means that $G^*$ has no cutvertices.

**Theorem 6.1.8** *Let $G = (\sigma, \mu)$ be a connected fuzzy graph such that the underlying graph $G^*$ has no cutvertices. Then $G$ is a block if and only if $\alpha_s(G)$ is a binary.*

*Proof* Let $G = (\sigma, \mu)$ be a connected fuzzy graph such that the underlying graph $G^*$ is a block. If $G$ is also a block, then by Theorem 6.1.6, $\alpha_s(G)$ is binary.

Conversely, suppose that, $\alpha_s(G)$ is a binary sequence. We have to prove that $G$ is a block. That is, we have to prove that $G$ has no fuzzy cutvertices. If possible, let $G$ has a fuzzy cutvertex, say, $v_i$. Then there exists two vertices $u$ and $w$ in $G$ such that $u \neq v_i \neq w$ and $CONN_{G-v_i}(u, w) < CONN_G(u, w)$. Because $G^*$ is a block, it has no cutvertices, and hence $v_i$ is not a cutvertex. Therefore, we can consider many possible $u - w$ paths not passing through the vertex $v_i$. Now, from the above inequality, clearly the weights of all edges in the $u - v_i - w$ path are strictly greater than weights of all edges in the possible $u - w$ paths, which are not passing through the vertex $v_i$. This means, all edges in the $u - v_i - w$ path are $\alpha$-strong, and hence $v_i$ is incident with at least 2 different $\alpha$-strong edges. Therefore, $n_i = 2$, which shows that $\alpha_s(G)$ is not binary. This is a contradiction. So our assumption is wrong. Hence, $G$ is a block. ■

Next we have a characterization of blocks using $\alpha$ strong sequences.

**Theorem 6.1.9** ([107]) *Let $G = (\sigma, \mu)$ be a connected fuzzy graph. Then $G$ is a block if and only if the following conditions are satisfied.*

*(i) $\alpha_s(G)$ is a binary sequence.*

*(ii) For any given pair of vertices $u$ $v$ in $\sigma^*$, there exists a cycle $C$, containing $u$ and $v$ such that $S_s(C)$ contains only entries which are at the least 2.*

*Proof* Let $G = (\sigma, \mu)$ be a connected fuzzy graph. First suppose that $G$ is a block. We have to prove conditions (i) and (ii). The proof of condition (i) is the same as that of Theorem 6.1.6. Now, we prove condition (ii) as follows. Let $u$ and $v$ be any two vertices of $G$. We have to prove that $u$ and $v$ lie on a common cycle $C$ such that all entries in $S_s(C)$ are at least 2. That is, we need to prove that $u$ and $v$ lie on a common strong cycle $C$. If possible, suppose the contrary. Let there be no common strong cycles containing both $u$ and $v$. There are two cases arise.

**Case 1**: $uv$ is a strong edge.

Because $u$ and $v$ are not on any strong cycle and edge $uv$ is strong, the edge $uv$ lies on every *MST* of $G$. Now, by Corollary 2.2.4 an edge belonging to every *MST* is a fuzzy bridge. Hence, $uv$ is a fuzzy bridge.

If $u$ is an end vertex of all maximum spanning trees, then clearly $v$ is the common vertex of at least 2 fuzzy bridges, and hence it will be a fuzzy cutvertex of $G$, which is a contradiction to our assumption that $G$ is free from fuzzy cutvertices as it is a block. On the other hand, if $v$ is an end vertex of all $MST's$, then $u$ will be a fuzzy cutvertex of $G$, contradiction to our assumption.

Now, suppose that $u$ is an end vertex of $MST$, $T_1$ and $v$ is an end vertex of $MST$, $T_2$. Because $T_1$ is a spanning tree, $v$ will be an internal vertex of $T_1$. Let $w$ be a strong neighbor of $u$ in $T_2$. Clearly, there appears a strong path $P$ in $T_1$ from $u$ to $w$ through $v$. This path $P$ together with the strong edge $uw$ forms a strong cycle $C$ in $G$, which is also a contradiction.

**Case 2**: $uv$ is a $\delta$-edge.

If $uv$ is a $\delta$-edge, then there exists a strong path between $u$ and $v$. Because there is a unique strong $u - v$ path $P$ in $G$, it belongs to all $MST's$ and all internal vertices of $P$ are fuzzy cutvertices of $G$, which is a contradiction.

So in all cases, our assumption is wrong. Hence, there exists a common strong cycle $C$, containing any given pair of vertices $u$ and $v$. Because $C$ is strong, $S_s(C)$ contains entries which are at least 2 only.

Conversely, suppose conditions (i) and (ii) hold. We have to prove that $G$ is a block. That is, we need to prove that $G$ has no fuzzy cutvertices. By condition (i), it is clear that at most one $\alpha$-strong edge is incident at every vertex. By condition (ii), any given pair of vertices $u$ and $v$ lies on a common strong cycle. So if we delete a vertex from $G$, then the strength of connectedness between every pair of vertices remains the same. Thus, no vertex of $G$ can be a fuzzy cutvertex and hence $G$ is a block. ■

There are several methods and algorithms for the identification of a fuzzy tree in the literature. In this section, we provide different characterizations for fuzzy trees. We use the concepts of sequences, maximum spanning trees, and sequence sums in these results. As explained before, these results can be used to determine the complete structure of certain types of fuzzy graphs. Also, these results link fuzzy graphs to sequence spaces and hence to functional analysis.

**Theorem 6.1.10** *If $G = (\sigma, \mu)$ is a fuzzy tree and $|\sigma^*| = p$, then $\alpha_s(G) \in (\mathbf{Z}^+)^p$.*

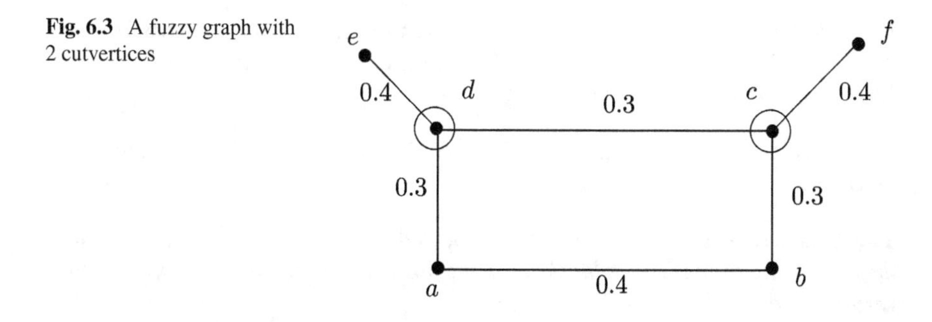

**Fig. 6.3** A fuzzy graph with 2 cutvertices

*Proof* By the definition of $\alpha_s(G)$, it is trivial that all of its elements are greater than or equal to zero. Here, we want to prove that all elements in $\alpha_s(G)$ are at least unity. Suppose the contrary. Let the $i$th element in $\alpha_s(G)$, say, $n_i$ be zero. Because $n_i = 0$, the corresponding vertex $v_i$ will not be incident with any $\alpha$-strong edge. This will result in the disconnection of the maximum spanning tree $F$ of $G$, which is a contradiction to the definition of $F$. So our assumption is wrong and hence all elements of $\alpha_s(G)$ are at least unity. ■

The condition in Theorem 6.1.10 is not sufficient as seen from the following example.

*Example 6.1.11* Consider the fuzzy graph $G = (\sigma, \mu)$ with $\sigma^* = \{a, b, c, d, e, f\}$, $\mu(ab) = \mu(cf) = \mu(de) = 0.4$ and $\mu(bc) = \mu(cd) = \mu(ad) = 0.3$ (Fig. 6.3). In this fuzzy graph, $\alpha_s(G) = (1, 1, 1, 1, 1, 1)$. But the fuzzy graph is not a fuzzy tree. Here vertices $c$ and $d$ are cutvertices of the fuzzy graph.

The next result helps us to obtain the number of fuzzy cutvertices of a fuzzy tree by looking at the $\alpha$-sequence of $G$.

**Theorem 6.1.12** *Let $G = (\sigma, \mu)$ be a connected fuzzy graph such that $|\sigma^*| = p$. Let $t$ be a positive integer such that $t \leq p$. If $\alpha_s(G)$ contains $t$ elements which are at least 2, then $G$ has exactly $t$ fuzzy cutvertices.*

*Proof* Let $G = (\sigma, \mu)$ be a fuzzy tree. Let $F$ be the spanning tree of $G$ with the property given in the definition of a fuzzy tree. Then by Theorem 2.3.16, the internal vertices of $F$ are the fuzzy cutvertices of $G$. Also, we know that, if a vertex is common to more than one $\alpha$-strong edges, then it is a fuzzy cutvertex (Theorems 2.2.11 and 3.2.9). So the vertex of $G$ corresponding to an entry in $\alpha_s(G)$ which is at least 2 must be a fuzzy cutvertex. ■

If the condition in the above theorem were also sufficient, then we would be able to identify the fuzzy cutvertices of $G$ with the information about the $\alpha$-sequence of $G$. But due to its nonsufficiency, we can get the number of fuzzy cutvertices of $G$ only. We illustrate this fact in the following example.

*Example 6.1.13* Consider the fuzzy graph given in Example 6.1.7 (Fig. 6.2). Here, the vertex $d$ is a cutvertex and hence a fuzzy cutvertex. But the entry in the $\alpha$-sequence corresponding to vertex $d$ is 0.

If we restrict the underlying graph $G^*$ of $G$ to be a block, then the non sufficiency of Theorem 6.1.12 can be avoided.

**Theorem 6.1.14** *Let $G = (\sigma, \mu)$ be a connected fuzzy graph such that $|V| = p$ and the underlying graph $G^*$ is a block. Then the fuzzy cutvertices of $G$ are exactly those vertices whose entry in $\alpha_s(G)$ is at least 2.*

**Definition 6.1.15** A **zero sequence** is a real sequence having all entries 0. It is denoted by (0).

The following theorem is a characterization of fuzzy trees, using the concept of $\beta$-sequence of a fuzzy graph.

**Theorem 6.1.16** *A connected fuzzy graph $G$ is a fuzzy tree if and only if $\beta_s(G) = (0)$.*

*Proof* Let $G = (\sigma, \mu)$ be a connected fuzzy graph. Suppose that $G$ is a fuzzy tree. Then all the strong edges of $G$ are fuzzy bridges. Now, an edge $e = uv$ in $G$ is a fuzzy bridge if and only if it is $\alpha$-strong (Theorem 3.2.9). Thus, all the strong edges of $G$ are $\alpha$-strong. Therefore, $G$ has no $\beta$ strong edges and hence $\beta_s(G) = (0)$.

If $G$ is a tree, there is nothing to prove. If $G$ is not a tree, then $G$ has a cycle, say, $C$. Because $G$ is a fuzzy tree, by Theorem 2.3.1, there exists an edge $e = uv$ such that $w(e) < CONN_{G-e}(u, v)$, where $G - e$ is the subgraph of $G$ obtained by deleting the edge $e$ from $G$. This means that $e$ is a $\delta$-edge. If $G - e$ is a fuzzy spanning tree of $G$, then all the edges in $G - e$ are $\alpha$. Hence, $\beta_s(G) = (0)$. If $G - e$ is not a fuzzy spanning tree of $G$, then continue the above procedure of deleting $\delta$ edges from $G - e$ until we get spanning tree.

Conversely, suppose that $\beta_s(G) = (0)$. We have to prove that $G$ is a fuzzy tree. If $G$ has no fuzzy cycles, then $G$ is a tree and hence a fuzzy tree. Suppose that $G$ has a cycle, say, $C$. Then $C$ contains only $\alpha$-strong and $\delta$-edges. Also, note that all the edges of $C$ cannot be $\alpha$-strong because otherwise it will contradict the definition of $\alpha$-strong edges. Thus, there exists at least one $\delta$-edge say $e$ in $C$. If we delete $e$ from $C$, we get a maximum spanning tree of $G$. Similarly, remove one $\delta$ edge each from the existing cycles of $G$. Finally, we get a unique maximum spanning tree of $G$. Thus, it follows that $G$ is a fuzzy tree. ■

We know that, if $G$ is a fuzzy tree and $F$, the spanning tree in the definition, then the edges of $F$ are the fuzzy bridges of $G$. That is, edges of $F$ are the $\alpha$-strong edges of $G$. Thus, $F$ has no $\beta$ strong edges and hence $\beta_s(F)$ contains only zeros. Also, because $F$ is a spanning tree of $G$, $\beta_s(G) = \beta_s(F)$.

In the following theorem, we present another necessary and sufficient condition for a fuzzy graph $G$ to be a fuzzy tree.

**Theorem 6.1.17** *A connected fuzzy graph $G = (\sigma, \mu)$ is a fuzzy tree if and only if $\alpha_s(G) = \alpha_s(F)$, where $F$ is a maximum spanning tree of $G$.*

*Proof* Suppose that $G = (\sigma, \mu)$ is a fuzzy tree. If $G$ itself is a tree, then $F$ and $G$ are isomorphic and hence $\alpha_s(G) = \alpha_s(F)$. If $G$ is not a tree, then $G$ contains a cycle $C$. Because $G$ is a fuzzy tree, there exists an edge $e = uv$ in $C$ such that $w(e) < CONN_{G-e}(u, v)$, where $G - e$ is the fuzzy subgraph obtained by deleting the edge $e$ from $G$. If $G - e$ is a fuzzy tree, then $G - e$ and $F$ are isomorphic and because $e$ is a $\delta$-edge, we get $\alpha_s(G) = \alpha_s(F.)$ If not, then continue the above procedure of deleting $\delta$-edges from cycles in $G - e$ until we get a maximum spanning tree $F$ of $G$ such that $\alpha_s(G) = \alpha_s(F)$.

Conversely, assume that $\alpha_s(G) = \alpha_s(F)$, where $F$ is a maximum spanning tree of $G$. We want to prove that $G$ is a fuzzy tree. If possible, let $G$ be not a fuzzy tree. Then there exists at least one $\beta$-strong edge in $G$. Let $uv$ be a $\beta$-strong edge in $G$. Then there exists at least one another $u - v$ path, say, $P$ in $G$ such that $\mu(xy) \geq \mu(uv)$ for every edge $xy$ in $P$. Now, the union of $P$ and edge $uv$ is a cycle in $G$. Suppose that the number of $\alpha$-strong edges incident at $u$ in $G$ is $t$. Now, to get $F$, delete the edge $uv$ from $G$, which has the least weight in $C$. Then the number of $\alpha$-strong edges incident at $u$ in $F$ is $t + 1$, which is a contradiction to our assumption that $\alpha_s(G) = \alpha_s(F)$. Hence, our assumption is wrong. Thus, $G$ is a fuzzy tree. ■

If a connected fuzzy graph $G$ is a tree (not a fuzzy tree), then $G$ has no cycles and every edge in it is an $\alpha$-strong edge. Therefore, the sum of the elements in $\alpha_s(G)$ is equal to twice the number of edges of $G$. The next theorem is another necessary and sufficient condition for a connected fuzzy graph $G$ to be a fuzzy tree. The uniqueness of the maximum spanning tree $F$ of $G$ is used in the proof.

**Theorem 6.1.18** *A connected fuzzy graph $G = (\sigma, \mu)$ is a fuzzy tree if and only if $\alpha_s(F)$ is the same for all maximum spanning trees $F$ of $G$.*

*Proof* We know by Theorem 2.3.19 that a connected fuzzy graph is a fuzzy tree if and only if it has a unique maximum spanning tree. Moreover the spanning tree $F$ in the definition of fuzzy tree, is a maximum spanning tree. ■

In the next theorem, we characterize fuzzy trees by using $\alpha$-sequences and number of disjoint cycles.

**Theorem 6.1.19** *Let $G = (\sigma, \mu)$ be a connected fuzzy graph with exactly $k$-edge disjoint cycles. Then $G$ is a fuzzy tree if and only if $\sum_{n_i \in \alpha_s(G)} n_i = 2(e - k)$, where $e$ is the total number of edges of $G$.*

*Proof* Let $G = (\sigma, \mu)$ be a connected fuzzy graph with exactly $k$-edge disjoint cycles. Suppose that $G$ is a fuzzy tree. Then it has no $\beta$-strong edges. This means all edges present in $G$ are either $\alpha$-strong or $\delta$ edges. Now, given that $G$ has $k$-edge disjoint cycles. Consider an arbitrary cycle, say, $C$. Let the minimum of the weights of all edges in $C$ be $w$ and let it be assigned to the edge $uv$. That is, $\mu(uv) = w$. Now, all the edges of $C$ have weights strictly greater than $w$. For if, let there be another edge $ab$ in $C$ with membership $w$. Then $CONN_{G-uv}(u, v) = w = \mu(uv)$, which implies that $uv$ is a $\beta$-strong edge. Hence, our assumption is wrong. So the minimum weight in a cycle is assigned to exactly one edge in $C$. So the edge $uv$ is the only $\delta$-edge in $C$ and all other edges are $\alpha$-strong. This situation is same in all the $k$ cycles in $G$, which are disjoint. Thus, the total number of $\delta$-edges of $G = k$, and hence total number of $\alpha$-strong edges of $G = e - k$. Therefore, $\sum_{n_i \in \alpha_s(G)} n_i = 2(e - k)$.

Conversely, assume that $\sum_{n_i \in \alpha_s(G)} n_i = 2(e - k)$, where $e$ is the total number of edges of $G$. We have to prove that $G$ is a fuzzy tree. It is enough if we prove that $G$ has no $\beta$-strong edges. Suppose the contrary. Let there be a $\beta$-strong edge, say, $e$ exists in $G$. Then clearly $e$ lies on a cycle, say, $C$. Then there will be at least one edge, say $e'$ in $C$, which is different from $e$ such that $\mu(e) = \mu(e')$. This means that $e'$ is also a $\beta$-strong edge. Thus, there will be at most $e - k - 1$ number of $\alpha$-strong edges in $G$. It is a contradiction to our assumption $\sum_{n_i \in \alpha_s(G)} n_i = 2(e - k)$. So $G$ is a fuzzy tree. ■

If $G = (\sigma, \mu)$ is a simple connected fuzzy graph such that the number of edges is at most number of vertices, then all cycles of $G$ will be edge disjoint. So we have the following results.

**Corollary 6.1.20** *Let $G = (\sigma, \mu)$ be a simple connected fuzzy graph such that $|\mu^*| \leq |\sigma^*|$ and have exactly k-cycles. Then G is a fuzzy tree if and only if $\sum_{n_i \in \alpha_s(G)} n_i = 2(e - k)$, where e is the total number of edges of G.*

**Corollary 6.1.21** *A connected fuzzy graph $G = (\sigma, \mu)$ is a tree, not a fuzzy tree if and only if $\sum_{n_i \in \alpha_s(G)} n_i = 2e$, where $e = |\mu^*|$.*

## 6.2 Saturation in Fuzzy Graphs

This section is based on [108]. Here, we discuss the concepts of vertex and edge saturation counts of fuzzy graphs. The vertex saturation count of a fuzzy graph gives a measure of the average strong degree of the fuzzy graph and the edge saturation count, the percentage of strong edges in the fuzzy graph. The following definitions are valid for different unweighted graph structures also. We consider fuzzy graphs $G = (\sigma, \mu)$ with $\sigma(u) = 1$ for all $u \in \sigma^*$ for convenience, unless otherwise specified.

**Definition 6.2.1** Let $G = (\sigma, \mu)$ be a connected fuzzy graph. Then the **strong vertex count** of $G$ is defined and denoted by

$$S_V(G) = \frac{\textit{number of strong edges of } G}{\textit{number of vertices of } G} = \frac{\textit{number of } \alpha \textit{ or } \beta \textit{ strong edges}}{|\sigma^*|},$$

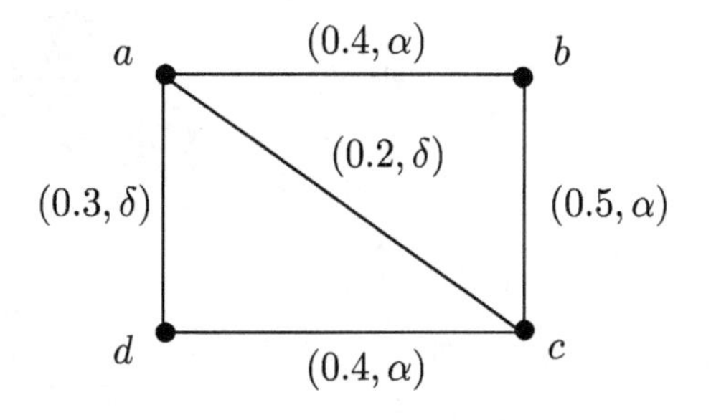


**Fig. 6.4** A graph with strong saturation count

and the **strong edge count** of $G$ is defined and denoted by

$$S_E(G) = \frac{number\ of\ strong\ edges\ of\ G}{number\ of\ edges\ of\ G} = \frac{number\ of\ \alpha\ or\ \beta\ strong\ edges}{|\mu^*|}.$$

*Remark 6.2.2* If we restrict the numerator in Definition 6.2.1 to the number of $\alpha$-strong edges only, then we get $\alpha$-vertex count $\alpha_V(G)$ and $\alpha$-edge count $\alpha_E(G)$, respectively. Similarly, replacing the numerator by $\beta$ strong edges, we get the $\beta$-vertex count $\beta_V(G)$ and the $\beta$-edge count $\beta_E(G)$, respectively.

*Example 6.2.3* Let $G = (\sigma, \mu)$ be the fuzzy graph with $\sigma^* = \{a, b, c, d\}$, $\mu(ab) = \mu(cd) = 0.4$, $\mu(bc) = 0.5$ and $\mu(ad) = 0.3$ (Fig. 6.4). Note that $\sigma(x) = 1$ for all $x \in \sigma^*$. The vertex counts and edge counts of $G$, are given in the following table.

| | |
|---|---|
| $\alpha_V(G) = \frac{3}{4}$, | $\alpha_E(G) = \frac{3}{5}$, |
| $\beta_V(G) = \frac{0}{4} = 0$, | $\beta_E(G) = \frac{0}{5} = 0$, |
| $S_V(G) = \frac{3}{4}$, | $SE(G) = \frac{3}{5}$. |

*Example 6.2.4* Let $G = (\sigma, \mu)$ be the fuzzy graph whose underlying graph is complete with $\sigma^* = \{a, b, c, d, e\}$, $\mu(ab) = \mu(ac) = \mu(ad) = \mu(be) = \mu(cd) = 0.5$, $\mu(ae) = \mu(bd) = 0.6$, $\mu(cd) = \mu(ce) = 0.8$ and $\mu(bc) = 0.9$ (Fig. 6.5). The edge classification of $G$ is given in the following table.

| ab | ac | ad | ae | bc | bd | be | cd | ce | de |
|---|---|---|---|---|---|---|---|---|---|
| $\delta$ | $\delta$ | $\beta$ | $\alpha$ | $\alpha$ | $\delta$ | $\delta$ | $\alpha$ | $\alpha$ | $\delta$ |

The saturation counts in $G$ are given below.

| | |
|---|---|
| $\alpha_V(G) = \frac{4}{5}$, | $\alpha_E(G) = \frac{4}{10}$, |
| $\beta_V(G) = \frac{1}{5}$, | $\beta_E(G) = \frac{1}{10}$, |
| $S_V(G) = \frac{5}{5} = 1$, | $S_E(G) = \frac{5}{10} = \frac{1}{2}$. |

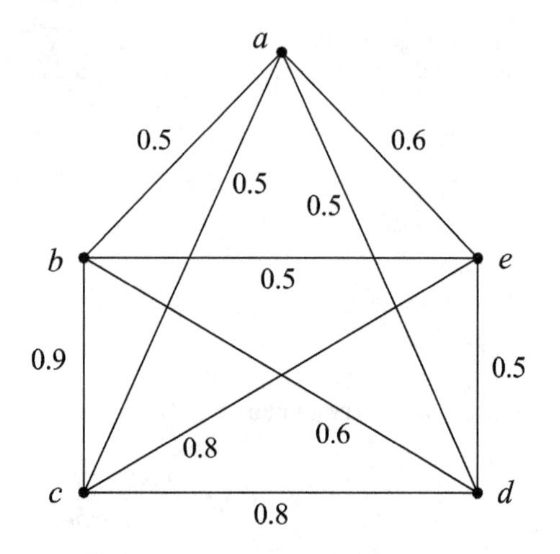


**Fig. 6.5** A graph with nonzero beta saturation count

As all the strong edges of a fuzzy tree $G$ are $\alpha$-strong, $\alpha_V(G) = \frac{n-1}{n}$ and $\alpha_E(G) = \frac{n-1}{n-1} = 1$. For any simple fuzzy graph structure, the number of $\alpha$-strong edges never exceed the number of vertices. In a complete fuzzy graph, by assigning equal membership values to all the vertices, we can make all the edges $\beta$-strong. For such a fuzzy graph $G$, $\beta_V(G) = \frac{^nC_2}{n}$ and $\beta_E(G) = \frac{^nC_2}{^nC_2} = 1$.

Based on all these observations, we have the following proposition.

**Proposition 6.2.5** *Let $G = (\sigma, \mu)$ be a connected fuzzy graph with $|\sigma^*| = n$. Then we have the following inequalities.*

*(i)* $0 \leq \alpha_V(G) \leq \frac{n-1}{n}$,
*(ii)* $0 \leq \beta_V(G) \leq \frac{^nC_2}{n}$,
*(iii)* $0 \leq S_V(G) \leq \frac{^nC_2}{n}$,
*(iv)* $0 \leq \alpha_E(G) \leq 1$,
*(v)* $0 \leq \beta_E(G) \leq 1$,
*(vi)* $0 \leq S_E(G) \leq 1$.

As the strong saturation count is simply the sum of $\alpha$ and $\beta$ saturation counts, the proofs of all the above inequalities are immediate and omitted.

In the following proposition, we make a comparison between the edge count and vertex count of a fuzzy graph.

**Proposition 6.2.6** *Let $G = (\sigma, \mu)$ be a connected fuzzy graph with $|\sigma^*| = n$. Then we have the following inequalities.*

*(i)* $0 \leq \beta_E(G) \leq \beta_V(G)$.
*(ii)* $0 \leq S_E(G) \leq S_V(G)$.
*Furthermore, if $G$ is any fuzzy graph other than a fuzzy tree, then*
*(iii)* $0 \leq \alpha_E(G) \leq \alpha_V(G)$.

**Proposition 6.2.7** *Let $G = (\sigma, \mu)$ be a fuzzy tree, then $0 \leq \alpha_V(G) < \alpha_E(G)$.*

For a fuzzy tree, it is obvious that, $\alpha_V(G) = \frac{n-1}{n}$ and $\alpha_E(G) = \frac{n-1}{n-1} = 1$.

In this section, saturation counts of fuzzy structures like fuzzy trees, cycles and blocks are discussed. Some useful characterizations for these structures are also provided. In the following theorem, a characterization for fuzzy trees whose support are trees is given in terms of its $\alpha$-vertex and $\alpha$-edge counts.

**Theorem 6.2.8** *Let $G = (\sigma, \mu)$ be a connected fuzzy graph with $|\sigma^*| = n$, where $n \geq 2$. Then the following statements are equivalent.*

*(i) $G$ is a tree.*

*(ii) $\alpha_V(G) = \frac{n-1}{n}$ and $\alpha_E(G) = 1$.*

*(iii) $n\ \alpha_V(G) = (n-1)\ \alpha_E(G)$.*

*Proof* (i)$\Rightarrow$ (ii).

Suppose that $G = (\sigma, \mu)$ is a fuzzy tree with $|\sigma^*| = n$, where $n \geq 2$. If $G$ is a tree, then the unique maximum spanning tree $F$ of $G$ is itself. Then $G$ is connected and has exactly $(n-1)$ strong edges. Note that any edge $e = uv$ in $G$ is the unique strong path between $u$ and $v$ in $G$. Hence, $e$ is $\alpha$-strong. Thus, all the $(n-1)$ edges of $G$ are $\alpha$-strong. Hence, $\alpha_V(G) = \frac{n-1}{n}$ and $\alpha_E(G) = 1$.

(ii)$\Rightarrow$ (iii).

Suppose that $\alpha_V(G) = \frac{n-1}{n}$ and $\alpha_E(G) = 1$. Then $n\alpha_V(G) = (n-1) = (n-1)1 = (n-1)\alpha_E(G)$ because $\alpha_E(G) = 1$.

(iii)$\Rightarrow$ (i).

Suppose that $n\alpha_V(G) = (n-1)\ \alpha_E(G)$, i.e., $\alpha_V(G) = \frac{n-1}{n}\ \alpha_E(G)$. We have to prove that $G$ is a tree. We have, $\frac{\alpha_V(G)}{\alpha_E(G)} = \frac{(n-1)}{n} = \alpha_V(G)$, which implies $\alpha_E(G) = 1$. Thus, all edges in $G$ are $\alpha$-strong edges, which is possible only if $G$ is connected and acyclic and hence $G$ is a tree. ■

In the following theorem, we present another characterization for fuzzy trees.

**Theorem 6.2.9** *Let $G = (\sigma, \mu)$ be a connected fuzzy graph. Then $G$ is a fuzzy tree if and only if $\alpha_V(G) = S_V(G)$ and $\alpha_E(G) = S_E(G)$.*

*Proof* Let $G = (\sigma, \mu)$ be a connected fuzzy graph. Suppose that $G$ is a fuzzy tree. Then by Proposition 2.3.4 and Theorem 3.2.15, it follows that all the edges of $G$ which are in $F$ are $\alpha$-strong. Other edges of $G$ will be $\delta$-edges by definition of a fuzzy tree. Therefore, $G$ has no $\beta$ strong edges and hence $\beta_V(G) = \frac{0}{|V|} = 0$ and $\beta_E(G) = \frac{0}{|E|} = 0$. Consequently, $\alpha_V(G) = S_V(G)$ and $\alpha_E(G) = S_E(G)$.

Conversely, suppose that $\alpha_V(G) = S_V(G)$ and $\alpha_E(G) = S_E(G)$. We have to prove that $G$ is a fuzzy tree. If $G$ has no cycles, then $G$ is a tree and hence is a fuzzy tree. Suppose that $G$ has a cycle, say, $C$. Then $C$ contains only $\alpha$-strong and $\delta$-edges. Also, note that all the edges of $C$ cannot be $\alpha$-strong because otherwise it will contradict the definition of $\alpha$-strong edges. Thus, there exists at least one $\delta$-edge in $C$. When we delete $e$ from $C$, if we get a unique maximum spanning tree of $G$, we are done. If not, remove $\delta$ edges from the existing cycles of $G$ one by one until we get a unique maximum spanning tree of $G$ proving that $G$ is a fuzzy tree. ■

Next result shows that the $\alpha$ vertex counts of a fuzzy tree and its corresponding maximum spanning tree are the same.

**Theorem 6.2.10** *A connected fuzzy graph $G = (\sigma, \mu)$ is a fuzzy tree if and only if $\alpha_V(G) = \alpha_V(F)$, where $F$ is the corresponding unique maximum spanning tree of $G$.*

*Proof* Suppose that $G = (\sigma, \mu)$ is a fuzzy tree. If $G$ is itself a tree, then $F$ and $G$ are isomorphic and,

$$\alpha_V(G) = \frac{\textit{number of } \alpha \textit{ strong edges in G}}{|V|} = \frac{\textit{number of } \alpha \textit{ strong edges in F}}{|V|} = \alpha_V(F).$$

If $G$ is not a tree, then $G$ contains a cycle $C$. Because $G$ is a fuzzy tree, there exists an edge $e = uv$ in $C$ such that $w(e) < CONN_{G-e}(u, v)$, where $G - e$ is the fuzzy subgraph obtained by deleting the edge $e$ from $G$. If $G - e$ is a tree, then $G - e$ and $F$ are isomorphic and because $e$ is a $\delta$ edge, we get $\alpha_V(G) = \alpha_V(F)$. If not, continue the above procedure of deleting $\delta$ edges from cycles in $G - e$ until we get a maximum spanning tree $F$ of $G$ such that $\alpha_V(G) = \alpha_V(F)$.

Conversely, assume that $\alpha_V(G) = \alpha_V(F)$, where $F$ is the corresponding maximum spanning tree of $G$. We want to prove $G$ is a fuzzy tree. If possible, suppose $G$ is not a fuzzy tree. Then there exists at least one $\beta$-strong edge in $G$. Let $uv$ be a $\beta$-strong edge in $G$. Then there exists at least one another $u - v$ path, say, $P$ in $G$ such that $\mu(xy) \geq \mu(uv)$ for every edge $xy$ in $P$. Now, the union of $P$ and edge $uv$ is a cycle in $G$. Suppose that the number of $\alpha$-strong edges incident at $u$ in $G$ is $t$. Now, to get $F$, delete the edge $uv$ from $G$, which has the least weight in $C$. Then the number of $\alpha$-strong edges incident at $u$ in $F$ is $t + 1$. Let $t'$ be the remaining number of $\alpha$-strong edges of $G$. Then $\alpha_V(G) = \frac{t+t'}{|V|}$ and $\alpha_V(F) = \frac{t+1+t'}{|V|}$, which is a contradiction to our assumption that $\alpha_V(G) = \alpha_V(F)$. Thus, our assumption is wrong. Hence, $G$ is a fuzzy tree. ■

Next theorem also is a characterization of fuzzy trees.

**Theorem 6.2.11** ([108]) *Let $G = (\sigma, \mu)$ be a connected fuzzy graph with $|\sigma^*| = n$, $(n > 2)$ and $|\mu^*| = q$. Then the following statements are equivalent.*

*(i) $G$ is a fuzzy tree.*

*(ii) $n\alpha_V(G) = q\alpha_E(G)$.*

*Proof* (i) $\Rightarrow$ (ii).

Suppose $G$ is a fuzzy tree. Then

$$\alpha_V(G) = \frac{\text{number of } \alpha \text{ strong edges}}{|V|} = \frac{n-1}{n}$$

and

$$\alpha_E(G) = \frac{\text{number of } \alpha \text{ strong edges}}{|E|} = \frac{n-1}{q}.$$

*(ii)* ⇒ *(i)*.

Suppose that $n\ \alpha_V(G) = q\ \alpha_E(G)$. We have to prove that $G$ is a fuzzy tree. If possible suppose the contrary. Let $G$ be not a fuzzy tree. We know that a fuzzy graph $G$ is a fuzzy tree if and only if it has no $\beta$-strong edges. Now, because $G$ is not a fuzzy tree, there exists at least one $\beta$-strong edge in $G$. Let $uv$ be a $\beta$-strong edge in $G$. Then there exists at least one another $u - v$ path, say, $P$ in $G$ such that $\mu(xy) \geq \mu(uv)$ for every edge $xy$ in $P$. Now, the union of $P$ and edge $uv$ is a cycle in $G$. Suppose that the number of $\alpha$-strong edges incident at $u$ in $G$ is $t$. Now, to get $F$ (the MST in the definition of fuzzy trees), delete the edge $uv$ from $G$, which has the least weight in $C$. Then the number of $\alpha$-strong edges incident at $u$ in $F$ is $t+1$. Let $t'$ be the remaining number of $\alpha$-strong edges of $G$. Then $\alpha_V(G) = \frac{t+t'}{|V|}$ and $\alpha_V(F) = \frac{t+1+t'}{|V|}$. Also, $\alpha_E(G) = \frac{t+t'}{q}$ and $\alpha_E(F) = \frac{t+t'+1}{n-1}$. Now, the assumption, $n\ \alpha_V(G) = q\ \alpha_E(G)$ becomes $n(\alpha_E(F) - \frac{1}{n}) = (n-1)(\alpha_E(F) - 1)$, where $\alpha_V(F) = \frac{(n-1)}{n}$ and $\alpha_E(F) = 1$. Therefore, we get $n = 2$, a contradiction to the given condition. So our assumption is wrong. Hence, $G$ is a fuzzy tree. ■

We know that the strong degree of a vertex in a fuzzy graph is greater than or equal to one. Because the strong edges are further classified into two, we can consider for further investigation how dense is the distribution of different types of strong edges in a fuzzy graph.

**Definition 6.2.12** Let $G = (\sigma, \mu)$ be a fuzzy graph. $G$ is said to be $\alpha$-**saturated**, if= at least one $\alpha$-strong edge is incident at every vertex $v \in \sigma^*$. $G$ is called $\beta$-**saturated**, if at least one $\beta$-strong edge is incident at every vertex.

As noted above, all fuzzy graphs are trivially strong saturated.

**Definition 6.2.13** Let $G = (\sigma, \mu)$ be a fuzzy graph. Then $G$ is called **saturated** if it is both $\alpha$-saturated and $\beta$-saturated. That is, at least one $\alpha$-strong edge and one $\beta$-strong edge is incident on every vertex $v \in \sigma^*$. Also, a fuzzy graph which is not saturated is called **unsaturated**.

*Example 6.2.14* Consider the fuzzy graph $G = (\sigma, \mu)$ such that $\sigma^* = \{a, b, c, d\}$, $\sigma(a) = \sigma(b) = \sigma(c) = \sigma(d) = 1$, $\mu(ab) = \mu(cd) = 0.4$, $\mu(ad) = \mu(bc) = 0.3$ and $\mu(ac) = 0.2$ (Fig. 6.6). Now, $G$ is both $\alpha$-saturated and $\beta$-saturated, and hence it is saturated.

*Example 6.2.15* Consider the fuzzy graph $G = (\sigma, \mu)$ given in Fig. 6.7 with 8 vertices and 12 edges. Even though the fuzzy graph is $\alpha$-saturated, it is unsaturated. Every vertex except $a$ is incident with both $\alpha$ and $\beta$ strong edges.

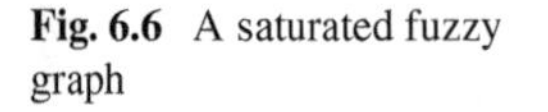

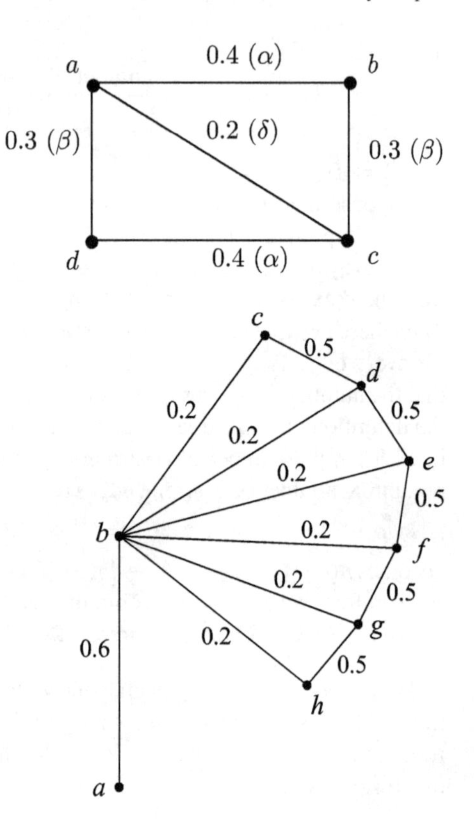


**Fig. 6.6** A saturated fuzzy graph

**Fig. 6.7** An unsaturated fuzzy graph

In the previous section, we have seen that a finite sequence of real numbers $\alpha_s(G) = (n_1, n_2, n_3, \ldots, n_p)$ is called the $\alpha$-sequence of $G = (\sigma, \mu)$ if $n_i =$ number of $\alpha$ strong edges incident on $v_i$ and 0, if no $\alpha$-strong edges are incident at $v_i$, where $G$ is a fuzzy graph with $|\sigma^*| = p$. Similarly, $\beta$ sequences were also defined. Note that

$$\sum_{n_i \in \alpha_S(G)} n_i + \sum_{n_i \in \beta_S(G)} n_i = \sum_{n_i \in S_S(G)} n_i.$$

In the next theorem, we discuss relationships between the sequences and saturation of a fuzzy graph.

**Theorem 6.2.16** *Let $G = (\sigma, \mu)$ be a fuzzy graph with $|\sigma^*| = p$ vertices. Then $G$ is*

*(i) $\alpha$-saturated if and only if the least entry in $\alpha_S(G)$ is unity. That is, $\sum_{n_i \in \alpha_S(G)} n_i$ is at the least $p$;*

*(ii) $\beta$-saturated if and only if the least entry in $\beta_S(G)$ is unity. That is, $\sum_{n_i \in \beta_S(G)} n_i$ is at the least $p$;*

*(iii) Saturated, only if the least entry in $S_S(G)$ is two. That is, if $\sum_{n_i \in S_S(G)} n_i$ is at the least $2p$.*

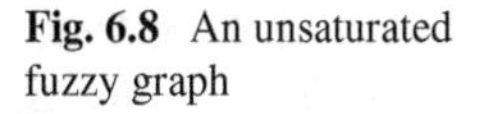
**Fig. 6.8** An unsaturated fuzzy graph

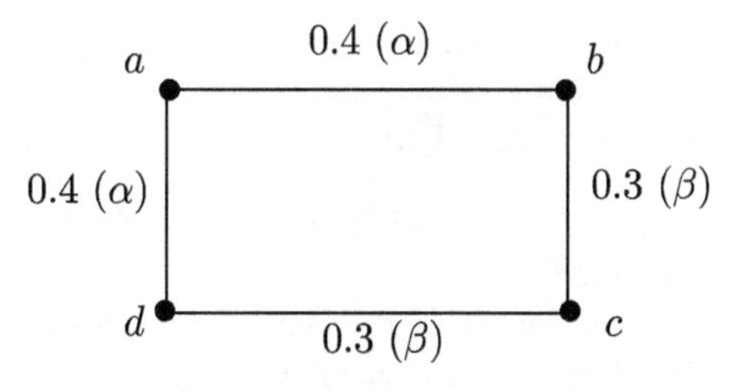


*Proof* Let $G = (\sigma, \mu)$ be a fuzzy graph with order $|\sigma^*| = p$. Let $\alpha_S(G)$, $\beta_S(G)$ and $S_S(G)$ be the $\alpha$, $\beta$ and strong sequences of $G$, respectively.

Suppose that $G$ is $\alpha$-saturated. Then all the vertices are incident with at least one $\alpha$-strong edge. This means unity is the least entry in $\alpha_S(G)$. Thus,

$$\sum_{n_i \in \alpha_S(G)} n_i \geq \underbrace{1 + 1 + 1 + \cdots + 1}_{p \text{ times}} = p.$$

Conversely, assume that the least entry in $\alpha_S(G)$ is unity. This means all the $p$ vertices of $G$ are incident with at least one $\alpha$ strong edge and hence $G$ is $\alpha$-saturated.

In a similar manner, we can prove the result for $\beta$-saturated fuzzy graphs.

Now, we have to prove the result for saturated fuzzy graphs. Suppose that $G$ is saturated. That is, $G$ is both $\alpha$-saturated and $\beta$-saturated. That is, each vertex of $G$ is incident with at least one $\alpha$-strong edge and with at least one $\beta$-strong edge. Thus, the least entry in $S_S(G)$ is at least two. ■

Note that the condition for saturated graphs is only necessary. There can be a situation in which a given vertex can be incident with 2 or more $\alpha$-strong edges only (no $\beta$-strong edges). In that case also, the least entry in $S_S(G)$ will be at least two, but the graph need not be saturated. This is clear from the following example.

*Example 6.2.17* Consider the fuzzy graph $G = (\sigma, \mu)$ in Fig. 6.8, whose underlying graph is $C_4$. $\sigma^* = \{a, b, c, d\}$, $\mu(ab) = \mu(ad) = 0.6$ and $\mu(bc) = \mu(cd) = 0.5$. The strong sequence of $G$ is $S_s(G) = (2, 2, 2, 2)$, but $G$ is unsaturated.

In the following theorem, upper bounds for saturation counts are provided.

**Theorem 6.2.18** *Let $G = (\sigma, \mu)$ be a fuzzy graph with $|\sigma^*| = n$. If $G$ is*
*(i) $\alpha$-saturated, then $\alpha_V(G) \geq 0.5$;*
*(ii) $\beta$-saturated, then $\beta_V(G) \geq 0.5$;*
*(iii) Saturated, then $S_V(G) \geq 1$.*

*Proof* Suppose that $G$ is $\alpha$-saturated. Then each vertex is incident with at least one $\alpha$-strong edge. This means $G$ contains at least $\frac{n}{2}$, $\alpha$-strong edges. Hence, $\alpha_V(G) \geq \frac{n/2}{n} = 0.5$. In the same manner, we can prove the result for $\beta$ saturated graphs. Now, we have to prove the result for saturated graphs. Suppose that $G$ is a saturated fuzzy graph. Then $G$ is both $\alpha$-saturated and $\beta$-saturated. That is, each vertex of $G$

is incident with at least one $\alpha$-strong edge and with at least one $\beta$-strong edge. Thus, the number of strong edges of $G =$ (number of $\alpha$-strong edges of $G$ + number of $\beta$-strong edges of $G) \geq \frac{n}{2} + \frac{n}{2} = n$. Hence, $S_V(G) \geq \frac{n}{n} = 1$. ■

Note that condition (*i*) in Theorem 6.2.18 is not sufficient, as it is clear from Example 6.2.17.

In the following theorem, we characterize saturated fuzzy cycles. We denote by $C_n$, a fuzzy cycle $C = (\tau, \nu)$ with $|\tau^*| = n$.

**Theorem 6.2.19** *Let $C_n$ be a fuzzy cycle. Then it is saturated if and only if the following two conditions are satisfied.*
*(i) $n = 2k$, where k is an integer;*
*(ii) $\alpha$-strong and $\beta$-strong edges appears alternatively on $C_n$.*

*Proof* Let $C_n$ be a fuzzy cycle. Then it has no $\delta$ edges. That is, all the edges appearing on $C_n$ are either $\alpha$ strong or $\beta$ strong. Suppose that $C_n$ is saturated. Then it is both $\alpha$-saturated and $\beta$-saturated. That is, each of its vertices are incident with at least one $\alpha$-strong edge and with at least one $\beta$-strong edge. This implies that the number of $\alpha$-strong edges $=$ the number of $\beta$-strong edges $= k$, where $k$ is a positive integer and $k + k = n$. Thus, $n = 2k$. Also, each vertex is incident with both $\alpha$-strong and $\beta$-strong edges, which happens only when they appear alternatively on $C_n$.

Conversely, suppose that $C_n$ is an even cycle and $\alpha$-strong and $\beta$-strong edges appear alternatively in $C_n$. This means all the vertices are incident with exactly one $\alpha$-strong and exactly with one $\beta$-strong edges. Then $C_n$ is both $\alpha$-saturated and $\beta$-saturated. Hence, $C_n$ is saturated. ■

*Example 6.2.20* Consider fuzzy graphs $G_1$ in Fig. 6.9 and $G_2$ in Fig. 6.10. All the edges with membership value 0.5 are $\alpha$-strong, and edges with value 0.2 are $\beta$-strong. In Fig. 6.9, the fuzzy cycle is a saturated cycle. Note that the number of vertices is even. In Fig. 6.10, there are odd number of vertices, and hence one vertex is incident with two $\alpha$-strong edges and no $\beta$-strong edges. Thus, the fuzzy cycle is $\alpha$-saturated, but not $\beta$-saturated. Hence, it is not saturated.

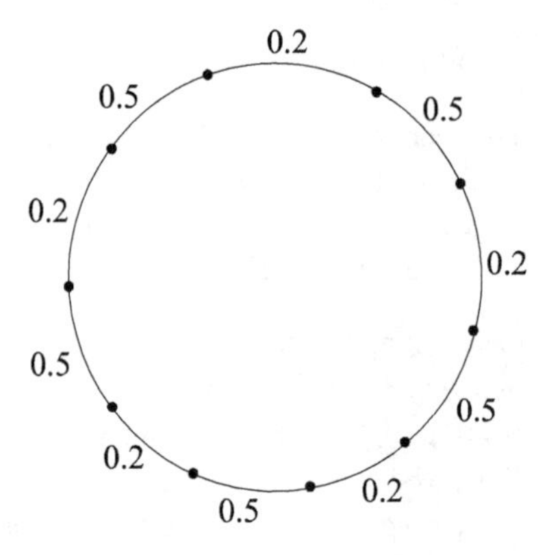


**Fig. 6.9** F-cycle on 10 vertices

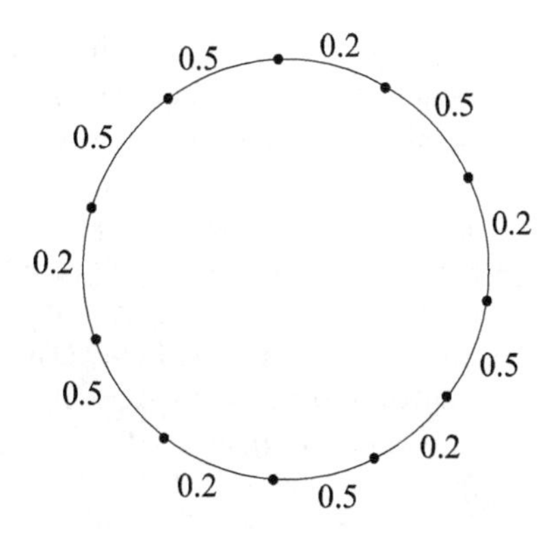


**Fig. 6.10** F-cycle on 11 vertices

Recall the definitions of strongest strong cycle and locamin cycle. In the following theorem, we present another characterization for saturated fuzzy cycles.

**Theorem 6.2.21** *Let $G = (\sigma, \mu)$ be a fuzzy cycle. Then the following are equivalent.*
*(i) G is either saturated or $\beta$-saturated.*
*(ii) G is a block.*
*(iii) G is a strongest strong cycle (SSC).*
*(iv) G is a locamin cycle.*

*Proof* The equivalence of conditions (ii), (iii) and (iv) are proved in Theorem 2.8.4. Now, we give only the equivalence between (i) and (ii).

*(i)* $\Rightarrow$ *(ii)*.

Suppose that the fuzzy cycle $G$ is saturated. Then it is both $\alpha$-saturated and $\beta$-saturated. That is, each vertex of $G$ is incident with at least one $\alpha$-strong edge and at least one $\beta$-strong edge. Because $G$ is a cycle, each vertex is incident with exactly two edges. Thus, exactly one $\alpha$-strong edge and exactly one $\beta$-strong edge is incident on every vertex. Hence, removal of any vertex from $G$ will not reduce the strength of connectedness between any other vertices. This implies that no vertex of $G$ is a fuzzy cutvertex, and hence $G$ is a block.

Also, suppose that the fuzzy cycle $G$ is $\beta$-saturated. That is, each vertex of $G$ is incident with at least one $\beta$-strong edge. We prove that $G$ is a block. We claim that $G$ has no fuzzy cutvertices. Suppose the contrary. Let $x$ be a fuzzy cutvertex of $G$. Then there exists two vertices $u$ and $v$ where $u \neq x \neq v$ such that $u - x - v$ path has more strength than $x - v$ path. This means all the edges in the $u - x - v$ path have weights more than the strength of the $x - y$ path. That is, all the edges in the $u - x - y$ path are $\alpha$-strong. Thus, both the edges, which are incident on $x$ will be $\alpha$-strong. It is a contradiction to our assumption that $G$ is $\beta$-saturated. Hence, our assumption is wrong and the claim is true. $G$ has no fuzzy cutvertices and $G$ is a block.

*(ii)* $\Rightarrow$ *(i)*.

Suppose that the fuzzy cycle $G$ is a block. We claim that $G$ has no $\delta$-edges. Suppose the contrary. Let $e = uv$ be a $\delta$-edge in $G$. Then all other edges in $G$ will be $\alpha$-strong and thereby $G$ will have exactly $n-2$ fuzzy cutvertices, which is a contradiction to our assumption that $G$ is a block. So our assumption is wrong and the claim is true. Thus, only $\alpha$-strong and $\beta$-strong edges are in $G$.

If $G$ contains both $\alpha$-strong and $\beta$-strong edges, then they must be appeared alternatively; otherwise the block structure will be lost. If the number of $\alpha$-strong edges = number of $\beta$-strong edges = $\frac{n}{2}$, then $G$ is both $\alpha$-saturated and $\beta$-saturated, and hence saturated. If the number of $\alpha$-strong edges < the number of $\beta$-strong edges, then $G$ will be $\beta$-saturated only. The case where the number of $\alpha$-strong edges > the number of $\beta$-strong edges, will not happen, as it looses the block structure. If all the edges in $G$ are $\beta$-strong, then it is $\beta$-saturated. Thus, in all the cases, $G$ is either saturated or $\beta$-saturated. ■

Let $G = (\sigma, \mu)$ be a saturated cycle or a $\beta$-saturated cycle. Then we have $\alpha_V(G) = 0.5$ and $\beta_V(G) = 0.5$ or $0.5 \leq \beta_V(G) \leq 1$. Thus, Theorem 6.2.21 may be rewritten as follows.

**Theorem 6.2.22** *Let $G = (\sigma, \mu)$ be fuzzy cycle. Then the following are equivalent.*
*(i) $\alpha_V(G) = 0.5$ and $\beta_V(G) = 0.5$ or $0.5 \leq \beta_V(G) \leq 1$.*
*(ii) $G$ is either saturated or $\beta$-saturated.*
*(iii) $G$ is a block.*
*(iv) $G$ is a strongest strong cycle (SSC).*
*(v) $G$ is a locamin cycle.*

In the following example, it is shown that all blocks in fuzzy graphs are not saturated and all saturated graphs are not blocks.

*Example 6.2.23* The fuzzy graph $G = (\sigma, \mu)$ given in Fig. 6.11 is an unsaturated fuzzy graph which is a block. Here $\sigma(s) = 1$ for all $s \in \sigma^*$ and $\mu(xy) = 0.5$ for every edge other than two edges of $\mu$ value 0.1. In $G$, edges with membership value 0.1 are $\delta$, and with 0.5 are $\beta$-strong. The graph is unsaturated and it is a block.

Now, we compute the vertex and edge saturation counts for a complete fuzzy graph (CFG).

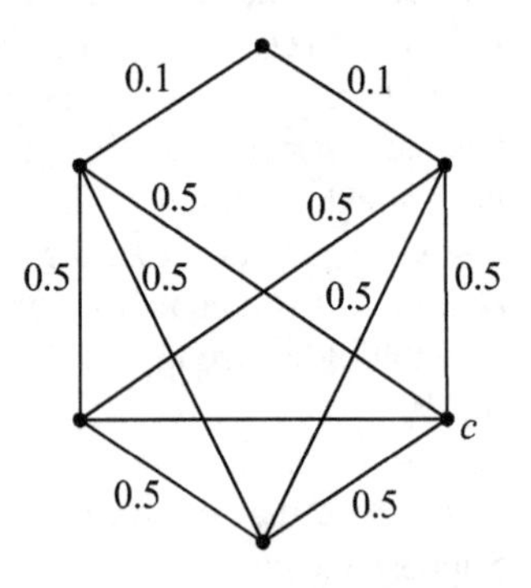


**Fig. 6.11** Unsaturated block

By Lemma 3.2.22, a CFG has no $\delta$ edges and at most one $\alpha$-strong edge. All CFG's are blocks in fuzzy graphs. Based on these observations, we have the following theorem.

**Theorem 6.2.24** *Let $G = (\sigma, \mu)$ be a CFG with $|\sigma^*| = n$. Then we have the following inequalities.*

*(i)* $0 \leq \alpha_V(G) \leq \frac{1}{n}$.

*(ii)* $\frac{n^2-n-2}{2n} \leq \beta_V(G) \leq \frac{n-1}{2}$.

*Proof* Let $G = (\sigma, \mu)$ be a CFG with $|\sigma^*| = n$. Then $G$ has at most one $\alpha$-strong edge. Hence, $0 \leq \alpha_V(G) \leq \frac{1}{n}$. Because $G$ has no $\delta$ edges, the minimum number of $\beta$-strong edges in $G$ is

$$^nC_2 - 1 = \frac{n(n-1)}{2} - 1.$$

Thus,

$$\beta_V(G) \geq \frac{\frac{n(n-1)}{2} - 1}{n} = \frac{n^2 - n - 2}{2n}.$$

Hence, $\frac{n^2-n-2}{2n} \leq \beta_V(G) \leq \frac{n-1}{2}$. ■

From the definition of CFG's, it is clear that all CFG's are unsaturated.

In the following theorem, an upper bound for $\alpha$-vertex count of a block is given.

**Theorem 6.2.25** *Let $G = (\sigma, \mu)$ be a block, then $\alpha_V(G) \leq \frac{1}{2}$.*

*Proof* Let $G = (\sigma, \mu)$ be a block on $n$ vertices. We also know that if a vertex $w$ is common for more than one $\alpha$-strong edges, then $w$ is a fuzzy cutvertex, and hence $G$ cannot be a block. Now,

$$\alpha_V(G) = \frac{\text{number of } \alpha\text{-strong edges in } G}{|V|} \leq \frac{n/2}{n} = \frac{1}{2}.$$

■

As a consequence of the above theorem, we have the following two corollaries.

**Corollary 6.2.26** *A block is $\alpha$-saturated if and only if $\alpha_V(G) = 0.5$.*

*Proof* Let $G = (\sigma, \mu)$ be a block. Let $G$ be $\alpha$-saturated. Then each vertex of $G$ is incident with at least one $\alpha$-strong edge. Also, as blocks in fuzzy graphs are free from fuzzy cutvertices, exactly one $\alpha$-strong edge is incident with each vertex. Therefore, $\alpha_V(G) = \frac{n/2}{n} = 0.5$. ■

**Corollary 6.2.27** *There exists no $\alpha$-saturated blocks of odd order.*

From the following example, one can see that only blocks with even order can be $\alpha$-saturated. Only the underlying graphs are given.

*Example 6.2.28* See Fig. 6.12.

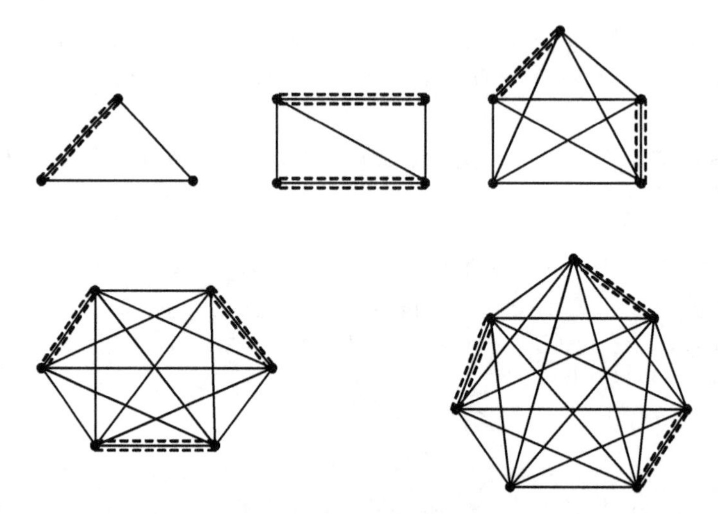

**Fig. 6.12** Saturated and unsaturated blocks in fuzzy graphs

## 6.3 Intervals in Fuzzy Graphs

The concept of distance and convexity are important in almost all branches of mathematics. We can view a fuzzy graph together with a metric as a metric space, and hence the concepts like convexity can be easily extended to fuzzy graphs. It was Mulder [132] who introduced the concept of an interval in graph theory. Several concepts like transit functions [55, 56] and median graphs were brought into the literature after that. These concepts have applications in the architecture and design of interconnection networks. Because these systems are fuzzy to some degree, it is necessary to extend these ideas to fuzzy graphs. The contents of this section are from [67]. We have the definition of interval by Mulder as follows.

Let $G(V, E)$ be a finite, connected, simple, loop less graph with a distance function $d$, where $d(u, v)$ is the length of the shortest $u - v$ path. The interval function $I$ on $G$ is defined as

$$I(u, v) = \{w \mid d(u, v) = d(u, w) + d(w, v)\},$$

for all $u, v \in V$. The set $I(u, v)$ is the interval between $u$ and $v$.

In a fuzzy graph there exists a strong path between any two vertices. Recall Definition 3.1.24. The geodesic distance or $g$-distance between vertices $u, v \in \sigma^*$, denoted by $d_g(u, v)$, is defined as the length of the shortest $u - v$ strong path. If $u$ and $v$ are not connected by a path, then $d_g(u, v) = \infty$. Hence, we have the following definition.

**Definition 6.3.1** Let $G = (\sigma, \mu)$ be a fuzzy graph. The **interval** between $u$ and $v$ with respect to the geodesic distance $d_g$ is defined as

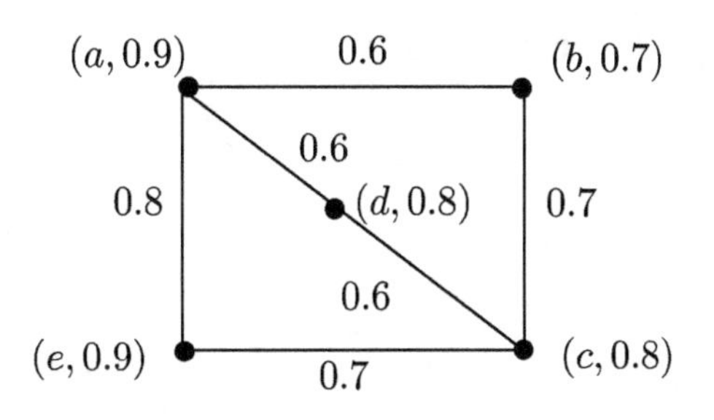


**Fig. 6.13** Intervals in a fuzzy graph $G$

$$I_g(u, v) = \{(x, \sigma(x)) \in \sigma^* \times [0, 1] \mid d_g(u, v) = d_g(u, x) + d_g(x, v)\}.$$

The support of the interval $I_g(u, v)$ is defined as

$$I_g^*(u, v) = \{x \in \sigma^* \mid (x, \sigma(x)) \in I_g(u, v), \sigma(x) > 0\}.$$

*Example 6.3.2* Consider the fuzzy graph $G = (\sigma, \mu)$ in Fig. 6.13. $\sigma^* = \{a, b, c, d, e\}$, $\sigma(a) = \sigma(e) = 0.9$, $\sigma(d) = \sigma(c) = 0.8$, $\sigma(b) = 0.9$, $\mu(ab) = 0.6$, $\mu(bc) = \mu(ce) = 0.7$, $\mu(ae) = 0.8$, $\mu(ad) = \mu(dc) = 0.6$. Here, all edges except $ab$ are strong, because $\mu(ab) < CONN_{G-ab}(a, b)$ and for all other edges $uv$, $\mu(uv) \geq CONN_{G-uv}(u, v)$. So $I_g(a, b) = \{(a, 0.9), (b, 0.7), (c, 0.8), (d, 0.8), (e, 0.9)\}, I_g(a, c) = \{(a, 0.9), (c, 0.8), (d, 0.8), (e, 0.9)\}$, $I_g(a, d) = \{(a, 0.9), (d, 0.8)\}$, $I_g(a, e) = \{(a, 0.9), (e, 0.9)\}$.

If $H = (\tau, \nu)$ is a partial fuzzy subgraph of $G = (\sigma, \mu)$, then for all $a, b \in \tau^*$, $I_{g(H)}(a, b) \subseteq I_{g(G)}(a, b)$. Here, $I_{g(H)}(a, b)$ and $I_{g(G)}(a, b)$ denote intervals in $H$ and $G$, respectively.

Consider a cycle $G = (\sigma, \mu)$. If $G$ is a fuzzy tree, then for all $x, y \in \sigma^*$ there exists a unique strong path in $G$. So $I_g^*(u, v)$ is the set of all vertices in the unique strong $u - v$ path. Moreover $I_g^*(u, v) = \sigma^*$ if and only if $uv$ is the unique weakest edge in $G$. If $G$ is a fuzzy cycle with $n$ vertices in $\sigma^*$, then for all $x, y \in \sigma^*$, $I_g^*(x, y)$ is the set of all vertices on shortest $x - y$ paths. If $n$ is even, then there exist $\frac{n}{2}$ pairs of diametrically opposite vertices $x, y$ in $\sigma^*$ such that $I_g^*(x, y) = \sigma^*$ and for all other pair of vertices $u, v$, $I_g^*(u, v) \subset \sigma^*$. If $n$ is odd, then there does not exist any pair of vertices $x, y$ in $\sigma^*$ such that $I_g^*(x, y) = \sigma^*$. If the edge $xy \in \mu^*$ is either $\alpha$-strong or $\beta$-strong, then $I_g^*(x, y) = \{x, y\}$. Otherwise, i.e., if $xy$ is a $\delta$-edge, then $I_g^*(x, y) = \sigma^*$.

Some of the properties of intervals in fuzzy graphs are given in the following theorem.

**Theorem 6.3.3** *Let $G = (\sigma, \mu)$ be a fuzzy graph. Then the following properties hold. Let $u, v, w, x \in \sigma^*$.*

*(i) $I_g(u, v) = \{(u, \sigma(u)), (v, \sigma(v))\}$ if and only if $uv$ is a strong edge.*

*(ii) $I_g(u, v) = I_g(v, u)$.*

*(iii) If $(w, \sigma(w)) \in I_g(u, v)$, then $I_g(u, w) \subseteq I_g(u, v)$ and $I_g(w, v) \subseteq I_g(u, v)$.*

*(iv) If $(w, \sigma(w)) \in I_g(u, v)$ and $(x, \sigma(x)) \in I_g(w, v)$, then $(x, \sigma(x)) \in I_g(u, v)$ and $(w, \sigma(w)) \in I_g(u, x)$.*

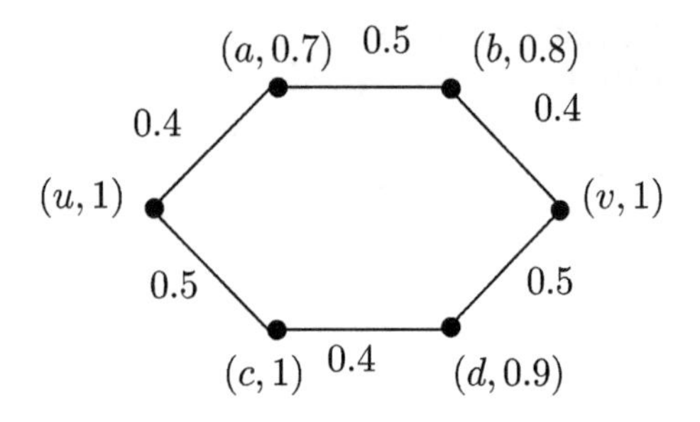


**Fig. 6.14** Fuzzy graph in Example 6.3.4

(*v*) *If* $(x, \sigma(x)) \in I_g(u, v)$, *then* $I_g(u, x) \cap I_g(x, v) = \{(x, \sigma(x))\}$ *and hence* $|I_g(u, x) \cap I_g(x, v)| = 1$.

(*vi*) *If* $(x, \sigma(x)) \in I_g(u, v)$, *then* $I_g(u, x) \cup I_g(x, v) \subseteq I_g(u, v)$.

(*vii*) *If* $|I_g(u, v)| = 2 = |I_g(x, y)|$, $v \in I_g^*(u, x)$ *and* $u \in I_g^*(v, y)$, *then* $x \in I_g^*(v, y)$.

*Proof* The proof of (*i*) to (*v*) are obvious.

(vi) Let $(x, \sigma(x)) \in I_g(u, v)$. Then $x$ is on a geodesic joining $u$ and $v$. Thus, every geodesic $P$ joining $u$ to $x$ followed by a geodesic $Q$ joining $x$ to $v$ is again a geodesic joining $u$ to $v$. So every vertex on $P \cup Q$ is in $I_g^*(u, v)$. Therefore, $I_g(u, x) \cup I_g(x, v) \subseteq I_g(u, v)$.

(viii) Let the assumption holds. Then $d_g(v, y) \le d_g(v, x) + d_g(x, y) = d_g(v, x) + 1$ because $|I_g(x, y)| = 2$. Thus, $d_g(v, y) = d_g(v, x)$ (because $v \in I_g(u, x)$) $\le d_g(u, y) + d_g(y, x) = d_g(u, y) + 1$ (because $|I_g(x, y)| = 2$) $= d_g(v, y)$ (because $u \in I_g(v, y)$) and hence $|I_g(u, v)| = 2$. Thus, $x \in I_g(v, y)$. ■

Equality does not hold always in Theorem 6.3.3(vi). For consider the fuzzy graph $G$ in Fig. 6.14.

*Example 6.3.4* Consider the fuzzy cycle $G$, given in Fig. 6.14. In $G$, $I_g(u, v) = \{(x, \sigma(x)) \mid \text{for all } x \in \sigma^*\}$. So $(a, 0.6) \in I_g(u, v)$. But $I_g(u, a) \cup I_g(a, v) = \{(u, 1), (a, 0.6), (b, 0.7), (v, 1)\} \neq I_g(u, v)$.

If $u$ and $v$ are joined by a unique strong path, then for any $(a, \sigma(a)) \in I_g(u, v)$, $I_g(u, a) \cup I_g(a, v) = I_g(u, v)$. Because there exists a unique strong path between any two vertices in a fuzzy tree, we have the following theorem.

**Theorem 6.3.5** *In a fuzzy tree* $G = (\sigma, \mu)$, $I_g(u, v) = I_g(u, x) \cup I_g(x, v)$ *for all* $x, u, v \in \sigma^*$ *such that* $(x, \sigma(x)) \in I_g(u, v)$.

*Proof* In a fuzzy tree $G = (\sigma, \mu)$, there exists a unique strong $u - v$ path between any two vertices $u, v \in \sigma^*$. Thus, $(x, \sigma(x)) \in I_g(u, v)$, which implies $I_g(u, v) = I_g(u, x) \cup I_g(x, v)$. ■

**Theorem 6.3.6** *If $x$ is a cutvertex, then there exists a pair of vertices $u, v$ in $\sigma^*$ such that* $I_g(u, v) = I_g(u, x) \cup I_g(x, v)$.

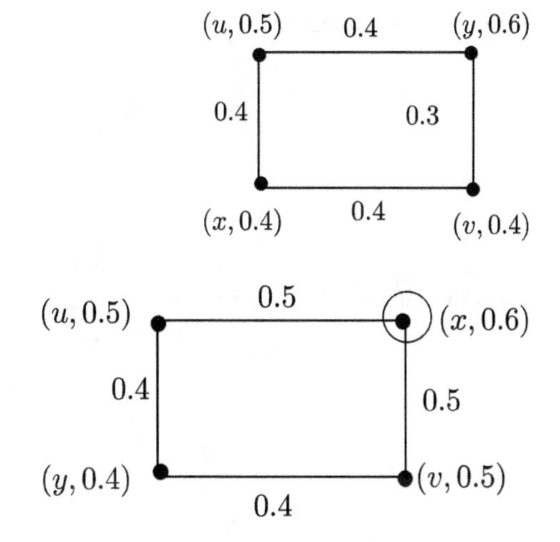


**Fig. 6.15** Fuzzy graph with a unique weak edge

**Fig. 6.16** A fuzzy graph with unique fuzzy cutvertex

*Proof* Because $x$ is a cutvertex, there exists a pair of vertices $u, v$ in $\sigma^*$ such that removal of $x$ from $G$ deletes all $u - v$ paths. This is because every $u - v$ path contains $x$. So every strong $u - v$ path contains $x$. Therefore, vertices on $u - v$ paths can be partitioned into two sets such that vertices before $x$ and after $x$. This gives $I_g(u, v) = I_g(u, x) \cup I_g(x, v)$. ■

The converse of Theorem 6.3.6 is not true. That is, the equation $I_g(u, v) = I_g(u, x) \cup I_g(x, v)$ does not imply that $x$ is a cutvertex of $G$, as seen from the following example.

*Example 6.3.7* Consider the fuzzy graph $G = (\sigma, \mu)$, $\sigma^* = \{u, v, x, y\}$, $\sigma(x) = \sigma(v) = 0.4$, $\sigma(u) = 0.5$, $\sigma(y) = 0.6$, $\mu(uy) = \mu(ux) = \mu(xv) = 0.4$ and $\mu(vy) = 0.3$ (Fig. 6.15). Clearly, $vy$ is the unique weak edge of $G$. Also, $I_g(u, v) = I_g(u, x) \cup I_g(x, v)$. However, $x$ is not a cutvertex of $G$.

If $x$ is a fuzzy cutvertex, then Theorem 6.3.6 need not be true. For example, consider the fuzzy graph in Fig. 6.16.

*Example 6.3.8* Consider the fuzzy graph $G = (\sigma, \mu)$, $\sigma^* = \{u, v, x, y\}$, $\sigma(x) = 0.6$, $\sigma(u) = \sigma(v) = 0.5$, $\sigma(y) = 0.4$, $\mu(ux) = 0.5 = \mu(xv)$ and $\mu(vy) = \mu(uy) = 0.4$ (Fig. 6.16). Clearly, $x$ is a fuzzy cutvertex of $G$. Now, $x$ is a fuzzy cutvertex, but there does not exist any pair $u, v$ in $\sigma^*$ such that $I_g(u, v) = I_g(u, x) \cup I_g(x, v)$. This is because every edge of $G$ is strong.

**Theorem 6.3.9** *In a complete fuzzy graph $G$, $I_g^*(u, v) = \{u, v\}$ for all $u, v \in \sigma^*$.*

*Proof* In a complete fuzzy graph, by Lemma 3.2.22, there are no $\delta$ edges and hence every two vertices are joined by a strong edge. So $I_g^*(u, v) = \{u, v\}$ for all $u$, $v \in \sigma^*$. ■

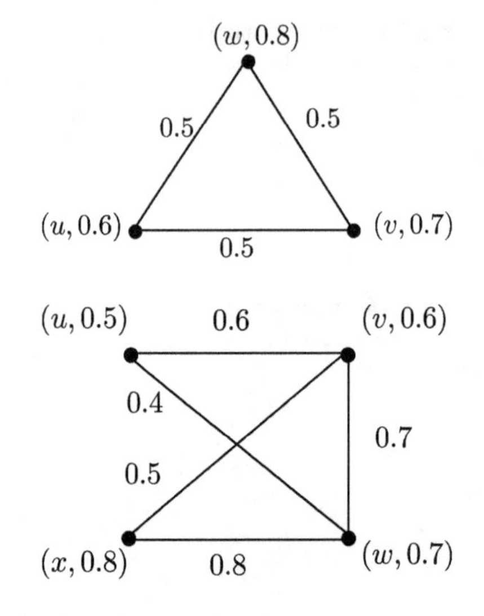


**Fig. 6.17** A non complete cycle

**Fig. 6.18** A fuzzy tree

Converse of the Theorem 6.3.9 is not true. Consider the following example.

*Example 6.3.10* Consider a cycle $G = (\sigma, \mu)$ with three vertices. $\sigma(u) = 0.6$, $\sigma(v) = 0.7$, $\sigma(w) = 0.8$ and $\mu(uv) = \mu(vw) = \mu(uw) = 0.5$ (Fig. 6.17). In G, every edge is strong. So $I_g^*(a, b) = \{a, b\}$, for all $a, b \in \sigma^*$. But $G$ is not a complete fuzzy graph.

**Theorem 6.3.11** *If a fuzzy tree $G = (\sigma, \mu)$ is a path or cycle, then there exists two vertices u and v such that $I_g^*(u, v) = \sigma^*$.*

*Proof* Consider the fuzzy tree $G$. If $G^*$ is a path. Then take $u$ and $v$ as the end vertices of $G$. Then $I_g^*(u, v) = \sigma^*$. If $G^*$ is a cycle, then because $G$ is a fuzzy tree, $G$ has exactly one weak edge (say) $uv$. Then $I_g^*(u, v) = \sigma^*$. ■

The converse of Theorem 6.3.11 is not true as seen from Example 6.3.12.

*Example 6.3.12* Consider the fuzzy graph $G = (\sigma, \mu)$ with $\sigma^* = \{u, v, w, x\}$, $\sigma(u) = 0.5$, $\sigma(v) = 0.6$, $\sigma(w) = 0.7$, $\sigma(x) = 0.8$. $\mu(uv) = 0.6$, $\mu(vw) = 0.7$, $\mu(wx) = 0.8$, $\mu(vx) = 0.5$ and $\mu(uw) = 0.4$ (Fig. 6.18). It is easy to see that $G$ is a fuzzy tree and $I_g^*(u, x) = \sigma^*$. However, it is neither a path nor a cycle.

**Theorem 6.3.13** *Let $G$ be a connected fuzzy graph and $u, v, x, y \in \sigma^*$. Then $I_g^*(u, x) \cup I_g^*(x, v) = I_g^*(u, v)$ if and only if $x$ is on every $u - v$ geodesic.*

*Proof* If $x$ is on every $u - v$ geodesic, then every $u - v$ geodesic can be considered as the union of a $u - x$ strong geodesic and a strong $x - v$ geodesic. Thus,

$$I_g^*(u, x) \cup I_g^*(x, v) = I_g^*(u, v). \tag{6.1}$$

Now, consider the converse. Assume that equation in the statement of the theorem is true. We want to prove that $x$ is on every $u - v$ geodesic. Suppose this is not true. Then there exists at least one $u - v$ geodesic $P$ which does not contain $x$. Let $P$ be $y_0 y_1 y_2 \cdots y_{n-1} y_n$, where $y_0 = u$ and $y_n = v$. Because $u \in I_g^*(u, x)$ and $v \in I_g^*(x, v)$, there exist $i \in \{1, 2, 3, \ldots, n-1\}$ such that $y_i \in I_g^*(u, x)$ and $y_{i+1} \in I_g^*(x, v)$ and $y_i y_{i+1}$ is a strong edge (because it is an edge of the strong path $P$). Now, $y_i \in I_g^*(u, x)$ implies

$$d_g(u, y_i) + d_g(y_i, x) = d_g(u, x). \text{ So we get } d_g(u, y_i) + 1 \le d_g(u, x). \quad (6.2)$$

Also, $y_{i+1} \in I_g^*(x, v)$ gives,

$$d_g(x, y_{i+1}) + d_g(y_{i+1}, v) = d_g(x, v). \text{ So we get } 1 + d_g(y_{i+1}, v) \le d_g(x, v). \quad (6.3)$$

Equation (6.1) and $x \in I_g^*(u, v)$ together gives

$$d_g(u, x) + d_g(x, v) = d_g(u, v). \quad (6.4)$$

Let $P_1$ be the $u - x$ geodesic containing $y_i$ and $P_2$, the $x - v$ geodesic containing $y_{i+1}$. Now, consider the strong $u - v$ path; $u - y_i$ part of $P_1$ followed by the strong edge $y_i y_{i+1}$ along with $y_{i+1} - v$ part of $P_2$, which has length $d_g(u, y_i) + d_g(y_i, y_{i+1}) + d_g(y_{i+1}, v) \le d_g(u, x) - 1 + 1 + d_g(x, v) - 1 = d_g(u, v) - 1$,

From (6.2)–(6.4), we get a contradiction. That is, $x$ is on every strong $u - v$ geodesic. ■

Because intervals and their supports in a fuzzy graph $G = (\sigma, \mu)$ are subsets of $\sigma^* \times [0, 1]$ and $\sigma^*$ respectively, their unions and intersections can be defined in terms of the usual union and usual intersection. Let $I_g(u, v)$ and $I_g(x, y)$ be two strong intervals in a fuzzy graph $G$. Then their union is defined as

$$I_g(u, v) \cup I_g(x, y) = \{(a, \sigma(a)) \mid (a, \sigma(a)) \in I_g(u, v) \text{ or } (a, \sigma(a)) \in I_g(x, y)\}.$$

Similarly, their intersection is defined as

$$I_g(u, v) \cap I_g(x, y) = \{(a, \sigma(a)) \mid (a, \sigma(a)) \in I_g(u, v) \text{ and } (a, \sigma(a)) \in I_g(x, y)\}.$$

The union of two or more strong intervals need not be a strong interval as seen from the following example.

*Example 6.3.14* Consider the fuzzy graph $G$ in Fig. 6.19. Here, $\sigma(a) = 1$, for all $a \in \{u, x, v, y\}$ and $\mu(ux) = \mu(xv) = \mu(xy) = 1$. Also, $I_g^*(u, v) = \{u, x, v\}$ and $I_g^*(u, y) = \{u, y, x\}$. But $I_g^*(u, v) \cup I_g^*(u, y) = \{u, x, v\} \cup \{u, y, x\} = \{u, v, x, y\} \ne I_g^*(a, b)$, for all $a, b \in \sigma^*$.

**Theorem 6.3.15** *If $x, y \in I_g^*(u, v)$, then $I_g^*(u, v) \cup I_g^*(x, y) = I_g^*(u, v)$.*

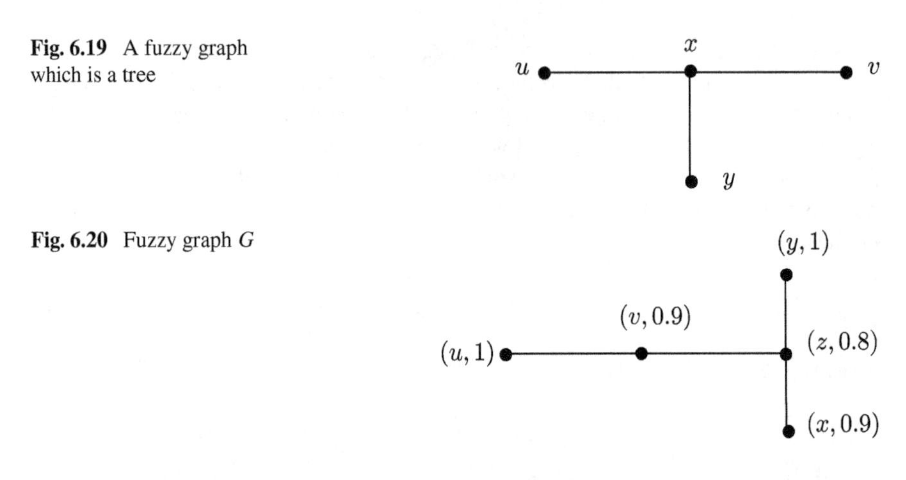


**Fig. 6.19** A fuzzy graph which is a tree

**Fig. 6.20** Fuzzy graph $G$

*Proof* $x, y \in I_g^*(u, v)$ implies $I_g^*(x, y) \subset I_g^*(u, v)$. So $I_g^*(x, y) \cup I_g^*(u, v) = I_g^*(u, v)$. ■

But If $I_g^*(u, v) \cap I_g^*(x, y) = \emptyset$, then $I_g(u, v) \cup I_g(x, y)$ need not be a strong interval as is clear from the following example.

*Example 6.3.16* Consider the fuzzy graph $G = (\sigma, \mu)$ such that $|\sigma^*| = 5$, given in Fig. 6.20. Here because $G^*$ is a tree, $G$ is a fuzzy tree irrespective of the $\mu$ values. Here, $I_g^*(u, v) = \{u, v\}$, $I_g^*(x, y) = \{x, z, y\}$ and $I_g^*(u, v) \cap I_g^*(x, y) = \emptyset$. But $I_g(u, v) \cup I_g(x, y)$ is not a strong interval.

**Theorem 6.3.17** *Every nonempty intersection of intervals is again an interval.*

*Proof* Consider two intervals $I_g(u, v)$ and $I_g(x, y)$ such that $I_g(u, v) \cap I_g(x, y) \neq \emptyset$. We want to prove that $I_g(u, v) \cap I_g(x, y)$ is again an interval. There are two cases.

**Case 1**: $I_g^*(u, v) \cap I_g^*(x, y) = \{a\}$ for some $a \in \sigma^*$.

Because $I_g^*(a, a) = \{a\}$ for any vertex $a \in \sigma^*$, we get $I_g^*(u, v) \cap I_g^*(x, y) = \{a\} = I_g^*(a, a)$. Thus, $I_g(u, v) \cap I_g(x, y)$ is again a strong interval.

**Case 2**: $I_g^*(u, v) \cap I_g^*(x, y)$ contains at least 2 vertices.

In this case, for any two vertices $a, b \in I_g^*(u, v) \cap I_g^*(x, y)$, $I_g^*(a, b) \subset I_g^*(u, v) \cap I_g^*(x, y)$. Now, choose $m, n \in I_g^*(u, v) \cap I_g^*(x, y)$ such that

$d_g(m, n) = \bigvee_{p,q \in I_g^*(u,v) \cap I_g^*(x,y)} d_g(p, q)$. We have $I_g^*(u, v) \cap I_g^*(x, y) = I_g^*(m, n)$. That is, $I_g(u, v) \cap I_g(x, y) = I_g(m, n)$. Hence, $I_g(u, v) \cap I_g(x, y)$ is again an interval. Thus, every nonempty intersection of strong intervals is again a strong interval. ■

**Theorem 6.3.18** *In a fuzzy tree $G = (\sigma, \mu)$, if the union of two disjoint intervals $I_g(u, v)$ and $I_g(x, y)$ is again an interval, then all the four vertices $u, v, x, y$ lie on a geodesic.*

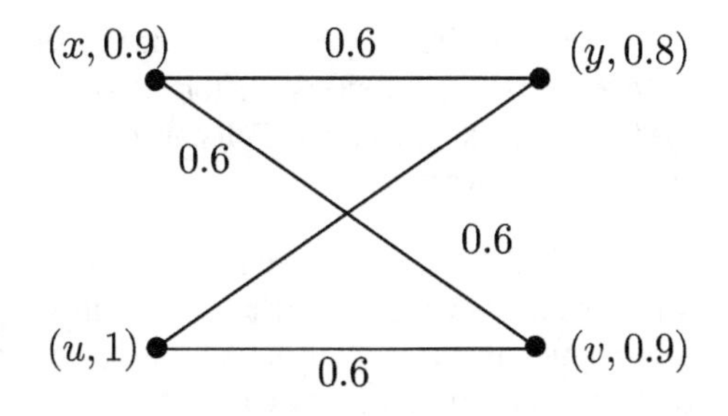


**Fig. 6.21** Uniqueness in Conjecture 6.3.19

*Proof* Given that $I_g(u, v) \cup I_g(x, y)$ is an interval. Let $I_g(u, v) \cup I_g(x, y) = I_g(a, b)$. Then clearly $u, v, x, y \in I_g^*(a, b)$. If possible assume that the vertices $u, v, x, y$ do not lie together on any $a - b$ geodesic. Consider two vertices $m, n \in \{u, v, x, y\}$ such that $m$ and $n$ lie on different $a - b$ geodesics, say $P_1$ and $P_2$ respectively. Then $P_1$ and $P_2$ together gives a fuzzy cycle. This is a contradiction because $G$ is a fuzzy tree. ■

Next we have a conjecture.

**Conjecture 6.3.19** *If $I_g(u, v) \cap I_g(x, y) = \emptyset$ and $I_g(u, v) \cup I_g(x, y)$ is an interval, then $I_g(u, v) \cup I_g(x, y) = I_g(a, b)$, where $a, b \in \{u, v, x, y\}$ such that $d_g(a, b) = \bigvee_{m,n \in \{u,v,x,y\}} d_g(m, n)$. Moreover, if pair $a, b$ is unique, then vertex $m \in \{u, v, x, y\}$ such that $m \neq a$ and $m \neq b$ is on every $a - b$ geodesic.*

In Conjecture 6.3.19, if $a, b$ is not unique, then vertex $m \in \{u, v, x, y\}$ such that $m \neq a$ and $m \neq b$ need not be on every $a - b$ geodesic.

This is clear from the following example.

*Example 6.3.20* Consider the fuzzy graph $H$ given in Fig. 6.21. Clearly, in $H$, $I_g(u, v) \cup I_g(x, y) = I_g(u, x) = I_g(v, y)$. But none of the vertices in $H$ is a fuzzy cutvertex.

## 6.4 Gates and Gated Sets in Fuzzy Graphs

This section is based on the work by Dhanymol and Mathew [66]. Interval functions are useful to study the properties of graphs which depend only on distances between vertices. Goldman and Witzgall first introduced the idea of gated sets in [80].

**Definition 6.4.1** For a subset $W$ of vertex set $V$ of a graph $G$ and a vertex $z \in V$, a vertex $x \in W$ is an $R$-**gate** for $z$ if $x$ lies in $R(z, w)$ for any $w$ in $W$. The set $W$ is called $R$-**gated**, if every vertex $z \in V$ has a unique $R$-gate in $W$.

Recall that the length of a geodesic between $u$ and $v$ is called the $g$-distance $d_g(u, v)$ and a geodesic is a shortest strong path between $u$ and $v$. In graphs, $g$-distance coincides with usual distance. So in graphs, $I_g(u, v)$ is the collection of all $u - v$ interior vertices.

**Definition 6.4.2** An induced fuzzy subgraph $H = (\tau, \nu)$ of $G = (\sigma, \mu)$ is said to be **gated** with respect to a metric $d$, if for all $v \in \sigma^* \setminus \tau^*$ and for all $u \in \tau^*$, $\exists$ a vertex $x \in \tau^*$ such that $d(u, v) = d(u, x) + d(x, v)$. Also, $x$ is known as a **gate** of $H$ in $G$ and $\tau^*$ is known as a gated set in $\sigma^*$.

Here, by an induced fuzzy subgraph $H = (\tau, \nu)$ of $G = (\sigma, \mu)$, we mean maximal fuzzy subgraph induced by $\tau^*$. That is, $\tau(x) = \sigma(x)$, for all $x \in \tau^*$ and $\nu(xy) = \mu(xy)$, for all $x, y \in \tau^*$. Moreover, a fuzzy subgraph of $G = (\sigma, \mu)$ is called an induced fuzzy subgraph of $G$ if it is a maximal fuzzy subgraph induced by some subsets of $\sigma^*$.

**Definition 6.4.3** A nontrivial induced fuzzy subgraph $H = (\tau, \nu)$ of $G = (\sigma, \mu)$ is said to be **gated** with respect to an interval function $I$, if for all $v \in \sigma^* \setminus \tau^*$ and for all $u \in \tau^*$, $\exists$ a vertex $x \in \tau^*$ such that $x \in I(u, v)$. Also, $x \in H$ is called a **gate of** $H$ **in** $G$ if and only if $x \in \bigcap_{u \in \tau^*} I(u, v)$, for all $v \in \sigma^* \setminus \tau^*$.

If a nontrivial induced fuzzy subgraph $H$ of $G$ is gated with respect to the $g$-distance $d_g$ (or strong interval function $I_g$), then $H$ is known as strongly gated and if $x$ is a gate of $H$ in $G$ with respect to the $g$-distance $d_g$ (or strong interval function $I_g$), then $x$ is known as a strong gate of $H$ in $G$.

All internal vertices of a path are strong gates for a path $P$: $u_0, u_1, \ldots, u_n$, where each vertex and each edge has membership value 1. Let, $u_1, u_2, \ldots, u_{n-1}$ be the internal vertices of $P$. Then for the fuzzy subgraph $P$ induced by $\{u_0, u_1, u_2, \ldots, u_i\}$, $u_i$ is a strong gate, where $i \in \{1, 2, 3, \ldots, n-1\}$.

**Definition 6.4.4** A vertex $x$ is known as a **gate of a fuzzy graph** $G$ if it is a gate of some nontrivial induced fuzzy subgraph $H$ of $G$. Also, it is clear that only connected graphs can have strong gates.

**Proposition 6.4.5** *A nontrivial induced fuzzy subgraph $H$ of a fuzzy graph $G$ has at most one strong gate in $G$.*

*Proof* Suppose $x$ and $y$ are strong gates of an induced fuzzy subgraph $H = (\tau, \nu)$ in $G = (\sigma, \mu)$. Then $x$ is a strong gate of $H$ implies $x \in I_g(y, v)$ for all $v \in \sigma^* \setminus \tau^*$. This gives,

$$d_g(y, v) = d_g(y, x) + d_g(x, v). \tag{6.5}$$

Similarly, $y$ is a strong gate of $H$ implies, $y \in I_g(x, v)$ for all $v \in \sigma^* \setminus \tau^*$. Therefore,

$$d_g(x, v) = d_g(x, y) + d_g(y, v). \tag{6.6}$$

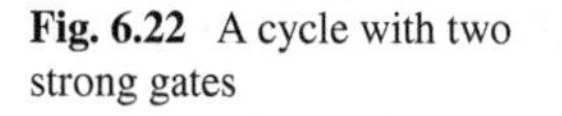
**Fig. 6.22** A cycle with two strong gates

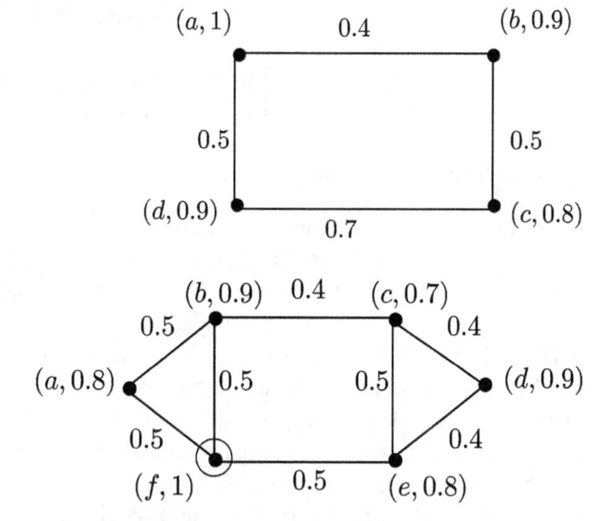


**Fig. 6.23** A fuzzy graph with a strong gate

Equations (6.5) and (6.6) together gives $d_g(x, y) = 0$, which implies $x = y$, because $d_g$ is a metric. ■

*Example 6.4.6* Consider the fuzzy graph $G = (\sigma, \mu)$ with $\sigma^* = \{a, b, c, d\}$, $\sigma(a) = 1, \sigma(b) = 0.9, \sigma(c) = 0.8, \sigma(d) = 0.9, \mu(ab) = 0.4, \mu(bc) = 0.5, \mu(cd) = 0.7$ and $\mu(ad) = 0.4$. $c$ and $d$ are strong gates. Here, $ab$ is the only weak edge. For the subgraph induced by $\{a, d\}$, $d$ is a strong gate in $G$ and for the subgraph induced by $\{b, c\}$, $c$ is a strong gate in $G$ (Fig. 6.22).

**Theorem 6.4.7** *Let G be a connected fuzzy graph and $H = (\tau, \nu)$ be a nontrivial induced fuzzy subgraph of $G = (\sigma, \mu)$. Then x is a strong gate of H in G if and only if every strong path from $\tau^*$ to $\sigma^* \setminus \tau^*$ passes through x.*

The proof is obvious. It can be obtained by contradiction.

$x$ is a strong gate of $H$ in $G$ does not imply that, all strongest paths from $\tau^*$ to $\sigma^* \setminus \tau^*$ pass through $x$. Consider the following example.

*Example 6.4.8* Consider the fuzzy graph given in Fig. 6.23.

In Fig. 6.23, $bc$ is the only $\delta$-edge of $G$ and $f$ is a strong gate of the subgraph induced by $\{a, b, f\}$ in $G$ because $(\bigcap_{u\in\{a,b,f\}} \bigcap_{v\in\{c,d,e\}} I_g(u, v)) \cap \{a, b, f\} = \{f\}$. Here, $abcd$ is a strongest $a - d$ path not containing $f$. The same case for $e$ also. $e$ is a strong gate of the fuzzy subgraph induced by $\{c, d, e\}$ because $(\bigcap_{v\in\{a,b,f\}} \bigcap_{u\in\{c,d,e\}} I_g(u, v)) \cap \{c, d, e\} = \{e\}$. But, $dcba$ is a strongest path from $d$ to $a$ and it does not pass through $e$.

Next theorem is about the number of strong gates of a fuzzy graph whose support is a cycle.

**Theorem 6.4.9** *Let $G$ be a fuzzy graph such that $G^*$ is a cycle with $|\sigma^*| = n$. Then*

$$\zeta(G) = \begin{cases} 0 & \text{if } G \text{ is a fuzzy cycle} \\ n-2 & \text{otherwise,} \end{cases}$$

*where $\zeta(G)$ denotes number of strong gates in $G$.*

*Proof* Disconnected subgraphs of $G$ have no strong gates. So, it is enough to consider connected subgraphs. Consider a fuzzy cycle $G = (\sigma, \mu)$ and a nontrivial connected induced subgraph $H = (\tau, \nu)$ of $G$. Then $H^*$ has at least two vertices. Clearly, no interior vertex of $H^*$ can be a strong gate of $H$ in $G$. Let $u$ and $v$ be pendent vertices of $H^*$ and let $w_1$ be the vertex in $\sigma^* \setminus \tau^*$ which is adjacent to $u$. Then $v \notin I_g(u, w_1)$ and $I_g(u, w_1) = \{u, w_1\}$. Also, let $w_2$ be the vertex in $\sigma^* \setminus \tau^*$ which is adjacent to $v$. Then $u \notin I_g(v, w_2)$ and $I_g(v, w_2) = \{v, w_2\}$. So $H$ has no strong gates in $G$. Because $H$ is arbitrary, it leads to the conclusion that $G$ has no strong gates.

Now, consider the case where $G$ is not a fuzzy cycle. Then $G$ has exactly one weakest edge. Let it be $xy$. Then any nontrivial connected induced fuzzy subgraph $H$ of $G$ containing $xy$ has no strong gates. Also, any nontrivial connected induced fuzzy subgraph, not containing both $x$ and $y$, has no strong gates. Now, consider a nontrivial connected fuzzy subgraph $H$ in $G$ containing either $x$ or $y$ (not both). Then that vertex will be one of the pendent vertices of $H^*$. Let $w$ be the other pendent vertex of $H^*$, different from $x$ and $y$. Then $w$ is a strong gate of $H$ in $G$. Also, $x$ and $y$ cannot be strong gates. That is, all vertices of $G^*$, except $x$ and $y$, are strong gates of $G$. Therefore, $\zeta(G) = n - 2$. ■

**Corollary 6.4.10** *No endvertex of a fuzzy tree is a strong gate.*

**Definition 6.4.11** A fuzzy subgraph is called ***I*-gate free** ($d$-gate free) if it has no $I$-gates ($d$-gates). Here ***I*-gates** means gates with respect to the interval function $I$. Similarly, ***d*-gates** means gates with respect to the metric $d$.

**Definition 6.4.12** A fuzzy graph $G$ is said to be **partially strong gated** if some connected induced fuzzy subgraphs of $G$ are gated. $G$ is said to be **fully strong gated** if every nontrivial connected induced fuzzy subgraphs of $G$ has a gate. $G$ is said to be **strongly restricted** if no connected induced fuzzy subgraph of $G$ is gated.

Every complete fuzzy graph is strongly restricted.

**Theorem 6.4.13** *Let $G = (\sigma, \mu)$ be any fuzzy graph and $H = (\tau, \nu)$ an induced connected fuzzy subgraph of $G$, which is strongly gated in $G$. Then the following properties hold.*

*(i) If $x$ is a strong gate of $H$ in $G$, then for any nontrivial connected induced fuzzy subgraph $K = (\tau', \nu')$ of $H$ containing $x$, having a strong gate in $G$, $x$ is a strong gate of $K$ in $G$.*

*(ii) If $x \in \tau^*$ is not a strong gate of $H$ in $G$, then there does not exist any nontrivial induced fuzzy subgraph $M$ of $G$ such that $H$ is a fuzzy subgraph of $M$ and $x$ is a strong gate of $M$ in $G$.*

*Proof* (*i*) Assume that $y$ is a strong gate of $K$. Consider the vertices $u$ and $v$ such that $u \in (\tau^{'})^*$ and $v \in \sigma^* \setminus \tau^*$. Then

$$d_g(u, v) = d_g(u, x) + d_g(x, v), \text{ because } x \text{ is the strong gate of } H. \tag{6.7}$$

$$d_g(u, v) = d_g(u, y) + d_g(y, v), \text{ because } y \text{ is the strong gate of } K. \tag{6.8}$$

Equation (6.7) implies

$$d_g(u, v) = d_g(u, x) + d_g(x, y) + d_g(y, v). \tag{6.9}$$

This is because $y$ is a strong gate of $K$ and $x \in K$. Thus,

$$d_g(u, v) = d_g(u, y) + d_g(y, x) + d_g(x, v), \text{ by Eq. (6.8).} \tag{6.10}$$

Equations (6.9) and (6.10)

$$d_g(u, x) + d_g(y, v) = d_g(u, y) + d_g(x, v). \tag{6.11}$$

By Eqs. (6.7) and (6.8)

$$d_g(u, x) + d_g(x, v) = d_g(u, y) + d_g(y, v). \tag{6.12}$$

Equations (6.11) and (6.12) imply

$$d_g(u, x) = d_g(u, y) \text{ and } d_g(y, v) = d_g(x, v). \tag{6.13}$$

Let $u$ be any vertex in $K^*$. Also, $x \in \tau^*$. So put $u = x$ in (6.13). Then $x = y$ because $d_g$ is a metric.

(*ii*) This is clear from the first part. ■

$x \in H^*$ is necessary in Theorem 6.4.13 as seen from Example 6.4.14.

*Example 6.4.14* Let $G$ be the fuzzy graph in Fig. 6.24. Then $b$ is a strong gate of the subgraph $H$ induced by $\{u, a, b\}$ and the subgraph $K$ induced by $\{u, a\}$ is also strongly gated. But, $b$ is not a strong gate of $K$.

For the second part of Theorem 6.4.13 to be valid, $H$ should be strongly gated in $G$. That is, even if $x \in H$ is not a strong gate of $H$, there exist subgraphs $M$ of $G$ such that $M$ contains $H$ and $x$ is a strong gate of $M$. It is clear from Fig. 6.25.

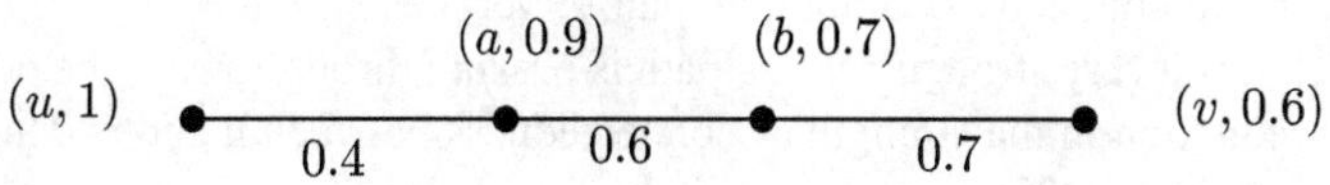

**Fig. 6.24** A gated path

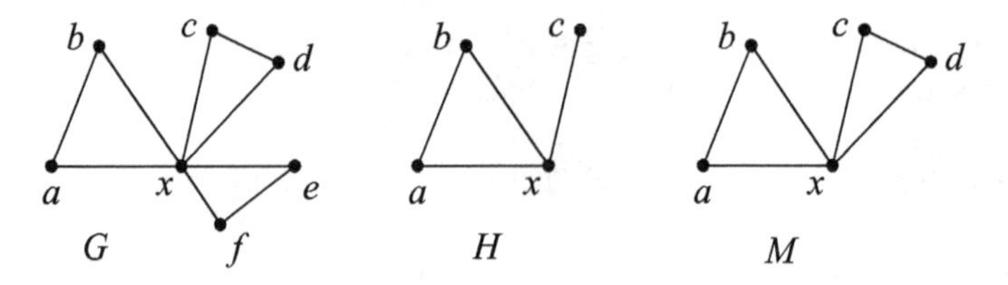


**Fig. 6.25** Fuzzy graph $G$ and its subgraphs $H$ and $M$

*Example 6.4.15* Let $G$ be the fuzzy graph, in Fig. 6.25, such that each edge and each vertex have membership value 1. Let $H$ and $M$ be fuzzy subgraphs of $G$. Then $x \in V(H^*)$ is not a strong gate of $H$. But $x$ is a strong gate of $M$ and $M$ contains $H$ as a subgraph.

**Theorem 6.4.16** *Every strong internal vertex (vertex whose strong degree greater than or equal to 2) of a fuzzy tree is a strong gate.*

*Proof* Between any two vertices, there exists a unique strong path and every interior vertex lies in the strong path joining any pair of its strong neighbors. So every strong interior vertex is a strong gate. ■

**Theorem 6.4.17** *Every strong gate in a fuzzy graph G is a fuzzy cutvertex.*

*Proof* Assume that $x$ is a strong gate in $G$. We want to prove that $x$ is a fuzzy cutvertex. Consider that $H$ is a nontrivial connected induced fuzzy subgraph of $G$ for which $x$ is a strong gate in $G$. Also, consider $u \in V(H^*)$ and $v \in V(G^*) \setminus V(H^*)$ such that $u, v$ has the maximum strength of connectedness among all pair of vertices $m, n \in V(H^*) \times (V(G^*) \setminus V(H^*))$. Assume that $CONN_G(u, v) = m$. Suppose there exists a strongest $u - v$ path $P$ which does not pass through $x$. Then strength of $P$, $s(P) = m$.

**Claim**: $P$ is not a strong path.

Assume, $P : a_1a_2a_3 \cdots a_k$ is a strong path, where $a_1 = u$ and $a_k = v$. Because $u \in V(H^*)$ and $v \in V(G^*) \setminus V(H^*)$, there exists $i \in \{1, 2, 3, \ldots, k-1\}$ such that $a_i \in V(H^*)$ and $a_{i+1} \in V(G^*) \setminus V(H^*)$. Because $a_ia_{i+1}$ is an edge in the strong path $P$, $a_ia_{i+1}$ is a strong edge. Therefore, $I_s(a_i, a_{i+1}) = \{a_i, a_{i+1}\}$. That is, $x \notin I_s(a_i, a_{i+1})$. This is a contradiction to the assumption that $x$ is a strong gate of $H$ in $G$. Therefore, $P$ is not a strong path.

Because $a_ia_{i+1}$ is a weak edge, $CONN_G(a_i, a_{i+1}) > \mu(a_ia_{i+1})$. Because $a_ia_{i+1} \in P$, we get $m \leq \mu(a_ia_{i+1})$. Both these equations together gives, $CONN_G(u, v) = m < CONN_G(a_i, a_{i+1})$. This is a contradiction to the assumption that $u, v$ has the maximum strength of connectivity among all pair of vertices $m, n \in V(H^*) \times (V(G^*) \setminus V(H^*))$. That is, every strongest $u - v$ path is passing through $x$. So, the deletion of $x$ makes a reduction in the strength of connectedness between $u$ and $v$. Therefore, $x$ is a fuzzy cutvertex. ■

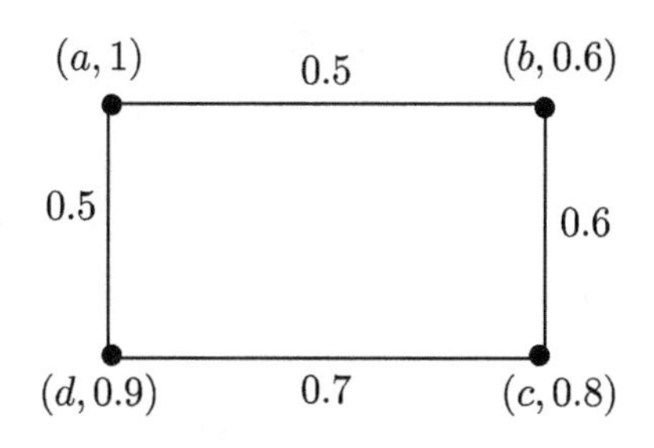


**Fig. 6.26** A strong cycle with a fuzzy cutvertex

*Example 6.4.18* Let $G$ be the fuzzy graph in Fig. 6.26. Now, $c$ is a partial cutvertex, but it is not a strong gate. Hence, the converse of Theorem 6.4.17 is not true.

Because fuzzy blocks have no fuzzy cutvertices, we have the following corollary.

**Corollary 6.4.19** *Every fuzzy block is strong gate free.*

Next we discuss sufficient conditions for a partial cutvertex to be a strong gate.

**Theorem 6.4.20** *Let $G$ be a fuzzy graph. A fuzzy cutvertex $w$ of $G$ is a strong gate if either one of the following statement is true.*
*(i) $w$ is a cutvertex of $G$*
*(ii) $w$ do not lie on any strong cycle.*

*Proof* (*i*) Consider a fuzzy cutvertex $w$ of $G$. Consider the case where $w$ is a cutvertex. $w$ is a cutvertex of $G$ implies $G - w$ is disconnected. Let $H$ be any component of $G - w$. Then $w$ is a strong gate of the fuzzy subgraph induced by $(H^*) \cup \{w\}$ in $G$.

(*ii*) Assume that $w$ do not lie on any strong cycle. Because $w$ is a fuzzy cutvertex, we can find two vertices $u$ and $v$ such that, $u$ and $v$ are strongly adjacent to $w$ and $CONN_{G-w}(u, v) < CONN_G(u, v)$. Also, $G$ does not have a strong $uv$ edge. So, $w$ do not lie on any strong cycle. Now, consider two sets $G_u$ and $G_v$ such that

$$G_u = \{x \in V(G^*) \mid x \text{ is strongly connected to } u \text{ in } G - w\}$$

and

$$G_v = \{y \in V(G^*) \mid y \text{ is strongly connected to } v \text{ in } G - w\}.$$

Here, two vertices are strongly connected means, there exists a strong path between them.

Then clearly $G_u \cap G_v = \emptyset$. If not; that is, if $a \in G_u \cap G_v$, there exists a strong $u - a$ path $P_1$ (say) and a strong $a - v$ path $P_2$ (say). Then the strong walk $P_1$ followed by $P_2$ contains a strong $u - v$ path $P$ (say) which does not contain the vertex $w$. $P$ followed by the strong path $vwu$ is a strong cycle which contains $w$. This gives a contradiction to our assumption that $w$ do not lie on any strong cycle. So, $G_u \cap G_v = \emptyset$. That is, every strong path from $G_u$ to $G \setminus G_u$ passes through $w$. So $w$ is a strong gate in $G$. ■

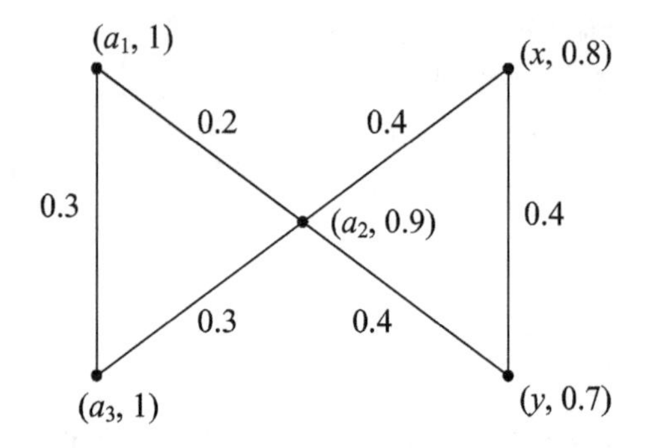


**Fig. 6.27** Fuzzy graph $G$

**Definition 6.4.21** A fuzzy graph $G = (\sigma, \mu)$ is said to be **strongly connected** if there is a strong path between any two vertices in $\sigma^*$. Maximal strongly connected fuzzy subgraphs of $G$ are called **strong components** of $G$. A fuzzy graph is known as **strongly disconnected** if it has at least two strong components. $\omega_s(G)$ denotes the number of strong components of $G$.

**Theorem 6.4.22** *In a connected fuzzy graph $G$, a vertex $x$ is a strong gate in $G$ if and only if $G - x$ is strongly disconnected.*

*Proof* Let $x$ be a strong gate of $H$ in $G$. This implies every strong path from $H$ to $G \setminus H$ passes through $x$. Thus, $G \setminus H$ is strongly disconnected.

Now, consider the case where $G$ is strongly connected and $G - x$ is strongly disconnected. Let $H$ be any strong component of $G - x$ and let $K$ be the fuzzy subgraph of $G$ induced by $H^*$ and $x$. Then $x$ is a strong gate of $K$ in $G$. ■

In a fuzzy tree all vertices of strong degree more than 1 are strong gates. However, this is not the case in a general fuzzy graph. That is, vertices of strong degree more than 1 cannot be strong gates.

*Example 6.4.23* Consider the fuzzy graph $G$ in Fig. 6.27. In $G$, both $x$ and $y$ have strong degree 2, but they are not strong gates in $G$. If we take $H$ as the fuzzy subgraph induced by $\{x, y\}$, then $a_2 \in \cap_{u \in V(G^*) \setminus V(H^*)} \cap_{v \in V(H^*)} I_s(u, v)$. However, $a_2$ is not a strong gate of $H$ because $a_2 \notin V(H^*)$.

Let $G = (V, E)$ be a graph. A vertex $u$ of $G$ is said to be a **pendant vertex** if $\deg_G(u) = 1$. A vertex $u$ of a fuzzy graph is called a **strong pendant vertex** if $d_s(u) = 1$.

**Theorem 6.4.24** ([66]) *A nonempty connected fuzzy graph is strongly restricted if and only if any two vertices lie on a strong cycle.*

*Proof* Consider a nonempty connected fuzzy graph $G = (\sigma, \mu)$ such that $G$ is strongly restricted. Now, we want to prove that any 2 vertices lie on a strong cycle. We will prove this result by mathematical induction on $d_g(u, v)$, where $u$ and $v$ are any two vertices in $V(G^*)$. Let $d_g(u, v) = 1$. That is, $uv$ is a strong edge in $G$. Because

$G$ is strongly restricted it has no strong gates. Thus, it has no strong pendent vertices. Therefore, strong degree of $u$ as well as $v$, greater than or equal to 2. So, consider a vertex $w \neq u$ such that $w$ is strongly adjacent to $v$. Then $uvw$ is a strong $u - w$ path.

**Claim**: There exists a strong $u - w$ path $P$ such that $P$ is internally disjoint from $uvw$.

Assume that all strong $u - w$ paths pass through $v$. Then $G - v$ is strongly disconnected. Let $\{V_u, V_w, V^{'}\}$ be a partition of $V((G - v)^*)$, defined as follows.

$$V_u = \{x \in V((G - v)^*) \mid x \text{ is strongly connected to } u\},$$

$$V_w = \{x \in V((G - v)^*) \mid x \text{ is strongly connected to } w\}$$

and

$$V^{'} = V((G - v)^*) \setminus (V_u \cup V_w).$$

Then $u \in V_u$ and $w \in V_w$. From the construction, it is clear that $V_u$, $V_w$ and $V'$ are pairwise disjoint and their union is $V((G - v)^*)$. Now, consider the fuzzy subgraph $H$ of $G$, induced by $V_u \cup v$. Then $v$ is a strong gate of $H$ in $G$. It is a contradiction, because $G$ is strongly restricted. So, there exists a strong $u - w$ path $P$ such that $P$ is internally disjoint to $uvw$. Then $P$ followed by $wvu$ is a strong cycle containing $u$ and $v$.

Now, assume that for all $u$, $v$ with $d_g(u, v) \leq n$, $u$ and $v$ lie on a strong cycle, where $n \geq 2$. Now, consider two vertices $x, y$ of $G$ such that $d_g(x, y) = n + 1$. Consider any $x - y$ strong geodesic. Let it be $u_0, u_1, u_2, \ldots, u_n, u_{n+1}$, where $u_0 = x$ and $u_{n+1} = y$. Because $d_g(x, u_n) = n$, by assumption there exist two internally disjoint $x - u_n$ strong paths, say $P_1$ and $P_2$. Let $w$ be the last vertex in $P_1$, strongly adjacent to $u_n$. Clearly $d_g(w, y) \leq 2$. Because $u_n$ is not a strong gate, there exists a strong $w - y$ path internally disjoint from $wu_ny$. Let it be $P_3$. $P_3$ can have common vertices with $P_1$ and $P_2$. Let $a$ be the last such vertex in $P_3$. Then there are two cases.

**Case 1**: $a$ lies on $P_1$. $x - a$ part of $P_1$ followed by $a - y$ part of $P_3$ is a strong $x - y$ path. This is internally disjoint from the strong $x - y$ path formed by $P_2$ followed by the strong edge $u_ny$. Now, we get two internally disjoint $x - y$ strong paths. They together form a strong cycle containing $x$ and $y$.

**Case 2**: $a$ lies on $P_2$. $P_1$ followed by the strong edge $u_ny$ is a strong $x - y$ path. Another $x - y$ strong path can be constructed as $x - a$ part of $P_2$ followed by $a - y$ part of $P_3$. These two strong paths are internally disjoint and they together form a strong cycle containing $x$ and $y$. By mathematical induction, we proved that any two vertices lie on a strong cycle if the graph is strongly restricted. Now, consider the converse. That is, we want to prove that $G$ is strongly restricted, if any two vertices lie on a strong cycle. Now, consider a graph $G$, which has the property that any two vertices lie on a strong cycle. To prove that $G$ is strongly restricted, it is enough to prove that $G$ has no strong gates. Because any two vertices lie on a strong cycle, for any vertex $x$ and for any pair of vertices $u$, $v$, there is a strong $u - v$ path not passing through $x$. So $x$ can not be a strong gate of $G$. Because $x$ is arbitrary vertex of $G^*$, it cannot have any strong gates. Hence, $G$ is strongly restricted. ■

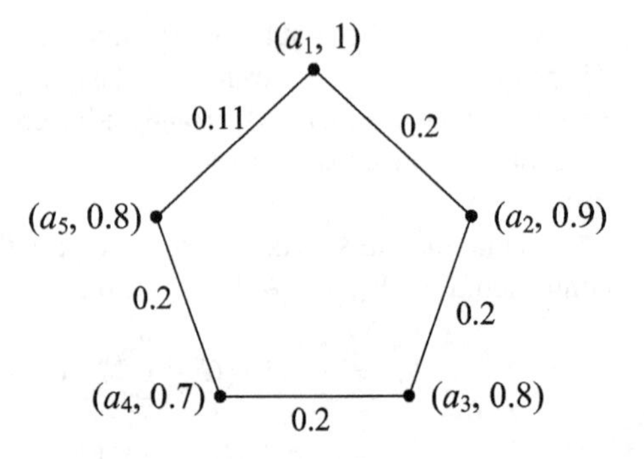


**Fig. 6.28** A fuzzy cycle with a unique weak edge

Let $G$ be a connected fuzzy graph and $x$ be a strong pendant vertex of $G$. Then the vertex $y$, which is strongly adjacent to $x$ is a strong gate of $H$ in $G$, where $H$ is the fuzzy subgraph of $G$, induced by $x$ and $y$.

Now, we check whether all strong gates are cutvertices in a fuzzy graphs. The answer is no. Consider the following example.

*Example 6.4.25* Consider the fuzzy graph in Fig. 6.28. In Fig. 6.28, $a_2$, $a_3$ and $a_4$ are strong gates, but they are not cutvertices. Thus, in a fuzzy graph $G$, set of cutvertices of $G \subseteq$ set of strong gates of $G \subseteq$ set of fuzzy cutvertices of $G$.

A fuzzy cycle has no strong gates. Thus, the minimum number of strong gates in a fuzzy graph is zero. Also, in any fuzzy graph, there exists at least 2 vertices which are not partial cutvertices. Moreover, in a path all internal vertices are strong gates. So, we have the following result.

**Proposition 6.4.26** *If $\varsigma(G)$ represents the number of strong gates in a graph $G$, then* $0 \leq \varsigma(G) \leq n - 2$.

# Chapter 7
# Interval-Valued Fuzzy Graphs

The results in this chapter are based mostly on the works in [5, 14, 19, 23]. In 1975, Zadeh [194] introduced the notion of interval-valued fuzzy sets as an extension of fuzzy sets [190] in which the values of the memberships degrees are intervals in [0, 1] instead of elements in [0, 1]. Interval-valued fuzzy sets provide a more adequate description of uncertainty than traditional fuzzy sets in some cases. It can therefore be important to use interval-valued fuzzy sets in applications, e.g., in fuzzy control. One of the most intensive parts of fuzzy control is defuzzification [121]. Interval-valued fuzzy sets have been widely studied and used, for example, in the work of Gorzalczany on approximate reasoning [81, 82], Roy and Biswas on medical diagnosis [155], Turksen on multivalued logic [177], and Mendel in intelligent control [121].

We define the operations of Cartesian product, composition, union, and join on interval-valued fuzzy graphs and investigate some of their properties. The work is based on [130]. We study isomorphisms (resp. weak isomorphism) between interval-valued fuzzy graphs in an equivalence relation (resp. partial order). We introduce the notion of interval-valued fuzzy complete graphs and present some properties of self-complementary and self-weak complementary interval-valued fuzzy complete graphs.

Other results and applications can be found in [1, 5, 8, 14, 18, 24, 26, 34, 37, 40, 41, 44, 50, 57, 76, 77, 83, 86, 103, 114, 122, 125, 128, 146, 151, 157, 164, 167, 186].

## 7.1 Interval-Valued Fuzzy Sets

Recall that a simple graph is a graph that has no loops and no more than one edge between two different vertices. A complete graph is a simple graph in which every pair of distinct vertices is connected by an edge. The complete graph on $n$ vertices has $n(n-1)/2$ edges. We consider only simple graphs.





S. Mathew et al., *Fuzzy Graph Theory*, Studies in Fuzziness and Soft Computing 363, https://doi.org/10.1007/978-3-319-71407-3_7

Let $G^* = (V, E)$ denote a graph. By a complementary graph $\overline{G^*}$ of a simple graph, we mean a graph having the same vertices as $G^*$ and such that two vertices are adjacent in $\overline{G^*}$ if and only if they are not adjacent in $G^*$.

Recall an isomorphism of the graphs $G_1^*$ and $G_2^*$ is a bijection $f$ between the vertex sets of $G_1^*$ and $G_2^*$ such that any two vertices $v_1$ and $v_2$ of $G_1^*$ are adjacent if and only if $f(v_1)$ and $f(v_2)$ are adjacent in $G_2^*$. Isomorphic graphs are denoted by $G_1^* \simeq G_2^*$. An **automorphism** of a graph is a graph isomorphism with itself.

Let $G_1^* = (V_1, E_1)$ and $G_2^* = (V_2, E_2)$ be two simple graphs. We can construct several new graphs. The **Cartesian product** of $G_1^*$ and $G_2^*$ is the graph $G_1^* \times G_2^* = (V, E)$ with $V = V_1 \times V_2$ and

$$E = \{(x, x_2)(y, y_2) \mid x \in V_1, x_2 y_2 \in E_2\} \cup \{(x_1, z)(y_1, z) \mid z \in V_2,\ x_1 y_1 \in E_1\}.$$

The **composition** of graphs $G_1^*$ with $G_2^*$ is the graph $G_1^*[G_2^*] = (V_1 \times V_2, E^0)$, where

$$E^0 = E \cup \{(x_1, x_2)(y_1, y_2) \mid x_1 y_1 \in E_1, x_2 \neq y_2\}.$$

Note that $G_1^*[G_2^*] \neq G_2^*[G_1^*]$.

The **union** of graphs $G_1^*$ and $G_2^*$ is defined as $G_1^* \cup G_2^* = (V_1 \cup V_2, E_1 \cup E_2)$.

The **join** of graphs $G_1^*$ and $G_2^*$ is the simple graph $G_1^* + G_2^* = (V_1 \cup V_2, E_1 \cup E_2 \cup E')$, where $E'$ is the set of all edges joining the nodes of $V_1$ and $V_2$; it is assumed that $V_1 \cap V_2 = \emptyset$.

An **interval number** $D$ is an interval $[a^-, a^+]$ with $0 \leq a^- \leq a^+ \leq 1$. The interval $[a, a]$ is identified with the number $a \in [0, 1]$. $D[0, 1]$ denotes the set of all interval numbers.

For interval numbers $D_1 = [a_1^-, b_1^+]$ and $D_2 = [a_2^-, b_2^+]$, we define
$D_1 \wedge D_2 = [a_1^- \wedge a_2^-, b_1^+ \wedge b_2^+]$,
$D_1 \vee D_2 = [a_1^- \vee a_2^-, b_1^+, \vee b_2^+]$,
$D_1 + D_2 = [a_1^- + a_2^- - a_1^- \cdot a_2^-, b_1^+ + b_2^+ - b_1^+ \cdot b_2^+]$,
$D_1 \leq D_2 \Longleftrightarrow a_1^- \leq a_2^-$ and $b_1^+ \leq b_2^+$,
$D_1 = D_2 \Longleftrightarrow a_1^- = a_2^-$ and $b_1^+ = b_2^+$,
$D_1 < D_2 \Longleftrightarrow D_1 \leq D_2$ and $D_1 \neq D_2$,
$kD_1 = k[a_1^-, b_1^+] = [ka_1^-, kb_1^+]$, where $0 \leq k \leq 1$.

It follows that $(D[0, 1], \leq, \vee, \wedge)$ is a complete lattice with $[0, 0]$ as the least element and $[1, 1]$ as the greatest.

The **interval-valued fuzzy set** $A$ in $V$ is defined by

$$A = \{(x, \mu_A^-(x), \mu_A^+(x)) \mid x \in V\},$$

where $\mu_A^-$ and $\mu_A^+$ are fuzzy subsets of $V$ such that $\mu_A^-(x) \leq \mu_A^+(x)$ for all $x \in V$. For any two interval-valued fuzzy sets $A = [\mu_A^-(x), \mu_A^+(x)]$ and $B = [\mu_B^-(x), \mu_B^+(x)]$ in $V$ we define:
$A \cup B = \{(x, \mu_A^-(x) \vee \mu_B^-(x), \mu_A^+(x) \vee \mu_B^+(x)) \mid x \in V\}$,
$A \cap B = \{(x, \mu_A^-(x) \wedge \mu_B^-(x), \mu_A^+(x) \wedge \mu_B^+(x)) \mid x \in V\}$.

If $G^* = (V, E)$ is a graph, then by an **interval-valued fuzzy relation** $B$ on a set $E$, we mean an interval-valued fuzzy set such that

$$\mu_B^-(xy) \le \mu_A^-(x) \wedge \mu_A^-(y),$$
$$\mu_B^+(xy) \le \mu_A^+(x) \wedge \mu_A^+(y)$$

for all $xy \in E$.

## 7.2 Operations on Interval-Valued Fuzzy Graphs

Throughout this section, $G^*$ is a crisp graph and $G$ is an interval-valued fuzzy graph.

**Definition 7.2.1** By an **interval-valued fuzzy graph** of a graph $G^* = (V, E)$, we mean a pair $G = (A, B)$, where $A = [\mu_A^-, \mu_A^+]$ is an interval-valued fuzzy set on $V$ and $B = [\mu_B^-, \mu_B^+]$ is an interval-valued fuzzy relation on $E$.

Let $\mu$ be a fuzzy subset of a set $X$. Suppose that $\mu(x) = a$. We sometimes write $\frac{x}{a}$ for $\mu(x) = a$ when $\mu$ is understood.

*Example 7.2.2* Consider a graph $G^* = (V, E)$ such that $V = \{x, y, z\}$ and $E = \{xy, yz, zx\}$. Let $A$ be an interval-valued fuzzy set of $V$ and let $B$ be an interval-valued fuzzy set of $E \subseteq V \times V$ defined by

$$A = \left\langle \left(\frac{x}{0.1}, \frac{y}{0.3}, \frac{z}{0.5}\right), \left(\frac{x}{0.4}, \frac{y}{0.6}, \frac{z}{0.5}\right\rangle,$$
$$B = \left\langle \left(\frac{xy}{0.1}, \frac{yz}{0.2}, \frac{zx}{0.1}\right), \left(\frac{xy}{0.3}, \frac{yz}{0.4}, \frac{zx}{0.4}\right)\right\rangle.$$

It is follows easily that $G = (A, B)$ is an interval-valued fuzzy graph of $G^*$.

**Definition 7.2.3** The **Cartesian product** $G_1 \times G_2$ of two interval-valued fuzzy graphs $G_1 = (A_1, B_1)$ and $G_2 = (A_2, B_2)$ of the graphs $G_1^* = (V_1, E_1)$ and $G_2^* = (V_2, E_2)$ is defined to be the pair $(A_1 \times A_2, B_1 \times B_2)$ such that

$$(\mu_{A_1}^- \times \mu_{A_2}^-)(x_1, x_2) = \mu_{A_1}^-(x_1) \wedge \mu_{A_2}^-(x_2),$$
$$(\mu_{A_1}^+ \times \mu_{A_2}^+)(x_1, x_2) = \mu_{A_1}^+(x_1) \wedge \mu_{A_2}^+(x_2),$$

for all $(x_1, x_2) \in V$,

$$(\mu_{B_1}^- \times \mu_{B_2}^-)(x, x_2)(x, y_2) = \mu_{A_1}^-(x) \wedge \mu_{B_2}^-(x_2y_2),$$
$$(\mu_{B_1}^+ \times \mu_{B_2}^+)(x, x_2)(x, y_2) = \mu_{A_1}^+(x) \wedge \mu_{B_2}^+(x_2y_2),$$

for all $x \in V_1$ and $x_2y_2 \in E_2$,

$$(\mu^-_{B_1} \times \mu^-_{B_2})(x_1, z)(y_1, z) = \mu^-_{B_1}(x_1y_1) \wedge \mu^-_{A_2}(z),$$
$$(\mu^+_{B_1} \times \mu^+_{B_2})(x_1, z)(y_1, z) = \mu^+_{B_1}(x_1y_1) \wedge \mu^+_{A_2}(z),$$

for all $z \in V_2$ and $x_1y_1 \in E_1$.

*Example 7.2.4* Let $G_1^* = (V_1, E_1)$ and $G_2^* = (V_2, E_2)$ be graphs such that $V_1 = \{a, b\}$, $V_2 = \{c, d\}$, $E_1 = \{ab\}$, $E_2 = \{cd\}$. Consider two interval-valued fuzzy graphs $G_1 = (A_1, B_1)$ and $G_2 = (A_2, B_2)$, where

$$A_1 = \left\langle \left(\frac{a}{0.2}, \frac{b}{0.3}\right), \left(\frac{a}{0.4}, \frac{b}{0.5}\right) \right\rangle, \quad B_1 = \left\langle \frac{ab}{0.1}, \frac{ab}{0.2} \right\rangle,$$
$$A_2 = \left\langle \left(\frac{c}{0.1}, \frac{d}{0.2}\right), \left(\frac{c}{0.4}, \frac{d}{0.6}\right) \right\rangle, \quad B_2 = \left\langle \frac{cd}{0.1}, \frac{cd}{0.3} \right\rangle.$$

Then it follows that

$$(\mu^-_{B_1} \times \mu^-_{B_2})(a, c)(a, d) = 0.1, \quad (\mu^+_{B_1} \times \mu^+_{B_2})(a, c)(a, d) = 0.3,$$
$$(\mu^-_{B_1} \times \mu^-_{B_2})(a, c)(b, c) = 0.1, \quad (\mu^+_{B_1} \times \mu^+_{B_2})(a, c)(b, c) = 0.2,$$
$$(\mu^-_{B_1} \times \mu^-_{B_2})(a, d)(b, d) = 0.1, \quad (\mu^+_{B_1} \times \mu^+_{B_2})(a, d)(b, d) = 0.2,$$
$$(\mu^-_{B_1} \times \mu^-_{B_2})(b, c)(b, d) = 0.1, \quad (\mu^+_{B_1} \times \mu^+_{B_2})(b, c)(b, d) = 0.3.$$

It follows that $G_1 \times G_2$ is an interval-valued fuzzy graph of $G_1^* \times G_2^*$.

**Proposition 7.2.5** *The Cartesian product $G_1 \times G_2 = (A_1 \times A_2, B_1 \times B_2)$ of two interval-valued fuzzy graph of $G_1^*$ and $G_2^*$ is an interval-valued fuzzy graph of $G_1^* \times G_2^*$.*

*Proof* Let $x_1, x_2 \in V$. Then $(\mu^-_{A_1} \times \mu^-_{A_2})(x_1, x_2) = \mu^-_{A_1}(x_1) \wedge \mu^-_{A_2}(x_2) \leq \mu^+_{A_1}(x_1) \wedge \mu^+_{A_2}(x_2) = (\mu^+_{A_1} \times \mu^+_{A_2})(x_1, x_2)$.

Let $x \in V_1$ and $x_2y_2 \in E_2$. Then

$$\begin{aligned}
(\mu^-_{B_1} \times \mu^-_{B_2})(x, x_2)(x, y_2) &= \mu^-_{A_1}(x) \wedge \mu^-_{B_2}(x_2y_2) \\
&\leq \mu^-_{A_1}(x) \wedge (\mu^-_{A_2}(x_2) \wedge \mu^-_{A_2}(y_2)) \\
&= (\mu^-_{A_1}(x) \wedge \mu^-_{A_2}(x_2)) \wedge (\mu^-_{A_1}(x) \wedge \mu^-_{A_2}(y_2) \\
&= (\mu^-_{A_1} \times \mu^-_{A_2})(x,\ x_2) \wedge (\mu^-_{A_1} \times \mu^-_{A_2})(x, y_2),
\end{aligned}$$

$$\begin{aligned}(\mu^+_{B_1} \times \mu^+_{B_2})(x, x_2)(x, y_2) &= \mu^+_{A_1}(x) \wedge \mu^+_{B_2}(x_2y_2)\\ &\le \mu^+_{A_1}(x) \wedge (\mu^+_{A_2}(x_2) \wedge \mu^+_{A_2}(y_2))\\ &= (\mu^+_{A_1}(x) \wedge \mu^+_{A_2}(x_2)) \wedge (\mu^+_{A_1}(x) \wedge \mu^+_{A_2}(y_2))\\ &= (\mu^+_{A_1} \times \mu^+_{A_2})(x, x_2) \wedge (\mu^+_{A_1} \times \mu^+_{A_2})(x, y_2).\end{aligned}$$

Similarly, for $z \in V_2$ and $x_1y_1 \in E_1$, we have

$$\begin{aligned}(\mu^-_{B_1} \times \mu^-_{B_2})(x_1, z)(y_1, z) &= \mu^-_{B_1}(x_1y_1) \wedge \mu^-_{A_2}(z)\\ &\le (\mu^-_{A_1}(x_1) \wedge \mu^-_{A_1}(y_1)) \wedge \mu^-_{A_2}(z)\\ &= (\mu^-_{A_1}(x_1) \wedge \mu^-_{A_2}(z)) \wedge (\mu^-_{A_1}(y_1) \wedge \mu^-_{A_2}(z))\\ &= (\mu^-_{A_1} \times \mu^-_{A_2})(x_1, z) \wedge (\mu^-_{A_1} \times \mu^-_{A_2})(y_1, z),\end{aligned}$$

$$\begin{aligned}(\mu^+_{B_1} \times \mu^+_{B_2})(x_1, z)(y_1, z) &= \mu^+_{B_1}(x_1y_1) \wedge \mu^+_{A_2}(z)\\ &\le (\mu^+_{A_1}(x_1) \wedge \mu^+_{A_1}(y_1)) \wedge \mu^+_{A_2}(z)\\ &= (\mu^+_{A_1}(x_1) \wedge \mu^+_{A_2}(z)) \wedge (\mu^+_{A_1}(y_1) \wedge \mu^+_{A_2}(z)\\ &= (\mu^+_{A_1} \times \mu^+_{A_2})(x_1, z) \wedge (\mu^+_{A_1} \times \mu^+_{A_2})(y_1, z).\end{aligned}$$

■

**Definition 7.2.6** The **composition** $G_1[G_2] = (A_1 \circ A_2, B_1 \circ B_1)$ of two interval-valued fuzzy graphs $G_1$ and $G_2$ of the graphs $G_1^*$ and $G_2^*$ is defined as follows:

$$\begin{aligned}(\mu^-_{A_1} \circ \mu^-_{A_2})(x_1, x_2) &= \mu^-_{A_1}(x_1) \wedge \mu^-_{A_2}(x_2),\\ (\mu^+_{A_1} \circ \mu^+_{A_2})(x_1, x_2) &= \mu^+_{A_1}(x_1) \wedge \mu^+_{A_2}(x_2),\end{aligned}$$

for all $(x_1, x_2) \in V$,

$$\begin{aligned}(\mu^-_{B_1} \circ \mu^-_{B_2})(x, x_2)(x, y_2) &= \mu^-_{A_1}(x) \wedge \mu^-_{B_2}(x_2y_2),\\ (\mu^+_{B_1} \circ \mu^+_{B_2})(x, x_2)(x, y_2) &= \mu^+_{A_1}(x) \wedge \mu^+_{B_2}(x_2y_2),\end{aligned}$$

for all $x \in V_1$ and $x_2y_2 \in E_2$,

$$\begin{aligned}(\mu^-_{B_1} \circ \mu^-_{B_2})(x_1, z)(y_1, z) &= \mu^-_{B_1}(x_1y_1) \wedge \mu^-_{A_2}(z),\\ (\mu^+_{B_1} \circ \mu^+_{B_2})(x_1, z)(y_1, z) &= \mu^+_{B_1}(x_1y_1) \wedge \mu^+_{A_2}(z),\end{aligned}$$

for all $z \in V_2$ and $x_1y_1 \in E_1$,

$$(\mu^-_{B_1} \circ \mu^-_{B_2})(x_1, x_2)(y_1, y_2) = \mu^-_{A_2}(x_2) \wedge \mu^-_{A_2}(y_2) \wedge \mu^-_{B_1}(x_1y_1),$$
$$(\mu^+_{B_1} \circ \mu^+_{B_2})(x_1, x_2)(y_1, y_2) = \mu^+_{A_2}(x_2) \wedge \mu^+_{A_2}(y_2) \wedge \mu^+_{B_1}(x_1y_1),$$

for all $(x_1, x_2)(y_1, y_2) \in E^0 - E$.

*Example 7.2.7* Let $G_1^*$ and $G_2^*$ be as in Example 7.2.4. Consider two interval-valued fuzzy graphs $G_1 = (A_1, B_1)$ and $G_2 = (A_2, B_2)$, where

$$A_1 = \left\langle \left(\frac{a}{0.2}, \frac{b}{0.3}\right), \left(\frac{a}{0.5}, \frac{b}{0.5}\right) \right\rangle, \ B_1 = \left\langle \frac{ab}{0.2}, \frac{ab}{0.4} \right\rangle,$$
$$A_2 = \left\langle \left(\frac{c}{0.1}, \frac{d}{0.3}\right), \left(\frac{c}{0.4}, \frac{d}{0.6}\right) \right\rangle, \ B_2 = \left\langle \frac{cd}{0.1}, \frac{cd}{0.3} \right\rangle.$$

It follows that

$$(\mu^-_{B_1} \circ \mu^-_{B_2})(a, c)(a, d) = 0.1, \quad (\mu^+_{B_1} \circ \mu^+_{B_2})(a, c)(a, d) = 0.3,$$
$$(\mu^-_{B_1} \circ \mu^-_{B_2})(b, c)(b, d) = 0.1, \quad (\mu^+_{B_1} \circ \mu^+_{B_2})(b, c)(b, d) = 0.3,$$
$$(\mu^-_{B_1} \circ \mu^-_{B_2})(a, c)(b, c) = 0.1, \quad (\mu^+_{B_1} \circ \mu^+_{B_2})(a, c)(b, c) = 0.4,$$
$$(\mu^-_{B_1} \circ \mu^-_{B_2})(a, d)(b, d) = 0.2, \quad (\mu^+_{B_1} \circ \mu^+_{B_2})(a, d)(b, d) = 0.4,$$
$$(\mu^-_{B_1} \circ \mu^-_{B_2})(a, c)(b, d) = 0.1, \quad (\mu^+_{B_1} \circ \mu^+_{B_2})(a, c)(b, d) = 0.4,$$
$$(\mu^-_{B_1} \circ \mu^-_{B_2})(b, c)(a, d) = 0.1, \quad (\mu^+_{B_1} \circ \mu^+_{B_2})(b, c)(a, d) = 0.4.$$

It follows easily that $G_1[G_2] = (A_1 \circ A_2, B_1 \circ B_1)$ is an interval-valued fuzzy graph of $G_1^*[G_2^*]$.

**Proposition 7.2.8** *The composition $G_1[G_2]$ of two interval-valued fuzzy graphs $G_1$ and $G_2$ of the graphs $G_1^*$ and $G_2^*$ is an interval-valued fuzzy graph of $G_1^*[G_2^*]$.*

*Proof* We verify the result only for $B_1 \circ B_2$. For $x \in V_1$ and $x_2y_2 \in E_2$, we have from Definition 7.2.6 that

$$\begin{aligned}(\mu^-_{B_1} \circ \mu^-_{B_2})(x, x_2)(x, y_2) &= \mu^-_{A_1}(x) \wedge \mu^-_{B_2}(x_2y_2) \\ &\leq \mu^-_{A_1}(x) \wedge ((\mu^-_{A_2}(x_2) \wedge \mu^-_{A_2}(y_2)) \\ &= (\mu^-_{A_1}(x) \wedge \mu^-_{A_2}(x_2)) \wedge (\mu^-_{A_1}(x) \wedge \mu^-_{A_2}(y_2) \\ &= \mu^-_{A_1} \circ \mu^-_{A_2}(x, x_2) \wedge \mu^-_{A_1} \circ \mu^-_{A_2}(x, y_2),\end{aligned}$$

$$\begin{aligned}(\mu^+_{B_1} \circ \mu^+_{B_2})(x, x_2)(x, y_2) &= \mu^+_{A_1}(x) \wedge \mu^+_{B_2}(x_2 y_2)\\ &\le \mu^+_{A_1}(x) \wedge (\mu^+_{A_2}(x_2) \wedge \mu^+_{A_2}(y_2))\\ &= (\mu^+_{A_1}(x) \wedge \mu^+_{A_2}(x_2)) \wedge (\mu^+_{A_1}(x) \wedge \mu^+_{A_2}(y_2))\\ &= (\mu^+_{A_1} \circ \mu^+_{A_2})(x,\ x_2) \wedge (\mu^+_{A_1} \circ \mu^+_{A_2})(x, y_2).\end{aligned}$$

The proof is similar for the case, $z \in V_2$ and $x_1 y_1 \in E_1$.
Suppose $(x_1, x_2)(y_1, y_2) \in E^0 \setminus E$. Then

$$\begin{aligned}(\mu^-_{B_1} \circ \mu^-_{B_2})(x_1, x_2)(y_1, y_2) &= \mu^-_{A_2}(x_2) \wedge \mu^-_{A_2}(y_2) \wedge \mu^-_{B_1}(x_1 y_1)\\ &\le \mu^-_{A_2}(x_2) \wedge \mu^-_{A_2}(y_2) \wedge \mu^-_{A_1}(x_1) \wedge \mu^-_{A_1}(y_1)\\ &= (\mu^-_{A_1}(x_1) \wedge \mu^-_{A_2}(x_2)) \wedge (\mu^-_{A_1}(y_1) \wedge \mu^-_{A_2}(y_2))\\ &= (\mu^-_{A_1} \circ \mu^-_{A_2})(x_1, x_2) \wedge (\mu^-_{A_1} \circ \mu^-_{A_2})(y_1, y_2),\end{aligned}$$

$$\begin{aligned}(\mu^+_{B_1} \circ \mu^+_{B_2})(x_1, x_2)(y_1, y_2) &= \mu^+_{A_2}(x_2) \wedge \mu^+_{A_2}(y_2) \wedge \mu^+_{B_1}(x_1 y_1)\\ &\le \mu^+_{A_2}(x_2) \wedge \mu^+_{A_2}(y_2) \wedge \mu^+_{A_1}(x_1) \wedge \mu^+_{A_1}(y_1)\\ &= (\mu^+_{A_1}(x_1) \wedge \mu^+_{A_2}(x_2)) \wedge (\mu^+_{A_1}(y_1) \wedge \mu^+_{A_2}(y_2))\\ &= (\mu^+_{A_1} \circ \mu^+_{A_2})(x_1, x_2) \wedge (\mu^+_{A_1} \circ \mu^+_{A_2})(y_1, y_2).\end{aligned}$$

■

**Definition 7.2.9** The **union** $G_1 \cup G_2 = (A_1 \cup A_2, B_1 \cup B_2)$ of two interval-valued fuzzy graphs $G_1$ and $G_2$ of the graphs $G_1^*$ and $G_2^*$ is defined as follows:

$$\begin{aligned}(\mu^-_{A_1} \cup \mu^-_{A_2})(x) &= \mu^-_{A_1}(x), \text{ if } x \in V_1 \text{ and } x \notin V_2,\\ (\mu^-_{A_1} \cup \mu^-_{A_2})(x) &= \mu^-_{A_2}(x), \text{ if } x \in V_2 \text{ and } x \notin V_1,\\ (\mu^-_{A_1} \cup \mu^-_{A_2})(x) &= \mu^-_{A_1}(x) \vee \mu^-_{A_2}(x), \text{ if } x \in V_1 \cap V_2,\end{aligned}$$

$$\begin{aligned}(\mu^+_{A_1} \cup \mu^+_{A_2})(x) &= \mu^+_{A_1}(x), \text{ if } x \in V_1 \text{ and } x \notin V_2,\\ (\mu^+_{A_1} \cup \mu^+_{A_2})(x) &= \mu^+_{A_1}(x), \text{ if } x \in V_2 \text{ and } x \notin V_1,\\ (\mu^+_{A_1} \cup \mu^+_{A_2})(x) &= \mu^+_{A_1}(x) \vee \mu^+_{A_2}(x), \text{ if } x \in V_1 \cap V_2,\end{aligned}$$

$$\begin{aligned}(\mu^-_{B_1} \cup \mu^-_{B_2})(xy) &= \mu^-_{B_1}(xy), \text{ if } xy \in E_1 \text{ and } xy \notin E_2,\\ (\mu^-_{B_1} \cup \mu^-_{B_2})(xy) &= \mu^-_{B_2}(xy), \text{ if } xy \in E_2 \text{ and } xy \notin E_1,\\ (\mu^-_{B_1} \cup \mu^-_{B_2})(xy) &= \mu^-_{B_1}(xy) \vee \mu^-_{B_2}(xy)\ , \text{ if } xy \in E_1 \cap E_2,\end{aligned}$$

$$(\mu^+_{B_1} \cup \mu^+_{B_2})(xy) = \mu^+_{B_1}(xy), \text{ if } xy \in E_1 \text{ and } xy \notin E_2,$$
$$(\mu^+_{B_1} \cup \mu^+_{B_2})(xy) = \mu^+_{B_2}(xy), \text{ if } xy \in E_2 \text{ and } xy \notin E_1,$$
$$(\mu^+_{B_1} \cup \mu^+_{B_2})(xy) = \mu^+_{B_1}(xy) \vee \mu^+_{B_2}(xy)\,, \text{ if } xy \in E_1 \cap E_2.$$

*Example 7.2.10* Let $G_1^* = (V_1, E_1)$ and $G_2^* = (V_2, E_2)$ be graphs such that $V_1 = \{a, b, c, d, e\}$, $E_1 = \{ab, bc, be, ce, ad, ed\}$, $V_2 = \{a, b, c, d, f\}$, $E_2 = \{ab, bc, cf, bf, bd\}$. Consider two interval-valued fuzzy graphs $G_1 = (A_1, B_1)$ and $G_2 = (A_2, B_2)$, where

$$A_1 = \left\langle \left(\frac{a}{0.2}, \frac{b}{0.4}, \frac{c}{0.3}, \frac{d}{0.3}, \frac{e}{0.2}\right), \left(\frac{a}{0.4}, \frac{b}{0.5}, \frac{c}{0.6}, \frac{d}{0.7}, \frac{e}{0.6}\right)\right\rangle,$$
$$B_1 = \left\langle \left(\frac{ab}{0.1}, \frac{bc}{0.2}, \frac{ce}{0.1}, \frac{be}{0.2}, \frac{ad}{0.1}, \frac{de}{0.1}\right), \left(\frac{ab}{0.3}, \frac{bc}{0.4}, \frac{ce}{0.5}, \frac{be}{0.5}, \frac{ad}{0.3}, \frac{de}{0.6}\right)\right\rangle,$$
$$A_2 = \left\langle \left(\frac{a}{0.2}, \frac{b}{0.2}, \frac{c}{0.3}, \frac{d}{0.2}, \frac{f}{0.4}\right), \left(\frac{a}{0.4}, \frac{b}{0.5}, \frac{c}{0.6}, \frac{d}{0.6}, \frac{f}{0.6}\right)\right\rangle,$$
$$B_2 = \left\langle \left(\frac{ab}{0.1}, \frac{bc}{0.2}, \frac{cf}{0.1}, \frac{bf}{0.1}, \frac{bd}{0.2}\right), \left(\frac{ab}{0.2}, \frac{bc}{0.4}, \frac{cf}{0.5}, \frac{bf}{0.2}, \frac{bd}{0.5}\right)\right\rangle.$$

Then according to the above definition:

$$(\mu^-_{A_1} \cup \mu^-_{A_2})(a) = 0.2, \quad (\mu^+_{A_1} \cup \mu^+_{A_2})(a) = 0.4,$$
$$(\mu^-_{A_1} \cup \mu^-_{A_2})(b) = 0.4, \quad (\mu^+_{A_1} \cup \mu^+_{A_2})(b) = 0.5,$$
$$(\mu^-_{A_1} \cup \mu^-_{A_2})(c) = 0.3, \quad (\mu^+_{A_1} \cup \mu^+_{A_2})(c) = 0.6,$$
$$(\mu^-_{A_1} \cup \mu^-_{A_2})(d) = 0.3, \quad (\mu^+_{A_1} \cup \mu^+_{A_2})(d) = 0.7,$$
$$(\mu^-_{A_1} \cup \mu^-_{A_2})(e) = 0.2, \quad (\mu^+_{A_1} \cup \mu^+_{A_2})(e) = 0.6,$$
$$(\mu^-_{A_1} \cup \mu^-_{A_2})(f) = 0.4, \quad (\mu^+_{A_1} \cup \mu^+_{A_2})(f) = 0.6,$$

$$(\mu^-_{B_1} \cup \mu^-_{B_2})(ab) = 0.1, \quad (\mu^+_{B_1} \cup \mu^+_{B_2})(ab) = 0.3,$$
$$(\mu^-_{B_1} \cup \mu^-_{B_2})(bc) = 0.2, \quad (\mu^+_{B_1} \cup \mu^+_{B_2})(bc) = 0.4,$$
$$(\mu^-_{B_1} \cup \mu^-_{B_2})(ce) = 0.1, \quad (\mu^+_{B_1} \cup \mu^+_{B_2})(ce) = 0.5,$$
$$(\mu^-_{B_1} \cup \mu^-_{B_2})(be) = 0.2, \quad (\mu^+_{B_1} \cup \mu^+_{B_2})(be) = 0.5,$$
$$(\mu^-_{B_1} \cup \mu^-_{B_2})(ad) = 0.1, \quad (\mu^+_{B_1} \cup \mu^+_{B_2})(ad) = 0.3,$$
$$(\mu^-_{B_1} \cup \mu^-_{B_2})(de) = 0.1, \quad (\mu^+_{B_1} \cup \mu^+_{B_2})(de) = 0.6,$$
$$(\mu^-_{B_1} \cup \mu^-_{B_2})(bd) = 0.2, \quad (\mu^+_{B_1} \cup \mu^+_{B_2})(bd) = 0.5,$$
$$(\mu^-_{B_1} \cup \mu^-_{B_2})(bf) = 0.1, \quad (\mu^+_{B_1} \cup \mu^+_{B_2})(bf) = 0.2.$$

Clearly, $G_1 \cup G_2 = (A_1 \cup A_2, B_1 \cup B_2)$ is interval-valued fuzzy graph of the graph $G_1^* \cup G_2^*$.

**Proposition 7.2.11** *The union of two interval-valued fuzzy graphs is an interval-valued fuzzy graph.*

*Proof* Let $G_1 = (A_1, B_1)$ and $G_2 = (A_2, B_2)$ be interval-valued fuzzy graphs of $G_1^*$ and $G_2^*$, respectively. We prove that $G_1 \cup G_2 = (A_1 \cup A_2, B_1 \cup B_2)$ is interval-valued fuzzy graphs of the graphs $G_1^* \cup G_2^*$. Because all conditions of $A_1 \cup A_2$ are automatically satisfied, we verify only the conditions of $B_1 \cup B_2$.

Suppose that $xy \in E_1 \cap E_2$. Then

$$\begin{aligned}(\mu_{B_1}^- \cup \mu_{B_2}^-)(xy) &= \mu_{B_1}^-(xy) \vee \mu_{B_2}^-(xy)\\ &\le (\mu_{A_1}^-(x) \wedge \mu_{A_1}^-(y)) \vee (\mu_{A_2}^-(x) \wedge \mu_{A_2}^-(y))\\ &= (\mu_{A_1}^-(x) \vee \mu_{A_2}^-(x)) \wedge (\mu_{A_1}^-(y) \vee \mu_{A_2}^-(y))\\ &= (\mu_{A_1}^- \cup \mu_{A_2}^-)(x) \wedge (\mu_{A_1}^- \cup \mu_{A_2}^-)(y),\end{aligned}$$

$$\begin{aligned}(\mu_{B_1}^+ \cup \mu_{B_2}^+)(xy) &= \mu_{B_1}^+(xy) \vee \mu_{B_2}^+(xy)\\ &\le (\mu_{A_1}^+(x) \wedge \mu_{A_1}^+(y)) \wedge (\mu_{A_2}^+(x) \wedge \mu_{A_2}^+(y))\\ &= (\mu_{A_1}^+(x) \vee \mu_{A_2}^+(x) \wedge (\mu_{A_1}^+(y) \vee \mu_{A_2}^+(y))\\ &= (\mu_{A_1}^+ \cup \mu_{A_2}^+)(x) \wedge (\mu_{A_1}^+ \cup \mu_{A_2}^+)(y).\end{aligned}$$

If $xy \in E_1$ and $xy \notin E_2$, then

$$\begin{aligned}(\mu_{B_1}^- \cup \mu_{B_2}^-)(xy) &= \mu_{B_1}^-(xy) \le \mu_{A_1}^-(x) \wedge \mu_{A_1}^-(y)\\ &\le (\mu_{A_1}^- \cup \mu_{A_2}^-)(x) \wedge (\mu_{A_1}^- \cup \mu_{A_2}^-)(y)\\ &\text{and similarly}\\ (\mu_{B_1}^+ \cup \mu_{B_2}^+)(xy) &\le (\mu_{A_1}^+ \cup \mu_{A_2}^+)(x) \wedge (\mu_{A_1}^+ \cup \mu_{A_2}^+)(y).\end{aligned}$$

If $xy \in E_2$ and $xy \notin E_1$, then

$$\begin{aligned}(\mu_{B_1}^- \cup \mu_{B_2}^-)(xy) &= \mu_{B_2}^-(xy) \le \mu_{A_2}^-(x) \wedge \mu_{A_2}^-(y)\\ &\le (\mu_{A_1}^- \cup \mu_{A_2}^-)(x) \wedge (\mu_{A_1}^- \cup \mu_{A_2}^-)(y)\\ &\text{and similarly}\\ (\mu_{B_1}^+ \cup \mu_{B_2}^+)(xy) &\le (\mu_{A_1}^+ \cup \mu_{A_2}^+)(x) \wedge (\mu_{A_1}^+ \cup \mu_{A_2}^+)(y).\end{aligned}$$

■

**Definition 7.2.12** The **join** $G_1 + G_2 = (A_1 + A_2, B_1 + B_1)$ of two interval-valued fuzzy graphs $G_1$ and $G_2$ of the graphs $G_1^*$ and $G_2^*$ is defined as follows:

$$\left.\begin{array}{l}(\mu^-_{A_1}+\mu^-_{A_2})(x)=(\mu^-_{A_1}\cup\mu^-_{A_2})(x)\\(\mu^+_{A_1}+\mu^+_{A_2})(x)=(\mu^+_{A_1}\cup\mu^+_{A_2})(x)\end{array}\right. \text{ if } x\in V_1\cup V_2,$$

$$\left.\begin{array}{l}(\mu^-_{B_1}+\mu^-_{B_2})(xy)=(\mu^-_{B_1}\cup\mu^-_{B_2})(xy)\\(\mu^+_{B_1}+\mu^+_{B_2})(xy)=(\mu^+_{B_1}\cup\mu^+_{B_2})(xy)\end{array}\right. \text{ if } xy\in E_1\cap E_2,$$

$$\left.\begin{array}{l}(\mu^-_{B_1}+\mu^-_{B_2})(xy)=\mu^-_{A_1}(x)\wedge\mu^-_{A_2}(y)\\(\mu^+_{B_1}+\mu^+_{B_2})(xy)=\mu^+_{A_1}(x)\wedge\mu^+_{A_2}(y)\end{array}\right. \text{ if } xy\in E',$$

where $E'$ is the set of all edges joining the nodes of $V_1$ and $V_2$.

**Proposition 7.2.13** *The join of two interval-valued fuzzy graphs is an interval-valued fuzzy graphs.*

*Proof* Let $G_1 = (A_1, B_1)$ and $G_2 = (A_2, B_2)$ be interval-valued fuzzy graphs of $G_1^*$ and $G_2^*$, respectively. We prove that $G_1 + G_2 = (A_1 + A_2, B_1 + B_2)$ is an interval-valued fuzzy graph of the graph $G_1^* + G_2^*$. In view of Proposition 7.2.11, it is sufficient to verify the case when $xy \in E'$. In this case, we have

$$\begin{aligned}(\mu^-_{B_1}+\mu^-_{B_2})(xy)&=\mu^-_{A_1}(x)\wedge\mu^-_{A_2}(y)\\&\leq(\mu^-_{A_1}\cup\mu^-_{A_2}(x))\wedge(\mu^-_{A_1}\cup\mu^-_{A_2})(y)\\&=(\mu^-_{A_1}+\mu^-_{A_2})(x)\wedge(\mu^-_{A_1}+\mu^-_{A_2})(y),\end{aligned}$$

$$\begin{aligned}(\mu^+_{B_1}+\mu^+_{B_2})(xy)&=\mu^+_{A_1}(x)\wedge\mu^+_{A_2}(y)\\&\leq(\mu^+_{A_1}\cup\mu^+_{A_2})(x)\wedge(\mu^+_{A_1}\cup\mu^+_{A_2})(y)\\&=(\mu^+_{A_1}+\mu^+_{A_2})(x)\wedge(\mu^+_{A_1}+\mu^+_{A_2})(y).\end{aligned}$$

■

**Proposition 7.2.14** *Let $G_1^* = (V_1, E_1)$ and $G_2^* = (V_2, E_2)$ be crisp graphs such that $V_1 \cap V_2 = \emptyset$. Let $A_1, A_2, B_1, B_2$ be interval-valued fuzzy subsets of $V_1, V_2, E_1, E_2$, respectively. Then $G_1 \cup G_2 = (A_1 \cup A_2, B_1 \cup B_2)$ is an interval-valued fuzzy graph of $G_1^* \cup G_2^*$ if and only if $G_1 = (A_1, B_1)$ and $G_2 = (A_2, B_2)$ are interval-valued fuzzy graphs of $G_1^*$ and $G_2^*$, respectively.*

*Proof* Suppose that $G_1 \cup G_2 = (A_1 \cup A_2, B_1 \cup B_2)$ is an interval-valued fuzzy graph of $G_1^* \cup G_2^*$. Let $xy \in E_1$. Then $xy \notin E_2$ and $x, y \in V_1 \setminus V_2$. Thus,

$$\begin{aligned}\mu^-_{B_1}(xy)&=(\mu^-_{B_1}\cup\mu^-_{B_2})(xy)\\&\leq(\mu^-_{A_1}\cup\mu^-_{A_2}(x))\wedge(\mu^-_{A_1}\cup\mu^-_{A_2})(y)\\&=\mu^-_{A_1}(x)\wedge\mu^-_{A_1}(y),\end{aligned}$$

$$\begin{aligned}\mu^+_{B_1}(xy) &= (\mu^+_{B_1} \cup \mu^+_{B_2})(xy)\\ &\le (\mu^+_{A_1} \cup \mu^+_{A_2})(x) \wedge (\mu^+_{A_1} \cup \mu^+_{A_2})(y)\\ &= \mu^+_{A_1}(x) \wedge \mu^+_{A_1}(y).\end{aligned}$$

This shows that $G_1 = (A_1, B_1)$ is an interval-valued fuzzy graph. Similarly, we can show that $G_2 = (A_2, B_2)$ is an interval-valued fuzzy graph.

The converse statement is given by Proposition 7.2.11. ■

As a consequence of Propositions 7.2.13 and 7.2.14, we obtain the following proposition.

**Proposition 7.2.15** *Let $G_1^* = (V_1, E_1)$ and $G_2^* = (V_2, E_2)$ be crisp graphs and let $V_1 \cap V_2 = \emptyset$. Let $A_1, A_2, B_1, B_2$ be interval-valued fuzzy subsets of $V_1, V_2, E_1, E_2$, respectively. Then $G_1 + G_2 = (A_1 + A_2, B_1 + B_2)$ is an interval-valued fuzzy graph of $G_1^* \cup G_2^*$ if and only if $G_1 = (A_1, B_1)$ and $G_2 = (A_2, B_2)$ are interval-valued fuzzy graphs of $G_1^*$ and $G_2^*$, respectively.*

## 7.3 Isomorphisms of Interval-Valued Fuzzy Graphs

In this section, we consider various types of (weak) isomorphisms of interval-valued fuzzy graphs.

**Definition 7.3.1** Let $G_1 = (A_1, B_1)$ and $G_2 = (A_2, B_2)$ be two interval-valued fuzzy graphs. A **homomorphism** $f : G_1 \to G_2$ is a mapping $f : V_1 \to V_2$ such that for all $x_1 \in V_1$, $x_1y_1 \in E_1$,

$(i)$ $\mu^-_{A_1}(x_1) \le \mu^-_{A_2}(f(x_1))$, $\mu^+_{A_1}(x_1) \le \mu^+_{A_2}(f(x_1))$,

$(ii)$ $\mu^-_{B_1}(x_1y_1) \le \mu^-_{B_2}(f(x_1)f(y_1))$, $\mu^+_{B_1}(x_1y_1) \le \mu^+_{B_2}(f(x_1)f(y_1))$.

A bijective homomorphism with the property

$(iii)$ $\mu^-_{A_1}(x_1) = \mu^-_{A_2}(f(x_1))$, $\mu^+_{A_1}(x_1) = \mu^+_{A_2}(f(x_1))$

is called a **weak isomorphism** and a **weak co-isomorphism** if

$(iv)$ $\mu^-_{B_1}(x_1y_1) = \mu^-_{B_2}(f(x_1)f(y_1))$, $\mu^+_{B_1}(x_1y_1) = \mu^+_{B_2}(f(x_1)f(y_1))$ for all $x_1, y_1 \in V_1$.

A bijective mapping $f : G_1 \to G_2$ satisfying $(iii)$ and $(iv)$ is called an **isomorphism**.

*Example 7.3.2* Let $G_1^* = (V_1, E_1)$ and $G_2^* = (V_2, E_2)$ be graphs such that $V_1 = \{a_1, b_1\}$, $V_2 = \{a_2, b_2\}$, $E_1 = \{a_1b_1\}$, $E_2 = \{a_2b_2\}$. Let $A_1$, $A_2$, $B_1$, and $B_2$ be interval-valued fuzzy subsets defined by

$$A_1 = \left\langle \left(\frac{a_1}{0.2}, \frac{b_1}{0.3}\right), \left(\frac{a_1}{0.5}, \frac{b_1}{0.6}\right) \right\rangle, \quad B_1 = \left\langle \frac{a_1b_1}{0.1}, \frac{a_1b_1}{0.3} \right\rangle,$$

$$A_2 = \left\langle \left(\frac{a_2}{0.3}, \frac{b_2}{0.2}\right), \left(\frac{a_2}{0.6}, \frac{b_2}{0.5}\right) \right\rangle, \quad B_2 = \left\langle \frac{a_2b_2}{0.1}, \frac{a_2b_2}{0.4} \right\rangle.$$

Then it follows that $G_1 = (A_1, B_1)$ and $G_2 = (A_2, B_2)$ are interval-valued fuzzy graphs of $G_1^*$ and $G_2^*$, respectively. The map $f : V_1 \to V_2$ defined by $f(a_1) = b_2$ and $f(b_1) = a_2$ is a weak isomorphism, but it is not an isomorphism.

*Example 7.3.3* Let $G_1^*$ and $G_2^*$ be as in Example 7.3.2 and let $A_1$, $A_2$, $B_1$, and $B_2$ be interval-valued fuzzy subsets defined by

$$A_1 = \left\langle \left(\frac{a_1}{0.2}\right) \frac{b_1}{0.3}), \left(\frac{a_1}{0.4}, \frac{b_1}{0.5}\right) \right\rangle, \quad B_1 = \left\langle \frac{a_1 b_1}{0.1}, \frac{a_1 b_1}{0.3} \right\rangle,$$
$$A_2 = \left\langle \left(\frac{a_2}{0.4}, \frac{b_2}{0.3}\right), \left(\frac{a_2}{0.5}, \frac{b_2}{0.6}\right) \right\rangle, \quad B_2 = \left\langle \frac{a_2 b_2}{0.1}, \frac{a_2 b_2}{0.3} \right\rangle.$$

Then $G_1 = (A_1, B_1)$ and $G_2 = (A_2, B_2)$ are interval-valued fuzzy graphs of $G_1^*$ and $G_2^*$, respectively. The map $f : V_1 \to V_2$ defined by $f(a_1) = b_2$ and $f(b_1) = a_2$ is a weak co-isomorphism, but it is not an isomorphism.

## 7.4 Strong Interval-Valued Fuzzy Graphs

In this section, we consider strong interval-valued fuzzy graphs.

**Definition 7.4.1** An interval-valued fuzzy graph $G = (A, B)$ is called **strong** if

$$\mu_{B_1}^-(xy) = \mu_{A_1}^-(x) \wedge \mu_{A_1}^-(y), \; \mu_{B_1}^+(xy) = \mu_{A_1}^+(x) \wedge \mu_{A_1}^+(y)$$

for all $xy \in E$.

*Example 7.4.2* Consider a graph $G^* = (V, E)$ such that $V = \{x, y, z\}$ and $E = \{xy, yz, zx\}$. If $A$ and $B$ are interval-valued fuzzy subsets defined by

$$A = \left\langle \left(\frac{x}{0.2}, \frac{y}{0.3}, \frac{z}{0.4}\right), \left(\frac{x}{0.4}, \frac{y}{0.5}, \frac{z}{0.5}\right) \right\rangle,$$
$$B = \left\langle \left(\frac{xy}{0.2}, \frac{yz}{0.3}, \frac{zx}{0.2}\right), \left(\frac{xy}{0.4}, \frac{yz}{0.5}, \frac{zx}{0.4}\right) \right\rangle,$$

then $G = (A, B)$ is a interval-valued complete fuzzy graph of $G^*$.

**Definition 7.4.3** The **complement** of an interval-valued fuzzy graph $G = (A, B)$ of $G^* = (V, E)$ is an interval-valued fuzzy graph $\overline{G} = (\overline{A}, \overline{B})$ on $\overline{G^*} = (V, \overline{E})$, where $\overline{A} = A = [\mu_A^-, \mu_A^+]$ and $\overline{B} = \left[\overline{\mu^-}_B, \overline{\mu^+}_B\right]$ is defined by

$$\overline{\mu^-}_B(xy) = \begin{cases} 0 & \text{if } \mu_B^-(xy) > 0, \\ \mu_A^-(x) \wedge \mu_A^-(y) & \text{if } \mu_B^-(xy) = 0. \end{cases}$$

$$\overline{\mu^+}_B(xy) = \begin{cases} 0 & \text{if } \mu_B^+(xy) > 0, \\ \mu_A^+(x) \wedge \mu_A^+(y) & \text{if } \mu_B^+(xy) = 0. \end{cases}$$

**Definition 7.4.4** An interval-valued fuzzy graph $G = (A, B)$ is called **self-complementary** if $G \simeq \overline{G}$.

*Example 7.4.5* Consider a graph $G^* = (V, E)$ such that $V = \{a, b, c, d\}$ and $E = \{ab, bc, cd\}$. Let $G = (A, B)$ be the interval-valued fuzzy graph such that

$$A = \left\langle \left( \frac{a}{0.2}, \frac{b}{0.2}, \frac{c}{0.1}, \frac{d}{0.1} \right), \left( \frac{a}{0.4}, \frac{b}{0.4}, \frac{c}{0.3}, \frac{d}{0.3} \right) \right\rangle,$$

$$B = \left\langle \left( \frac{ab}{0.1}, \frac{bc}{0.1}, \frac{cd}{0.1} \right), \left( \frac{ab}{0.3}, \frac{bc}{0.3}, \frac{cd}{0.3} \right) \right\rangle.$$

Then $\overline{G} = (\overline{A}, \overline{B})$ is as follows:

$$\overline{A} = \left\langle \left( \frac{a}{0.2}, \frac{b}{0.2}, \frac{c}{0.1}, \frac{d}{0.1} \right), \left( \frac{a}{0.4}, \frac{b}{0.4}, \frac{c}{0.3}, \frac{d}{0.3} \right) \right\rangle,$$

$$\overline{B} = \left\langle \left( \frac{ad}{0.1}, \frac{ac}{0.1}, \frac{bd}{0.1} \right), \left( \frac{ad}{0.3}, \frac{ac}{0.3}, \frac{bd}{0.3} \right) \right\rangle.$$

It follows that the function $f : V \to V$ defined by $f(a) = c$, $f(b) = a$, $f(c) = d$, $f(d) = b$ yields an isomorphism of $G$ onto $\overline{G}$.

**Proposition 7.4.6** *Suppose $G = (A, B)$ is a strong self-complementary interval-valued fuzzy graph. Then the following properties hold.*

(*i*) $\sum_{x \neq y} \mu_B^-(xy) = \sum_{x \neq y} \mu_A^-(x) \wedge \mu_A^-(y)$,
(*ii*) $\sum_{x \neq y} \mu_B^+(xy) = \sum_{x \neq y} \mu_A^+(x) \wedge \mu_A^+(y)$.

*Proof* Let $G = (A, B)$ be a strong self-complementary interval-valued fuzzy graph. Then there exists an isomorphism $f : V \to V$ such that $\mu_A^-(f(x)) = \mu_A^-(x)$, $\mu_A^+(f(x)) = \mu_A^+(x)$, $\overline{\mu_B^-}(f(x)f(y)) = \mu_B^-(xy)$ and $\overline{\mu_B^+}(f(x)f(y)) = \mu_B^+(xy)$ for all $x, y \in V$. Thus, for all $x, y \in V$, we have that

$$\mu_B^-(xy) = \overline{\mu_B^-}(f(x)f(y)) = \mu_A^-(f(x)) \wedge \mu_A^-(f(y)) = \mu_A^-(x) \wedge \mu_A^-(y).$$

Thus, (*i*) holds. The proof of (*ii*) follows in a similar manner. ■

**Proposition 7.4.7** *Let $G_1 = (A_1, B_1)$ and $G_2 = (A_2, B_2)$ be interval-valued fuzzy strong graphs. Then $G_1 \simeq G_2$ if and only if $\overline{G_1} \simeq \overline{G_2}$.*

*Proof* Assume that $G_1$ and $G_2$ are isomorphic. Then there exists a bijective map $f : V_1 \to V_2$ satisfying

$$\mu^-_{A_1}(x) = \mu^-_{A_2}(f(x)),\ \mu^+_{A_1}(x) = \mu^+_{A_2}(f(x)) \text{ for all } x \in V_1,$$

$$\mu^-_{B_1}(xy) = \mu^-_{B_2}(f(x)f(y)),\ \mu^+_{B}(xy) = \mu^+_{B_2}(f(x)f(y)) \text{ for all } xy \in E_1.$$

By the definition of complement, we have

$$\overline{\mu^-}_{B_1}(xy) = \mu^-_{A_1}(x) \wedge \mu^-_{A_1}(y) = \mu^-_{A_2}(f(x)) \wedge \mu^-_{A_2}(f(y)) = \overline{\mu^-}_{B_2}(f(x)f(y)),$$

$$\overline{\mu^+}_{B_1}(xy) = \mu^+_{A_1}(x) \wedge \mu^+_{A_1}(y) = \mu^+_{A_2}(f(x)) \wedge \mu^+_{A_2}(f(y)) = \overline{\mu^+}_{B_2}(f(x)f(y))$$

for all $xy \notin E_1$. Hence, $\overline{G_1} \simeq \overline{G_2}$.

The proof of the converse follows immediately because $\overline{G_1}$ and $\overline{G_2}$ are strong. ■

## 7.5 Interval-Valued Fuzzy Line Graphs

In this section, we present the work of [9]. The work is based on [125]. We prove a necessary and sufficient condition for an interval-valued fuzzy graph to be isomorphic to its corresponding interval-valued fuzzy line graph. We determine when an isomorphism between two interval-valued fuzzy graphs follows from an isomorphism of their corresponding fuzzy line graphs.

Recall isomorphic graphs are denoted by $G_1^* \simeq G_2^*$. In graph theory, the line graph $L(G^*)$ of a simple graph is a graph that represents the adjacencies between edges of $G^*$. Given a graph $G^*$, its line graph $L(G^*)$ is a graph such that each vertex of $L(G^*)$ represents and edge of $G^*$; and two vertices of $L(G^*)$ are adjacent if and only if there corresponding edges share a common endpoint in $G^*$.

Let $G^* = (V, E)$ be an undirected graph, where $V = \{v_1, v_2, \ldots, v_n\}$. Let $S_i = \{v_i, x_{i1}, x_{i2}, \ldots, x_{iq_i}\}$, where $x_{ij} \in E$ has vertex $v_i$, $i = 1, 2, \ldots, n$, $j = 1, 2, \ldots, q_i$. Let $S = \{S_1, S_2, \ldots, S_n\}$. Let $T = \{S_iS_j \mid S_i, S_j \in S, S_i \cap S_j \neq \emptyset, i \neq j\}$. Then $P(S) = (S, T)$ is an intersection graph and $P(S) \simeq G^*$. The line graph $L(G^*)$ of a simple graph $G^*$ is by definition the intersection graph $P(E)$. That is, $L(G^*) = (Z, W)$, where $Z = \{\{x\} \cup \{u_x, v_x\} \mid x \in E,\ u_x, v_x \in V,\ x = u_xv_x\}$ and $W = \{S_xS_y \mid S_x \cap S_y \neq \emptyset, x, y \in E, x \neq y\}$, and $S_x = \{\{x\} \cup \{u_x, v_x\}\}$, $x \in E$.

Throughout this section, $G^*$ is a crisp simple graph, and $G$ is an interval-valued fuzzy graph.

**Definition 7.5.1** Let $P(S) = (S, T)$ be an intersection graph of a simple graph $G^* = (V, E)$. Let $G = (A_1, B_1)$ be an interval-valued fuzzy graph of $G^*$. We define an **interval-valued fuzzy intersection graph** $P(G) = (A_2, B_2)$ of $P(S)$ as follows:

$(i)$ $A_2$, $B_2$ are interval-valued fuzzy subsets of $S$ and $T$, respectively,

$(ii)$ $\mu^-_{A_2}(S_i) = \mu^-_{B_1}(v_i)$, $\mu^+_{A_2}(S_i) = \mu^+_{B_1}(v_i)$,

$(iii)$ $\mu^-_{B_2}(S_iS_j) = \mu^-_{B_1}(v_iv_j)$, $\mu^+_{B_2}(S_iS_j) = \mu^+_{B_1}(v_iv_j)$

for all $S_i, S_j \in S$, $S_iS_j \in T$. That is, any interval-valued fuzzy graph of $P(S)$ is called an interval-valued fuzzy intersection graph.

The following proposition is clear.

**Proposition 7.5.2** *Let $G = (A_1, B_1)$ be an interval-valued fuzzy graph of $G^*$. Then $P(G) = (A_2, B_2)$ is an interval-valued fuzzy graph of $P(S)$ and $G \simeq P(G)$.*

This proposition shows that any interval-valued fuzzy graph is isomorphic to an interval-valued fuzzy intersection graph.

**Definition 7.5.3** Let $L(G^*) = (Z, W)$ be a line graph of a simple graph $G^* = (V, E)$. Let $G = (A_1, B_1)$ be an interval-valued fuzzy graph of $G^*$. We define an **interval-valued fuzzy line graph** $L(G) = (A_2, B_2)$ of $G$ as follows:

$(i)$ $A_2$ and $B_2$ are interval-valued fuzzy subsets of $Z$ and $W$, respectively,
$(ii)$ $\mu^-_{A_2}(S_x) = \mu^-_{B_1}(x) = \mu^-_{B_1}(u_xv_x)$,
$(iii)$ $\mu^+_{A_2}(S_x) = \mu^+_{B_1}(x) = \mu^+_{B_1}(u_xv_x)$,
$(iv)$ $\mu^-_{B_2}(S_xS_y) = \mu^-_{B_1}(x) \wedge \mu^-_{B_1}(y))$,
$(v)$ $\mu^+_{B_2}(S_xS_y) = \mu^+_{B_1}(x) \wedge \mu^+_{B_1}(y)$,
for all $S_x, S_y \in Z$, $S_xS_y \in W$.

*Example 7.5.4* Consider a graph $G^* = (V, E)$ such that $V = \{v_1, v_2, v_3, v_4\}$ and $E = \{x_1 = v_1v_2, x_2 = v_2v_3, x_3 = v_3v_4, x_4 = v_4v_1\}$. Let $A_1$ be an interval-valued fuzzy subset of $V$ and let $B_1$ be an interval-valued fuzzy subset of $E$ defined by

| | $v_1$ | $v_2$ | $v_3$ | $v_4$ |
|---|---|---|---|---|
| $\mu^-_{A_1}$ | 0.2 | 0.3 | 0.4 | 0.2 |
| $\mu^+_{A_1}$ | 0.5 | 0.4 | 0.5 | 0.3 |

| | $v_1v_2$ | $v_2v_3$ | $v_3v_4$ | $v_4v_1$ |
|---|---|---|---|---|
| $\mu^-_{B_1}$ | 0.1 | 0.2 | 0.1 | 0.1 |
| $\mu^+_{B_1}$ | 0.2 | 0.3 | 0.2 | 0.2 |

It is easily shown that $G$ is an interval-valued fuzzy graph. Consider a line graph $L(G^*) = (Z, W)$ such that

$$Z = \{S_{x_1}, S_{x_2}, S_{x_3}, S_{x_4}\}$$

and

$$W = \{S_{x_1}S_{x_2}, S_{x_2}S_{x_3}, S_{x_3}S_{x_4}, S_{x_4}S_{x_1}\}.$$

Let $A_2 = [\mu^-_{A_2}, \mu^+_{A_2}]$ and $B_2 = [\mu^-_{B_2}, \mu^+_{B_2}]$ be interval-valued fuzzy sets of $Z$ and $W$, respectively. Then it follows that

$$\begin{array}{ll}
\mu^-_{A_2}(S_{x_1}) = 0.1 & \mu^+_{A_2}(S_{x_1}) = 0.2 \\
\mu^-_{A_2}(S_{x_2}) = 0.2 & \mu^+_{A_2}(S_{x_2}) = 0.3 \\
\mu^-_{A_2}(S_{x_3}) = 0.1 & \mu^+_{A_2}(S_{x_3}) = 0.2 \\
\mu^-_{A_2}(S_{x_4}) = 0.1 & \mu^+_{A_2}(S_{x_4}) = 0.2 \\
\mu^-_{B_2}(S_{x_1}S_{x_2}) = 0.1 & \mu^+_{B_2}(S_{x_1}S_{x_2}) = 0.2 \\
\mu^-_{B_2}(S_{x_2}S_{x_3}) = 0.1 & \mu^+_{B_2}(S_{x_2}S_{x_3}) = 0.2 \\
\mu^-_{B_2}(S_{x_3}S_{x_4}) = 0.1 & \mu^+_{B_2}(S_{x_3}S_{x_4}) = 0.2 \\
\mu^-_{B_2}(S_{x_4}S_{x_1}) = 0.1 & \mu^+_{B_2}(S_{x_4}S_{x_1}) = 0.2.
\end{array}$$

Clearly, $L(G)$ is an interval-valued fuzzy line graph.

The following proposition is immediate.

**Proposition 7.5.5** *$L(G)$ is an interval-valued fuzzy line graph corresponding to an interval-valued fuzzy graph.*

**Proposition 7.5.6** *If $L(G)$ is an interval-valued fuzzy line graph of an interval-valued fuzzy graph, then $L(G^*)$ is the line graph of $G^*$.*

*Proof* Because $G = (A_1, B_1)$ is an interval-valued fuzzy graph of $G^*$ and $L(G) = (A_2, B_2)$ is an interval-valued fuzzy line graph of $L(G^*)$.

$$\mu^-_{A_2}(S_x) = \mu^-_{B_1}(x),\ \mu^+_{A_2}(S_x) = \mu^+_{B_1}(x) \text{ for all } x \in E,$$

Thus, $S_x \in Z \Leftrightarrow x \in E$. Also,

$$\mu^-_{B_2}(S_xS_y) = \mu^-_{B_1}(x) \wedge \mu^-_{B_1}(y)$$
$$\mu^+_{B_2}(S_xS_y) = \mu^+_{B_1}(x) \wedge \mu^+_{B_1}(y)$$

for all $S_xS_y \in W$. Hence,

$$W = \{S_xS_y \mid S_x \cap S_y \neq \emptyset,\ x, y \in E, x \neq y\}.$$

■

**Proposition 7.5.7** *Let $L(G) = (A_2, B_2)$ is an interval-valued fuzzy line graph of $L(G^*)$. Then $L(G)$ is an interval-valued fuzzy line graph of some interval-valued fuzzy line graph $G$ if and only if*

$$\mu^-_{B_2}(S_xS_y) = \mu^-_{A_2}(S_x) \wedge \mu^-_{A_2}(S_y) \textit{ for all } S_xS_y \in W,$$

$$\mu^+_{B_2}(S_xS_y) = \mu^+_{A_2}(S_x) \wedge \mu^+_{A_2}(S_y) \textit{ for all } S_xS_y \in W.$$

*Proof* Suppose that $\mu^-_{B_2}(S_xS_y) = \mu^-_{A_2}(S_x) \wedge \mu^-_{A_2}(S_y)$ for all $S_xS_y \in W$. For all $x \in E$, define $\mu^-_{A_1}(x) = \mu^-_{A_2}(S_x)$. Then

$$\mu^-_{B_2}(S_xS_y) = \mu^-_{A_2}(S_x) \wedge \mu^-_{A_2}(S_y) = \mu^-_{A_1}(x) \wedge \mu^-_{A_1}(y),$$

$$\mu^+_{B_2}(S_xS_y) = \mu^+_{A_2}(S_x) \wedge \mu^+_{A_2}(S_y) = \mu^+_{A_1}(x) \wedge \mu^+_{A_1}(y).$$

An interval-valued fuzzy set $A_1 = [\mu^-_{A_1}, \mu^+_{A_1}]$ that yields the property

$$\mu^-_{B_1}(xy) \leq \mu^-_{A_1}(x) \wedge \mu^-_{A_1}(y),$$

$$\mu^+_{B_1}(xy) \leq \mu^+_{A_1}(x) \wedge \mu^+_{A_1}(y)$$

will suffice.

The converse follows easily. ■

**Proposition 7.5.8** *$L(G)$ is an interval-valued fuzzy line graph if and only if $L(G^*)$ is a line graph and*

$$\mu^-_{B_2}(uv) \leq \mu^-_{A_2}(u) \wedge \mu^-_{A_2}(v),$$

$$\mu^+_{B_2}(uv) \leq \mu^+_{A_2}(u) \wedge \mu^+_{A_2}(v)$$

*for all $uv \in W$.*

*Proof* The proof follows from Propositions 7.5.6 and 7.5.7. ■

**Proposition 7.5.9** *Let $G_1$ and $G_2$ be interval-valued fuzzy graphs. If $f$ is a weak isomorphism of $G_1$ onto $G_2$, then $f$ is an isomorphism of $G_1^*$ onto $G_2^*$.*

*Proof* $v \in V_1 \Leftrightarrow f(v) \in V_2$ and $uv \in E_1 \Leftrightarrow f(u)f(v) \in E_2$. ■

**Theorem 7.5.10** ([9]) *Let $L(G) = (A_2, B_2)$ be the interval-valued fuzzy line graph corresponding to interval-valued fuzzy graph $G = (A_1, B_1)$. Suppose that $G^* = (V, E)$ is connected. Then the following properties hold.*

*$(i)$ There exists a weak isomorphism of $G$ onto $L(G)$ if and only if $G^*$ is a cycle and for all $v \in V$, $x \in E$, $\mu^-_{A_1}(v) = \mu^-_{B_1}(x)$, $\mu^+_{A_1}(v) = \mu^+_{B_1}(x)$, i.e., $A_1 = \left[\mu^-_{A_1}, \mu^+_{A_1}\right]$ and $B_1 = \left[\mu^-_{B_1}, \mu^+_{B_1}\right]$ are constant functions on $V$ and $E$, respectively, taking on the same value.*

*$(ii)$ If $f$ is a weak isomorphism of $G$ onto $L(G)$, then $f$ is an isomorphism.*

*Proof* Assume that $f$ is a weak isomorphism of $L(G)$ onto $G$. From Proposition 7.5.8, it follows that $G^* = (V, E)$ is a cycle.

Let $V = \{v_1, v_2, \ldots, v_n\}$ and $E = \{x_1 = v_1v_2, x_2 = v_2v_3, \ldots, x_n = v_nv_1\}$, where $v_1v_2v_3 \cdots v_n$ is a cycle. Define interval-valued fuzzy sets

$$\mu^-_{A_1}(x_i) = s_i, \ \mu^+_{A_1}(x_i) = s'_i$$

and

$$\mu^-_{B_1}(v_iv_{i+1}) = r_i, \ \mu^+_{B_1}(v_iv_{i+1}) = r'_i, \ i = 1, 2, \ldots, n, \ v_{n+1} = v_1.$$

Then $s_{n+1} = s_1, s'_{n+1} = s'_1$.

$$\begin{cases} r_i \leq s_i \wedge s_{i+1} \\ r'_i \leq s'_i \wedge s'_{i+1}, \; i = 1, 2, \ldots, n. \end{cases} \tag{7.1}$$

Now, $Z = \{S_{x_1}, S_{x_2}, \ldots, S_{x_n}\}$ and $W = \{S_{x_1}S_{x_2}, S_{x_2}S_{x_3}, \ldots, S_{x_n}S_{x_1}\}$. Also, for $r'_{n+1} = r'_1$,

$$\begin{aligned}
&\mu^-_{A_2}(S_{x_i}) = \mu^-_{A_1}(x_i) = \mu^-_{A_1}(v_i v_{i+1}) = r_i, \\
&\mu^+_{A_2}(S_{x_i}) = \mu^+_{A_1}(x_i) = \mu^+_{A_1}(v_i v_{i+1}) = r'_i, \\
&\mu^-_{B_2}(S_{x_i}S_{x_{i+1}}) = \mu^-_{B_1}(x_i) \wedge \mu^-_{B_1}(x_{i+1}) = \\
&\mu^-_{B_1}(v_i v_{i+1}) \wedge \mu^-_{B_1}(v_{i+1} v_{i+2}) = r_i \wedge r_{i+1}, \\
&\mu^+_{B_2}(S_{x_i}S_{x_{i+1}}) = \mu^+_{B_1}(x_i) \wedge \mu^+_{B_1}(x_{i+1}) = \\
&\mu^+_{B_1}(v_i v_{i+1}) \wedge \mu^+_{B_1}(v_{i+1} v_{i+2}) = r'_i \wedge r'_{i+1}
\end{aligned}$$

for $i = 1, 2, \ldots, n$, $v_{n+1} = v_1$, $v_{n+2} = v_2$. Because $f$ is an isomorphism of $G^*$ onto $L(G^*)$, $f$ maps $V$ one-to-one and onto $Z$. Also, $f$ preserves adjacency. Hence, $f$ induces a permutation $\pi$ of $\{1, 2, \ldots, n\}$ such that

$$f(v_i) = S_{x_{\pi(i)}} = S_{x_{\pi(i)}} S_{x_{\pi(i+1)}}$$

and

$$x_i = v_i v_{i+1} \rightarrow f(v_i) f(v_{i+1}) = S_{v_{\pi(i)}} S_{v_{\pi(i+1)}} S_{v_{\pi(i+2)}}, \; i = 1, 2, \ldots, n-1.$$

Now,

$$\begin{aligned}
s_i &= \mu^-_{A_1}(v_i) \leq \mu^-_{A_2}(f(v_i)) = \mu^-_{A_2}(S_{v_{\pi(i)}} S_{v_{\pi(i+1)}}) = r_{\pi(i)}, \\
s'_i &= \mu^+_{A_1}(v_i) \leq \mu^+_{A_2}(f(v_i)) = \mu^+_{A_2}(S_{v_{\pi(i)}} S_{v_{\pi(i+1)}}) = r'_{\pi(i)},
\end{aligned}$$

$$\begin{aligned}
r_i &= \mu^-_{B_1}(v_i v_{i+1}) \leq \mu^-_{B_2}(f(v_i) f(v_{i+1})) = \mu^-_{B_2}(S_{v_{\pi(i)}} S_{v_{\pi(i+1)}} S_{v_{\pi(i+)+1}}) \\
&= r_{\pi(i)} \wedge r_{\pi(i+1)}, \\
r'_i &= \mu^+_{B_1}(v_i v_{i+1}) \leq \mu^+_{B_2}(f(v_i) f(v_{i+1})) = \mu^+_{B_2}(S_{v_{\pi(i)}} S_{v_{\pi(i+1)}} S_{v_{\pi(i+)+1}}) \\
&= r'_{\pi(i)} \wedge r'_{\pi(i+1)}
\end{aligned}$$

for $i = 1, 2, \ldots, n$. That is,

$$s_i \leq r_{\pi(i)}, \; s'_i \leq r'_{\pi(i)}$$

and

$$\begin{cases} r_i \leq r_{\pi(i)} \wedge r_{\pi(i+1)} \\ r'_i \leq r'_{\pi(i)} \wedge r'_{\pi(i+1)}. \end{cases} \tag{7.2}$$

By (7.2) we have $r_i \le r_{\pi(i)}$, $r'_i \le r'_{\pi(i)}$ for $i = 1, 2, \ldots, n$ and so $r_{\pi(i)} \le r_{\pi(\pi(i))}$, $r'_{\pi(i)} \le r'_{\pi(\pi(i))}$ for $i = 1, 2, \ldots, n$. Continuing, we have

$$r_i \le r_{\pi(i)} \le r_{\pi(\pi(i))} \le \cdots \le r_{\pi^j(i)} \le r_i,$$
$$r'_i \le r'_{\pi(i)} \le r'_{\pi(\pi(i))} \le \cdots \le r'_{\pi^j(i)} \le r'_i$$

and so $r_i = r_{\pi(i)}$, $r'_i = r'_{\pi(i)}$ for $i = 1, 2, \ldots, n$, where $\pi^{j+1}$ is the identity map. Again, by (7.2), we have

$$r_i \le r_{\pi(i+1)} = r_{i+1},\ i = 1, 2, \ldots, n,\ r_{n+1} = r_1,$$
$$r'_i \le r'_{\pi(i+1)} = r'_{i+1},\ i = 1, 2, \ldots, n,\ r_{n+1} = r_1.$$

Hence, by (7.1) and (7.2),

$$r_1 = \cdots = r_n = s_1 = \cdots = s_n,$$
$$r'_1 = \cdots = r'_n = s'_1 = \cdots = s'_n.$$

Thus, we have not only proved the conclusion about $A_1$ and $B_1$ being constant functions, but we have also shown that (ii) holds.

The converse follows easily. ■

We state the following theorem without proof.

**Theorem 7.5.11** *Let G and H be interval-valued fuzzy graphs of $G^*$ and $H^*$, respectively, such that $G^*$ and $H^*$ are connected. Let $L(G)$ and $L(H)$ be the interval-valued fuzzy line graphs corresponding to G and H, respectively. Suppose that it is not the case that one of $G^*$ and $H^*$ is complete graph $K_3$ and the other is bipartite complete graph $K_{1,3}$. If $L(G)$ and $L(H)$ are isomorphic, then G and H are line-isomorphic.*

*Example 7.5.12* Consider an interval-valued fuzzy line graph given in Example 7.5.4. By routine calculations, it is easy to see that $L(G)$ is a strong interval-valued fuzzy graph.

The proof of the following proposition follows easily.

**Proposition 7.5.13** *An interval-valued fuzzy line graph is a strong interval-valued fuzzy graph.*

## 7.6 Balanced Interval-Valued Fuzzy Graphs

We next consider certain types of interval-valued fuzzy graphs including balanced interval-valued fuzzy graphs, neighborly irregular interval-valued fuzzy graphs, neighborly total irregular interval-valued fuzzy graphs, highly irregular interval-valued fuzzy graphs, and highly total irregular interval-valued fuzzy graphs. The work is due to [19]. We present necessary and sufficient conditions under which neighborly irregular and highly irregular interval-valued fuzzy graphs are equivalent.

In [169], the definition of complement of a fuzzy graph was modified so that the complement of the complement is the original fuzzy graph, which agrees with the crisp graph case. Ju and Wang gave the definition of interval-valued fuzzy graph in [88]. Akram et al. [5, 6, 9, 13, 14] introduced many new concepts including bipolar fuzzy graphs, interval-valued fuzzy line graphs, strong intuitionistic fuzzy graphs. We propose certain types of interval-valued fuzzy graphs including balanced interval-valued fuzzy graphs, neighborly irregular interval-valued fuzzy graphs, neighborly total irregular interval-valued fuzzy graphs, highly irregular interval-valued fuzzy graphs, and highly total irregular interval-valued fuzzy graphs. Some properties associated with these new interval-valued fuzzy graphs are investigated, and necessary and sufficient conditions under which neighborly irregular and highly irregular interval-valued fuzzy graphs are equivalent are obtained. We also describe the relationship between intuitionistic fuzzy graphs and interval-valued fuzzy graphs.

Recall that a regular graph is a graph where each vertex has the same number of neighbors, that is, all the vertices have the same closed neighborhood degree. A connected graph is **highly irregular** if each of its vertices is adjacent only to vertices with distinct degrees. Equivalently, a graph $G^*$ is highly irregular if every two vertices of $G^*$ connected by a path of length 2 have distinct degrees. A connected graph is said to be **neighborly irregular** if no two adjacent vertices of $G^*$ have the same degree. Equivalently, a connected graph $G^*$ is called neighborly irregular if every two adjacent vertices of $G^*$ have distinct degree.

One of the best known classes of graphs is the class of regular graphs. These graphs have been studied extensively in various contexts. Regular graphs of degree $r$ and order $n$ exist with only limited, but natural, restrictions. Indeed, for integers $r$ and $n$ with $0 \leq r \leq n-1$, an $r$ -regular graph of order $n$ exists if and only if $nr$ is even. A graph that is not regular is called **irregular**. It is well known that all nontrivial graphs, regular or irregular, must contain at least vertices of the same degree [90]. In a regular graph, every vertex is adjacent only to vertices having the same degree. On the other hand, it is possible for a vertex in an irregular graph to be adjacent only to vertices with distinct degrees. We consider graphs that are opposite, in a certain sense, to regular graphs.

**Definition 7.6.1** Let $G$ be an interval-valued fuzzy graph. The **neighborhood degree** of a vertex $x$ in $G$ is defined by $\deg(x) = [\deg_{\mu^-}(x), \deg_{\mu^+}(x)]$, where $\deg_{\mu^-}(x) = \sum_{y \in N(x)} \mu_A^-(y)$ and $\deg_{\mu^+}(x) = \sum_{y \in N(x)} \mu_A^+(y)$.

In Definition 7.6.1, $\mu_B^-(xy) > 0$ and $\mu_B^+(xy) > 0$ for $xy \in E$, and $\mu_B^-(xy) = \mu_B^+(xy) = 0$ for $xy \notin E$.

**Definition 7.6.2** Let $G = (A, B)$ be an interval-valued fuzzy graph on $G^*$. If all the vertices have the same open neighborhood degree $n$, then $G$ is called an $n$-**regular interval-valued fuzzy graph**. The **open neighborhood degree** of $x$ in $G$ is defined by $\deg(x) = [\deg_{\mu^-}(x), \deg_{\mu^+}(x)]$, where $\deg_{\mu^-}(x) = \sum_{y\in N(x)} \mu_A^-(y)$ and $\deg_{\mu^+}(x) = \sum_{y\in N(x)} \mu_A^+(y)$.

**Definition 7.6.3** Let $G$ be an interval-valued fuzzy graph. The **closed neighborhood degree** of $x$ in $G$ is defined by $\deg[x] = [\deg_{\mu^-}[x], \deg_{\mu^+}[x]]$, where

$$\deg_{\mu^-}[x] = \deg_{\mu^-}(x) + \mu_A^-(x)$$
$$\deg_{\mu^+}[x] = \deg_{\mu^+}(x) + \mu_A^+(x).$$

If all the vertices have the same closed neighborhood degree $m$, then $m$ is called a $m$-**totally regular interval-valued fuzzy graph**.

*Example 7.6.4* Consider a graph $G^*$ such that $V = \{a, b, c, d\}$, $E = \{ab, bc, cd, ad\}$, and

| | $a$ | $b$ | $c$ | $d$ |
|---|---|---|---|---|
| $\mu_A^-$ | 0.3 | 0.3 | 0.3 | 0.3 |
| $\mu_A^+$ | 0.5 | 0.5 | 0.5 | 0.5 |

| | $ab$ | $bc$ | $cd$ | $ad$ |
|---|---|---|---|---|
| $\mu_B^-$ | 0.1 | 0.1 | 0.1 | 0.1 |
| $\mu_B^+$ | 0.2 | 0.4 | 0.2 | 0.4 |

It is easily shown that the interval-valued fuzzy graph $G$ is both regular and totally regular.

*Example 7.6.5* Consider a graph $G^*$ such that $V = \{v_1, v_2, v_3\}$, $E = \{v_1v_2, v_1v_3\}$. Let $A$ be an interval-valued fuzzy subset of $V$ and let $B$ be an interval-valued fuzzy subset of $E$ defined by

| | $v_1$ | $v_2$ | $v_3$ |
|---|---|---|---|
| $\mu_A^-$ | 0.4 | 0.7 | 0.6 |
| $\mu_A^+$ | 0.4 | 0.8 | 0.7 |

| | $v_1v_2$ | $v_1v_3$ |
|---|---|---|
| $\mu_B^-$ | 0.2 | 0.2 |
| $\mu_B^+$ | 0.3 | 0.4 |

It is easily shown that the interval-valued fuzzy graph $G$ is neither regular nor totally regular.

**Definition 7.6.6** We define the **order** $O(G)$ and the **size** of $S(G)$ of an interval-valued fuzzy graph $G = (A, B)$ by

$$O(G) = \sum_{x\in V} \frac{1+\mu_A^+(x)-\mu_A^-(x)}{2}$$
$$Z(G) = \sum_{xy\in E} \frac{1+\mu_B^+(xy)-\mu_B^-(xy)}{2}$$

*Example 7.6.7* Consider a graph $G^*$ such that $V = \{x, y, z\}$, $E = \{xy, yz, zx\}$. Let $A$ be an interval-valued fuzzy subset of $V$ and let $B$ be an interval-valued fuzzy subset of $E$ defined by

| | $x$ | $y$ | $z$ |
|---|---|---|---|
| $\mu_A^-$ | 0.3 | 0.4 | 0.5 |
| $\mu_A^+$ | 0.5 | 0.7 | 0.6 |

| | $xy$ | $yz$ | $zx$ |
|---|---|---|---|
| $\mu_B^-$ | 0.3 | 0.4 | 0.3 |
| $\mu_B^+$ | 0.5 | 0.6 | 0.5 |

It is easy to show that $G$ is both strong and totally regular interval-valued fuzzy graph, but $G$ is not regular because $\deg(x) \neq \deg(z) \neq \deg(y)$.

**Theorem 7.6.8** *Every complete interval-valued fuzzy graph is totally regular.*

**Theorem 7.6.9** *Let $G = (A, B)$ be an interval-valued fuzzy graph of a graph $G^*$. Then $A = [\mu_A^-, \mu_A^+]$ is a constant function if and only if the following statements are equivalent.*

*$(i)$ $G$ is a regular interval-valued fuzzy graph.*

*$(ii)$ $G$ is a totally regular interval-valued fuzzy graph.*

*Proof* Suppose that $A = [\mu_A^-, \mu_A^+]$ is a constant function. Let $\mu_A^-(x) = c_1$ and $\mu_A^+(x) = c_2$ for all $x \in V$.

$(i) \Rightarrow (ii)$ : Assume that $G$ is an $n$-regular interval-valued fuzzy graph. Then $\deg_{\mu^-}(x) = n_1$ and $\deg_{\mu^+}(x) = n_2$ for all $x \in V$. So

$$\deg_{\mu^-}[x] = \deg_{\mu^-}(x) + \mu_A^-(x)$$
$$\deg_{\mu^+}[x] = \deg_{\mu^+}(x) + \mu_A^+(x)$$

for all $x \in V$. Thus,

$$\deg_{\mu^-}[x] = \deg_{\mu^-}(x) + \mu_A^-(x) = n_1 + c_1$$
$$\deg_{\mu^+}[x] = \deg_{\mu^+}(x) + \mu_A^+(x) = n_2 + c_2$$

for all $x \in V$. Hence, $G$ is a totally regular interval-valued fuzzy graph.

$(ii) \Rightarrow (i)$ : Suppose that $G$ is a totally regular interval-valued fuzzy graph. Then

$$\deg_{\mu^-}[x] = k_1 \text{ and } \deg_{\mu^+}[x] = k_2 \text{ for all } x \in V,$$

or

$$\deg_{\mu^-}(x) + \mu_A^-(x) = k_1$$
$$\deg_{\mu^+}(x) + \mu_A^+(x) = k_2$$

for all $x \in V$, or

$$\deg_{\mu^-}(x) + c_1 = k_1$$
$$\deg_{\mu^+}(x) + c_2 = k_2$$

for all $x \in V$, or

$$\deg_{\mu^-}(x) = k_1 - c_1$$
$$\deg_{\mu^+}(x) = k_2 - c_2$$

for all $x \in V$. Thus, $G$ is a regular interval-valued fuzzy graph. Hence, $(i)$ and $(ii)$ are equivalent.

The converse follows easily. ■

**Theorem 7.6.10** ([19]) *Let $G$ be an interval-valued fuzzy graph, where the crisp graph $G^*$ is an odd cycle. Then $G$ is a regular interval-valued fuzzy graph if and only if $B$ is a constant function.*

*Proof* If $B = [\mu_B^-, \mu_B^+]$ is a constant function, say $\mu_B^-(xy) = c_1$ and $\mu_B^+(xy) = c_2$ for all $xy \in E$, then $\deg_{\mu^-}(x) = 2c_1$ and $\deg_{\mu^+}(x) = 2c_2$ for all $x \in V$. Hence, $G$ is a regular interval-valued fuzzy graph.

Conversely, suppose that $G$ is a $(k_1, k_2)$-regular interval-valued fuzzy graph. Let $e_1, e_2, \ldots, e_{2n+1}$ be the edges of $G$ in that order. Let $\mu_B^-(e_1) = c_1$, $\mu_B^-(e_2) = k_1 - c_1$, $\mu_B^-(e_3) = k_1 - (k_1 - c_1) = c_1$, $\mu_B^-(e_4) = k_1 - c_1$, and so on. Then

$$\mu_B^-(e_i) = \begin{cases} c_1 & \text{if } i \text{ is odd} \\ k_1 - c_1 & \text{if } i \text{ is even.} \end{cases}$$

Thus, $\mu_B^-(e_1) = \mu_B^-(e_{2n+1}) = c_1$. So if $e_1$ and $e_{2n+1}$ are incident at a vertex $v_1$, then $\deg_{\mu^-}(v_1) = k_1$, $\deg_{\mu^-}(e_1) + \deg_{\mu^-}(e_{2n+1}) = k_1$, $c_1 + c_1 = k_1$, i.e., $2c_1 = k_1$, and so $c_1 = k_1/2$. This shows that $\mu_B^-$ is a constant function.

Similarly, let $\mu_B^+(e_1) = c_2$, $\mu_B^+(e_2) = k_2 - c_2$, $\mu_B^+(e_3) = k_2 - (k_2 - c_2) = c_2$, $\mu_B^+(e_4) = k_2 - c_2$, and so on. Therefore,

$$\mu_B^+(e_i) = \begin{cases} c_2 & \text{if } i \text{ is odd} \\ k_2 - c_2 & \text{if } i \text{ is even.} \end{cases}$$

Hence, $\mu_B^+(e_1) = \mu_B^+(e_{2n+1}) = c_2$. So if $e_1$ and $e_{2n+1}$ are incident at a vertex $v_1$, then $\deg_{\mu^+}(v_1) = k_1$, $\deg_{\mu^+}(e_1) + \deg_{\mu^+}(e_{2n+1}) = k_2$, $c_2 + c_2 = k_2$, i.e., $2c_2 = k_2$, and so $c_2 = k_2/2$. This shows that $\mu_B^+$ is a constant function. Thus, $B = [\mu_B^-, \mu_B^+]$ is a constant function. ■

The following result also holds.

**Theorem 7.6.11** *Let $G$ be an interval-valued fuzzy graph, where the crisp graph $G^*$ is an even cycle. Then $G$ is a regular interval-valued fuzzy graph if and only if either $B = [\mu_B^-, \mu_B^+]$ is a constant function or the alternate edges have the same membership values.* ■

**Definition 7.6.12** The **density** of an interval-valued fuzzy graph $D$ is $D(G) = (D^-(G), D^+(G))$, where

$$D^-(G) = \frac{2\sum_{x,y \in V} \mu_B^-(xy)}{\sum_{x,y \in V}(\mu_A^-(x) \wedge \mu_A^-(y))}$$

for $x, y \in V$ and

$$D^+(G) = \frac{2\sum_{x,y \in V} \mu_B^+(xy)}{\sum_{x,y \in V}(\mu_A^+(x) \wedge \mu_A^+(y))}.$$

An interval-valued fuzzy graph $G$ is **balanced** if $D(H) \le D(G)$, i.e., $D^-(H) \le D^-(G)$ and $D^+(H) \le D^+(G)$ for all subgraphs $H$ of $G$. An interval-valued fuzzy graph is **strictly balanced** if for all $x, y \in V$, $D(H) = D(G)$ for all nonempty subgraphs $H$.

*Example 7.6.13* Consider the regular interval-valued fuzzy graph $G$ given in Example 7.6.4. It follows that $D^-(G) = 0.67$ and $D^+(G) = 1.2$. Thus, $D(G) = (0.67, 1.2)$. Consider that $H_1 = \{a, b, c\}$, $H_2 = \{a, c\}$, $H_3 = \{a, d\}$ and $H_4 = \{b, c\}$ are nonempty subgraphs of $G$. Then $D(H_1) = (0.67, 1.2)$, $D(H_2) = (0, 0)$, $D(H_3) = (0.67, 1.6)$, $D(H_4) = (0.67, 1.6)$. It is easy to see that regular interval-valued fuzzy graph is not balanced.

It's worth noting that every regular interval-valued graph may not be balanced.

*Example 7.6.14* Consider the regular interval-valued fuzzy graph $G$ given in Example 7.6.7. It follows that $D^-(G) = 2$ and $D^+(G) = 2$. Thus, $D(G) = (2, 2)$. Consider that $H_1 = \{x, y\}$, $H_2 = \{x, z\}$ and $H_3 = \{y, z\}$ are nonempty subgraphs of $G$. Then $D(H_1) = (2, 2)$, $D(H_2) = (2, 2)$, and $D(H_3) = (2, 2)$. It is easy to see that regular interval-valued fuzzy graph is balanced and also strictly balanced.

**Proposition 7.6.15** *Any strong interval-valued fuzzy graph is balanced.*

**Proposition 7.6.16** *Let $G$ be a self-complementary interval-valued fuzzy graph. Then $D(G) = (1, 1)$.*

**Proposition 7.6.17** *Let $G_1$ and $G_2$ be two balanced interval-valued fuzzy graphs. Then $G_1 \times G_2$ is balanced if and only if $D(G_1) = D(G_2) = D(G_1 \times G_2)$.*

**Theorem 7.6.18** *Let $G$ be a strictly balanced interval-valued fuzzy graph and let $\overline{G}$ be its complement. Then $D(G) + D(\overline{G}) = (2n^2, 2n^2)$, where $|V| = n$.*

*Proof* Let $G$ be a strictly balanced interval-valued fuzzy graph and $\overline{G}$ be its complement. Let $H$ be a nonempty subgraph of $G$. Because $G$ is strictly balanced, $D(G) = D(H)$ for all $H \subseteq G$ and $x, y \in V$. In $\overline{G}$,

$$\overline{\mu_B^-(xy)} = \mu_A^-(x) \wedge \mu_A^-(y) - \mu_B^-(xy). \tag{7.3}$$

$$\overline{\mu_B^+(xy)} = \mu_A^+(x) \wedge \mu_A^+(y) - \mu_B^+(xy). \tag{7.4}$$

for all $x, y \in V$. Dividing (7.3) by $\mu_A^-(x) \wedge \mu_A^-(y)$, we get

$$\frac{\overline{\mu_B^-(xy)}}{\mu_A^-(x) \wedge \mu_A^-(y)} = 1 - \frac{\mu_B^-(xy)}{\mu_A^-(x) \wedge \mu_A^-(y)}$$

for all $x, y \in V$. Dividing (7.4) by $\mu_A^+(x) \wedge \mu_A^+(y)$, we get

$$\frac{\overline{\mu_B^+(xy)}}{\mu_A^+(x) \wedge \mu_A^+(y)} = 1 - \frac{\mu_B^+(xy)}{\mu_A^+(x) \wedge \mu_A^+(x)}$$

for all $x, y \in V$. Thus,

$$\sum_{x,y \in V} \frac{\overline{\mu_B^-(xy)}}{\mu_A^-(x) \wedge \mu_A^-(y)} = n^2 - \sum_{x,y \in V} \frac{\mu_B^-(xy)}{\mu_A^-(x) \wedge \mu_A^-(y)}, \tag{7.5}$$

and

$$\sum_{x,y \in V} \frac{\overline{\mu_B^+(xy)}}{\mu_A^+(x) \wedge \mu_A^+(y)} = n^2 - \sum_{x,y \in V} \frac{\mu_B^+(xy)}{\mu_A^+(x) \wedge \mu_A^+(y)}. \tag{7.6}$$

Multiplying both sides of (7.5) and (7.6) by 2, we obtain

$$2 \sum_{x,y \in V} \frac{\overline{\mu_B^-(xy)}}{\mu_A^-(x) \wedge \mu_A^-(y)} = 2n^2 - 2 \sum_{x,y \in V} \frac{\mu_B^-(xy)}{\mu_A^-(x) \wedge \mu_A^-(y)}$$

$$2 \sum_{x,y \in V} \frac{\overline{\mu_B^+(xy)}}{\mu_A^+(x) \wedge \mu_A^+(y)} = 2n^2 - 2 \sum_{x,y \in V} \frac{\mu_B^+(xy)}{\mu_A^+(x) \wedge \mu_A^+(y)}.$$

Hence, $D^-(\overline{G}) = 2n^2 - D^-(G)$ and $D^+(\overline{G}) = 2n^2 - D^+(G)$.

Now,

$$\begin{aligned} D(G) + D(\overline{G}) &= (D^-(G), D^+(G)) + (D^-(\overline{G}), D^+(\overline{G})) \\ &= (D^-(G) + D^-(\overline{G}), D^+(G) + D^+(\overline{G})) \\ &= (2n^2, 2n^2). \end{aligned}$$

■

**Corollary 7.6.19** *The complement of a strictly balanced interval-valued fuzzy graph is strictly balanced.*

**Theorem 7.6.20** *Let $G_1$ and $G_2$ be isomorphic interval-valued fuzzy graphs. If $G_2$ is balanced, the $G_1$ is balanced.*

## 7.7 Irregularity in Interval-Valued Fuzzy Graphs

**Definition 7.7.1** Let $G$ be an interval-valued fuzzy graph on $G^*$. If there is a vertex which is adjacent to vertices with distinct neighborhood degrees, then $G$ is called an **irregular interval-valued fuzzy graph**.

*Example 7.7.2* Consider a graph $G^*$ such that

$$V = \{v_1, v_2, v_3\}, \ E = \{v_1v_2, v_2v_3, v_1v_3\}.$$

Let $A$ be an interval-valued fuzzy subset of $V$ and let $B$ be an interval-valued fuzzy subset of $E \subseteq V \times V$ defined by

| | $v_1$ | $v_2$ | $v_3$ |
|---|---|---|---|
| $\mu_A^-$ | 0.2 | 0.2 | 0.3 |
| $\mu_A^+$ | 0.6 | 0.7 | 0.4 |

| | $v_1v_2$ | $v_1v_3$ | $v_2v_3$ |
|---|---|---|---|
| $\mu_B^-$ | 0.1 | 0.1 | 0.2 |
| $\mu_B^+$ | 0.2 | 0.2 | 0.3 |

It is easily shown that $\deg(v_1) = [0.5, 1.1]$, $\deg(v_2) = [0.5, 1.0]$, and $\deg(v_3) = [0.4, 1.3]$. Clearly $G$ is an irregular interval-valued fuzzy graph.

**Definition 7.7.3** Let $G$ be an interval-valued fuzzy graph. If there is a vertex which is adjacent to vertices with distinct closed neighborhood degrees, then $G$ is called a **totally irregular interval-valued fuzzy graph**.

*Example 7.7.4* Let $G$ be an interval-valued fuzzy graph $G$ such that

$$V = \{v_1, v_2, v_3, v_4, v_5\},\ E = \{v_1v_2, v_2v_3, v_2v_4, v_3v_1, v_3v_4, v_4v_1, v_4v_5\}.$$

and

| | $v_1$ | $v_2$ | $v_3$ | $v_4$ | $v_5$ |
|---|---|---|---|---|---|
| $\mu_A^-$ | 0.4 | 0.3 | 0.3 | 0.4 | 0.2 |
| $\mu_A^+$ | 0.6 | 0.5 | 0.7 | 0.6 | 0.2 |

| | $v_1v_2$ | $v_2v_3$ | $v_2v_4$ | $v_3v_1$ | $v_3v_4$ | $v_4v_1$ | $v_4v_5$ |
|---|---|---|---|---|---|---|---|
| $\mu_B^-$ | 0.2 | 0.2 | 0.2 | 0.1 | 0.3 | 0.1 | 0.1 |
| $\mu_B^+$ | 0.3 | 0.2 | 0.4 | 0.4 | 0.5 | 0.3 | 0.2 |

Clearly, $\deg[v_1] = [1.4, 2.4]$, $\deg[v_2] = [1.4, 2.4]$, $\deg[v_3] = [1.4, 2.4]$, $\deg[v_4] = [1.6, 2.6]$, $\deg[v_5] = [0.6, 0.8]$. It is also clear that $G$ is a totally irregular interval-valued fuzzy graph.

**Definition 7.7.5** A connected interval-valued fuzzy graph $G$ is said to be **neighborly irregular** if every two adjacent vertices of $G$ have distinct open neighborhood degree.

*Example 7.7.6* Let $G$ be an interval-valued fuzzy graph $G$ such that

$$V = \{v_1, v_2, v_3, v_4\},\ E = \{v_1v_2, v_2v_3, v_3v_4, v_4v_1\}$$

and

| | $v_1$ | $v_2$ | $v_3$ | $v_4$ |
|---|---|---|---|---|
| $\mu_A^-$ | 0.2 | 0.3 | 0.4 | 0.5 |
| $\mu_A^+$ | 0.6 | 0.7 | 0.4 | 0.5 |

| | $v_1v_2$ | $v_2v_3$ | $v_3v_4$ | $v_4v_1$ |
|---|---|---|---|---|
| $\mu_B^-$ | 0.1 | 0.1 | 0.1 | 0.1 |
| $\mu_B^+$ | 0.2 | 0.4 | 0.2 | 0.4 |

It follows that $\deg(v_1) = [0.8, 1.2]$, $\deg(v_2) = [0.6, 1.0]$, $\deg(v_3) = [0.8, 1.2]$, and $\deg(v_4) = [0.6, 1.0]$. Hence, $G$ is neighborly irregular.

**Definition 7.7.7** A connected interval-valued fuzzy graph $G$ is said to be **neighborly totally irregular** if every two adjacent vertices of $G$ have distinct closed neighborhood degree.

*Example 7.7.8* Let $G$ be an interval-valued fuzzy graph $G$ such that

$$V = \{v_1, v_2, v_3, v_4\}, \ E = \{v_1v_2, v_2v_3, v_3v_4, v_4v_1\}$$

and

| | $v_1$ | $v_2$ | $v_3$ | $v_4$ |
|---|---|---|---|---|
| $\mu_A^-$ | 0.3 | 0.4 | 0.2 | 0.4 |
| $\mu_A^+$ | 0.6 | 0.5 | 0.7 | 0.5 |

| | $v_1v_2$ | $v_2v_3$ | $v_3v_4$ | $v_4v_1$ |
|---|---|---|---|---|
| $\mu_B^-$ | 0.1 | 0.1 | 0.1 | 0.1 |
| $\mu_B^+$ | 0.2 | 0.4 | 0.2 | 0.4 |

It is easily shown that $\deg[v_1] = [1.1, 1.6]$, $\deg[v_2] = [0.9, 1.8]$, $\deg[v_3] = [1.0, 1.7]$, and $\deg[v_4] = [0.9, 1.8]$. Hence, $G$ is neighborly totally irregular.

**Definition 7.7.9** Let $G$ be a connected interval-valued fuzzy graph. $G$ is called **highly irregular** if every vertex of $G$ is adjacent to vertices with distinct neighborhood degrees.

*Example 7.7.10* Let $G$ be an interval-valued fuzzy graph $G$ such that

$$V = \{v_1, v_2, v_3, v_4, v_5, v_6\},$$

$$E = \{v_1v_2, v_2v_3, v_2v_6, v_3v_4, v_3v_5, v_4v_5, v_5v_1\}$$

and

| | $v_1$ | $v_2$ | $v_3$ | $v_4$ | $v_5$ | $v_6$ |
|---|---|---|---|---|---|---|
| $\mu_A^-$ | 0.2 | 0.1 | 0.3 | 0.5 | 0.3 | 0.1 |
| $\mu_A^+$ | 0.6 | 0.4 | 0.7 | 0.5 | 0.4 | 0.4 |

| | $v_1v_2$ | $v_2v_3$ | $v_2v_6$ | $v_3v_4$ | $v_3v_5$ | $v_4v_5$ | $v_5v_1$ |
|---|---|---|---|---|---|---|---|
| $\mu_B^-$ | 0.2 | 0.2 | 0.1 | 0.2 | 0.2 | 0.2 | 0.1 |
| $\mu_B^+$ | 0.2 | 0.2 | 0.4 | 0.4 | 0.3 | 0.4 | 0.3 |

It follows that $\deg(v_1) = [0.4, 0.8]$, $\deg(v_2) = [0.6, 1.7]$, $\deg(v_3) = [0.9, 1.3]$, $\deg(v_4) = [0.6, 1.1]$, $\deg(v_5) = [1.0, 1.8]$ and $\deg(v_6) = [0.1, 0.4]$. Clearly, $G$ is a highly irregular.

*Example 7.7.11* Let $G$ be an interval-valued fuzzy graph $G$ such that

$$V = \{v_1, v_2, v_3, v_4\}, \ E = \{v_1v_2, v_2v_3, v_3v_4, v_4v_1\}$$

and

| | $v_1$ | $v_2$ | $v_3$ | $v_4$ |
|---|---|---|---|---|
| $\mu_A^-$ | 0.3 | 0.2 | 0.5 | 0.4 |
| $\mu_A^+$ | 0.4 | 0.5 | 0.5 | 0.5 |

| | $v_1v_2$ | $v_2v_3$ | $v_3v_4$ | $v_4v_1$ |
|---|---|---|---|---|
| $\mu_B^-$ | 0.1 | 0.1 | 0.1 | 0.1 |
| $\mu_B^+$ | 0.2 | 0.4 | 0.2 | 0.4 |

By routine computations, we have $\deg(v_1) = [0.6, 1.0]$, $\deg(v_2) = [0.8, 0.9]$, $\deg(v_3) = [0.6, 1.0]$, and $\deg(v_4) = [0.8, 0.9]$. Clearly, $G$ is a neighborly irregular, but not highly irregular.

We note that a neighborly irregular interval-valued fuzzy graph may not be highly irregular.

**Theorem 7.7.12** *An interval-valued fuzzy graph $G$ is highly irregular and neighborly irregular interval-valued fuzzy graph if and only if the neighborhood degrees of all the vertices of $G$ are distinct.*

*Proof* Let $G$ be an interval-valued fuzzy graph with $n$ vertices $v_1, v_2, \ldots, v_n$. Assume that $G$ is highly irregular and neighborly irregular. Let $\deg(v_i) = [k_i, l_i]$, $i = 1, 2, \ldots, n$. Let the adjacent vertices of $v_1$ be $v_2, v_3, \ldots, v_n$ with neighborhood degrees $[k_2, l_2], [k_3, l_3], \ldots, [k_n, l_n]$, respectively. Then $k_2 \neq k_3 \neq \cdots \neq k_n$ and $l_2 \neq l_3 \neq \cdots \neq l_n$ because $G$ is highly irregular. Also, $k_1 \neq k_2 \neq k_3 \neq \cdots \neq k_n$ and $l_1 \neq l_2 \neq l_3 \neq \cdots \neq l_n$ because $G$ is neighborly irregular. Hence, the neighborhood degree of all the vertices of $G$ is distinct.

Conversely, assume that the neighborhood degree of all the vertices of $G$ is distinct. Let $\deg(v_i) = [k_i, l_i]$, $i = 1, 2, \ldots, n$. Given that $k_1 \neq k_2 \neq k_3 \neq \cdots \neq k_n$ and $l_1 \neq l_2 \neq l_3 \neq \cdots \neq l_n$, which implies that every two adjacent vertices have distinct neighborhood degrees and to every vertex, the adjacent vertices have distinct neighborhood degrees. Thus, $G$ is highly irregular and neighborly irregular interval-valued fuzzy graph. ■

**Theorem 7.7.13** *An interval-valued fuzzy graph $G$ of $G^*$, where $G^*$ is a cycle with 3 vertices that is neighborly irregular and highly irregular if and only if the lower and upper membership values of the vertices between every pair of vertices are all distinct.*

*Proof* Assume that the lower and upper membership values of the vertices are all distinct.

Let $v_i, v_j, v_k \in V$. Given that $\mu_A^-(v_i) \neq \mu_A^-(v_j) \neq \mu_A^-(v_k)$ and $\mu_A^+(v_i) \neq \mu_A^+(v_j) \neq \mu_A^+(v_k)$, which implies that $\sum_{x \in N(x)} \mu_A^-(v_i) \neq \sum_{x \in N(x)} \mu_A^-(v_j) \neq \sum_{x \in N(x)} \mu_A^-(v_k)$ and $\sum_{x \in N(x)} \mu_A^+(v_i) \neq \sum_{x \in N(x)} \mu_A^+(v_j) \neq \sum_{x \in N(x)} \mu_A^+(v_k)$. That is, $\deg(v_i) \neq \deg(v_j) \neq \deg(v_k)$. Hence, $G$ is neighborly irregular and highly irregular.

Conversely, assume that $G$ is neighborly irregular and highly irregular. Let $\deg(v_i) = [k_i, l_i]$, $i = 1, 2, \ldots, n$. Suppose that lower and upper membership values of any two vertices are the same. Let $v_1, v_2 \in V$. Let $\mu_A^-(v_1) = \mu_A^-(v_2)$ and $\mu_A^+(v_1) = \mu_A^+(v_2)$. Then $\deg(v_1) = \deg(v_2)$ because $G^*$ is a cycle, which is a contradiction to the fact that $G$ is neighborly irregular and highly irregular interval-valued fuzzy graphs Hence, the lower and upper membership values of the vertices are all distinct. ■

*Example 7.7.14* Consider an interval-valued fuzzy graph $G$ such that

$$V = \{v_1, v_2, v_3\},\ E = \{v_1v_2, v_2v_3, v_1v_3\}.$$

Let $A$ be an interval-valued fuzzy subset of $V$ and let $B$ be an interval-valued fuzzy subset of $E \subseteq V \times V$ defined by

| | $v_1$ | $v_2$ | $v_3$ |
|---|---|---|---|
| $\mu_A^-$ | 0.4 | 0.4 | 0.2 |
| $\mu_A^+$ | 0.6 | 0.6 | 0.7 |

| | $v_1v_2$ | $v_1v_3$ | $v_2v_3$ |
|---|---|---|---|
| $\mu_B^-$ | 0.4 | 0.2 | 0.2 |
| $\mu_B^+$ | 0.6 | 0.6 | 0.6 |

It follows that $\deg(v_1) = [0.6, 1.3]$, $\deg(v_2) = [0.6, 1.3]$, and $\deg(v_3) = [0.8, 1.2]$. We see that neighborhood degree of $v_1$ and $v_2$ are not distinct. Hence, $G$ is not neighborly irregular, but it is complete.

Note that a neighborly total irregular interval-valued fuzzy graph may not be neighborly irregular.

*Example 7.7.15* Let $G$ be an interval-valued fuzzy graph $G$ such that

$$V = \{v_1, v_2, v_3, v_4\},\ E = \{v_1v_2, v_2v_3, v_3v_4, v_4v_1\}.$$

and

| | $v_1$ | $v_2$ | $v_3$ | $v_4$ |
|---|---|---|---|---|
| $\mu_A^-$ | 0.4 | 0.3 | 0.4 | 0.4 |
| $\mu_A^+$ | 0.6 | 0.4 | 0.5 | 0.5 |

| | $v_1v_2$ | $v_2v_3$ | $v_3v_4$ | $v_4v_1$ |
|---|---|---|---|---|
| $\mu_B^-$ | 0.1 | 0.1 | 0.1 | 0.1 |
| $\mu_B^+$ | 0.2 | 0.4 | 0.2 | 0.4 |

We have that $\deg(v_1) = [0.7, 0.9]$, $\deg(v_2) = [0.8, 1.1]$, $\deg(v_3) = [0.7, 0.9]$, $\deg(v_4) = [0.8, 1.1]$, $\deg[v_1] = [1.1, 1.5]$, $\deg[v_2] = [1.1, 1.5]$, $\deg[v_3] = [1.1, 1.4]$, and $\deg[v_4] = [1.2, 1.6]$. We see that $\deg[v_1] = \deg[v_2]$. Thus, $G$ is neighborly irregular, but not neighborly total irregular.

**Proposition 7.7.16** *If an interval-valued fuzzy graph $G$ is neighborly irregular and $A = [\mu_A^-, \mu_A^+]$ is a constant function, then it is neighborly totally irregular.*

*Proof* Assume that $G$ is a neighborly irregular interval-valued fuzzy graph. Then the open neighborhood degrees of every two adjacent vertices are distinct. Let $v_i, v_j \in V$ be adjacent vertices with distinct open neighborhood degrees $\deg(v_i) = [k_1, l_1]$ and $\deg(v_j) = [k_2, l_2]$, where $k_1 \neq k_2$ and $l_1 \neq l_2$. Let us assume that $(\mu_1(v_i), \nu_1(v_i)) = (\mu_1(v_j), \nu_1(v_j)) = [c_1, c_2]$, where $c_1$ and $c_2$ are constants and $c_1, c_2 \in [0, 1]$. Therefore, $\deg_{\mu^-}[v_i] = \deg_{\mu^-}(v_i) + \mu_1(v_i) = k_1 + c_1$, $\deg_{\mu^+}[v_i] = \deg_{\mu^+}(v_i) + \nu_1(v_i) = l_1 + c_2$, $\deg_{\mu^-}[v_j] = \deg_{\mu^-}(v_j) + \mu_1(v_j) = k_2 + c_1$, $\deg_{\mu^+}[v_j] = \deg_{\mu^+}(v_j) + \nu_1(v_j) = l_2 + c_2$.

We show that $\deg_{\mu^-}[v_i] \neq \deg_{\mu^-}[v_j]$ and $\deg_{\mu^+}[v_i] \neq \deg_{\mu^+}[v_j]$. Suppose that $\deg_{\mu^-}[v_i] = \deg_{\mu^-}[v_j]$ and $\deg_{\mu^+}[v_i] = \deg_{\mu^+}[v_j]$. Suppose that

$$\deg_{\mu^-}[v_i] = \deg_{\mu^-}[v_j] \text{ so}$$
$$k_1 + c_1 = k_2 + c_1$$
$$k_1 - k_2 = c_1 - c_1 = 0,$$

Then $k_1 = k_2$, which contradicts that $k_1 \neq k_2$. Therefore, $\deg_{\mu^-}[v_i] \neq \deg_{\mu^-}[v_j]$.
Suppose that

$$\deg_{\mu^+}[v_i] = \deg_{\mu^+}[v_j] \text{ so}$$
$$l_1 + c_2 = l_2 + c_2$$
$$l_1 - l_2 = c_2 - c_2 = 0.$$

Then $l_1 = l_2$, which contradicts $l_1 \neq l_2$. Therefore, $\deg_{\mu^+}[v_i] \neq \deg_{\mu^+}[v_j]$.
Hence, $G$ is a neighborly totally irregular interval-valued fuzzy graph. ∎

**Theorem 7.7.17** *If an interval-valued fuzzy graph G is neighborly totally irregular and* $A = [\mu_A^-, \mu_A^+]$ *is a constant function, then it is a neighborly irregular interval-valued fuzzy graph.*

*Proof* Assume that $G$ is a neighborly totally irregular interval-valued fuzzy graph. Then the closed neighborhood degree of every two adjacent vertices is distinct. Let $v_i, v_j \in V$ and $\deg(v_i) = [k_1, l_1]$ and $\deg(v_j) = [k_2, l_2]$, where $k_1 \neq k_2$and $l_1 \neq l_2$. Assume that $(\mu_1(v_i), \nu_1(v_i)) = (\mu_1(v_j), \nu_1(v_j)) = [c_1, c_2]$, where $c_1$ and $c_2$ are constants and $c_1, c_2 \in [0, 1]$ and $\deg[v_i] \neq \deg[v_j]$.
Suppose that $\deg(v_i) \neq \deg(v_j)$. Then $\deg[v_i] \neq \deg[v_j]$. Thus, $\deg_{\mu^-}[v_i] \neq \deg_{\mu^-}[v_j]$ and $\deg_{\mu^+}[v_i] \neq \deg_{\mu^+}[v_j]$. Now,

$$\deg_{\mu^-}[v_i] \neq \deg_{\mu^-}[v_j]$$
$$k_1 + c_1 \neq k_2 + c_1$$
$$k_1 \neq k_2$$

and

$$\deg_{\mu^+}[v_i] \neq \deg_{\mu^+}[v_j]$$
$$l_1 + c_2 \neq l_2 + c_2$$
$$l_1 \neq l_2.$$

That is, the neighborhood degrees of adjacent vertices of $G$ are distinct. Hence, neighborhood degree of every pair of adjacent vertices is distinct in $G$. ∎

**Proposition 7.7.18** *If an interval-valued fuzzy graph G is neighborly irregular and neighborly totally irregular, then* $[\mu_A^-, \mu_A^+]$ *need not be a constant function.* ∎

Note that if $G$ is a neighborly irregular interval-valued fuzzy graph, then interval-valued subgraph $H = (A', B')$ if $G$ may not be neighborly irregular.
Note also that if $G$ is totally irregular interval-valued fuzzy graph, then interval-valued fuzzy subgraph $H = (A', B')$ of $G$ may not be totally irregular.

## 7.8 Self-centered Interval-Valued Fuzzy Graphs

The work in this section is due to [23]. We present some metric-aspects of interval-valued graphs. We also discuss some properties of self centered interval-valued fuzzy graphs.

A **complete graph** is a simple graph in which every pair of distinct vertices is connected by an edge. The complete graph on $n$ vertices has $n$ vertices and $n(n-1)/2$ edges. For a pair of vertices $u, v$ in a connected graph $G^*$, the **distance** $d(u, v)$ between $u$ and $v$ is the length of a shortest path connecting $u$ and $v$. The **eccentricity** $e(v)$ if a vertex $v$ in a graph $G^*$ is the distance from $v$ to a vertex furthest from $v$, that is $e(v) = \vee\{d(u, v) \mid u \in V\}$. The **radius** of a connected graph (or weighted graph) $G^*$ is defined as $rad(G^*) = \wedge\{e(v) \mid v \in V\}$. The **diameter** of a connected graph (or weighted graph) $G^*$ is defined as $diam(G^*) = \vee\{e(v) \mid v \in V\}$. The eccentric set $S$ of a graph is the set of eccentricities. The **center** $C(G^*)$ of a graph $G^*$ is the set of vertices with minimum eccentricity. A graph is **self-centered** if all its vertices lie in the center. Thus, the eccentricity of a self-centered graph contains only one element, that is, all the vertices have the same eccentricity. Equivalently, a self-centered graph is a graph whose diameter equals its radius.

**Definition 7.8.1** An interval-valued graph $G$ is called **complete** if

$$\mu_2^-(xy) = \mu_1^-(x) \wedge \mu_1^-(y) \text{ and } \mu_2^+(xy) = \mu_1^+(x) \wedge \mu_1^+(y)$$

for each $x, y \in V$.

**Definition 7.8.2** A **path** $P$ in an interval-valued fuzzy graph $G$ is a sequence of distinct vertices $v_1, v_2, \ldots, v_n$ such that either one of the following conditions is satisfied:

$(i)$ $\mu_{2ij}^- > 0$ and $\mu_{2ij}^+ = 0$ for some $i, j$.
$(ii)$ $\mu_{2ij}^- = 0$ and $\mu_{2ij}^+ > 0$ for some $i, j$.
A path $P : v_1 v_2 \cdots v_{n+1}$ in $G$ is called a **cycle** if $v_1 = v_{n+1}$ and $n \geq 3$.

**Definition 7.8.3** Let $P : v_0, v_1, v_2, \ldots, v_n$ be a path in an interval-valued fuzzy graph $G$. The $\mu^-$**-strength of the paths** connecting two vertices $v_i$ and $v_j$ is defined as $\vee(\mu_2^-(v_i, v_j))$ and is denoted by $(\mu_{2ij}^-)^\infty$. The $\mu^+$**-strength of the paths** connecting two vertices $v_i$ and $v_j$ is defined as $\vee(\mu_2^+(v_i, v_j))$ and is denoted by $(\mu_{2ij}^+)^\infty$. If some edge possesses both the $\mu^-$-strength and $\mu^+$-strength values, then it is the strength of the strongest path $P$ and it is denoted by $S_P = [(\mu_{2ij}^-)^\infty, (\mu_{2ij}^+)^\infty]$ for all $i, j = 1, 2, \ldots, n$.

**Definition 7.8.4** An interval-valued fuzzy graph $G$ is **connected** if any two vertices are joined by a path. That is, an interval-valued fuzzy graph is connected if $(\mu_{2ij}^-)^\infty > 0$ and $(\mu_{2ij}^+)^\infty > 0$.

*Example 7.8.5* Consider an interval-valued fuzzy connected graph such that $V = \{a, b, c, d\}$, $E = \{(a, b), (a, d), (b, d), (b, c), (c, d)\}$.

| | $a$ | $b$ | $c$ | $d$ |
|---|---|---|---|---|
| $\mu_1^-$ | 0.1 | 0.3 | 0.2 | 0.1 |
| $\mu_1^+$ | 0.4 | 0.5 | 0.5 | 0.4 |

| | $(a, b)$ | $(a, d)$ | $(b, d)$ | $(b, c)$ | $(c, d)$ |
|---|---|---|---|---|---|
| $\mu_2^-$ | 0.1 | 0.1 | 0.1 | 0.2 | 0.2 |
| $\mu_2^+$ | 0.3 | 0.4 | 0.4 | 0.4 | 0.3 |

It follows easily that $ad$ is a path of length 1 and the strength is $[0.1, 0.4]$, $abd$ is a path of length 2 and the strength is $[0.1, 0.4]$, and $abcd$ is a path of length 3 and the strength is $[0.2, 0.4]$.

**Definition 7.8.6** Let $G$ be a connected interval-valued fuzzy graph. The $\mu^-$**-length** of a path $P : v_1 v_2 \cdots v_n$ in $G$, $l_{\mu^-}(P)$, is defined as

$$l_{\mu^-}(P) = \sum_{i=1}^{n-1} \mu_2^-(v_i, v_{i+1}).$$

The $\mu^+$**-length** of a path $P : v_1 v_2 \cdots v_n$ in $G$, $l_{\mu^+}(P)$, is defined as

$$l_{\mu^+}(P) = \sum_{i=1}^{n-1} \mu_2^+(v_i, v_{i+1}).$$

The $\mu^-\mu^+$**-length** of a path $P : v_1 v_2 \cdots v_n$ in $G$, $l_{\mu^-\mu^+}(P)$, is defined as

$$l_{\mu^-\mu^+}(P) = [l_{\mu^-}(P),\ l_{\mu^+}(P)].$$

**Definition 7.8.7** Let $G$ be a connected interval-valued fuzzy graph. The $\mu^-$**-distance**, $\delta_{\mu^-}(v_i, v_j)$, is the smallest $\mu^-$-length of any $v_i - v_j$ path $P$ in $G$, where $v_i, v_j \in V$. That is, $\delta_{\mu^-}(v_i, v_j) = \wedge(l_{\mu^-}(P))$. The $\mu^+$**-distance,** $\delta_{\mu^+}(v_i, v_j)$, is the largest $\mu^+$-length of any $v_i - v_j$ path $P$ in $G$, where $v_i, v_j \in V$. That is, $\delta_{\mu^+}(v_i, v_j) = \vee(l_{\mu^+}(P))$. The **distance** $\delta(v_i, v_j)$, is defined as $\delta(v_i, v_j) = [\delta_{\mu^-}(v_i, v_j), \delta_{\mu^+}(v_i, v_j)]$.

**Definition 7.8.8** Let $G$ be a connected interval-valued fuzzy graph. For each $v_i \in V$, the $\mu^-$**-eccentricity** of $v_i$, denoted by $e_{\mu^-}(v_i)$, is defined as, $e_{\mu^-}(v_i) = \vee\{\delta_{\mu^-}(v_i, v_j) \mid v_j \in V,\ v_i \neq v_j\}$. For each $v_i \in V$, the $\mu^+$**-eccentricity** of $v_i$, denoted by $e_{\mu^+}(v_i)$, is defined as, $e_{\mu^+}(v_i) = \vee\{\delta_{\mu^+}(v_i, v_j) \mid v_j \in V,\ v_i \neq v_j\}$. For each $v_i \in V$, the **eccentricity** of $v_i$, denoted by $e(v_i)$, is defined as $e(v_i) = [e_{\mu^-}(v_i), e_{\mu^+}(v_i)]$.

**Definition 7.8.9** Let $G$ be a connected interval-valued fuzzy graph. The $\mu^-$**-radius** of $G$, denoted by $r_{\mu^-}(G)$, is defined as, $r_{\mu^-}(G) = \wedge\{e_{\mu^-}(v_i) \mid v_i \in V\}$. The $\mu^+$-**radius** of $G$, denoted by $r_{\mu^+}(G)$, is defined as, $r_{\mu^+}(G) = \wedge\{e_{\mu^+}(v_i) \mid v_i \in V\}$. The **radius** of $G$, denoted by $r(G)$, is defined as $r(G) = [r_{\mu^-}(G), r_{\mu^+}(G)]$.

**Definition 7.8.10** Let $G$ be a connected interval-valued fuzzy graph. The $\mu^-$-**diameter** of $G$, denoted by $d_{\mu^-}(G)$, is defined as, $d_{\mu^-}(G) = \vee\{e_{\mu^-}(v_i) \mid v_i \in V\}$. The $\mu^+$**-diameter** of $G$, denoted by $d_{\mu^+}(G)$, is defined as, $d_{\mu^+}(G) = \vee\{e_{\mu^+}(v_i) \mid v_i \in V\}$. The **diameter** of $G$, denoted by $d(G)$, is defined as $d(G) = [d_{\mu^-}(G), d_{\mu^+}(G)]$.

*Example 7.8.11* Consider the interval-valued fuzzy connected graph such that $V = \{a, b, c, d\}$, $E = \{(a, b), (a, c), (a, d), (b, c), (c, d)\}$.

| | $a$ | $b$ | $c$ | $d$ |
|---|---|---|---|---|
| $\mu_1^-$ | 0.1 | 0.3 | 0.2 | 0.1 |
| $\mu_1^+$ | 0.3 | 0.5 | 0.5 | 0.3 |

| | $(a, b)$ | $(a, c)$ | $(a, d)$ | $(b, c)$ | $(c, d)$ |
|---|---|---|---|---|---|
| $\mu_2^-$ | 0.1 | 0.1 | 0.1 | 0.1 | 0.1 |
| $\mu_2^+$ | 0.2 | 0.2 | 0.2 | 0.2 | 0.2 |

It follows easily that
$(i)$

$$\delta_{\mu^-}(a, b) = 0.1,\ \delta_{\mu^-}(a, c) = 0.1,\ \ \delta_{\mu^-}(a, d) = 0.1,$$
$$\delta_{\mu^-}(b, c) = 0.1,\ \delta_{\mu^-}(b, d) = 0.19,\ \delta_{\mu^-}(c, d) = 0.1,$$
$$\delta_{\mu^+}(a, b) = 0.2,\ \delta_{\mu^+}(a, c) = 0.2,\ \ \delta_{\mu^+}(a, d) = 0.2,$$
$$\delta_{\mu^+}(b, c) = 0.2,\ \delta_{\mu^+}(b, d) = 0.36,\ \delta_{\mu^+}(c, d) = 0.2.$$

The distance $\delta(v_i, v_j)$ is

$$\delta(a, b) = [0.1, 0.2],\ \delta(a, c) = [0.1, 0.2],\ \ \ \delta(a, d) = [0.1, 0.2].$$
$$\delta(b, c) = [0.1, 0.2],\ \delta(b, d) = [0.19, 0.36].\ \delta(c, d) = [0.1, 0.2].$$

$(ii)$ The $t$-eccentricity and $f$-eccentricity of each vertex is

$$e_{\mu^-}(a) = 0.1,\ e_{\mu^-}(b) = 0.19,\ e_{\mu^-}(c) = 0.1,\ e_{\mu^-}(d) = 0.1,$$
$$e_{\mu^+}(a) = 0.2,\ e_{\mu^+}(b) = 0.36,\ e_{\mu^+}(c) = 0.2,\ e_{\mu^+}(d) = 0.2.$$

The eccentricity of each vertex is

$$e(a) = [0.1, 0.2],\ e(b) = [0.19, 0.36],\ e(c) = [0.1, 0.2],\ e(d) = [0.1, 0.2].$$

$(iii)$ The radius of $G$ is $[0.1, 0.2]$ and the diameter of $G$ is $[0.19, 0.36]$.

**Definition 7.8.12** A vertex $v_i \in V$ is called a **central vertex** of a connected interval-valued fuzzy graph $G$ if $r_{\mu^-}(G) = e_{\mu^-}(v_i)$ and $r_{\mu^+}(G) = e_{\mu^+}(v_i)$ and the set of all central vertices of an interval-valued fuzzy graph is denoted by $C(G)$.

**Definition 7.8.13** A connected interval-valued fuzzy graph $G$ is **self-centered**, if every vertex of $G$ is a central vertex, that is, $r_{\mu^-}(G) = e_{\mu^-}(v_i)$ and $r_{\mu^+}(G) = e_{\mu^+}(v_i)$ for all $v_i \in V$.

*Example 7.8.14* Consider a connected interval-valued fuzzy graph $G$ such that $V = \{a, b, c\}$ and $E = \{ab, bc, ca\}$.

| | $a$ | $b$ | $c$ |
|---|---|---|---|
| $\mu_1^-$ | 0.3 | 0.4 | 0.2 |
| $\mu_1^+$ | 0.5 | 0.7 | 0.76 |

| | $(a, b)$ | $(b, c)$ | $(c, a)$ |
|---|---|---|---|
| $\mu_2^-$ | 0.1 | 0.1 | 0.1 |
| $\mu_2^+$ | 0.5 | 0.5 | 0.5 |

By routine computations, it is easy to see that the following properties hold.

$(i)$ The distance is

$$\delta(a, b) = [0.1, 0.5],\ \delta(a, c) = [0.1, 0.5],\ \delta(b, c) = [0.1, 0.5].$$

$(ii)$ The eccentricity of each vertex is $[0.1, 0.5]$.

$(iii)$ The radius of $G$ is $[0.1, 0.5]$. Hence, $G$ is a self-centered interval-valued fuzzy graph.

**Definition 7.8.15** A **path cover** of an interval-valued fuzzy graph $G$ is a set $P$ of paths such that every vertex $G$ is incident with some path in $P$.

**Definition 7.8.16** An **edge cover** of an interval-valued fuzzy graph $G$ is a set $L$ of edges such that every vertex $G$ is incident to some edge in $L$.

**Definition 7.8.17** An interval-valued fuzzy graph $G$ is said to be **bipartite** if the vertex set $V$ can be partitioned into two nonempty subsets $V_1$ and $V_2$ such that the following properties hold.

$(i)$ $\mu_2^-(v_i, v_j) = 0$ and $\mu_2^+(v_i, v_j) = 0$ if $v_i, v_j \in V_1$ or $v_i, v_j \in V_2$.

$(ii)$ $\mu_2^-(v_i, v_j) > 0$ and $\mu_2^+(v_i, v_j) > 0$ if $v_i \in V_1$ or $v_j \in V_2$ for some $i$ and $j$, (or)

$$\mu_2^-(v_i, v_j) = 0 \text{ and } \mu_2^+(v_i, v_j) > 0 \text{ if } v_i \in V_1 \text{ or } v_j \in V_2 \text{ for some } i \text{ and } j,$$

(or)

$$\mu_2^-(v_i, v_j) > 0 \text{ and } \mu_2^+(v_i, v_j) = 0 \text{ if } v_i \in V_1 \text{ or } v_j \in V_2 \text{ for some } i \text{ and } j.$$

**Theorem 7.8.18** *In an interval-valued fuzzy graph $G$ for which $A = [\mu_2^-, \mu_2^+] : V \times V \to D[0, 1]$ is not constant map, an edge $(v_i, v_j)$ for which $\mu_{2ij}^-$ is minimum and $\mu_{2ij}^+$ is maximum. Therefore, it is a bridge of $G$.*

**Theorem 7.8.19** *If $G$ is an interval-valued fuzzy bipartite graph then it has no strong cycle of odd length.*

*Proof* Let $G$ be an interval-valued fuzzy bipartite graph with interval-valued fuzzy bipartition $V_1$ and $V_2$. Suppose that is contains a string cycle of length, say $v_1, v_2, \ldots, v_n, v_1$ for some odd $n$. Without loss of generality, let $v_1 \in V_1$. Because $(v_i, v_{i+1})$ is strong for $i = 1, 2, \ldots, n-1$ and the nodes appear alternatively in $V_1$ and $V_2$, we have $v_n, v_1 \in V_1$. But this implies that $(v_n, v_1)$ is an edge in $V_1$, which contradicts the assumption that $G$ is an interval-valued fuzzy bipartite graph. Hence, an interval-valued fuzzy bipartite graph has no strong cycle of odd length. ■

**Theorem 7.8.20** ([23]) *Every complete interval-valued fuzzy graph $G$ is a self centered interval-valued fuzzy graph and $r_{\mu^-}(G) = \frac{1}{\mu^-_{1i}}$ and $r_{\mu^+}(G) = \frac{1}{\mu^+_{1i}}$, where is the $\mu^-_{1i}$ least and $\mu^+_{1i}$ is the greatest.*

*Proof* Let $G$ be a complete interval-valued fuzzy graph. To prove $G$ is a self centered interval-valued fuzzy graph, we have to show that every vertex is a central vertex. First we claim that $G$ is a $\mu^-$-self centered interval-valued fuzzy graph and $r_{\mu^-}(G) = \frac{1}{\mu^-_{1i}}$ is the least. Now, fix a vertex $v_i \in V$ such that $\mu^-_{1i}$is the least vertex membership of $G$.

**Case 1**: Consider all the $v_i - v_j$ paths $P$ of length $n$ in $G$, for all $v_j \in V$.

$(i)$ If $n = 1$, then $\mu^-$-length of $P = l_{\mu^-}(P) = \frac{1}{\mu^-_{1i}}$.

$(ii)$ If $n > 1$, then one of edges of $P$ possesses the $\mu^-$-strength $\mu^-_{1i}$ and hence, $\mu^-$-length of a $v_i - v_j$ path will exceed $\frac{1}{\mu^-_{1i}}$. That is $\mu^-$-length of $P = l_{\mu^-}(P) > \frac{1}{\mu^-_{1i}}$. Hence,

$$\delta^-_{\mu}(v_i, v_j) = \wedge(l_{\mu^-}(P)) = \frac{1}{\mu^-_{1i}}, \text{ for all } v_j \in V. \tag{7.7}$$

**Case 2**: Let $v_k \neq v_i \in V$. Consider all $v_k - v_j$ path $Q$ of length $n$ in $G$ for all $v_j \in V$.

$(i)$ If $n = 1$, then $\mu^-_2(v_k, v_j) = \mu^-_{1k} \wedge \mu^-_{1j} \geq \mu^-_{1i}$ because $\mu^-_{1i}$ is the least. Hence, then $\mu^-$-length of $Q = l_{\mu^-}(Q) = \frac{1}{\mu^-_2(v_k, v_j)} \leq \frac{1}{\mu^-_{1i}}$.

$(ii)$ If $n = 2$, then $l_{\mu^-}(Q) = \frac{1}{\mu^-_2(v_k, v_{k+1})} + \frac{1}{\mu^-_2(v_{k+1}, v_j)} \leq \frac{2}{\mu^-_{1i}}$ because $\mu^-_{1i}$ is the least.

$(iii)$ If $n > 2$, then $l_{\mu^-}(Q) \leq \frac{n}{\mu^-_{1i}}$ because $\mu^-_{1i}$ is the least. Thus,

$$\delta^-_{\mu}(v_k, v_j) = \wedge(l_{\mu^-}(Q)) \leq \frac{1}{\mu^-_{1i}} \text{ for all } v_k, v_j \in V. \tag{7.8}$$

From Eqs. (7.7) and (7.8), we have

$$e_{\mu^-}(v_i) = \wedge(\delta^-_{\mu}(v_i, v_j)) = \frac{1}{\mu^-_{1i}} \text{ for all } v_i \in V. \tag{7.9}$$

Hence, $G$ is $\mu^-$-self centered interval-valued fuzzy graph.

Now, $r_{\mu^-}(G) = \wedge(e_{\mu^-}(v_i)) = \frac{1}{\mu^-_{1i}}$ because by (7.9) $r_{\mu^-}(G) = \frac{1}{\mu^-_{1i}}$, $\mu^-_{1i}$ is the least.

Next we claim that $G$ is a $\mu^+$-self centered interval-valued fuzzy graph and $r_{\mu^+}(G) = \frac{1}{\mu^+_{1i}}$ is the greatest. Choose some vertex $v_i \in V$ such that $\mu^+_{1i}$ is the greatest vertex membership of $G$.

First, consider all the $v_i - v_j$ paths $P$ of length $n$ in $G$, for all $v_j \in V$.

$(i)$ If $n = 1$, then $\mu^+_{2ij} = \mu^+_{1i} \wedge \mu^+_{1j} = \mu^+_{1i}$. Therefore, $\mu^+$-length of $P = l_{\mu^+}(P) = \frac{1}{\mu^+_{1i}}$.

$(ii)$ If $n > 1$, then one of the edges of $P$ possesses the $\mu^+$ -strength $\mu^+_{1i}$ and hence, $\mu^=$-length of $P$ will exceed $\frac{1}{\mu^+_{1i}}$. That is, $\mu^+$-length of $P = l_{\mu+}(P) > \frac{1}{\mu^+_{1i}}$. Thus,

$$\delta^+_{\mu}(v_i, v_j) = \wedge(l_{\mu^+}(P)) = \frac{1}{\mu^+_{1i}} \text{ for all } v_j \in V. \tag{7.10}$$

Second, let $v_k \neq v_i \in V$. Consider all $v_k - v_j$ path $Q$ of length $n$ in $G$ for all $v_j \in V$.

$(i)$ If $n = 1$, then $\mu^+_2(v_k, v_j) = \mu^+_{1k} \wedge \mu^+_{1j} \leq \mu^+_{1i}$ because $\mu^+_{1i}$ is the greatest. Hence, then $\mu^+$-length of $Q = l_{\mu^+}(Q) = \frac{1}{\mu^+_2(v_k, v_j)} \geq \frac{1}{\mu^+_{1i}}$.

$(ii)$ If $n = 2$, then $l_{\mu^+}(Q) = \frac{1}{\mu^+_2(v_k, v_{k+1})} + \frac{1}{\mu^+_2(v_{k+1}, v_j)} \geq \frac{2}{\mu^-_{1i}}$ because $\mu^+_{1i}$ is the greatest.

$(iii)$ If $n > 2$, then $l_{\mu^+}(Q) \geq \frac{n}{\mu^+_{1i}}$ because $\mu^+_{1i}$ is the greatest. Hence,

$$\delta^+_{\mu}(v_k, v_j) = \wedge(l_{\mu^+}(Q)) \geq \frac{1}{\mu^+_{1i}}, \text{ for all } v_k, v_j \in V. \tag{7.11}$$

From Eqs. (7.10) and (7.11), we have

$$e_{\mu^+}(v_i) = \wedge(\delta^+_{\mu}(v_i, v_j)) = \frac{1}{\mu^+_{1i}}, \text{ for all } v_i \in V. \tag{7.12}$$

Hence, $G$ is $\mu^+$-self centered interval-valued fuzzy graph.

Now, $r_{\mu^+}(G) = \wedge(e_{\mu^+}(v_i)) = \frac{1}{\mu^+_{1i}}$ because by (7.10) $r_{\mu^+}(G) = \frac{1}{\mu^+_{1i}}$, $\mu^+_{1i}$ is the greatest.

From Eqs. (7.9) and (7.12), every vertex of $G$ is a central vertex. Thus, $G$ is a self centered interval-valued fuzzy graph. ∎

**Lemma 7.8.21** *An interval-valued fuzzy graph G is a self centered interval-valued fuzzy graph if and only if* $r_{\mu^-}(G) = D_{\mu^-}(G)$ *and* $r_{\mu^+}(G) = d_{\mu^+}(G)$.

**Theorem 7.8.22** *If G is a complete interval-valued fuzzy graph, then for at least one edge* $(\mu^-_2)^{\infty}(v_i, v_j) = \mu^-_2(v_i, v_j)$ *and* $(\mu^+_2)^{\infty}(v_i, v_j) = \mu^+_2(v_i, v_j)$.

*Proof* Let $G$ be a complete interval-valued fuzzy graph. Consider a vertex $v_i$, whose membership value is $\mu^-_{1i}$ and non-membership value is $\mu^+_{1i}$.

Let $\mu_{1i}^-$ be the least and $\mu_{1i}^+$ be the greatest in the vertex $v_i \in V$. Let $v_i, v_j \in V$. Then $(\mu_{2ij}^-, \mu_{2ij}^+) = (\mu_{1i}^-, \mu_{1i}^+)$ and $((\mu_{2ij}^-)^\infty, (\mu_2^+)^\infty) = (\mu_{1i}^-, \mu_{1i}^+)$. The strength of all the edges which are incident on the vertex $v_i$ is $(\mu_{1i}^-, \mu_{1i}^+)$ because $G$ is a complete interval-valued fuzzy graph.

Now, let $\mu_{1i}^-$ be the least and $\mu_{1k}^+$ be the greatest, where $v_i \neq v_k$. Then $(\mu_{2ik}^-, \mu_{2ik}^+) = (\mu_{1i}^-, \mu_{1k}^+)$. Because it is a complete interval-valued fuzzy graph, there will be an edge between $v_i$ and $v_k$. Therefore, $(\mu_{2ik}^-)^\infty = \mu_{1i}^-$ and $(\mu_{2ik}^+)^\infty = \mu_{1k}^+$. ■

**Theorem 7.8.23** *Let $G$ be a complete interval-valued fuzzy graph with path covers $P_1$ and $P_2$ of $G$. Then a necessary and sufficient condition for an interval-valued fuzzy graph to be self centered interval-valued fuzzy graph is*

$$\delta_\mu^-(v_i, v_j) = d_{\mu^-}(G) \textit{ for all } (v_i, v_j) \in P_1 \textit{ and} \tag{7.13}$$

$$\delta_\mu^+(v_i, v_j) = r_{\mu^+}(G) \textit{ for all } (v_i, v_j) \in P_2. \tag{7.14}$$

*Proof* Assume that $G$ is a self-centered interval-valued fuzzy graph. Then we must prove that Eqs. (7.13) and (7.14) hold. Suppose Eqs. (7.13) and (7.14) do not hold. Then we have $\delta_\mu^-(v_i, v_j) \neq d_{\mu^-}(G)$ for some $(v_i, v_j) \in P_1$ and $\delta_\mu^+(v_i, v_j) \neq r_{\mu^+}(G)$ for some $(v_i, v_j) \in P_2$.

By Lemma 7.8.21, the above inequality becomes $\delta_\mu^-(v_i, v_j) \neq r_{\mu^-}(G)$ for some $(v_i, v_j) \in P_1$ and $\delta_\mu^+(v_i, v_j) \neq r_{\mu^+}(G)$ for some $(v_i, v_j) \in P_2$. Thus, $e_{\mu^-}(v_i) \neq r_{\mu^-}(G)$, $e_{\mu^+}(v_i) \neq r_{\mu^+}(G)$ for some $v_i \in V$, which implies that $G$ is not a self-centered interval-valued fuzzy graph, a contradiction. Hence, $\delta_\mu^-(v_i, v_j) = d_{\mu^-}(G)$ for all $(v_i, v_j) \in P_1$ and $\delta_\mu^+(v_i, v_j) = r_{\mu^+}(G)$ for all $(v_i, v_j) \in P_2$.

Conversely, assume that Eqs. (7.13) and (7.14) hold. Then we must prove that $G$ is a self-centered interval-valued fuzzy graph. By Eqs. (7.13) and (7.14), we have that $e_{\mu^-}(v_i) = \delta_\mu^-(v_i, v_j)$ for all $(v_i, v_j) \in P_1$ and $e_{\mu^+}(v_i) = \delta_\mu^+(v_i, v_j)$ for all $(v_i, v_j) \in P_2$. Thus, $e_{\mu^-}(v_i) = r_{\mu^-}(G)$, $e_{\mu^+}(v_i) = r_{\mu^+}(G)$ for all $v_i \in V$. Hence, is a self-centered interval-valued fuzzy graph. ■

**Corollary 7.8.24** *If $G$ is connected complete interval-valued fuzzy graph with an edge cover $L$ of $G$, then a necessary and sufficient condition for an interval-valued fuzzy graph to be self-centered is*

$$\delta_{\mu^-}(v_i, v_j) = d_{\mu^-}(G) \textit{ for all } (v_i, v_j) \in L \textit{ and}$$

$$\delta_{\mu^+}(v_i, v_j) = r_{\mu^+}(G) \textit{ for all } (v_i, v_j) \in L.$$

**Theorem 7.8.25** *Let $H$ be a connected $\mu^-\mu^+$-self centered interval-valued fuzzy graph. Then there exists a connected interval-valued fuzzy graph $G$ such that $\langle C(G)\rangle$ is isomorphic to $H$. Also, $d_{\mu^-}(G) = 2r_{\mu^-}(G)$ and $d_{\mu^+}(G) = 2r_{\mu^+}(G)$.*

*Proof* Let $d_{\mu^-}(H) = l$ and $d_{\mu^+}(H) = m$. We construct $G$ from $H$ as follows:

Take two vertices $v_i, v_j \in V$ with $\mu_1^-(v_i) = \mu_1^-(v_j) = \frac{1}{l}$, $\mu_1^+(v_i) = \mu_1^+(v_j) = \frac{1}{2m}$ and join all the vertices of $H$ to both $v_i$ and $v_j$ with $\mu_2^-(v_i, v_k) = \mu_2^-(v_j, v_k) = \frac{1}{l}$,

$\mu_2^+(v_i, v_k) = \mu_2^+(v_j, v_k) = \frac{1}{2m}$ for all $v_k \in V'$. Put $\mu_1^- = (\mu^-)'_1$ and $\mu_1^+ = (\mu^+)'_1$ for all vertices in $H$ and $\mu_2^- = (\mu^-)'_2$ and $\mu_2^+ = (\mu^+)'_2$ for all edges in $H$.

We first show that $G$ is an interval-valued fuzzy graph. First note that $\mu_1^-(v_i) \leq \mu_1^-(v_k)$ for all $v_k \in H$. If possible, let $\mu_1^-(v_i) > \mu_1^-(v_k)$ for at least one vertex $v_k \in H$. Then $\frac{1}{l} > \mu_1^-(v_k)$, i.e., $l < \frac{1}{\mu_1^-(v_k)} \leq \frac{1}{\mu_2^-(v_k, v_i)}$, where the last inequality holds for all $v_i \in V'$ because $H$ is an interval-valued fuzzy graph. That is, $\frac{1}{\mu_2^-(v_k, v_i)} > l$ for all $v_k \in H$, which is a contradicts that $d_{\mu^-}(H) = l$. Therefore, $\mu_1^-(v_i) \leq \mu_1^-(v_k)$ for all $v_k \in V'$ and $\mu_2^-(v_i, v_k) \leq \mu_{1i}^- \wedge \mu_{1k}^- = \frac{1}{l}$. Similarly, $\mu_2^-(v_j, v_k) \leq \mu_{1j}^- \vee \mu_{1k}^- = \frac{1}{l}$ for all $v_k \in V'$. Note that $\mu_1^+(v_i) \leq \mu_1^+(v_k)$, $\mu_1^+(v_j) \leq \mu_1^+(v_k)$ for all $v_k \in V'$ because $d_{\mu^+}(H) = m$. Therefore, $\mu_2^+(v_i, v_k) \leq \mu_{1i}^+ \wedge \mu_{1k}^+ = \frac{1}{2m}$. Similarly, $\mu_2^+(v_j, v_k) \leq \mu_{1j}^+ \wedge \mu_{1k}^+ = \frac{1}{2m}$. Hence, $G$ is an interval-valued fuzzy graph.

Also, $e_{\mu^-}(v_k) = l$ for all $v_k \in V'$ and $e_{\mu^-}(v_i) = e_{\mu^-}(v_j) = \frac{1}{\mu_2^+(v_i, v_k)} + \frac{1}{\mu_2^+(v_k, v_j)} = 2l$, $r_{\mu^-}(G) = l$, $d_{\mu^-}(G) = 2l$. Next $e_{\mu^+}(v_k) = m$ for all $v_k \in V'$ and $e_{\mu^+}(v_i) = e_{\mu^+}(v_j) = \frac{1}{\mu_2^+(v_j, v_k)} = 2m$ for all $v_k \in V'$. Therefore, $r_{\mu^+}(G) = m$, $d_{\mu^+}(G) = 2m$. Hence, $\langle C(G) \rangle$ is isomorphic to $H$. ■

**Theorem 7.8.26** *An interval-valued fuzzy graph G is self-centered if and only if* $\delta_{\mu^-}(v_i, v_j) \leq r_{\mu^-}(G)$, $\delta_{\mu^+}(v_i, v_j) \geq r_{\mu^-}(G)$ *for all* $v_i, v_j \in G$.

*Proof* Assume that $G$ is a self-centered interval-valued fuzzy graph. That is, $e_{\mu^-}(v_i) = e_{\mu^-}(v_j)$ and $e_{\mu^+}(v_i) = e_{\mu^+}(v_j)$ for all $v_i, v_j \in V$, $r_{\mu^-}(G) = e_{\mu^-}(v_i)$, $r_{\mu^+}(G) = e_{\mu^+}(v_i)$ for all $v_i \in V$. Now, we wish to show that $\delta_{\mu^-}(v_i, v_j) \leq r_{\mu^-}(G)$ and $\delta_{\mu^+}(v_i, v_j) \geq r_{\mu^+}(G)$ for all $v_i, v_j \in V$. By the definition of the eccentricity, we obtain $\delta_{\mu^-}(v_i, v_j) \leq e_{\mu^-}(v_i)$, $\delta_{\mu^+}(v_i, v_j) \geq e_{\mu^+}(v_i)$ for all $v_i, v_j \in V$. This is possible only when $e_{\mu^-}(v_i) = e_{\mu^-}(v_j)$ and $e_{\mu^+}(v_i) = e_{\mu^+}(v_j)$ for all $v_i, v_j \in V$. Because $G$ is a self-centered interval-valued fuzzy graph, the above inequality becomes $\delta_{\mu^-}(v_i, v_j) \leq r_{\mu^-}(G)$, $\delta_{\mu^+}(v_i, v_j) \geq r_{\mu^-}(G)$ for all $v_i, v_j \in G$.

Conversely, suppose that $\delta_{\mu^-}(v_i, v_j) \leq r_{\mu^-}(G)$, $\delta_{\mu^+}(v_i, v_j) \geq r_{\mu^-}(G)$ for all $v_i, v_j \in G$. Then we must prove that $G$ is a self-centered interval-valued fuzzy graph. Suppose that $G$ is not a self-centered interval-valued fuzzy graph. Then $e_{\mu^-}(v_i) \neq r_{\mu^-}(G)$ and $e_{\mu^+}(v_i) \neq r_{\mu^+}(G)$ for some $v_i \in V$. Assume that $e_{\mu^-}(v_i)$ and $e_{\mu^+}(v_i)$ are the least values among all other eccentricities. That is,

$$r_{\mu^-}(G) = e_{\mu^-}(v_i) \text{ and } r_{\mu^+}(G) = e_{\mu^+}(v_i), \tag{7.15}$$

where $e_{\mu^-}(v_i) < e_{\mu^-}(v_j)$ and $e_{\mu^+}(v_i) < e_{\mu^+}(v_j)$ for some $v_i, v_j \in V$ and

$$\delta_{\mu^-}(v_i, v_j) = e_{\mu^-}(v_j) > e_{\mu^-}(v_i), \tag{7.16}$$

$$\delta_{\mu^+}(v_i, v_j) = e_{\mu^+}(v_j) > e_{\mu^+}(v_i), \tag{7.17}$$

for some $v_i, v_j \in V$.

Hence, from Eqs. (7.15)–(7.17), we have $\delta_{\mu^-}(v_i, v_j) > r_{\mu^-}(G)$, $\delta_{\mu^+}(v_i, v_j) < r_{\mu^+}(G)$ for some $v_i, v_j \in V$, which is a contradiction to the fact that $\delta_{\mu^-}(v_i, v_j) \leq$

$r_{\mu^-}(G)$, $\delta_{\mu^+}(v_i, v_j) \geq r_{\mu^+}(G)$ for all $v_i, v_j \in V$. Hence, $G$ is a self-centered interval-valued fuzzy graph. ■

**Theorem 7.8.27** *Let G be an interval-valued fuzzy graph. If the graph G is a complete bipartite interval-valued fuzzy graph, then the complement of G is a self-centered interval-valued fuzzy graph.*

*Proof* A bipartite interval-valued fuzzy graph $G$ is said to be complete if

$$\mu_2^-(v_i, v_j) = \mu_1^-(v_i) \wedge \mu_1^-(v_j)$$
$$\mu_2^+(v_i, v_j) = \mu_1^+(v_i) \wedge \mu_1^+(v_j)$$

for all $v_i \in V_1$ and $v_j \in V_2$, and

$$\mu_2^-(v_i, v_j) = 0, \ \mu_2^+(v_i, v_j) = 0 \tag{7.18}$$

for all $v_i, v_j \in V_1$ or $v_i, v_j \in V_2$.

Now,

$$\mu_2^-(v_i, v_j) = \mu_1^-(v_i) \wedge \mu_1^-(v_j) - \mu_{2ij}^-,$$
$$\mu_2^+(v_i, v_j) = \mu_1^+(v_i) \wedge \mu_1^+(v_j) - \mu_{2ij}^+.$$

By (7.18), we have that

$$\mu_2^-(v_i, v_j) = \mu_1^-(v_i) \wedge \mu_1^-(v_j) \tag{7.19}$$

$$\mu_2^+(v_i, v_j) = \mu_1^+(v_i) \wedge \mu_1^+(v_j) \tag{7.20}$$

for all $v_i, v_j \in V_1$ or $v_i, v_j \in V_2$.

Thus, from Eqs. (7.18)–(7.20), the complement of $G$ has two components and each component is a complete interval-valued fuzzy graph and is clearly a self-centered interval-valued fuzzy graph. ■

An interval-valued fuzzy set is a generalization of the notion of a fuzzy set. Because interval-valued fuzzy models give more precision, flexibility and compatibility to the system as compared to the fuzzy models, we have introduced the concept of self centered interval-valued fuzzy graphs in this paper. The concept of interval-valued graph can be applied in various domains of engineering and computer science.

# Chapter 8
# Bipolar Fuzzy Graphs

## 8.1 Bipolar Fuzzy Sets

In 1994, Zhang [195, 196] introduced the concept of bipolar fuzzy sets as a generalization of the notion of Zadeh's fuzzy sets. A bipolar fuzzy subset of a set is a pair of functions one from the set into the interval $[0, 1]$ and the other into the interval $[-1, 0]$. In a bipolar fuzzy set, the membership degree 0 of an element can be interpreted that the element is irrelevant to the corresponding property, the membership degree in $(0, 1]$ of an element indicates the intensity that the element satisfies the property, and the membership degree in $[-1, 0)$ of an element indicates the element does not satisfy the property. Fuzzy and possibilistic formalisms for bipolar information have been proposed in [71] because bipolarity exists when dealing with spatial information in image processing or in spatial reasoning applications.

The positive degree membership $\mu_B^P$ is used to denote the satisfaction degree of an element $x$ to the property corresponding to a bipolar fuzzy set $B$, and the negative degree membership $\mu_B^N$ is used to denote the satisfaction degree of an element $x$ to some implicit counter-property corresponding to a bipolar fuzzy set $B$. If $\mu_B^P(x) \neq 0$ and $\mu_B^N(x) = 0$, it is the situation that $x$ is regarded as having only positive satisfaction for $B$. If $\mu_B^P(x) = 0$ and $\mu_B^N(x) \neq 0$, it is the situation that $x$ does not satisfy the property of $B$, but somewhat satisfies the counter-property of $B$. If $\mu_B^P(x) \neq 0$ and $\mu_B^N(x) \neq 0$, then there is both a positive intensity for which $x$ satisfies the property and negative intensity with which it doesn't.

We note that results involving $\mu_B^N$ are analogous to those of $\mu_B^P$. Define the function $f : [-1, 0] \to [0, 1]$ by for all $x \in [-1, 0]$, $f(x) = -x$. Then clearly $f$ is a one-to-one function of $[-1, 0]$ onto $[0, 1]$. Now, for all $x, y \in [-1, 0]$, $f(x \wedge y) = -(x \wedge y) = -x \vee -y = f(x) \vee f(y)$ and $f(x \vee y) = -(x \vee y) = -x \wedge -y = f(x) \wedge f(y)$. Thus, the algebraic structures $([-1, 0], \wedge, \vee)$ and $([0, 1], \vee, \wedge)$ are isomorphic. However, we supply full proofs at times for the sake of completeness and due to the fact that applications involve different interpretations of $\mu_B^P$ and $\mu_B^N$ at times.



S. Mathew et al., *Fuzzy Graph Theory*, Studies in Fuzziness
and Soft Computing 363, https://doi.org/10.1007/978-3-319-71407-3_8

**Definition 8.1.1** Let $X$ be a non-empty set. A **bipolar fuzzy set** $B$ in $X$ is a set of triples

$$B = \{(x, \mu_B^P(x), \mu_B^N(x)) \mid x \in X\},$$

where $\mu_B^P : X \to [0, 1]$ and $\mu_B^N : X \to [-1, 0]$.

We often write $B = (\mu_B^P, \mu_B^N)$ for the bipolar fuzzy set $B = \{(x, \mu_B^P(x), \mu_B^N(x)) \mid x \in X\}$. We also use the notation $B = \{(x, m^+(x), m^-(x)) \mid x \in X\}$ or $B = (m^+, m^-)$ for a bipolar fuzzy set.

**Definition 8.1.2** Let $B = \{(x, m^+(x), m^-(x) \mid x \in X\}$ be a bipolar set on a non-empty set $X$. The **height** of $B$, written $h(B)$, is defined by $h(B) = \vee\{m^+(x) \mid x \in X\}$ and the **depth** of $B$, written $d(B)$, is defined as $d(B) = \wedge\{m^-(x) \mid x \in X\}$.

**Definition 8.1.3** Let $B_1 = \{(x, m_1^+(x), m_1^-(x) \mid x \in X\}$ and $B_2 = \{(x, m_2^+(x), m_2^-(x) \mid x \in X\}$ be bipolar fuzzy sets in $X$. Then $B_1 \subseteq B_2$ if $m_1^+(x) \leq m_2^+(x)$ for all $x \in X$ and $m_1^-(x) \geq m_2^-(x)$ for all $x \in X$.

**Definition 8.1.4** Let $B = \{(x, m^+(x), m^-(x) \mid x \in X\}$ be a bipolar set on a non-empty set $X$. The **support** of $B$, written Supp$(B)$, is defined by Supp$(B) = \{x \in X \mid m^+(x) \neq 0$ or $m^-(x) \neq 0\}$. The **upper core** of $B$, written $\overline{c}(B)$, is defined by $\overline{c}(B) = \{x \in X \mid m^+(x) = 1\}$ and the **lower core** of $B$, written $\underline{c}(B)$, is defined by $\underline{c}(B) = \{x \in X \mid m^-(x) = -1\}$.

**Definition 8.1.5** Let $B = (m^+, m^-)$ be a bipolar fuzzy set. Let $t_1 \in (0, 1]$ and $t_2 \in [-1, 0)$. Define the $\{t_1, t_2\}$ **cut level set** of $B$ to be the set $B_{t_2}^{t_1} = \{x \in \text{Supp}(B) \mid m^+(x) \geq t_1$ and $m^-(x) \leq t_2\}$.

**Theorem 8.1.6** *Let $A$ and $B$ be bipolar fuzzy sets. Then $A \subseteq B$ if and only if $A_k^t \subseteq B_k^t$ for all $t \in (0, 1], k \in [-1, 0)$.*

*Proof* Let $A = (m_1^+, m_1^-)$ and $B = (m_2^+, m_2^-)$. Suppose $A \subseteq B$. Then $m_1^+(x) \leq m_2^+(x)$ for all $x \in X$ and $m_1^-(x) \geq m_2^-(x)$ for all $x \in X$. Thus, $m_1^+(x) \geq t$ implies $m_2^+(x) \geq t$ and $m_1^-(x) \leq k$ implies $m_2^-(x) \leq k$ for all $x \in X$ and for all $t \in (0, 1]$ and for all $k \in [-1, 0)$. Hence, $A_k^t \subseteq B_k^t$ for all $t \in (0, 1], k \in [-1, 0)$.

Conversely, suppose $A_k^t \subseteq B_k^t$ for all $t \in (0, 1], k \in [-1, 0)$. Then to show $A \subseteq B$, it suffices to show $m_1^+(x) \leq m_2^+(x)$ for all $x \in X$ and $m_1^-(x) \geq m_2^-(x)$ for all $x \in X$. If $x \in X$ is such that $m_1^+(x) = t$ and $m_1^-(x) = k$, then $x \in A_k^t$ and so $x \in B_k^t$. Hence, $m_2^+(x) \geq t$ and $m_2^-(x) \leq k$. ∎

**Definition 8.1.7** Let $A = (\mu_A^P, \mu_A^N)$ and $B = (\mu_B^P, \mu_B^N)$ be bipolar fuzzy sets in $X$. We define the bipolar fuzzy sets $A \cap B$ and $A \cup B$ as follows: for all $x \in X$,

$$(A \cap B)(x) = (\mu_A^P(x) \wedge \mu_B^P(x), \mu_A^N(x) \vee \mu_B^N(x)),$$
$$(A \cup B)(x) = (\mu_A^P(x) \vee \mu_B^P(x), \mu_A^N(x) \wedge \mu_B^N(x)).$$

**Theorem 8.1.8** *Let $B_1$ and $B_2$ be bipolar fuzzy sets. Then $(B_1 \cap B_2)_{t_2}^{t_1} = (B_1)_{t_2}^{t_1} \cap (B_2)_{t_2}^{t_1}$ for all $t_1 \in (0, 1]$ and for all $t_2 \in [-1, 0)$.*

*Proof* Let $B_1 = (m_1^+, m_1^-)$ and $B_2 = (m_2^+, m_2^-)$. Then $B_1 \cap B_2$ is a bipolar fuzzy set on $X$. Let $B_1 \cap B_2 = (m^+, m^-)$, where $m^+ = m_1^+ \cap m_2^+$ and $m^- = m_1^- \cup m_2^-$. Then $x \in (B_1 \cap B_2)_{t_2}^{t_1} \Leftrightarrow m^+(x) \geq t_1$ and $m^-(x) \leq t_2 \Leftrightarrow (m_1^+ \cap m_2^+)(x) \geq t_1$ and $(m_1^- \cup m_2^-)(x) \leq t_2 \Leftrightarrow (m_1^+)(x) \geq t_1$ and $(m_2^+)(x) \geq t_1$ and $(m_1^-)(x) \leq t_2$ and $(m_2^-)(x) \leq t_2 \Leftrightarrow x \in (B_1)_{t_2}^{t_1} \cap (B_2)_{t_2}^{t_1}$. ■

*Example 8.1.9* Let $X = \{x\}$. Let $B_1 = (m_1^+, m_1^-)$ and $B_2 = (m_2^+, m_2^-)$ be bipolar sets defined on $X$ as follows:

$$m_1^+(x) = 3/4, m_1^-(x) = -1/4, m_2^+(x) = 1/4, m_2^-(x) = -3/4.$$

Let $t_1 = 1/2$ and $t_2 = -1/2$. Then $m_1^+(x) \nleq 1/2$ and $m_2^-(x) \ngeq -1/2$. Thus, $(B_1)_{-1/2}^{1/2} = \emptyset$ and $(B_2)_{-1/2}^{1/2} = \emptyset$. However, $m_1^+(x) \geq 1/2$ or $m_2^+(x) \geq 1/2$ and $m_1^-(x) \leq -1/2$ or $m_2^-(x) \leq -1/2$. That is, $m_1^+(x) \vee m_2^+(x) \geq 1/2$ and $m_1^-(x) \wedge m_2^-(x) \leq -1/2$. Thus, $(B_1 \cup B_2)_{-1/2}^{1/2} = \{x\}$. Hence, $(B_1)_{-1/2}^{1/2} \cup (B_2)_{-1/2}^{1/2} \subset (B_1 \cup B_2)_{-1/2}^{1/2}$.

**Theorem 8.1.10** *Let $B_1$ and $B_2$ be bipolar fuzzy sets. Then $(B_1)_{t_2}^{t_1} \cup (B_2)_{t_2}^{t_1} \subseteq (B_1 \cup B_2)_{t_2}^{t_1}$ for all $t_1 \in (0, 1]$ and for all $t_2 \in [-1, 0)$.*

*Proof* The result is immediate from the fact that $B_i \subseteq B_1 \cup B_2, i = 1, 2$. ■

## 8.2 Bipolar Fuzzy Graphs

In this section, we introduce the notion of bipolar fuzzy graphs. Most of the results in the next few sections are due to Akram [5]. We also include the work of [184]. We describe various methods of their constructions and introduce the concept of isomorphism of bipolar fuzzy graphs. We introduce the notion of strong bipolar fuzzy graphs. We also examine the notions of self-complimentary and self weak complementary strong bipolar fuzzy graphs.

Results and applications not mentioned here can be found in [1, 7, 10, 14, 24, 33, 71, 98, 125].

**Definition 8.2.1** A **bipolar fuzzy graph** with an underlying set $V$ is defined to be a pair $G = (A, B)$, where $A = (\mu_A^P, \mu_A^N)$ is a bipolar fuzzy set in $V$ and $B = (\mu_B^P, \mu_B^N)$ is a bipolar fuzzy set in $E$ such that

$$\mu_B^P(xy) \leq \mu_A^P(x) \wedge \mu_A^P(y) \text{ and } \mu_B^N(xy) \geq \mu_A^N(x) \vee \mu_A^N(y) \text{ for all } xy \in E,$$
$$\mu^P(xy) = \mu_B^N(xy) = 0 \text{ for all } xy \in \mathcal{E} \backslash E.$$

We call $A$ the **bipolar fuzzy vertex set** of $V$ and $B$ the **bipolar fuzzy edge set** of $E$, respectively.

The requirement that $\mu^P(xy) = \mu_B^N(xy) = 0$ for all $xy \in \mathcal{E}\setminus E$ in Definition 8.2.1 is needed when we introduce the notion of the complement of $G$.

*Example 8.2.2* Consider a graph $G^* = (V, E)$, where $V = \{a, b, c\}$, $E = \{ab, bc, ca\}$. Let $A = (\mu_A^P, \mu_A^N)$ is a bipolar fuzzy subset of $V$ and $B = (\mu_B^P, \mu_B^N)$ is a bipolar fuzzy subset of $E$ defined by

| | $a$ | $b$ | $c$ |
|---|---|---|---|
| $\mu_A^P$ | 0.5 | 0.6 | 0.4 |
| $\mu_A^N$ | −0.7 | −0.5 | −0.7 |

| | $ab$ | $bc$ | $ca$ |
|---|---|---|---|
| $\mu_B^P$ | 0.4 | 0.3 | 0.3 |
| $\mu_B^N$ | −0.3 | −0.2 | −0.1 |

It follows easily that $G = (A, B)$ is a bipolar fuzzy graph of $G^*$.

We next define certain operations of bipolar fuzzy graphs. Let $G_1^* = (V_1, E_1)$ and $G_2^* = (V_2, E_2)$ be graphs in this section.

**Definition 8.2.3** Let $A_1 = (\mu_{A_1}^P, \mu_{A_1}^N)$ and $A_2 = (\mu_{A_2}^P, \mu_{A_2}^N)$ be bipolar fuzzy subsets of $V_1$ and $V_2$ and $B_1 = (\mu_{B_1}^P, \mu_{B_1}^N)$ and $B_2 = (\mu_{B_2}^P, \mu_{B_2}^N)$ be bipolar fuzzy subsets of $E_1$ and $E_2$, respectively. Then we denote the **Cartesian product** of two bipolar fuzzy graphs $G_1$ and $G_2$ of the graphs $G_1^*$ and $G_2^*$ by $G_1 \times G_2 = (A_1 \times A_2, B_1 \times B_2)$ and defined by

(*i*)
$$(\mu_{A_1}^P \times \mu_{A_2}^P)(x_1, x_2) = \mu_{A_1}^P(x_1) \wedge \mu_{A_2}^P(x_2)$$

and
$$(\mu_{A_1}^N \times \mu_{A_2}^N)(x_1, x_2) = \mu_{A_1}^N(x_1) \vee \mu_{A_2}^N(x_2)$$

for all $(x_1, x_2) \in V \times V$.

(*ii*)
$$(\mu_{B_1}^P \times \mu_{B_2}^P)((x, x_2)(x, y_2)) = \mu_{A_1}^P(x) \wedge \mu_{B_2}^P(x_2y_2)$$

and
$$(\mu_{B_1}^N \times \mu_{B_2}^N)((x, x_2)(x, y_2)) = \mu_{A_1}^N(x) \vee \mu_{B_2}^N(x_2y_2)$$

for all $x \in V_1$, for all $x_2y_2 \in E_2$.

(*iii*)
$$(\mu_{B_1}^P \times \mu_{B_2}^P)((x_1, z)(y_1, z)) = \mu_{B_1}^P(x_1y_1) \wedge \mu_{A_2}^P(z)$$

and
$$(\mu_{B_1}^N \times \mu_{B_2}^N)((x_1, z)(y_1, z)) = \mu_{B_1}^N(x_1y_1) \vee \mu_{A_2}^N(z)$$

for all $z \in V_2$, for all $x_1y_1 \in E_1$.

In the previous definition, the set of edges of $G_1 \times G_2$ is the set $E = \{(x, x_2)(x, y_2) \mid x \in V_1, x_2y_2 \in E_2\} \cup \{(x_1, z)(y_1, z) \mid z \in V_2, x_1y_1 \in E_1\}$.

**Proposition 8.2.4** *If $G_1$ and $G_2$ are bipolar fuzzy graphs, then $G_1 \times G_2$ is a bipolar fuzzy graph.*

*Proof* Let $x \in V_1$ and $x_2y_2 \in E_2$. Then

$$\begin{aligned}
(\mu^P_{B_1} \times \mu^P_{B_2})((x, x_2)(x, y_2)) &= \mu^P_{A_1}(x) \wedge \mu^P_{B_2}(x_2y_2) \\
&\le \wedge\{\mu^P_{A_1}(x), \mu^P_{A_2}(x_2) \wedge \mu^P_{A_2}(y_2)\} \\
&= \wedge\{(\mu^P_{A_1}(x) \wedge \mu^P_{A_2}(x_2)), \mu^P_{A_1}(x) \wedge \mu^P_{A_2}(y_2)\} \\
&= (\mu^P_{A_1} \times \mu^P_{A_2})(x, x_2) \wedge (\mu^P_{A_1} \times \mu^P_{A_2})(x, y_2),
\end{aligned}$$

$$\begin{aligned}
(\mu^N_{B_1} \times \mu^N_{B_2})((x, x_2)(x, y_2)) &= \mu^N_{A_1}(x) \vee \mu^N_{B_2}(x_2y_2) \\
&\ge \vee\{\mu^N_{A_1}(x), (\mu^N_{A_2}(x_2) \vee \mu^N_{A_2}(y_2)\} \\
&= \vee\{\mu^N_{A_1}(x) \vee \mu^N_{A_2}(x_2), \mu^N_{A_1}(x) \vee \mu^N_{A_2}(y_2)\} \\
&= (\mu^N_{A_1} \times \mu^N_{A_2})(x, x_2) \vee (\mu^N_{A_1} \times \mu^N_{A_2})(x, y_2).
\end{aligned}$$

Let $z \in V_2$, and $x_1y_1 \in E_1$. Then

$$\begin{aligned}
&(\mu^P_{B_1} \times \mu^P_{B_2})((x_1, z)(y_1, z)) \\
&= \mu^P_{B_1}(x_1y_1) \wedge \mu^P_{A_2}(z) \\
&\le \wedge\{(\mu^P_{A_1}(x_1) \wedge \mu^P_{A_1}(y_1)), \mu^P_{A_2}(z)\} \\
&= \wedge\{\mu^P_{A_1}(x_1) \wedge \mu^P_{A_2}(z), \mu^P_{A_1}(y_1) \wedge \mu^P_{A_2}(z)\} \\
&= (\mu^P_{A_1} \times \mu^P_{A_2})(x_1, z) \wedge (\mu^P_{A_1} \times \mu^P_{A_2}(y_1, z)),
\end{aligned}$$

$$\begin{aligned}
(\mu^N_{B_1} \times \mu^N_{B_2})((x_1, z)(y_1, z)) &= \mu^N_{B_1}(x_1y_1) \vee \mu^N_{A_2}(z) \\
&\ge \vee\{\mu^N_{A_1}(x_1) \vee \mu^N_{A_1}(y_1), \mu^N_{A_2}(z)\} \\
&= \vee\{\mu^N_{A_1}(x_1) \vee \mu^N_{A_2}(z), \mu^N_{A_1}(y_1) \vee \mu^N_{A_2}(z)\} \\
&= (\mu^N_{A_1} \times \mu^N_{A_2})(x_1, z) \vee (\mu^N_{A_1} \times \mu^N_{A_2})(y_1, z).
\end{aligned}$$

■

Let $E = \{(x, x_2)(x, y_2) \mid x \in V_1, x_2y_2 \in E_2\} \cup \{(x_1, z)(y_1, z) \mid z \in V_2, x_1y_1 \in E_1\}$ and let $E^{(0)} = E \cup (x_1, x_2)(y_1, y_2) \mid x_1y_1 \in E_1$ and $x_2, y_2 \in V_2, x_2 \neq y_2\}$.

Let $E$ be defined as following Definition 8.2.3 and let $E^{(0)} = E \cup \{(x_1, x_2)(y_1, y_2) \mid x_1y_1 \in E_1, x_2 \neq y_2\}$.

**Definition 8.2.5** Let $A_1 = (\mu^P_{A_1}, \mu^N_{A_1})$ and $A_2 = (\mu^P_{A_2}, \mu^N_{A_2})$ be bipolar fuzzy subsets of $V_1$ and $V_2$ and $B_1 = (\mu^P_{B_1}, \mu^N_{B_1})$ and $B_2 = (\mu^P_{B_2}, \mu^N_{B_2})$ be bipolar fuzzy subsets of

$E_1$ and $E_2$, respectively. We denote the **composition** of two bipolar fuzzy graphs $G_1$ and $G_2$ of the graphs $G_1^*$ and $G_2^*$ by $G_1[G_2] = (A_1 \circ A_2, B_1 \circ B_2)$ and defined by

(*i*)

$$(\mu_{A_1}^P \circ \mu_{A_2}^P)(x_1, x_2) = \mu_{A_1}^P(x_1) \wedge \mu_{A_2}^P(x_2)$$

and

$$(\mu_{A_1}^N \circ \mu_{A_2}^N)(x_1, x_2) = \mu_{A_1}^N(x_1) \vee \mu_{A_2}^N(x_2)$$

for all $(x_1, x_2) \in V$.

(*ii*)

$$(\mu_{B_1}^P \circ \mu_{B_2}^P)((x, x_2)(x, y_2)) = \mu_{A_1}^P(x) \wedge \mu_{B_2}^P(x_2y_2)$$

and

$$(\mu_{B_1}^N \circ \mu_{B_2}^N)((x, x_2)(x, y_2)) = \mu_{A_1}^N(x) \vee \mu_{B_2}^N(x_2y_2)$$

for all $x \in V_1$, for all $x_2y_2 \in E_2$.

(*iii*)

$$(\mu_{B_1}^P \circ \mu_{B_2}^P)((x_1, z)(y_1, z)) = \mu_{B_1}^P(x_1y_1) \wedge \mu_{A_2}^P(z)$$

and

$$(\mu_{B_1}^N \circ \mu_{B_2}^N)((x_1, z)(y_1, z)) = \mu_{B_1}^N(x_1y_1) \vee \mu_{A_2}^N(z)$$

for all $z \in V_2$, for all $x_1y_1 \in E_1$.

(*iv*)

$$(\mu_{B_1}^P \circ \mu_{B_2}^P)((x_1, x_2)(y_1, y_2)) = \wedge\{\mu_{A_2}^P(x_2),\ \mu_{A_2}^P(y_2), \mu_{B_1}^P(x_1y_1)\}$$

and

$$(\mu_{B_1}^N \circ \mu_{B_2}^N)((x_1, x_2)(y_1, y_2)) = \vee\{\mu_{A_2}^N(x_2),\ \mu_{A_2}^N(y_2),\ \mu_{B_1}^N(x_1y_1)\}$$

for all $(x_1, x_2)(y_1, y_2) \in E^0 \setminus E$.

**Proposition 8.2.6** *If $G_1$ and $G_2$ are bipolar fuzzy graphs, then $G_1[G_2]$ is a bipolar fuzzy graph.*

*Proof* Let $x \in V_1$ and $x_2y_2 \in E_2$. Then

$$\begin{aligned}
&(\mu_{B_1}^P \circ \mu_{B_2}^P)((x, x_2)(x, y_2)) \\
&= \mu_{A_1}^P(x) \wedge \mu_{B_2}^P(x_2y_2) \\
&\le \wedge\{\mu_{A_1}^P(x), \mu_{A_2}^P(x_2) \wedge \mu_{A_2}^P(y_2)\} \\
&= \wedge\{\mu_{A_1}^P(x) \wedge \mu_{A_2}^P(x_2), \mu_{A_1}^P(x) \wedge \mu_{A_2}^P(y_2)\} \\
&= (\mu_{A_1}^P \circ \mu_{A_2}^P)(x, x_2) \wedge (\mu_{A_1}^P \circ \mu_{A_2}^P(x, y_2)),
\end{aligned}$$

$$\begin{aligned}(\mu^N_{B_1}\circ\mu^N_{B_2})((x,x_2)(x,y_2)) &= \mu^N_{A_1}(x)\vee\mu^N_{B_2}(x_2y_2)\\ &\geq \vee\{\mu^N_{A_1}(x),\mu^N_{A_2}(x_2)\vee\mu^N_{A_2}(y_2)\}\\ &= \vee\{\mu^N_{A_1}(x)\vee\mu^N_{A_2}(x_2),\mu^N_{A_1}(x)\vee\mu^N_{A_2}(y_2)\}\\ &= (\mu^N_{A_1}\circ\mu^N_{A_2})(x,x_2)\vee(\mu^N_{A_1}\circ\mu^N_{A_2})(x,y_2).\end{aligned}$$

Let $z\in V_2$, and $x_1y_1\in E_1$. Then

$$\begin{aligned}&(\mu^P_{B_1}\circ\mu^P_{B_2})((x_1,z)(y_1,z))\\ &= \mu^P_{B_1}(x_1y_1)\wedge\mu^P_{A_2}(z)\\ &\leq \wedge\{\mu^P_{A_1}(x_1),\ \mu^P_{A_1}(y_1),\mu^P_{A_2}(z)\}\\ &= \wedge\{\mu^P_{A_1}(x_1)\wedge\mu^P_{A_2}(z),\mu^P_{A_1}(y_1)\wedge\mu^P_{A_2}(z)\}\\ &= (\mu^P_{A_1}\circ\mu^P_{A_2}(x_1,z))\wedge(\mu^P_{A_1}\circ\mu^P_{A_2}(y_1,z)),\end{aligned}$$

$$\begin{aligned}(\mu^N_{B_1}\circ\mu^N_{B_2})((x_1,z)(y_1,z)) &= \mu^N_{B_1}(x_1y_1)\vee\mu^N_{A_2}(z)\\ &\geq \vee\{(\mu^N_{A_1}(x_1),\ \mu^N_{A_1}(y_1)),\mu^N_{A_2}(z)\}\\ &= \vee\{\mu^N_{A_1}(x_1)\vee\mu^N_{A_2}(z),\mu^N_{A_1}(y_1)\vee\mu^N_{A_2}(z)\}\\ &= (\mu^N_{A_1}\circ\mu^N_{A_2})(x_1,z)\vee(\mu^N_{A_1}\circ\mu^N_{A_2})(y_1,z).\end{aligned}$$

Let $(x_1,x_2)(y_1,y_2)\in E^0\backslash E$. Then $x_1y_1\in E_1$, $x_2\neq y_2$. Thus,

$$\begin{aligned}&(\mu^P_{B_1}\circ\mu^P_{B_2})((x_1,x_2)(y_1,y_2))\\ &= \wedge\{\mu^P_{A_2}(x_2),\ \mu^P_{A_2}(y_2),\ \mu^P_{B_1}(x_1y_1)\}\\ &\leq \wedge\{\mu^P_{A_2}(x_2),\ \mu^P_{A_2}(y_2),\ \mu^P_{A_1}(x_1)\wedge\mu^P_{A_1}(y_1)\}\\ &= \wedge\{\mu^P_{A_1}(x_1)\wedge\mu^P_{A_2}(x_2),\ \mu^P_{A_1}(y_1)\wedge\mu^P_{A_2}(y_2)\}\\ &= (\mu^P_{A_1}\circ\mu^P_{A_2})(x_1,x_2)\wedge(\mu^P_{A_1}\circ\mu^P_{A_2})(y_1,y_2).\end{aligned}$$

■

**Definition 8.2.7** Let $A_1=(\mu^P_{A_1},\mu^N_{A_1})$ and $A_2=(\mu^P_{A_2},\mu^N_{A_2})$ be bipolar fuzzy subsets of $V_1$ and $V_2$, respectively, and let $B_1=(\mu^P_{B_1},\mu^N_{B_1})$ and $B_2=(\mu^P_{B_2},\mu^N_{B_2})$ be bipolar fuzzy subsets of $E_1$ and $E_2$, respectively. Then we denote the **union** of two bipolar fuzzy graphs $G_1$ and $G_2$ of the graphs $G_1^*$ and $G_2^*$ by $G_1\cup G_2=(A_1\cup A_2,B_1\cup B_2)$ and defined as follows:

$$(\mu^P_{A_1}\cup\mu^P_{A_2})(x)=\begin{cases}\mu^P_{A_1}(x) & \text{if } x\in V_1\cap\overline{V_2}\\ \mu^P_{A_2}(x) & \text{if } x\in V_2\cap\overline{V_1}\\ \mu^P_{A_1}(x)\vee\mu^P_{A_2}(x) & \text{if } x\in V_1\cap V_2\end{cases}$$

$$(\mu^N_{A_1} \cup \mu^N_{A_2})(x) = \begin{cases} \mu^N_{A_1}(x) & \text{if } x \in V_1 \cap \overline{V_2} \\ \mu^N_{A_2}(x) & \text{if } x \in V_2 \cap \overline{V_1} \\ \mu^N_{A_1}(x) \wedge \mu^N_{A_2}(x) & \text{if } x \in V_1 \cap V_2 \end{cases}$$

$$(\mu^P_{B_1} \cup \mu^P_{B_2})(xy) = \begin{cases} \mu^P_{B_1}(xy) & \text{if } xy \in E_1 \cap \overline{E_2} \\ \mu^P_{B_2}(xy) & \text{if } xy \in E_2 \cap \overline{E_1} \\ \mu^P_{B_1}(xy) \vee \mu^P_{B_2}(xy) & \text{if } xy \in E_1 \cap E_2 \end{cases}$$

$$(\mu^N_{B_1} \cup \mu^N_{B_2})(xy) = \begin{cases} \mu^N_{B_1}(xy) & \text{if } xy \in E_1 \cap \overline{E_2} \\ \mu^N_{B_2}(xy) & \text{if } xy \in E_2 \cap \overline{E_1} \\ \mu^N_{B_1}(xy) \wedge \mu^N_{B_2}(xy) & \text{if } xy \in E_1 \cap E_2. \end{cases}$$

*Example 8.2.8* Consider the bipolar fuzzy graphs $G_1$ and $G_2$, where $V_1 = \{a, b, c\}$, $E_1 = \{ab, ac, bd\}$ and $V_2 = \{a, b, d\}$, $E_2 = \{ab, ad, bd\}$ such that

| | $a$ | $b$ | $c$ | | $a$ | $b$ | $d$ |
|---|---|---|---|---|---|---|---|
| $\mu^P_{A_1}$ | 0.3 | 0.5 | 0.4 | $\mu^P_{A_2}$ | 0.4 | 0.4 | 0.3 |
| $\mu^N_{A_1}$ | −0.4 | −0.6 | −0.5 | $\mu^N_{A_2}$ | −0.5 | −0.3 | −0.2 |

and

| | $ab$ | $ac$ | $bc$ | | $ab$ | $ad$ | $bd$ |
|---|---|---|---|---|---|---|---|
| $\mu^P_{B_1}$ | 0.2 | 0.3 | 0.4 | $\mu^P_{B_2}$ | 0.4 | 0.3 | 0.3 |
| $\mu^N_{B_1}$ | −0.4 | −0.4 | −0.5 | $\mu^N_{B_2}$ | −0.3 | −0.2 | −0.2 |

Then the following tables yield $G_1 \cup G_2$.

| | $a$ | $b$ | $c$ | $d$ |
|---|---|---|---|---|
| $\mu^P_{A_1} \cup \mu^P_{A_2}$ | 0.4 | 0.5 | 0.4 | 0.3 |
| $\mu^N_{A_1} \cup \mu^N_{A_2}$ | −0.5 | −0.6 | −0.5 | −0.2 |

| | $ab$ | $ac$ | $ad$ | $bc$ | $bd$ |
|---|---|---|---|---|---|
| $\mu^P_{B_1} \cup \mu^P_{B_2}$ | 0.4 | 0.3 | 0.3 | 0.4 | 0.3 |
| $\mu^N_{B_1} \cup \mu^N_{B_2}$ | −0.4 | −0.4 | −0.2 | −0.5 | −0.2 |

**Proposition 8.2.9** *If $G_1$ and $G_2$ are bipolar fuzzy graphs, then $G_1 \cup G_2$ is a bipolar fuzzy graph.*

*Proof* Let $xy \in E_1 \cap E_2$. Then

$$\begin{aligned} (\mu^P_{B_1} \cup \mu^P_{B_2})(xy) &= \mu^P_{B_1}(xy) \vee \mu^P_{B_2}(xy) \\ &\leq \vee\{\mu^P_{A_1}(x) \wedge \mu^P_{A_1}(y), \mu^P_{A_2}(x) \wedge \mu^P_{A_2}(y)\} \\ &= \wedge\{\mu^P_{A_1}(x) \vee \mu^P_{A_2}(x), \mu^P_{A_1}(y) \vee \mu^P_{A_2}(y)\} \\ &= (\mu^P_{A_1} \cup \mu^P_{A_2})(x) \wedge (\mu^P_{A_1} \cup \mu^P_{A_2})(y). \end{aligned}$$

$$
\begin{aligned}
(\mu^N_{B_1} \cup \mu^N_{B_2})(xy) &= \mu^N_{B_1}(xy) \wedge \mu^N_{B_2}(xy) \\
&\geq \wedge\{\mu^N_{A_1}(x) \vee \mu^N_{A_1}(y), \mu^N_{A_2}(x) \vee \mu^N_{A_2}(y)\} \\
&= \vee\{\mu^N_{A_1}(x) \wedge \mu^N_{A_2}(x), \mu^N_{A_1}(y) \wedge \mu^N_{A_2}(y)\} \\
&= (\mu^N_{A_1} \cup \mu^N_{A_2}(x)) \vee (\mu^N_{A_1} \cup \mu^N_{A_2})(y).
\end{aligned}
$$

Similarly, for $xy \in E_1 \cap \overline{E_2}$, we have

$$
\begin{aligned}
(\mu^P_{B_1} \cup \mu^P_{B_2})(xy) &\leq (\mu^P_{A_1} \cup \mu^P_{A_2}(x)) \wedge (\mu^P_{A_1} \cup \mu^P_{A_2})(y) \\
(\mu^N_{B_1} \cup \mu^N_{B_2})(xy) &\geq (\mu^N_{A_1} \cup \mu^N_{A_2})(x) \vee (\mu^N_{A_1} \cup \mu^N_{A_2})(y).
\end{aligned}
$$

If $xy \in E_2 \cap \overline{E_1}$, then

$$
\begin{aligned}
(\mu^P_{B_1} \cup \mu^P_{B_2})(xy) &\leq (\mu^P_{A_1} \cup \mu^P_{A_2})(x) \wedge (\mu^P_{A_1} \cup \mu^P_{A_2})(y) \\
(\mu^N_{B_1} \cup \mu^N_{B_2})(xy) &\geq (\mu^N_{A_1} \cup \mu^N_{A_2})(x) \vee (\mu^N_{A_1} \cup \mu^N_{A_2})(y).
\end{aligned}
$$

Thus, $G_1 \cup G_2$ is a bipolar fuzzy graph ■

**Proposition 8.2.10** *Let $\{G_i \mid i \in I\}$ be a family of bipolar fuzzy graphs with the underlying set $V$. Then $\cap_{i\in I} G_i$ is a bipolar fuzzy graph.*

*Proof* For any $x, y \in V$, we have that

$$
\begin{aligned}
\cap_{i\in I}\mu^P_B(xy) &= \wedge_{i\in I}\mu^P_{B_i}(xy) \leq \wedge_{i\in I}\{\mu^P_{A_i}(x) \wedge \mu^P_{A_i}(y)\} \\
&= (\wedge_{i\in I}\mu^P_{A_i}(x)) \wedge (\wedge_{i\in I}\mu^P_{A_i}(y)) \\
&= (\cap_{i\in I}\mu^P_{A_i}(x)) \wedge (\cap_{i\in I}\mu^P_{A_i}(y)),
\end{aligned}
$$

$$
\begin{aligned}
\cap_{i\in I}\mu^N_B(xy) &= \vee_{i\in I}\mu^N_{B_i}(xy) \geq \vee_{i\in I}(\mu^N_{A_i}(x) \vee \mu^N_{A_i}(y)) \\
&= (\vee_{i\in I}\mu^N_{A_i}(x)) \vee (\vee_{i\in I}\mu^N_{A_i}(y)) \\
&= \cap_{i\in I}\mu^N_{A_i}(x) \vee \cap_{i\in I}\mu^N_{A_i}(y).
\end{aligned}
$$

Hence, $\cap_{i\in I} G_i$ is a bipolar fuzzy graph. ■

**Definition 8.2.11** Let $A_1 = (\mu^P_{A_1}, \mu^N_{A_1})$ and $A_2 = (\mu^P_{A_2}, \mu^N_{A_2})$ be bipolar fuzzy subsets of $V_1$ and $V_2$ and let $B_1 = (\mu^P_{B_1}, \mu^N_{B_1})$ and $B_2 = (\mu^P_{B_2}, \mu^N_{B_2})$ be bipolar fuzzy subsets of $E_1$ and $E_2$, respectively. We denote the **join** of two bipolar fuzzy graphs $G_1 = (A_1, B_1)$ and $G_2 = (A_2, B_2)$ of the graphs $G^*_1 = (V_1, E_1)$ and $G^*_2 = (V_2, E_2)$, respectively, by $G_1 + G_2 = (A_1 + A_2, B_1 + B_2)$ and defined as follows:

(*i*) $(\mu^P_{A_1} + \mu^P_{A_2})(x) = (\mu^P_{A_1} \cup \mu^P_{A_2})(x)$ and $(\mu^N_{A_1} + \mu^N_{A_2})(x) = (\mu^N_{A_1} \cap \mu^N_{A_2})(x)$ if $x \in V_1 \cup V_2$;

(*ii*) $(\mu^P_{B_1} + \mu^P_{B_2})(xy) = (\mu^P_{B_1} \cup \mu^P_{B_2})(xy) = \mu^P_{B_1}(xy)$ and $(\mu^N_{B_1} + \mu^N_{B_2})(xy) = (\mu^N_{B_1} \cap \mu^N_{B_2})(xy) = \mu^N_{B_1}(xy)$ if $xy \in E_1 \cap E_2$;

(*iii*) $(\mu^P_{B_1} + \mu^P_{B_2})(xy) = \mu^P_{A_1}(x) \vee \mu^P_{A_2}(y)$ and $(\mu^N_{B_1} + \mu^N_{B_2})(xy) = \mu^N_{A_1}(x) \wedge \mu^N_{A_2}(y)$ if $xy \in E$, where $E$ is the set of all edges joining the vertices $V_1$ and $V_2$.

**Proposition 8.2.12** *If $G_1$ and $G_2$ are bipolar fuzzy graphs, then $G_1 + G_2$ is a bipolar fuzzy graph.*

*Proof* Let $xy \in E'$. Then

$$\begin{aligned}(\mu^P_{B_1} + \mu^P_{B_2})(xy) &= \mu^P_{A_1}(x) \vee \mu^P_{A_2}(y)\\ &\le (\mu^P_{A_1} \cup \mu^P_{A_2})(x) \vee (\mu^P_{A_1} \cup \mu^P_{A_2})(y)\\ &= (\mu^P_{A_1} + \mu^P_{A_2})(x) \vee (\mu^P_{A_1} + \mu^P_{A_2})(y)\end{aligned}$$

and

$$\begin{aligned}(\mu^N_{B_1} + \mu^N_{B_2})(xy) &= \mu^N_{A_1}(x) \wedge \mu^N_{A_2}(y)\\ &\ge (\mu^N_{A_1} \cap \mu^N_{A_2})(x) \wedge (\mu^N_{A_1} \cap \mu^N_{A_2})(y)\\ &= (\mu^N_{A_1} + \mu^N_{A_2})(x) \wedge (\mu^N_{A_1} + \mu^N_{A_2})(y).\end{aligned}$$

If $xy \in E_1 \cup E_2$, then the result follows from Proposition 8.2.9. ■

**Proposition 8.2.13** *Let $G_1^* = (V_1, E_1)$ and $G_2^* = (V_2, E_2)$ be crisp graphs and let $V_1 \cap V_2 = \emptyset$. Let $A_1, A_2, B_1$, and $B_2$ be bipolar fuzzy subsets of $V_1, V_2, E_1$, and $E_2$, respectively. Then $G_1 \cup G_2 = (A_1 \cup A_2, B_1 \cup B_2)$ is a bipolar fuzzy graph of $G^* = G_1^* \cup G_2^*$ if and only if $G_1 = (A_1, B_1)$ and $G_2 = (A_2, B_2)$ are bipolar fuzzy graphs of $G_1^*$ and $G_2^*$, respectively.*

*Proof* Suppose that $G_1 \cup G_2 = (A_1 \cup A_2, B_1 \cup B_2)$ is a bipolar fuzzy graph. Let $xy \in E_1$. Then $xy \notin E_2$ and $x, y \in V_1 \backslash V_2$. Thus,

$$\begin{aligned}\mu^P_{B_1}(xy) &= (\mu^P_{B_1} \cup \mu^P_{B_2})(xy) \le (\mu^P_{A_1} \cup \mu^P_{A_2})(x) \wedge (\mu^P_{A_1} \cup \mu^P_{A_2})(y)\\ &= \mu^P_{A_1}(x) \wedge \mu^P_{A_1}(y),\end{aligned}$$

$$\begin{aligned}\mu^N_{B_1}(xy) &= (\mu^N_{B_1} \cup \mu^N_{B_2})(xy) \ge (\mu^N_{A_1} \cap \mu^N_{A_2})(x) \vee (\mu^N_{A_1} \cap \mu^N_{A_2})(y)\\ &= \mu^N_{A_1}(x) \vee \mu^N_{A_1}(y).\end{aligned}$$

Hence, $G_1 = (A_1, B_1)$ is a bipolar fuzzy graph. Similarly, we can show that $G_2 = (A_2, B_2)$ is a bipolar fuzzy graph.

The converse follows by Proposition 8.2.9. ■

As a consequence of Propositions 8.2.12 and 8.2.13, we have the following.

**Proposition 8.2.14** *Let $G_1^* = (V_1, E_1)$ and $G_2^* = (V_2, E_2)$ be crisp graphs and let $V_1 \cap V_2 = \emptyset$. Let $A_1, A_2, B_1$, and $B_2$ be bipolar fuzzy subsets of $V_1, V_2, E_1$, and $E_2$, respectively. Then $G_1 + G_2 = (A_1 + A_2, B_1 + B_2)$ is a bipolar fuzzy graph of $G^* = G_1^* + G_2^*$ if and only if $G_1 = (A_1, B_1)$ and $G_2 = (A_2, B_2)$ are bipolar fuzzy graphs of $G_1^*$ and $G_2^*$, respectively.*

## 8.3 Isomorphisms of Bipolar Fuzzy Graphs

In this section, we consider the concept of an isomorphism of bipolar fuzzy graphs.

**Definition 8.3.1** Let $G_1$ and $G_2$ be bipolar fuzzy graphs. A **homomorphism** $f : G_1 \to G_2$ is a bijective mapping $f : V_1 \to V_2$ which satisfies the following conditions:

$(i)$ $\mu_{A_1}^P(x_1) \le \mu_{A_2}^P(f(x_1)), \mu_{A_1}^N(x_1) \ge \mu_{A_2}^N(f(x_1))$ for all $x_1 \in V_1$.

$(ii)$ $\mu_{B_1}^P(x_1y_1) \le \mu_{B_2}^P(f(x_1)f(y_1)), \mu_{B_1}^N(x_1y_1) \ge \mu_{B_2}^N(f(x_1)f(y_1))$ for all $x_1 \in V_1$ and $x_1y_1 \in E_1$.

**Definition 8.3.2** Let $G_1$ and $G_2$ be bipolar fuzzy graphs. An **isomorphism** $f : G_1 \to G_2$ is a bijective mapping $f : V_1 \to V_2$ which satisfies the following conditions:

$(i)$ $\mu_{A_1}^P(x_1) = \mu_{A_2}^P(f(x_1)), \mu_{A_1}^N(x_1) = \mu_{A_2}^N(f(x_1))$.

$(ii)$ $\mu_{B_1}^P(x_1y_1) = \mu_{B_2}^P(f(x_1)f(y_1)), \mu_{B_1}^N(x_1y_1) = \mu_{B_2}^N(f(x_1)f(y_1))$ for all $x_1 \in V_1$ and $x_1y_1 \in \mathcal{E}_1$.

**Definition 8.3.3** Let $G_1$ and $G_2$ be bipolar fuzzy graphs. A **weak isomorphism** $f : G_1 \to G_2$ is a bijective mapping $f : V_1 \to V_2$ which satisfies the following conditions:

$(i)$ $f$ is a homomorphism,

$(ii)$ $\mu_{A_1}^P(x_1) = \mu_{A_2}^P(f(x_1)), \mu_{A_1}^N(x_1) = \mu_{A_2}^N(f(x_1))$ for all $x_1 \in V_1$.

It follows that a weak isomorphism preserves the weights of the vertices, but not necessarily the weights of the edges.

*Example 8.3.4* Let $V_1 = \{a_1, b_1\}$, $V_2 = \{a_2, b_2\}$, $E_1 = \{a_1b_1\}$, and $E_2 = \{a_2b_2\}$. Consider bipolar fuzzy graphs $G_1 = (A_1, B_1)$ and $G_2 = (A_2, B_2)$ of $G_1^* = (V_1, E_1)$ and $G_2^* = (V_2, E_2)$, respectively, defined as follows:

$\mu_{A_1}^P(a_1) = 0.2$, $\mu_{A_1}^N(a_1) = -0.5$, $\mu_{A_1}^P(b_1) = 0.3$, $\mu_{A_1}^N(b_1) = -0.6$, $\mu_{B_1}^P(a_1b_1) = 0.1$, $\mu_{B_1}^N(a_1b_1) = -0.4$; $\mu_{A_2}^P(a_2) = 0.3_{B_1}$, $\mu_{A_2}^N(a_2) = -0.6$, $\mu_{A_2}^P(b_2) = 0.2$, $\mu_{A_2}^N(b_2) = -0.5$, $\mu_{B_2}^P(a_2b_2) = 0.2$, $\mu_{B_2}^N(a_2, b_2) = -0.4$.

Define the function $f : V_1 \to V_2$ by $f(a_1) = b_2$ and $f(b_1) = a_2$. Then we have: $\mu_{A_1}^P(a_1) = \mu_{A_2}^P(b_2)$, $\mu_{A_1}^N(a_1) = \mu_{A_2}^N(b_2)$, $\mu_{A_1}^P(b_1) = \mu_{A_2}^P(a_2)$, $\mu_{A_1}^N(b_1) = \mu_{A_2}^N(a_2)$, and

$\mu^N_{B_1}(a_1b_1) = \mu^N_{B_2}(a_2b_2)$, but $\mu^P_{B_1}(a_1b_1) \neq \mu^P_{B_2}(f(a_1)f(b_1)) = \mu^P_{B_2}(a_2b_2)$. Hence, the map is weak isomorphism, but not an isomorphism.

**Definition 8.3.5** Let $G_1$ and $G_2$ be bipolar fuzzy graphs. A **co-weak isomorphism** $f : G_1 \to G_2$ is a bijective mapping $f : V_1 \to V_2$ which satisfies the following conditions:

$(i)$ $f$ is a homomorphism,

$(ii)$ $\mu^P_{B_1}(x_1y_1) = \mu^P_{B_2}(f(x_1)f(y_1)), \mu^N_{B_1}(x_1y_1) = \mu^N_{B_2}(f(x_1)f(y_1))$ for all $x_1, y_1 \in V_1$.

We see that a weak isomorphism preserves the weights of the edges, but not necessarily the weights of the vertices.

*Example 8.3.6* Let $V_1 = \{a_1, b_1\}$, $V_2 = \{a_2, b_2\}$, $E_1 = \{a_1b_1\}$, and $E_2 = \{a_2b_2\}$. Consider bipolar fuzzy graphs $G_1 = (A_1, B_1)$ and $G_2 = (A_2, B_2)$ of $G_1^* = (V_1, E_1)$ and $G_2^* = (V_2, E_2)$, respectively, defined as follows:

$\mu^P_{A_1}(a_1) = 0.2$, $\mu^N_{A_1}(a_1) = -0.4$, $\mu^P_{A_1}(b_1) = 0.3$, $\mu^N_{A_1}(b_1) = -0.5$, $\mu^P_{B_1}(a_1b_1) = 0.1$, $\mu^N_{B_1}(a_1b_1) = -0.3$; $\mu^P_{A_2}(a_2) = 0.4_{B_1}$, $\mu^N_{A_2}(a_2) = -0.5$, $\mu^P_{A_2}(b_2) = 0.3$, $\mu^N_{A_2}(b_2) = -0.6$, $\mu^P_{B_2}(a_2b_2) = 0.1$, $\mu^N_{B_2}(a_2, b_2) = -0.3$.

The function $f : V_1 \to V_2$ defined by $f(a_1) = a_2$ and $f(b_1) = b_2$ can be shown to be a co-weak isomorphism, but not an isomorphism because $\mu^P_{A_1}(a_1) \neq \mu^P_{A_2}(b_2)$, $\mu^N_{A_1}(a_1) \neq \mu^N_{A_2}(b_2)$.

We next make some definitions and observations.

(1) If $G_1 = G_2 = G$, then the homomorphism $f$ of $G$ into $G$ itself is called an **endomorphism**. An isomorphism $f$ over $G$ is called an **automorphism**.

(2) Let $A = (\mu^P_A, \mu^N_A)$ be a bipolar fuzzy graph with an underlying set $V$. Let $Aut(G)$ be the set of all bipolar automorphism of $G$. Let $e : G \to G$ be a map defined by $e(x) = x$ for all $x \in V$. Clearly, $e \in Aut(G)$.

(3) If $G_1 = G_2$, then the weak and co-weak isomorphisms actually becomes isomorphic.

**Definition 8.3.7** A bipolar fuzzy set $A = (\mu^P_A, \mu^N_A)$ in a semigroup $S$ is called a **bipolar subsemigroup** of $S$ it satisfies: $\mu^P_A(xy) \geq \mu^P_A(x) \wedge \mu^P_A(y)$ and $\mu^N_A(xy) \leq \mu^N_A(x) \vee \mu^N_A(y)$ for all $x, y \in S$.

A bipolar fuzzy set $A = (\mu^P_A, \mu^N_A)$ in a group $G$ is called a **bipolar subgroup** of $G$ if it is a bipolar fuzzy subsemigroup of $G$ and satisfies: $\mu^P_A(x^{-1}) = \mu^P_A(x)$ and $\mu^N_A(x^{-1}) = \mu^N_A(x)$ for all $x \in G$.

We now show how to associate a bipolar fuzzy group with a bipolar fuzzy graph.

**Proposition 8.3.8** *Let $G = (A, B)$ be a bipolar fuzzy graph and let $Aut(G)$ be the set of all automorphisms of $G$. Then $(Aut(G), \circ)$ forms a group.*

*Proof* Let $\phi, \psi \in Aut(G)$ and let $x, y \in V$. Then

$$\mu_B^P((\phi \circ \psi)(x)(\phi \circ \psi)(y)) = \mu_B^P((\phi(\psi(x))\phi(\psi(y)) = \mu_B^P(\psi(x)\psi(y)) = \mu_B^P(xy),$$

$$\mu_B^N((\phi \circ \psi)(x)(\phi \circ \psi)(y)) = \mu_B^N((\phi(\psi(x))\phi(\psi(y)) = \mu_B^N(\psi(x)\psi(y)) = \mu_B^N(xy),$$

$$\mu_B^P((\phi \circ \psi)(x)) = \mu_B^P((\phi(\psi(x))) = \mu_B^P((\psi(x)) = \mu_B^P(x),$$

$$\mu_B^N((\phi \circ \psi)(x)) = \mu_B^N((\phi(\psi(x))) = \mu_B^N((\psi(x)) = \mu_B^N(x).$$

Thus, $\phi \circ \psi \in Aut(G)$. Clearly $Aut(G)$ satisfies associativity under the operation $\circ$, $\phi \circ e = \phi = e \circ \phi$, $\mu_A^P(\phi^{-1}) = \mu_A^P(\phi)$, $\mu_A^N(\phi^{-1}) = \mu_A^N(\phi)$ for all $\phi \in Aut(G)$. Hence, $(Aut(G), \circ)$ forms a group. ■

We state the two propositions without proofs.

**Proposition 8.3.9** *Let $G = (A, B)$ be a bipolar fuzzy graph and let $Aut(G)$ be the set of all automorphisms of $G$. Let $g = (\mu_g^P, \mu_g^N)$ be a bipolar fuzzy set in $Aut(G)$ defined by*

$$\mu_g^P(\phi) = \vee \left\{\mu_B^P(\phi(x), \phi(y)) \mid (x, y) \in V \times V\right\},$$
$$\mu_g^N(\phi) = \wedge\{\mu_B^N(\phi(x), \phi(y)) \mid (x, y) \in V \times V\}$$

*for all $\phi \in Aut(G)$. Then $g = (\mu_g^P, \mu_g^N)$ is a bipolar fuzzy group on $Aut(G)$.*

**Proposition 8.3.10** *Every bipolar fuzzy group has an embedding into the bipolar fuzzy group of the group of automorphisms of some bipolar fuzzy group.*

**Proposition 8.3.11** *Let $G_1$, $G_2$, and $G_3$ be bipolar fuzzy graphs. Then the following properties hold.*

*(i) If $f$ is an isomorphism of $G_1$ onto $G_2$, then $f^{-1}$ is an isomorphism of $G_2$ onto $G_1$.*

*(ii) If $f$ is an isomorphism of $G_1$ onto $G_2$ and $g$ is an isomorphism of $G_2$ onto $G_3$, then $g \circ f$ is an isomorphism of $G_1$ onto $G_3$.*

*Proof* (*i*) Now, $f : V_1 \to V_2$. Let $x_1 \in V_1$. Then we have $\mu_{A_1}^P(x_1) = \mu_{A_2}^P(f(x_1))$, $\mu_{A_1}^N(x_1) = \mu_{A_2}^N(f(x_1))$ for all $x_1 \in V_1$ and $\mu_{B_1}^P(x_1y_1) = \mu_{B_2}^P(f(x_1)f(y_1))$, $\mu_{B_1}^N(x_1y_1) = \mu_{B_2}^N(f(x_1)f(y_1))$ for all $x_1, y_1 \in E_1$.

Because $f$ is bijective, it follows that $f^{-1}(x_2) = x_1$, where $f(x_1) = x_2 \in V_2$. Thus, $\mu_{A_1}^P(f^{-1}(x_2)) = \mu_{A_2}^P(x_2)$, $\mu_{A_1}^N(f^{-1}(x_2)) = \mu_{A_2}^N(x_2)$ for all $x_2 \in V_2$ and $\mu_{B_1}^P(f^{-1}(x_2y_2)) = \mu_{B_2}^P(x_2y_2)$, $\mu_{B_1}^N(f^{-1}(x_2y_2)) = \mu_{B_2}^N(x_2y_2)$ for all $x_2, y_2 \in E_2$.

Hence, the bijective map $f^{-1} : V_2 \to V_1$ is an isomorphism from $G_2$ onto $G_1$.

$(ii)$ We have $f : V_1 \rightarrow V_2$ and $g : V_2 \rightarrow V_3$. Then $g \circ f : V_1 \rightarrow V_3$ is bijective map from $V_1$ onto $V_3$, where $(g \circ f)(x_1) = g(f(x_1))$ for all $x_1 \in V_1$. Let $x_2 = f(x_1)$. Thus,

$$\mu^P_{A_1}(x_1) = \mu^P_{A_2}(f(x_1)) = \mu^P_{A_2}(x_2), \tag{8.1}$$
$$\mu^N_{A_1}(x_1) = \mu^N_{A_2}(f(x_1)) = \mu^N_{A_2}(x_2) \tag{8.2}$$

for all $x_1 \in V_1$.

$$\mu^P_{B_1}(x_1y_1) = \mu^P_{B_2}(f(x_1)f(y_1)) = \mu^P_{B_2}(x_2y_2), \tag{8.3}$$
$$\mu^N_{B_1}(x_1y_1) = \mu^N_{B_2}(f(x_1)f(y_1)) = \mu^N_{B_2}(x_2y_2) \tag{8.4}$$

for all $x_1y_1 \in E_1$.

Let $x_3 = g(x_2) = x_3$. Then

$$\mu^P_{A_2}(x_2) = \mu^P_{A_3}(g(x_2)) = \mu^P_{A_3}(x_3), \tag{8.5}$$
$$\mu^N_{A_2}(x_2) = \mu^N_{A_3}(g(x_2)) = \mu^N_{A_3}(x_3) \tag{8.6}$$

for all $x_2 \in V_1$.

$$\mu^P_{B_2}(x_2y_2) = \mu^P_{B_3}(g(x_2)g(y_2)) = \mu^P_{B_3}(x_3y_3), \tag{8.7}$$
$$\mu^N_{B_2}(x_2y_2) = \mu^N_{B_3}(g(x_2)g(y_2)) = \mu^N_{B_3}(x_3y_3) \tag{8.8}$$

for all $x_2y_2 \in E_3$.

From (8.1), (8.2), (8.5), (8.6), and $f(x_1) = x_2$, $x_1 \in V_1$, we have

$$\mu^P_{A_1}(x_1) = \mu^P_{A_2}(f(x_1)) = \mu^P_{A_2}(x_2) = \mu^P_{A_2}(g(x_2)) = \mu^P_{A_2}(g(f(x_1))),$$

$$\mu^N_{A_1}(x_1) = \mu^N_{A_2}(f(x_1)) = \mu^N_{A_2}(x_2) = \mu^N_{A_2}(g(x_2)) = \mu^N_{A_2}(g(f(x_1)))$$

for all $x_1 \in V_1$.

From (8.3), (8.4), (8.7), (8.8), we have

$$\begin{aligned}\mu^P_{B_1}(x_1y_1) &= \mu^P_{B_2}(f(x_1)f(y_1)) = \mu^P_{B_2}(x_2y_2)\\ &= \mu^P_{B_3}(g(x_2)g(y_2)) = \mu^P_{B_3}(g(f(x_1))g(f(y_1))),\end{aligned}$$

and

$$\begin{aligned}\mu^N_{B_1}(x_1y_1) &= \mu^N_{B_2}(f(x_1)f(y_1)) = \mu^N_{B_2}(x_2y_2)\\ &= \mu^N_{B_3}(g(x_2)g(y_2)) = \mu^N_{B_3}(g(f(x_1))g(f(y_1)))\end{aligned}$$

for all $x_1y_1 \in E$. Therefore, $g \circ f$ is an isomorphism between $G_1$ and $G_3$. ■

Let $f : V_1 \to V_2$ be a weak isomorphism of $G_1$ onto $G_2$. Let $x_1 \in V_1$. Then $\mu^P_{A_1}(x_1) = \mu^P_{A_2}(f(x_1)), \mu^N_{A_1}(x_1) = \mu^N_{A_2}(f(x_1))$,

$$\mu^P_{B_1}(x_1y_1) \le \mu^P_{B_2}(f(x_1)f(y_1)),\ \mu^N_{B_1}(x_1y_1) \ge \mu^N_{B_2}(f(x_1)f(y_1)) \tag{8.9}$$

for all $x_1, y_1 \in E_1$.

Let $g : V_2 \to V_1$ be a weak isomorphism of $G_2$ onto $G_1$. Then $g$ is a bijective map defined by $g(x_2) = x_1, x_2 \in V_2$, satisfying $\mu^P_{A_2}(x_2) = \mu^P_{A_1}(g(x_2)), \mu^N_{A_2}(x_2) = \mu^N_{A_1}(g(x_2))$ for all $x_2 \in V_2$ and

$$\mu^P_{B_2}(x_2y_2) \le \mu^P_{B_1}(g(x_2)g(y_2)),\ \mu^N_{B_2}(x_2y_2) \ge \mu^N_{B_1}(g(x_2)g(y_2)) \tag{8.10}$$

for all $x_2y_2 \in E_2$.

The inequalities (8.9) and (8.10) hold on the finite sets $V_1$ and $V_2$ only when $G_1$ and $G_2$ have the same number of edges and the corresponding edges have the same weights. Hence, $G_1$ and $G_2$ are identical.

**Proposition 8.3.12** *Let $G_1, G_2$, and $G_3$ be bipolar fuzzy graphs. If $f$ is a weak isomorphism of $G_1$ onto $G_2$ and $g$ is a weak isomorphism of $G_2$ onto $G_3$, then $g \circ f$ is a weak isomorphism of $G_1$ onto $G_3$.*

*Proof* Let $f : V_1 \to V_2$ and $g : V_2 \to V_3$ be weak isomorphisms of $G_1$ onto $G_2$ and $G_2$ onto $G_3$, respectively. Then $g \circ f : V_1 \to V_3$ is bijective map from $V_1$ onto $V_3$, where $(g \circ f)(x_1) = g(f(x_1))$ for all $x_1 \in V_1$.

Let $x_1 \in V_1$ and $x_2 = f(x_1)$. Then

$$\mu^P_{A_1}(x_1) = \mu^P_{A_2}(f(x_1)) = \mu^P_{A_2}(x_2), \tag{8.11}$$
$$\mu^N_{A_1}(x_1) = \mu^N_{A_2}(f(x_1)) = \mu^N_{A_2}(x_2) \tag{8.12}$$

for all $x_1 \in V_1$.

$$\begin{aligned}\mu^P_{B_1}(x_1y_1) &\le \mu^P_{B_2}(f(x_1)f(y_1)) = \mu^P_{B_2}(x_2y_2)\mu^N_{B_1}(x_1y_1)\\ &\ge \mu^N_{B_2}(f(x_1)f(y_1)) = \mu^N_{B_2}(x_2y_2)\end{aligned} \tag{8.13}$$

for all $x_1y_1 \in E_1$.

Let $x_3 = g(x_2)$. Then we have,

$$\mu^P_{A_2}(x_2) = \mu^P_{A_3}(g(x_2)) = \mu^P_{A_3}(x_3), \tag{8.14}$$
$$\mu^N_{A_2}(x_2) = \mu^N_{A_3}(g(x_2)) = \mu^N_{A_3}(x_3) \tag{8.15}$$

for all $x_2 \in V_1$.

$$\mu^P_{B_2}(x_2y_2) \le \mu^P_{B_3}(g(x_2)g(y_2)) = \mu^P_{B_3}(x_3y_3), \tag{8.16}$$
$$\mu^N_{B_2}(x_2y_2) \ge \mu^N_{B_3}(g(x_2)g(y_2)) = \mu^N_{B_3}(x_3y_3) \tag{8.17}$$

for all $x_2y_2 \in E_3$.

From (8.11), (8.12), (8.14), (8.15), and $f(x_1) = x_2$, $x_1 \in V_1$, we have $\mu^P_{A_1}(x_1) = \mu^P_{A_2}(f(x_1)) = \mu^P_{A_2}(x_2) = \mu^P_{A_2}(g(x_2)) = \mu^P_{A_2}(g(f(x_1)))$, $\mu^N_{A_1}(x_1) = \mu^N_{A_2}(f(x_1)) = \mu^N_{A_2}(x_2) = \mu^N_{A_2}(g(x_2)) = \mu^N_{A_2}(g(f(x_1)))$ for all $x_1 \in V_1$.

From (8.13), (8.16), and (8.17), we have, $\mu^P_{B_1}(x_1y_1) \le \mu^P_{B_2}(f(x_1)f(y_1)) = \mu^P_{B_2}(x_2y_2) = \mu^P_{B_3}(g(x_2)g(y_2)) = \mu^P_{B_3}(g(f(x_1))g(f(y_1)))$ and $\mu^N_{B_1}(x_1y_1) \ge \mu^N_{B_2}(f(x_1)f(y_1)) = \mu^N_{B_2}(x_2y_2) = \mu^N_{B_3}(g(x_2)g(y_2)) = \mu^N_{B_3}(g(f(x_1))g(f(y_1)))$ for all $x_1, y_1 \in E$. Therefore, $g \circ f$ is an isomorphism between $G_1$ and $G_3$. ■

## 8.4 Strong Bipolar Fuzzy Graphs

**Definition 8.4.1** A bipolar fuzzy graph $G = (A, B)$ is called **strong** if $\mu^P_B(xy) = \mu^P_A(x) \wedge \mu^P_A(y)$ and $\mu^N_B(xy) = \mu^N_A(x) \vee \mu^N_A(y)$ for all $xy \in E$.

*Example 8.4.2* Consider a graph $G^*$ such that $V = \{x, y, z\}$ and $E = \{xy, yz, zx\}$. Let $A$ be a bipolar fuzzy subset of $V$ and $B$ be a bipolar fuzzy subset of $E$ defined by

| | $x$ | $y$ | $z$ |
|---|---|---|---|
| $\mu^P_A$ | 0.2 | 0.3 | 0.4 |
| $\mu^N_A$ | −0.4 | −0.5 | −0.5 |

| | $xy$ | $yz$ | $zx$ |
|---|---|---|---|
| $\mu^P_B$ | 0.2 | 0.3 | 0.2 |
| $\mu^N_B$ | −0.4 | −0.5 | −0.4 |

It follows easily that $G = (A, B)$ is a strong bipolar fuzzy graph of $G^*$.

**Proposition 8.4.3** *If $G_1$ and $G_2$ are strong bipolar fuzzy graphs, then $G_1 \times G_2$, $G_1[G_2]$, and $G_1 + G_2$ are strong bipolar fuzzy graphs.*

*Proof* The proof follows from Propositions 8.2.4, 8.2.6, and 8.2.12. ■

We next show that the union of two strong bipolar fuzzy graphs is not necessarily a strong bipolar fuzzy graph.

*Example 8.4.4* Let $V_1 = \{a, b, c\} = V_2$, $E_1 = \{ab, bc\}$, and $E_2 = \{bc, ac\}$. Let $G_1 = (A_1, B_1)$ and $G_2 = (A_2, B_2)$ be bipolar fuzzy graphs defined as follows:

$\mu^P_{A_1}(a) = 0.6$, $\mu^N_{A_1}(a) = -0.5$, $\mu^P_{A_1}(b) = 0.4$, $\mu^N_{A_1}(b) = -0.7$, $\mu^P_{A_1}(c) = 0.5$, $\mu^N_{A_1}(c) = -0.6$, $\mu^P_{B_1}(ab) = 0.4$, $\mu^N_{B_1}(ab) = -0.5$, $\mu^P_{B_1}(bc) = 0.4$, $\mu^N_{B_1}(bc) = -0.6$;

$\mu^P_{A_2}(a) = 0.2$, $\mu^N_{A_2}(a) = -0.5$, $\mu^P_{A_2}(b) = 0.3$, $\mu^N_{A_2}(b) = -0.4$, $\mu^P_{A_2}(c) = 0.7$, $\mu^N_{A_2}(c) = -0.6$, $\mu^P_{B_2}(bc) = 0.3$, $\mu^N_{B_2}(bc) = -0.4$, $\mu^P_{B_2}(ac) = 0.2$, $\mu^N_{B_2}(ac) = -0.5$.

Then we have that $(\mu^P_{A_1} \cup \mu^P_{A_2}(a)) \vee (\mu^P_{A_1} \cup \mu^P_{A_2}(c)) = 0.7 > 0.2 = (\mu^P_{B_1} \cup \mu^P_{B_2})(ac)$.

**Proposition 8.4.5** *If $G_1 \times G_2$ is a strong bipolar fuzzy graph, then at least one of $G_1$ or $G_2$ must be strong.*

*Proof* Suppose $G_1$ and $G_2$ are not strong. Then there exists $x_1y_1 \in E_1$ and $x_2y_2 \in E_2$ such that $\mu^P_{B_1}(x_1y_1) < \mu^P_{A_1}(x_1) \wedge \mu^P_{A_1}(y), \mu^P_{B_2}(x_1y_1) < \mu^P_{A_2}(x_1) \wedge \mu^P_{A_2}(y)$ $\mu^N_{B_1}(x_1y_1) > \mu^N_{A_1}(x_1) \vee \mu^N_{A_1}(y)$, $\mu^N_{B_2}(x_1y_1) > \mu^N_{A_2}(x_1) \vee \mu^N_{A_2}(y)$.

Suppose that $\mu^P_{B_2}(x_2y_2) \leq \mu^P_{B_1}(x_1y_1) < \mu^P_{A_1}(x_1) \wedge \mu^P_{A_1}(y) \leq \mu^P_{A_1}(x_1)$. Let $E = \{(x, x_2)(x, y_2) \mid x \in V_1, x_2y_2 \in E_2\} \cup \{(x_1, z)(y_1, z) \mid z \in V_2, x_1y_1 \in E_1\}$. Let $(x, x_2)(x, y_2) \in E$. Then we have that $(\mu^P_{B_1} \times \mu^P_{B_2})((x, x_2)(x, y_2)) = \mu^P_{A_1}(x) \wedge \mu^P_{B_2}(x_2y_2) < \wedge\{\mu^P_{A_1}(x), \mu^P_{A_2}(x_2) \wedge \mu^P_{A_2}(y_2)\}$ and $(\mu^P_{A_1} \times \mu^P_{A_2})(x, x_2) = \mu^P_{A_1}(x) \wedge \mu^P_{A_2}(x_2)$, $(\mu^P_{A_1} \times \mu^P_{A_2})(x, y_2) = \mu^P_{A_1}(x) \wedge \mu^P_{A_2}(y_2)$.

Thus, $(\mu^P_{A_1} \times \mu^P_{A_2})(x, x_2) \wedge (\mu^P_{A_1} \times \mu^P_{A_2})(x, y_2) = \wedge\{\mu^P_{A_1}(x), \mu^P_{A_2}(x_2), \mu^P_{A_2}(y_2)\}$ and hence, $(\mu^P_{B_1} \times \mu^P_{B_2})((x, x_2)(x, y_2)) < (\mu^P_{A_1} \times \mu^P_{A_2})(x, x_2) \wedge (\mu^P_{A_1} \times \mu^P_{A_2})(x, y_2)$.

Similarly, it follows that $(\mu^N_{B_1} \times \mu^N_{B_2})((x, x_2)(x, y_2)) < (\mu^N_{A_1} \times \mu^N_{A_2})(x, x_2) \wedge (\mu^N_{A_1} \times \mu^N_{A_2})(x, y_2)$. That is, $G_1 \times G_2$ is not a strong bipolar fuzzy graph, a contradiction. Hence, if $G_1 \times G_2$ is a strong bipolar fuzzy graph, then at least $G_1$ or $G_2$ must be a strong bipolar fuzzy graph. ■

The following result is immediate.

**Proposition 8.4.6** *If $G_1[G_2]$ is a strong bipolar fuzzy graph, then at least one of $G_1$ or $G_2$ must be strong.*

**Proposition 8.4.7** *Let $G = (A, B)$ be a strong bipolar fuzzy graph of a graph $G^* = (V, E)$. If $\overline{G} = (\overline{A}, \overline{B})$ satisfies $\overline{A} = A$ and $\overline{B} = (\overline{\mu^P_B}, \overline{\mu^N_B})$ defined by for all $xy \in \mathcal{E}$,*

$$\overline{\mu^P_B}(xy) = \begin{cases} 0 & \text{if } 0 < \mu^P_B(xy) \leq 1, \\ \mu^P_A(x) \wedge \mu^P_A(y) & \text{if } \mu^N_B(xy) = 0, \end{cases}$$

$$\overline{\mu^N_B}(xy) = \begin{cases} 0 & \text{if } -1 \leq \mu^N_B(xy) < 0, \\ \mu^P_A(x) \vee \mu^P_A(y) & \text{if } \mu_B(xy) = 0. \end{cases}$$

*Then $\overline{G}$ is a strong bipolar fuzzy graph of $\overline{G} = (V, \mathcal{E} \backslash E)$.*

*Proof* Clearly, the bipolar sets $\overline{A}$ and $\overline{B}$ satisfy $\overline{\mu^N_B}(xy \leq \overline{\mu^P_A}(x) \wedge \overline{\mu^P_A}(y)$ and $\overline{\mu^N_B}(xy) \geq \overline{\mu^N_A}(x) \vee \overline{\mu^N_A}(y)$ for all $xy \in \mathcal{E}$. Let $xy \in \mathcal{E} \backslash (\mathcal{E} \backslash E) = E$. Then $\mu^N_B(xy = \mu^P_A(x) \wedge \mu^P_A(y)$ because $G$ is strong. If $\overline{\mu^P_B}(xy) = 0$, then by Definition 8.2.1, we have $\overline{\mu^P_B}(xy) = \mu^P_A(x) \wedge \mu^P_A(y) = \mu^P_B(xy) = 0$. Suppose $0 < \mu^P_B(xy) \leq 1$. Then we have $\overline{\mu^P_B}(xy) = 0$. Thus, for all $xy \in \mathcal{E} \backslash (\mathcal{E} \backslash E) = E$, $\overline{\mu^P_B}(xy) = 0$. Similarly, we can show that for all $xy \in \mathcal{E}\backslash(\mathcal{E}\backslash E) = E$, $\overline{\mu_B^N}(xy) = 0$. Hence, $\overline{G}$ is a bipolar fuzzy graph of $\overline{G^*} = (V, \mathcal{E}\backslash E)$.

We now show $\overline{G}$ is strong. By definition, $\mu_B^P(xy) = \mu_B^N(xy) = 0$ for all $xy \in \mathcal{E}\backslash E$. Then $\overline{\mu_B^P}(xy) = \mu_A^P(x) \wedge \mu_A^P(y) = \overline{\mu_A^P}(x) \wedge \overline{\mu_A^P}(y)$ and $\overline{\mu_B^N}(xy) = \mu_A^N(x) \vee \mu_A^N(y) = \overline{\mu_A^N}(x) \vee \overline{\mu_A^N}(y)$. Hence, $\overline{G}$ is a strong bipolar fuzzy graph of $\overline{G^*} = (V, \mathcal{E}\backslash E)$. ■

**Definition 8.4.8** The **complement** of a strong bipolar fuzzy graph $G = (A, B)$ of $G^* = (V, E)$ is a strong bipolar fuzzy graph $\overline{G} = (\overline{A}, \overline{B})$ on $\overline{G^*}$, where $\overline{A} = (\overline{\mu_A^P}, \overline{\mu_A^N})$ and $\overline{B} = (\overline{\mu_B^P}, \overline{\mu_B^N})$ are defined by

(i) $\overline{V} = V$,

(ii) $\overline{\mu_A^P}(x) = \mu_A^P(x)$, $\overline{\mu_A^N}(x) = \mu_A^N(x)$ for all $x \in V$,

(iii)

$$\overline{\mu_B^P}(xy) = \begin{cases} 0 & \text{if } \mu_B^P(xy) > 0 \\ \mu_A^P(x) \wedge \mu_A^P(y) & \text{if } \mu_B^P(xy) = 0 \end{cases}$$

$$\overline{\mu_B^N}(xy) = \begin{cases} 0 & \text{if } \mu_B^N(xy) > 0 \\ \mu_A^N(x) \vee \mu_A^N(y) & \text{if } \mu_B^N(xy) = 0. \end{cases}$$

**Definition 8.4.9** A strong bipolar fuzzy graph $G$ is called **self complementary** if $G$ and $\overline{G}$ are isomorphic.

We next give an example of a self complementary strong bipolar fuzzy graph.

*Example 8.4.10* Let $G^* = (V, E)$ be a graph, where $V = \{a, b, c, d\}$ and $E = \{ab, ac, cd\}$. Let $G = (A, B)$ the strong bipolar graph fuzzy graph defined as follows:

$$\mu_A^P(a) = \mu_A^P(b) = \mu_A^P(c) = \mu_A^P(d) = 0.1,$$
$$\mu_A^N(a) = \mu_A^N(b) = \mu_A^N(c) = \mu_A^N(d) = -0.2.$$

$$\mu_B^P(ab) = \mu_B^P(ac) = \mu_B^P(cd) = 0.1, \mu_B^P(ad) = \mu_B^P(bc) = \mu_B^P(bd) = 0,$$
$$\mu_B^N(ab) = \mu_B^N(ac) = \mu_B^N(cd) = -0.2, \mu_B^N(ad) = \mu_B^N(bc) = \mu_B^N(bd) = 0.$$

Then the complement $\overline{G} = (\overline{A}, \overline{B})$ of $G$ is given as follows: $\overline{A} = A$ and

$$\overline{\mu_B^P}(bc) = \overline{\mu_B^P}(ad) = \overline{\mu_B^P}(bd) = 0.1, \overline{\mu_B^P}(cd) = \overline{\mu_B^P}(ac) = \overline{\mu_B^P}(ab) = 0,$$
$$\overline{\mu_B^N}(bc) = \overline{\mu_B^N}(ad) = \overline{\mu_B^N}(bd) = -0.2, \overline{\mu_B^N}(cd) = \overline{\mu_B^N}(ac) = \overline{\mu_B^N}(ab) = 0.$$

Define the function $f : V \to V$ by $f(a) = b$, $f(b) = c$, $f(c) = d$, $f(d) = a$. Then $f$ is bijective and satisfies the following properties:

$\mu_B^P(ab) = 0.1 = \overline{\mu_B^P}(bc) = \overline{\mu_B^P}(f(a)f(b))$, $\mu_B^N(ab) = -0.2 = \overline{\mu_B^N}(bc) = \overline{\mu_B^N}(f(a)f(b))$, $\mu_B^P(cd) = 0.1 = \overline{\mu_B^P}(ad) = \overline{\mu_B^P}(f(c)f(d))$, $\mu_B^N(cd) = -0.2 =$

$\overline{\mu_B^N}(ad) = \overline{\mu_B^N}(f(c)f(d))$, $\mu_B^P(ac) = 0.1 = \overline{\mu_B^P}(bd) = \overline{\mu_B^P}(f(a)f(c))$, $\mu_B^N(ac) = -0.2 = \overline{\mu_B^N}(bd) = \overline{\mu_B^N}(f(a)f(c))$, $\mu_B^P(bc) = 0 = \overline{\mu_B^P}(cd) = \overline{\mu_B^P}(f(b)f(c))$, $\mu_B^N(bc) = 0 = \overline{\mu_B^N}(cd) = \overline{\mu_B^N}(f(b)f(c))$, $\mu_B^P(bd) = 0 = \overline{\mu_B^P}(ac) = \overline{\mu_B^P}(f(b)f(d))$, $\mu_B^N(bd) = 0 = \overline{\mu_B^N}(ac) = \overline{\mu_B^N}(f(b)f(d))$, $\mu_B^P(ad) = 0 = \overline{\mu_B^P}(ab) = \overline{\mu_B^P}(f(a)f(d))$, $\mu_B^N(ad) = 0 = \overline{\mu_B^N}(ab) = \overline{\mu_B^N}(f(a)f(d))$. Thus, $G$ and $\overline{G}$ are isomorphic.

**Proposition 8.4.11** *Let* $G = (A, B)$ *be a strong bipolar graph of* $G^* = (V, E)$ *and* $\overline{G} = (\overline{A}, \overline{B})$ *be the complement of* $G$. *Then*

$$\overline{\mu_B^P}(xy) = \mu_A^P(x) \wedge \mu_A^P(y) - \mu_B^P(xy) \textit{ for all } xy \in \mathcal{E},$$
$$\overline{\mu_B^N}(xy) = \mu_A^N(x) \vee \mu_A^N(y) - \mu_B^N(xy) \textit{ for all } xy \in \mathcal{E}.$$

*Proof* Let $xy \in \mathcal{E}$. If $0 < \mu_B^P(xy) \leq 1$, then $xy \in E$. Because $G$ is strong, it follows that $\mu_A^P(x) \wedge \mu_A^P(y) - \mu_B^P(xy) = 0 = \overline{\mu_B^P}(xy)$. Suppose that $\mu_B^P(xy) = 0$. Then $\mu_A^P(x) \wedge \mu_A^P(y) - \mu_B^P(xy) = \mu_A^P(x) \wedge \mu_A^P(y) = \overline{\mu_B^P}(xy)$. Therefore, $\overline{\mu_B^P}(xy) = \mu_A^P(x) \wedge \mu_A^P(y) - \mu_B^P(xy)$ for all $xy \in \mathcal{E}$.

It follows analogously that $\overline{\mu_B^N}(xy) = \mu_A^N(x) \vee \mu_A^N(y) - \mu_B^N(xy)$ for all $xy \in \mathcal{E}$. ■

**Definition 8.4.12** ([45]) The **complement** of a fuzzy graph $G = (\sigma, \mu)$ is a fuzzy graph $\overline{G} = (\overline{\sigma}, \overline{\mu})$, where $\overline{\sigma} = \sigma$ and $\overline{\mu}(xy) = \wedge\{\sigma(x) \wedge \sigma(y) - \mu(xy) \mid xy \in \mathcal{E}\}$.

**Proposition 8.4.13** *Let* $G$ *be a self complementary strong bipolar fuzzy graph. Then*

(*i*) $\sum \mu_B^P(xy) = \frac{1}{2} \sum_{x \neq y} \mu_A^P(x) \wedge \mu_B^P(y)$,

(*ii*) $\sum \mu_B^N(xy) = \frac{1}{2} \sum_{x \neq y} \mu_A^N(x) \vee \mu_B^N(y)$.

*Proof* Because $G$ is a self complementary strong fuzzy bipolar graph, $\mu_B^P(xy) = \mu_A^P(x) \wedge \mu_A^P(y)$ and $\mu_B^N(xy) = \mu_A^N(x) \vee \mu_A^N(y)$ for all $xy \in \mathcal{E}$. Also, there exists an isomorphism $f$ of $G$ to $\overline{G}$ such that $\mu_A^P(x) = \overline{\mu_A^P}(f(x))$ and $\mu_A^N(x) = \overline{\mu_A^N}(f(x))$ for all $x \in V$ and $\mu_B^P(xy) = \overline{\mu_B^P}(f(x)f(y))$ and $\mu_B^N(xy) = \overline{\mu_B^N}(f(x)f(y))$ for all $xy \in \mathcal{E}$.

(*i*) For all $xy \in \mathcal{E}$, we have that $\overline{\mu_B^P}(f(x)f(y)) = \mu_A^P(f(x)) \wedge \mu_A^P(f(y)) - \mu_B^P(f(x)f(y))$. Hence, $\mu_B^P(xy) = \mu_A^P(f(x)) \wedge \mu_A^P(f(y)) - \mu_B^P(f(x)f(y))$. Thus,

$$\begin{aligned}\sum_{x \neq y} \mu_B^P(xy) + \sum_{x \neq y} \mu_B^P(f(x)f(y)) &= \sum_{x \neq y} \mu_A^P(f(x)) \wedge \mu_A^P(f(y) \\ &= \sum_{x \neq y} \mu_A^P(x) \wedge \mu_A^P(y).\end{aligned}$$

Hence, $2\sum_{x \neq y} \mu_B^P(xy) = \sum_{x \neq y} \mu_A^P(x) \wedge \mu_A^P(y)$ and so the desired result is immediate.

$(ii)$ For all $xy \in \mathcal{E}$, we have that $\overline{\mu_B^N}(f(x)f(y)) = \mu_A^N(f(x)) \vee \mu_A^N(f(y)) - \mu_B^N(f(x)f(y))$. Hence, $\mu_B^N(xy) = \mu_A^N(f(x)) \vee \mu_A^N(f(y)) - \mu_B^N(xy)$. Thus,

$$\sum_{x \neq y} \mu_B^N(xy) + \sum_{x \neq y} \mu_B^N(f(x)f(y)) = \sum_{x \neq y} \mu_A^N(f(x)) \vee \mu_A^N(f(y)$$
$$= \sum_{x \neq y} \mu_A^N(x) \vee \mu_A^N(y).$$

Hence, $2\sum_{x \neq y} \mu_B^N(xy) = \sum_{x \neq y} \mu_A^N(x) \vee \mu_A^N(y)$ and so the desired result is immediate. ■

**Proposition 8.4.14** *Suppose $G = (A, B)$ is a strong bipolar fuzzy graph of $G^* = (V, E)$. If $\mu_B^P(xy) = \frac{1}{2}(\mu_A^P(x) \wedge \mu_A^P(y))$ and $\mu_B^N(xy) = \frac{1}{2}(\mu_A^N(x) \vee \mu_A^N(y))$ for all $xy \in \mathcal{E}$, then $G$ is self complementary.*

*Proof* Clearly, the identity map $I : V \to V$ is an isomorphism from $G$ to $\overline{G}$. In fact by Proposition 8.4.11, we have for all $xy \in \mathcal{E}$,

$$\begin{aligned}
\overline{\mu_B^P}(I(x)I(y)) &= \overline{\mu_B^P}(xy) \\
&= \mu_A^P(x) \wedge \mu_A^P(y) - \mu_B^P(xy) \\
&= \mu_A^P(x) \wedge \mu_A^P(y) - \frac{1}{2}(\mu_A^P(x) \wedge \mu_A^P(y)) \\
&= \frac{1}{2}(\mu_A^P(x) \wedge \mu_A^P(y)) \\
&= \mu_B^P(xy).
\end{aligned}$$

Similarly, $\overline{\mu_B^P}(I(x)I(y)) = \mu_B^P(xy)$. Thus, $G$ and $\overline{G}$ are isomorphic. ■

**Corollary 8.4.15** *Suppose $G = (A, B)$ is a strong bipolar fuzzy graph of $G^* = (V, E)$. Then $G$ is self complementary if and only if $\mu_B^P(xy) = \frac{1}{2}(\mu_A^P(x) \wedge \mu_A^P(y))$ and $\mu_B^N(xy) = \frac{1}{2}(\mu_A^N(x) \vee \mu_A^N(y))$ for all $xy \in \mathcal{E}$.*

**Proposition 8.4.16** *Let $G_1$ and $G_2$ be strong bipolar fuzzy graphs. Then $G_1$ and $G_2$ are isomorphic if and only if $\overline{G_1}$ and $\overline{G_2}$ are isomorphic.*

*Proof* Suppose $G_1$ and $G_2$ are isomorphic. Then there exists a bijective function $f : V_1 \to V_2$ satisfying

$$\mu_{A_1}^P(x) = \mu_{A_2}^P(f(x)) \text{ and } \mu_{A_1}^N(x) = \mu_{A_2}^N(f(x)) \text{ for all } x \in V_1,$$
$$\mu_{B_1}^P(xy) = \mu_{B_2}^P(f(x)f(y)) \text{ and } \mu_{B_1}^P(xy) = \mu_{B_2}^P(f(x)f(y)) \text{ for all } xy \in \mathcal{E}_1.$$

Let $xy \in \mathcal{E}_1$. If $\mu^P_{B_1}(xy) = 0$, then $\mu^P_{B_2}(f(x)f(y)) = 0$. Now, $\overline{\mu^P_{B_1}}(xy) = \mu^P_{A_1}(x) \wedge \mu^P_{A_1}(y) = \mu^P_{A_2}(f(x)) \wedge \mu^P_{A_2}(f(y)) = \overline{\mu^P_{B_2}}(f(x)f(y))$. Suppose $0 < \mu^P_{B_1}(xy) \leq 1$. Then $0 < \mu^P_{B_2}(f(x)f(y)) \leq 1$. Thus, $\mu^P_{B_1}(xy) = 0 = \overline{\mu^P_{B_2}}(f(x)f(y))$. Analogously, $\mu^N_{B_1}(xy) = \overline{\mu^N_{B_2}}(f(x)f(y))$ for all $xy \in \mathcal{E}_1$. Hence, $\overline{G_1}$ and $\overline{G_2}$ are isomorphic.

Conversely, suppose $\overline{G_1}$ and $\overline{G_2}$ are isomorphic. Then there exists a bijective function $f : V_1 \to V_2$ satisfying

$$\overline{\mu^P_{A_1}}(x) = \overline{\mu^P_{A_2}}(f(x)) \text{ and } \overline{\mu^N_{A_1}}(x) = \overline{\mu^N_{A_2}}(f(x)) \text{ for all } x \in V_1,$$
$$\overline{\mu^P_{B_1}}(xy) = \overline{\mu^P_{B_2}}(f(x)f(y)) \text{ and } \overline{\mu^N_{B_1}}(xy) = \overline{\mu^N_{B_2}}(f(x)f(y)) \text{ for all } xy \in \mathcal{E}_1.$$

Let $xy \in \mathcal{E}_1$. Suppose that $\mu^P_{B_1}(xy) = 0$. Then

$$\begin{aligned}\overline{\mu^P_{B_2}}(f(x)f(y)) &= \overline{\mu^P_{B_1}}(xy)\\ &= \mu^P_{A_1}(x) \wedge \mu^P_{A_1}(y)\\ &= \overline{\mu^P_{A_1}}(x) \wedge \overline{\mu^P_{A_1}}(y)\\ &= \overline{\mu^P_{A_2}}(f(x)) \wedge \overline{\mu^P_{A_2}}(f(y)\\ &= \mu^P_{A_2}(f(x)) \wedge \mu^P_{A_2}(f(y)).\end{aligned}$$

Thus, $\mu^P_{B_2}(f(x)f(y)) = 0 = \mu^P_{B_1}(xy)$. Suppose that $0 < \mu^P_{B_1}(xy) \leq 1$. Then $\overline{\mu^P_{B_2}}(f(x)f(y)) = \overline{\mu^P_{B_1}}(xy) = 0$. Hence,

$$\begin{aligned}\mu^P_{B_2}(f(x)f(y)) &= \mu^P_{A_2}(f(x)) \wedge \mu^P_{A_2}(f(y))\\ &= \overline{\mu^P_{A_2}}(f(x)) \wedge \overline{\mu^P_{A_2}}(f(y)\\ &= \mu^P_{A_1}(x) \wedge \mu^P_{A_1}(y)\\ &= \mu^P_{B_1}(xy).\end{aligned}$$

It follows analogously that $\mu^N_{B_1}(xy) = \mu^N_{B_2}(f(x)f(y))$. Therefore, $G_1$ and $G_2$ are isomorphic. ■

The following result is immediate.

**Proposition 8.4.17** *Let $G_1$ and $G_2$ be strong bipolar fuzzy graphs. If there is a co-weak isomorphism between $G_1$ and $G_2$, then there is a co-weak isomorphism between $\overline{G_1}$ and $\overline{G_2}$.*

## 8.5 Regular Bipolar Fuzzy Graphs

The results in this section are due to those in [16]. We present the concepts of regular and totally regular bipolar fuzzy graphs. We prove necessary and sufficient conditions for which regular bipolar fuzzy graphs and totally bipolar fuzzy line graphs are equivalent. We introduce the notion of bipolar fuzzy line graphs and present some of their properties. We give a necessary and sufficient condition for a bipolar fuzzy graph to be isomorphic to its corresponding bipolar fuzzy line graph. We examine when an isomorphism between two bipolar fuzzy graphs follows from an isomorphism of their corresponding bipolar fuzzy line graphs. Sufficient conditions for a bipolar fuzzy graph to be isomorphic to its corresponding bipolar fuzzy line graph are provided.

Recall that for given a graph $G^* = (V, E)$, two vertices $x, y \in V$ are said to be **neighbors**, or **adjacent vertices**, if $xy \in E$. The **neighborhood** of a vertex $v$ in a graph $G^*$ is the induced subgraph of $G^*$ consisting of all vertices adjacent to $v$ and all edges connecting two such vertices. The neighborhood is denoted by $N(v)$. The degree $\deg(v)$ of vertex $v$ is the number of edges incident on $v$ or equivalently, $\deg(v) = |N(v)|$ .The set of neighbors, called an **open neighborhood** $N(v)$ for a vertex $v$ in a graph $G^*$ consists of all vertices adjacent two $v$ but not including $v$, that is $N(v) = \{u \in V \mid vu \in E\}$. When $v$ is also included, it is called a **closed neighborhood** $N[v]$, that is, $N[v] = N(v) \cup \{v\}$. A **regular graph** is a graph where each vertex has the same number of neighbors, i.e., all the vertices have the same closed neighborhood degree. A **complete graph** is a simple graph in which every pair of distinct vertices is connected by an edge.

An **isomorphism** of graphs $G_1^*$ and $G_2^*$ is a bijection between the vertex sets of $G_1^*$ and $G_2^*$ such that any two vertices $v_1$ and $v_2$ of $G_1^*$ are adjacent in $G_1^*$ if and only if $f(v_1)$ and $f(v_2)$ are adjacent in $G_2^*$. Isomorphic graphs are denoted by $G_1^* \simeq G_2^*$.

In graph theory, the line graph $L(G^*)$ of a simple graph $G^*$ is a graph that represents the adjacencies between edges of $G^*$. Given a graph $G^*$, its line graph $L(G^*)$ is a graph such that each vertex of $L(G^*)$ represents an edge of $G^*$; and two vertices of $L(G^*)$ are adjacent if and only if their corresponding edges share a common endpoint in $G^*$.

Let $G^* = (V, E)$ be an undirected graph, where $V = \{v_1, v_2, \ldots, v_{n-1}, v_n\}$. Let $S_i = \{v_i, x_{i1}, \ldots, x_{iq_i}\}$, where $x_{ij} = v_i v_j \in E$, $i = 1, 2, \ldots, n$, $j = 1, 2, \ldots, q_i$. Let $S = \{S_1, S_2, \ldots, S_n\}$. Let $T = \{S_i S_j \mid S_i, S_j \in S, S_i \cap S_j \neq \emptyset, i \neq j\}$. Then $P(S) = (S, T)$ is an intersection graph and $P(S) = G^*$. The line graph $L(G^*)$ is by definition the intersection graph $P(E)$. That is, $L(G^*) = (Z, W)$, where $Z = \{\{x\} \cup \{u_x, v_x\} \mid x \in E, u_x, v_x \in V, x = u_x v_x\}$ and $W = \{S_x S_y \mid S_x \cap S_y \neq \emptyset$, $x, y \in E, x \neq y\}$, and $S_x = \{x\} \cup \{u_x, v_x\}$, $x \in E$.

Recall that a graph is called **regular** if all vertices have the same degree.

**Proposition 8.5.1** *If $G$ is regular of degree $k$, then the line graph $L(G)$ is regular of degree $2k - 2$.*

**Definition 8.5.2** Let $G = (A, B)$ be a bipolar fuzzy graph on $G^* = (V, E)$. The open neighborhood degree of a vertex $x$ in $G$ is denoted by

$$\deg(x) = (\deg^P(x), \deg^N(x)),$$

where

$$\deg^P(x) = \sum_{y \in N(x)} \mu_A^P(y)$$

and

$$\deg^N(x) = \sum_{y \in N(x)} \mu_A^N(y).$$

If all the vertices have the same open neighborhood degree $n$, then $G$ is called an $n$-**regular bipolar fuzzy graph**.

**Definition 8.5.3** Let $G = (A, B)$ be a regular bipolar fuzzy graph. If each vertex of $G$ has the same closed neighborhood degree $m$, then $G$ is called a **totally regular bipolar fuzzy graph**. The closed neighborhood degree of a vertex $x$ is defined by $\deg[x] = \deg^P[x] + \deg^N[x]$, where

$$\deg^P[x] = \deg^P(x) + \mu_A^P(x),$$

$$\deg^N[x] = \deg^N(x) + \mu_A^N(x).$$

The following examples show that there is no relationship between $n$-regular bipolar fuzzy graphs and $m$-totally regular bipolar fuzzy graphs.

*Example 8.5.4* Consider a graph $G^*$ such that $V = \{a, b, c, d\}$ and $E = \{ab, bc, cd, ad\}$. Let $A$ be a bipolar fuzzy subset of $A$ and let $B$ be a bipolar fuzzy subset of $E$ defined by

| | $a$ | $b$ | $c$ | $d$ |
|---|---|---|---|---|
| $\mu_A^P$ | 0.5 | 0.5 | 0.5 | 0.5 |
| $\mu_A^N$ | −0.3 | −0.3 | −0.3 | −0.3 |

| | $ab$ | $bc$ | $cd$ | $ad$ |
|---|---|---|---|---|
| $\mu_B^P$ | 0.2 | 0.4 | 0.2 | 0.4 |
| $\mu_B^N$ | −0.1 | −0.1 | −0.1 | −0.1 |

It is easily shown that the bipolar fuzzy graph $G$ is both regular and totally regular.

*Example 8.5.5* Consider a $G^*$ such that $V = \{v_1, v_2, v_3\}$ and $E = \{v_1v_2, v_1v_3\}$. Let $A$ be a bipolar fuzzy subset of $V$ and let $B$ be a bipolar fuzzy subset of $E$ defined by

| | $v_1$ | $v_2$ | $v_3$ |
|---|---|---|---|
| $\mu_A^P$ | 0.4 | 0.8 | 0.7 |
| $\mu_A^N$ | −0.4 | −0.7 | −0.6 |

| | $v_1v_2$ | $v_1v_3$ |
|---|---|---|
| $\mu_B^P$ | 0.3 | 0.4 |
| $\mu_B^N$ | −0.2 | −0.2 |

It follows easily that the bipolar fuzzy graph $G$ is neither totally regular nor regular.

**Definition 8.5.6** A bipolar fuzzy graph $G = (A, B)$ is called **complete** if

$$\mu_B^P(xy) = \mu_A^P(x) \wedge \mu_A^P(y) \text{ and } \mu_B^N(xy) = \mu_A^N(x) \vee \mu_A^N(y)$$

for all $x, y \in V$.

*Example 8.5.7* Consider a $G^*$ such that $V = \{x, y, z\}$ and $E = \{xy, yz, zx\}$. Let $A$ be a bipolar fuzzy subset of $V$ and let $B$ be a bipolar fuzzy subset of $E$ defined by

| | $x$ | $y$ | $z$ |
|---|---|---|---|
| $\mu_A^P$ | 0.5 | 0.7 | 0.6 |
| $\mu_A^N$ | −0.3 | −0.4 | −0.5 |

| | $xy$ | $yz$ | $zx$ |
|---|---|---|---|
| $\mu_B^P$ | 0.5 | 0.6 | 0.5 |
| $\mu_B^N$ | −0.3 | −0.4 | −0.3 |

It is easily shown that $G$ is a both complete and totally regular bipolar fuzzy graph, but $G$ is not regular because $\deg(x) \neq \deg(z) \neq \deg(y)$.

**Theorem 8.5.8** *Every complete bipolar fuzzy graph is a totally regular bipolar fuzzy graph.*

Let $G = (A, B)$ be a bipolar fuzzy graph of a graph $G^*$. Then $A = (\mu_A^P, \mu_A^N)$ is a constant function if $\exists c_1 \in [0, 1]$ such that for all $x \in V$, $\mu_A^P(x) = c_1$ and $\exists c_2 \in V$ such that for all $x \in V$, $\mu_A^N(x) = c_1$.

**Theorem 8.5.9** ([14]) *Let $G = (A, B)$ be a bipolar fuzzy graph of a graph $G^*$. Then $A = (\mu_A^P, \mu_A^N)$ is a constant function if and only if the following conditions are equivalent:*

*$(i)$ $G$ is a regular bipolar fuzzy graph.*

*$(ii)$ $G$ is a totally regular bipolar fuzzy graph.*

*Proof* Suppose that $A = (\mu_A^P, \mu_A^N)$ is a constant function. Let $\mu_A^P(x) = c_1$ and $\mu_A^N(x) = c_2$ for all $x \in V$.

$(i) \Rightarrow (ii)$: Assume that $G$ is $n$-regular bipolar fuzzy graph. Then $\deg^P(x) = n_1$ and $\deg^N(x) = n_2$ for all $x \in V$. So $\deg^P[x] = \deg^P(x) + \mu_A^P(x)$ and $\deg^N[x] = \deg^N(x) + \mu_A^N(x)$ for all $x \in V$. Thus, $\deg^P[x] = n_1 + c_1$ and $\deg^N[x] = n_2 + c_2$ for all $x \in V$. Hence, $G$ is a totally regular bipolar fuzzy graph.

$(i) \Rightarrow (ii)$: Suppose that $G$ is a totally regular bipolar fuzzy graph. Then $\deg^P[x] = n_1$ and $\deg^N[x] = n_2$ for all $x \in V$ or $\deg^P(x) + \mu_A^P(x) = k_1$ and $\deg^N[x] + \mu_A^N(x) = k_2$ for all $x \in V$ or $\deg^P(x) + c_1 = k_1$ and $\deg^N[x] + c_2 = k_2$ for all $x \in V$ or $\deg^P(x) = k_1 - c_1$ and $\deg^N[x] = k_2 - c_2$ for all $x \in V$. Thus, $G$ is a regular bipolar fuzzy graph. The converse is immediate. ■

**Proposition 8.5.10** *If a bipolar fuzzy graph is both regular and totally regular, then $A = (\mu_A^P, \mu_A^N)$ is a constant function.*

*Proof* Let $G$ be a regular and totally regular fuzzy graph. Then

$$\deg^P(x) = n_1,\ \deg^N(x) = n_2 \text{ for all } x \in V,$$

$$\deg^P[x] = k_1 \text{ and } \deg^N[x] = k_2 \text{ for all } x \in V.$$

Now, $\deg^P[x] = k_1 \Leftrightarrow \deg^P(x) + \mu_A^P(x) = k_1 \Leftrightarrow n_1 + \mu_A^P(x) = k_1 \Leftrightarrow \mu_A^P(x) = k_1 - n_1$ for all $x \in V$. Similarly, $\mu_A^N(x) = k_2 - n_2$ for all $x \in V$. Hence, $A = (\mu_A^P, \mu_A^N)$ is a constant function. ■

The converse of Proposition 8.5.10 is not true, in general.

We state the following theorem without proof.

**Theorem 8.5.11** *Let $G$ be a bipolar fuzzy graph, where the crisp graph $G^*$ is an odd cycle. Then $G$ is a regular bipolar fuzzy graph if and only if $B$ is a constant function.*

## 8.6 Bipolar Fuzzy Line Graphs

We consider bipolar fuzzy line graphs in this section.

**Definition 8.6.1** Let $P(S) = (S, T)$ be an intersection graph of a simple graph $G^* = (V, E)$. Let $G = (A_1, B_1)$ be a bipolar fuzzy graph of $G^*$. We define a bipolar fuzzy intersection graph $P(G) = (A_2, B_2)$ of $P(S)$ as follows:

($i$) $A_2$ and $B_2$ are bipolar fuzzy subsets of $S$ and $T$, respectively.
($ii$) $\mu_{A_2}^P(S_i) = \mu_{B_1}^P(v_i)$, $\mu_{A_2}^N(S_i) = \mu_{B_1}^N(v_i)$.
($iii$) $\mu_{B_2}^P(S_iS_j) = \mu_{B_1}^P(v_iv_j)$, $\mu_{A_2}^N(S_iS_j) = \mu_{B_1}^N(v_iv_j)$ for all $S_i, S_j \in S$, $S_iS_j \in T$.

A bipolar fuzzy graph of $P(S)$ is called a **bipolar fuzzy intersection graph**.

The following proposition is immediate.

**Proposition 8.6.2** *Let $G = (A_1, B_1)$ be a bipolar fuzzy graph of $G^*$. Then $P(G) = (A_2, B_2)$ is a bipolar fuzzy graph of $P(S)$ and $G \simeq P(G)$.*

This proposition shows that any bipolar fuzzy graph is isomorphic to a bipolar fuzzy intersection graph.

**Definition 8.6.3** Let $L(G^*) = (Z, W)$ be a line graph of a simple graph $G^* = (V, E)$. Let $G = (A_1, B_1)$ be a bipolar fuzzy graph of $G^*$. We define a **bipolar fuzzy line graph** $L(G) = (A_2, B_2)$ of $G$ as follows:

($i$) $A_2$ and $B_2$ are bipolar fuzzy subsets of $Z$ and $W$, respectively.

$(ii)\ \mu^P_{A_2}(S_x) = \mu^P_{B_1}(x) = \mu^P_{B_1}(u_x v_x)$.

$(iii)\ \mu^N_{A_2}(S_x) = \mu^N_{B_1}(x) = \mu^N_{B_1}(u_x v_x)$.

$(iv)\ \mu^P_{B_2}(S_x S_y) = \mu^P_{B_1}(x) \wedge \mu^P_{B_1}(y)$.

$(v)\ \mu^P_{B_2}(S_x S_y) = \mu^N_{B_1}(x) \vee \mu^N_{B_1}(y)$ for all $S_x, S_y \in Z$, $S_x S_y \in W$.

*Example 8.6.4* Consider a graph $G^* = (V, E)$ such that $V = \{v_1, v_2, v_3, v_4\}$ and $E = \{x_1 = v_1 v_2\ x_2 = v_2 v_3, x_3 = v_3 v_4, x_4 = v_4 v_1\}$. Let $A_1$ be a bipolar fuzzy subset of $V$ and let $B_1$ be a bipolar fuzzy subset of $E$ defined by

| | $v_1$ | $v_2$ | $v_3$ | $v_4$ |
|---|---|---|---|---|
| $\mu^P_{A_1}$ | 0.2 | 0.3 | 0.4 | 0.1 |
| $\mu^N_{A_1}$ | −0.5 | −0.4 | −0.5 | −0.3 |

| | $x_1$ | $x_2$ | $x_3$ | $x_4$ |
|---|---|---|---|---|
| $\mu^P_{B_1}$ | 0.1 | 0.2 | 0.1 | 0.1 |
| $\mu^N_{B_1}$ | −0.2 | −0.3 | −0.2 | −0.2 |

It follows easily that $G$ is a bipolar fuzzy graph.

Consider a line graph $L(G^*) = (Z, W)$, where $Z = \{S_{x_1}, S_{x_2}, S_{x_3}, S_{x_4}\}$ and $W = \{S_{x_1} S_{x_2}, S_{x_2} S_{x_3}, S_{x_3} S_{x_4}, S_{x_4} S_{x_1}\}$. Let $A_2 = (\mu^P_{A_2}, \mu^N_{A_2})$ and $B_2 = (\mu^P_{B_2}, \mu^N_{B_2})$ be bipolar fuzzy subsets of $Z$ and $W$, respectively. Then it follows that,

$\mu^P_{A_2}(S_{x_1}) = 0.1$, $\mu^P_{A_2}(S_{x_2}) = 0.2$, $\mu^P_{A_2}(S_{x_3}) = 0.1$, $\mu^P_{A_2}(S_{x_4}) = 0.1$, $\mu^N_{A_2}(S_{x_1}) = -0.2$, $\mu^N_{A_2}(S_{x_2}) = -0.3$, $\mu^N_{A_2}(S_{x_3}) = -0.2$, $\mu^N_{A_2}(S_{x_4}) = -0.2$, $\mu^P_{B_2}(S_{x_1} S_{x_2}) = 0.1$, $\mu^P_{B_2}(S_{x_2} S_{x_3}) = 0.1$, $\mu^P_{B_2}(S_{x_3} S_{x_4}) = 0.1$, $\mu^P_{B_2}(S_{x_4} S_{x_1}) = 0.1$, $\mu^N_{B_2}(S_{x_1} S_{x_2}) = -0.2$, $\mu^N_{B_2}(S_{x_2} S_{x_3}) = -0.2$, $\mu^N_{B_2}(S_{x_3} S_{x_4}) = -0.2$, $\mu^N_{B_2}(S_{x_4} S_{x_1}) = -0.2$.

It follows easily that $L(G)$ is a bipolar fuzzy line graph. It is neither a regular bipolar fuzzy line graph nor a totally regular bipolar fuzzy line graph.

**Proposition 8.6.5** *If $L(G)$ is a bipolar fuzzy line graph of $G$, then $L(G^*)$ is a line graph of $G^*$.*

*Proof* Because $G = (A_1, B_1)$ be a bipolar fuzzy graph and $L(G)$ is a bipolar fuzzy line graph, $\mu^P_{A_1}(S_x) = \mu^P_{B_1}(x)$, $\mu^N_{A_1}(S_x) = \mu^N_{B_1}(x)$ for all $x \in E$. Thus, $S_x \in Z \Leftrightarrow x \in E$. Also, $\mu^N_{B_2}(S_x S_y) = \mu^N_{B_1}(x) \vee \mu^N_{B_1}(y)$ for all $S_x, S_y \in Z$. Hence, $W = \{S_x S_y \mid S_x \cap S_y \neq \emptyset, x, y \in E, x \neq y\}$. ■

**Proposition 8.6.6** *$L(G)$ is a bipolar fuzzy line graph of some bipolar fuzzy graph $G$ if and only if $\mu^P_{B_2}(S_x S_y) = \mu^P_{A_2}(S_x) \wedge \mu^P_{A_2}(S_y)$ for all $S_x S_y \in W$ and $\mu^N_{B_2}(S_x S_y) = \mu^N_{A_2}(S_x) \vee \mu^N_{A_2}(S_y)$ for all $S_x S_y \in W$.*

*Proof* Assume that $\mu^P_{B_2}(S_x S_y) = \mu^P_{A_2}(S_x) \wedge \mu^P_{A_2}(S_y)$ for all $S_x S_y \in W$. Define $\mu^P_{A_1}(x) = \mu^P_{A_2}(S_x)$ for all $x \in V$. Then $\mu^P_{B_2}(S_x S_y) = \mu^P_{A_2}(S_x) \wedge \mu^P_{A_2}(S_y) = \mu^P_{A_1}(x) \wedge \mu^P_{A_1}(y)$ and $\mu^N_{B_2}(S_x S_y) = \mu^N_{A_2}(S_x) \vee \mu^N_{A_2}(S_y) = \mu^N_{A_1}(x) \vee \mu^n_{A_1}(y)$. A bipolar fuzzy set $A_1 = (\mu^P_{A_1}, \mu^N_{A_1})$ that yields that the property $\mu^P_{B_1}(xy) = \mu^P_{A_1}(x) \wedge \mu^P_{A_1}(y)$, and $\mu^N_{B_1}(xy) = \mu^N_{A_1}(x) \vee \mu^N_{A_1}(y)$ will suffice.

The converse follows easily. ■

**Proposition 8.6.7** *$L(G)$ is a bipolar fuzzy line graph if and only if $L(G^*)$ is a line graph and $(i)$ $\mu^P_{B_2}(uv) \le \mu^P_{A_2}(u) \wedge \mu^P_{A_2}(v)$ for all $uv \in W$ $(ii)$ $\mu^N_{B_2}(uv) \ge \mu^N_{A_2}(u) \vee \mu^N_{A_2}(v)$ for all $uv \in W$.*

*Proof* The result follows from Propositions 8.6.5 and 8.6.6. ■

Let $G_1$ and $G_2$ be two bipolar fuzzy graphs. Recall that a homomorphism of $f : G_1 \to G_2$ is a mapping $f : V_1 \to V_2$ such that,

$(i)$ $\mu^P_{A_1}(x_1) \le \mu^P_{A_2}(f(x_1))$, $\mu^N_{A_1}(x_1) \ge \mu^N_{A_2}(f(x_1))$,

$(ii)$ $\mu^P_{B_1}(x_1y_1) \le \mu^P_{B_2}(f(x_1)f(y_1))$, $\mu^N_{B_1}(x_1y_1) \le \mu^N_{B_2}(f(x_1)f(y_1))$ for all $x_1 \in V_1$, $x_1y_1 \in E_1$.

A bijective homomorphism with the property $\mu^P_{A_1}(x_1) = \mu^P_{A_2}(f(x_1))$, $\mu^N_{A_1}(x_1) = \mu^N_{A_2}(f(x_1))$ is called a (**weak**) **vertex-isomorphism**.

If $f$ is a (weak) vertex-isomorphism and a (weak) line-isomorphism of $G_1$ onto $G_2$, then $f$ is called a (**weak**) **isomorphism** of $G_1$ and $G_2$.

**Proposition 8.6.8** *Let $G_1$ and $G_2$ be bipolar fuzzy graphs. If $f$ is a weak isomorphism of $G_1$ and $G_2$, then $f$ is an isomorphism of $G_1^*$ and $G_2^*$.*

**Theorem 8.6.9** *Let $L(G) = (A_2, B_2)$ be the bipolar fuzzy line graph corresponding to bipolar fuzzy graph $G_1 = (A_1, B_1)$. Suppose that $G^* = (V, E)$ is connected. Then the following properties hold.*

*$(i)$ There exists a weak isomorphism of $G$ onto $L(G)$ if and only if $G^*$ is a cycle and if for all $v \in V$, $x \in E$, $\mu^P_{A_1}(v) = \mu^P_{B_1}(x)$, $\mu^N_{A_1}(v) = \mu^N_{B_1}(x)$, i.e., $A_1 = (\mu^P_{A_1}, \mu^N_{A_1})$ and $B_1 = (\mu^P_{B_1}, \mu^N_{B_1})$ are constant functions on $V$ and $E$, respectively, taking on the same value.*

*$(ii)$ If $f$ is a weak isomorphism of $G$ onto $L(G)$, then $f$ is an isomorphism.*

*Proof* Assume that $f$ is a weak isomorphism of $G$ onto $L(G)$. It follows that $G^* = (V, E)$ is a cycle.

Let $V = \{v_1, v_2, \ldots, v_{n-1}, v_n\}$ and $E = \{x_1 = v_1v_2, x_2 = v_2v_3, \ldots, x_n = v_nv_1\}$, where $v_1v_2 \cdots v_nv_1$ is a cycle. Define the bipolar fuzzy subsets

$$\mu^P_{A_1}(v_i) = s_i,\ \mu^N_{A_1}(v_i) = s_i'$$

and

$$\mu^P_{B_1}(v_iv_{i+1}) = r_i,\ \mu^N_{B_1}(v_iv_{i+1}) = r_i',,\ i = 1, 2, \ldots, n,\ v_{n+1} = v_1.$$

Then for $s_{n+1} = s_1$, $s'_{n+1} = s'_1$, we have that

$$r_i \leq s_i \wedge s_{i+1}, \quad r'_i \geq s'_i \vee s'_{i+1}, \quad i = 1, 2, \ldots, n. \tag{8.18}$$

Now, $Z = \{S_{x_1}, S_{x_2}, \ldots, S_{x_n}\}$ and $W = \{S_{x_1}S_{x_2}, S_{x_2}S_{x_3}, \ldots, S_{x_n}S_{x_1}\}$. Also, for $r_{n+1} = r_1$, we have that

$$\mu^P_{A_2}(S_{x_i}) = \mu^P_{A_1}(x_i) = \mu^P_{A_1}(v_i v_{i+1}) = r_i,$$

$$\mu^N_{A_2}(S_{x_i}) = \mu^N_{A_1}(x_i) = \mu^N_{A_1}(v_i v_{i+1}) = r'_i,$$

$$\begin{aligned}\mu^P_{B_2}(S_{x_i}S_{x_{i+1}}) &= \mu^P_{B_1}(x_i) \wedge \mu^P_{B_1}(x_{i+1}) \\ &= \mu^P_{B_1}(v_i v_{i+1}) \wedge \mu^P_{B_1}(v_{i+1}v_{i+2}) \\ &= r_i \wedge r_{i+1},\end{aligned}$$

$$\begin{aligned}\mu^N_{B_2}(S_{x_i}S_{x_{i+1}}) &= \mu^N_{B_1}(x_i) \vee \mu^N_{B_1}(x_{i+1}) \\ &= \mu^N_{B_1}(v_i v_{i+1}) \vee \mu^N_{B_1}(v_{i+1}v_{i+2}) \\ &= r'_i \vee r'_{i+1}\end{aligned}$$

for $i = 1, 2, \ldots, n$, $v_{n+1} = v_1$, $v_{i+2} = v_2$. Because $f$ is an isomorphism of $G^*$ onto $L(G^*)$, $f$ maps $V$ one-to-one and onto $Z$. Also, $f$ preserves adjacency. Hence, $f$ induces a permutation $\pi$ of $\{1, 2, \ldots, n\}$ such that

$$f(v_i) = S_{x_{\pi(i)}} = S_{x_{\pi(i)}}S_{x_{\pi(i+1)}}$$

and

$$x_i = v_i v_{i+1} \rightarrow f(v_i)f(v_{i+1}) = S_{v_{\pi(i)}}S_{v_{\pi(i+1)}}S_{v_{\pi(i+2)}}, \; i = 1, 2, \ldots, n-1.$$

Now,

$$s_i = \mu^P_{A_1}(v_i) \leq \mu^P_{A_2}(f(v_i)) = \mu^P_{A_2}(S_{v_{\pi(i)}}S_{v_{\pi(i+1)}}) = r_{\pi(i)},$$

$$s'_i = \mu^N_{A_1}(v_i) \geq \mu^P_{A_2}(f(v_i)) = \mu^N_{A_2}(S_{v_{\pi(i)}}S_{v_{\pi(i+1)}}) = r'_{\pi(i)},$$

$$\begin{aligned}r_i &= \mu^P_{B_1}(v_i v_{i+1}) \leq \mu^P_{B_2}(f(v_i)f(v_{i+1})) \\ &= \mu^P_{B_2}(S_{v_{\pi(i)}}S_{v_{\pi(i)+1}}S_{v_{\pi(i+1)+1}}) \\ &= \mu^P_{B_1}(v_{\pi(i)}v_{\pi(i)+1} \wedge \mu^P_{B_1}(v_{\pi(i)+1}v_{\pi(i+1)+1}) \\ &= r_{\pi(i)} \wedge r_{\pi(i+1)},\end{aligned}$$

$$\begin{aligned} r_i' &= \mu_{B_1}^N(v_i v_{i+1}) \geq \mu_{B_2}^N(f(v_i)f(v_{i+1})) \\ &= \mu_{B_2}^N(S_{v_{\pi(i)}} S_{v_{\pi(i)+1}} S_{v_{\pi(i+1)+1}}) \\ &= \mu_{B_1}^N(v_{\pi(i)} v_{\pi(i)+1} \vee \mu_{B_1}^N(v_{\pi(i)+1} v_{\pi(i+1)+1}) \\ &= r'_{\pi(i)} \vee r'_{\pi(i+1)} \end{aligned}$$

for $i = 1, 2, \ldots, n$, That is,

$$s_i \leq r_{\pi(i)}, \ s_i' \geq r'_{\pi(i)}$$

and

$$r_i \leq r_{\pi(i)} \wedge r_{\pi(i+1)}, \quad r_i' \geq r'_{\pi(i)} \vee r'_{\pi(i+1)} \tag{8.19}$$

By (8.19), we have $r_i \leq r_{\pi(i)}, r_i' \geq r'_{\pi(i)}$ for $i = 1, 2, \ldots, n$ and so $r_\pi(i) \leq r_{\pi(\pi(i))}$, $r'_\pi(i) \geq r'_{\pi(\pi(i))}$ for $i = 1, 2, \ldots, n$. Continuing, we have

$$r_i \leq r_{\pi(i)} \leq \cdots \leq r_{\pi^j(i)} \leq r_i$$

$$r_i' \geq r'_{\pi(i)} \geq \cdots \geq r'_{\pi^j(i)} \geq r_i'$$

and so $r_i = r_{\pi(i)}$, $i = 1, 2, \ldots, n$, where $\pi^{j+1}$ is the identity map. Again, by (8.19), we have

$$r_i \leq r_{\pi(i+1)} = r_{i+1}, \ i = 1, 2, \ldots, n, \ r_{n+1} = r_1,$$

$$r_i' \geq r'_{\pi(i+1)} = r'_{i+1}, \ i = 1, 2, \ldots, n, \ r'_{n+1} = r_1'.$$

Hence, by (8.18) and (8.19),

$$r_1 = \cdots = r_n = s_1 = \cdots = s_n$$

$$r_1' = \cdots = r_n' = s_1' = \cdots = s_n'.$$

Thus, $A_1$ and $B_1$ are constant functions. Thus, $(ii)$ holds.

The converse follows easily. ■

**Theorem 8.6.10** *Let $G$ and $H$ be bipolar fuzzy graphs of $G^*$ and $H^*$, respectively, such that $G^*$ and $H^*$ are connected. Let $L(G)$ and $L(H)$ be the bipolar fuzzy line graphs corresponding to $G$ and $H$, respectively. Suppose that it is not the case that one of $G^*$ and $H^*$ is complete $K_3$ and other is bipartite $K_{1,3}$. If $L(G)$ and $L(H)$ are isomorphic, then $G$ and $H$ are line-isomorphic.*

## 8.7 Connectivity in Bipolar Fuzzy Graphs

In this section, we introduce some connectivity concepts in bipolar fuzzy graphs. The work is based primarily on that by Mathew, Sunitha, and Anjali [117]. Analogous to fuzzy cutvertices and fuzzy bridges in fuzzy graphs, bipolar fuzzy cut vertices and bipolar fuzzy bridges are introduced and characterized. Also, the concepts of gain and loss for paths between pairs of vertices and connectivity in complete bipolar fuzzy graphs are discussed.

This definition of a bipolar fuzzy graph given in this section is slightly different form Definition 8.2.1, where "bipolar fuzzy graph of a graph $G = (V, E)$" is defined. Because bipolar fuzzy graph is a generalization of a fuzzy graph, and hence that of a graph, definition independent of a graph is more appropriate.

$V$ may be called the underlying set of $G = (A, B)$. $A$ is said to be a bipolar fuzzy vertex set of $G$ and $B$, bipolar fuzzy edge set of $G$. Let us denote $\{x, y\}$ by $xy$.

**Definition 8.7.1** ([158]) The underlying crisp graph of a bipolar fuzzy graph $G = (A, B)$, is the graph $G = (V', E')$, where $V' = \{v \in V \mid \mu_A^P(v) > 0 \text{ or } \mu_A^N(v) < 0\}$ and $E' = \{\{x, y\} \mid \mu_B^P(x, y) > 0 \text{ or } \mu_B^N(x, y) < 0\}$. $V'$ is called the vertex set and $E'$ is called the edge set. A bipolar fuzzy graph may be also denoted as $G = (V', E')$.

**Definition 8.7.2** ([158]) A bipolar fuzzy graph $G = (A, B)$ is **connected** if the underlying crisp graph $G = (V', E')$ is connected.

**Definition 8.7.3** A **partial bipolar fuzzy** subgraph of a bipolar fuzzy graph $G = (A, B)$ is a bipolar fuzzy graph $H = (A', B')$ such that $\mu_{A'}^P(v_i) \leq \mu_A^P(v_i)$ and $\mu_{A'}^N(v_i) \geq \mu_A^N(v_i)$ for all $v_i \in V$ and $\mu_{B'}^P(v_i v_j) \leq \mu_B^P(v_i v_j)$ and $\mu_{B'}^N(v_i v_j) \geq \mu_B^P(v_i v_j)$ for all $v_i, v_j \in V$.

**Definition 8.7.4** A **bipolar fuzzy subgraph** of a bipolar fuzzy graph $G = (A, B)$ is a bipolar fuzzy graph $H = (A', B')$ such that $\mu_{A'}^P(v_i) = \mu_A^P(v_i)$ and $\mu_{A'}^N(v_i) = \mu_A^N(v_i)$ for all $v_i \in V$ and $\mu_{B'}^P(v_i v_j) = \mu_B^P(v_i v_j)$ and $\mu_{B'}^P(v_i v_j) = \mu_B^P(v_i v_j)$ for all $v_i, v_j \in V$.

*Example 8.7.5* In Fig. 8.1, the first graph $G$ is a bipolar fuzzy graph. The second is a partial bipolar fuzzy subgraph and the third, a bipolar fuzzy subgraph of $G$.

**Notation 8.7.6** We use the following notations to denote the conditions in the Definition 8.2.1.

$$\mu_{2ij}^P = \mu_2^P(v_i v_j) \leq \mu_{1i}^P \wedge \mu_{1j}^+,$$
$$\mu_{2ij}^N = \mu_2^N(v_i v_j) \geq \mu_{1i}^N \vee \mu_{1j}^N.$$

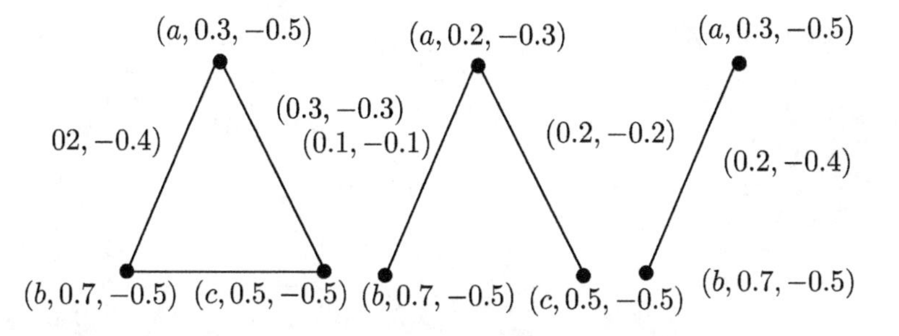


**Fig. 8.1** A bipolar fuzzy graph and two types of subgraphs

**Fig. 8.2** A bipolar fuzzy graph which is not a path

(0.1, −0.2) (0.3, −0.4)

a(0.2, −0.3) b(0.3, −0.5) c(1, −1)

**Definition 8.7.7** ([17]) A **path** $P$ in a bipolar fuzzy graph is a sequence of distinct vertices $v_1, v_2, \ldots, v_n$ such that either one of the following conditions is satisfied.

$(i)$ $\mu^P_{2ij} > 0$ and $\mu^N_{2ij} = 0$ for some $i$ and $j$.

$(ii)$ $\mu^P_{2ij} = 0$ and $\mu^N_{2ij} < 0$ for some $i$ and $j$.

According to the above definition of path, the bipolar fuzzy graph in Fig. 8.2 is not a path. So, we define a $b$-path.

**Definition 8.7.8** A sequence of distinct vertices $v_1, v_2, \ldots, v_n$ is called a **bipolar path** or $b$**-path** if at least one of $\mu^P_{2i(i+1)}$ or $\mu^N_{2i(i+1)}$is different from zero for $i = 1, 2, \ldots, n-1$.

Clearly, a bipolar fuzzy graph is connected if and only if every pair of vertices is joined by a $b$-path.

**Definition 8.7.9** A sequence of vertices $v_1, v_2, \ldots, v_n$, not necessarily distinct is called a **bipolar walk** or $b$**-walk** if at least one of $\mu^P_{2i(i+1)}$ or $\mu^N_{2i(i+1)}$is different from zero for $i = 1, 2, \ldots, n-1$. As in graphs, where every walk contains a path, every $b$-walk also contains a $b$-path. Hereafter, by a path we refer to a $b$-path and by a walk, we refer to a $b$-walk.

The concept of loss and gain are very important in several problems in economics, operations research and computer organization. We associate these concepts to a bipolar fuzzy graph in the following definitions.

**Definition 8.7.10** Let $G = (V', E')$ be a bipolar fuzzy graph. For a $u - v$ path $P$ : $u = u_1, u_2, \ldots, u_n = v$ in $G$, we define $\{\mu^P_2(u_1u_2) \wedge \mu^P_2(u_2u_3) \wedge \cdots \wedge \mu^P_2(u_{n-1}u_n)\}$ as the **gain** of $P$, denoted by $g(P)$ and $|\mu^N_2(u_1u_2)| \vee |\mu^N_2(u_2u_3)| \vee \cdots \vee |\mu^N_2(u_{n-1}u_n)|$ as the **loss** of $P$, denoted by $l(P)$.

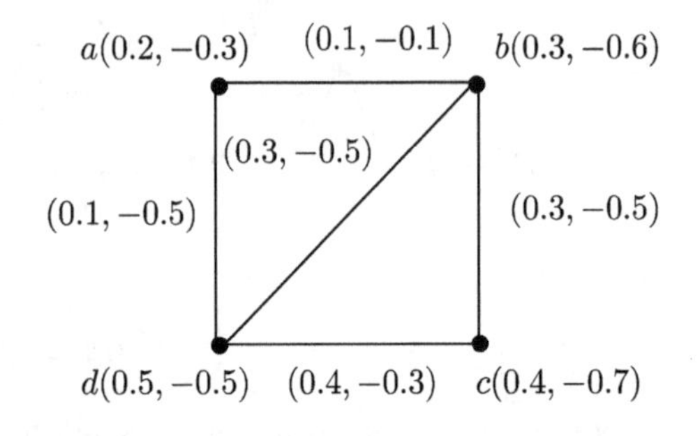


**Fig. 8.3** Gain paths and loss paths

In Fig. 8.2, the gain of the path $P : abc = g(P) = 0.1 \wedge 0.3 = 0.1$ and the loss of the path $P : abc = l(P) = 0.2 \vee 0.4 = 0.4$.

Note that if $e = uv$ is an edge, then its gain, denoted by $g(e) = \mu_2^P(uv)$ and loss of $e$, denoted by $l(e) = \left|\mu_2^N(uv)\right|$. In Fig. 8.2, $g(ab) = 0.1$, $l(ab) = 0.2$.

**Definition 8.7.11** A path $P$ is said to be a **gain path** if $g(P) > l(P)$ and a **loss path**, otherwise.

Similarly, gain edges and loss edges can be defined. In Fig. 8.2, path $P : abc$ is a loss path because $l(P) = 0.4 > 0.1 = g(P)$.

**Definition 8.7.12** Let $u, v$ be any two vertices in a connected bipolar fuzzy graph. Among all $u - v$ paths in $G$, a path whose gain is greater than or equal to that of any other $u - v$ path in $G$, is said to be a **maximum**$u - v$ **gain path** (max$(u - v)g$**-path**). Similarly, a $u - v$ path whose loss is less than or equal to that of any other $u - v$ path in $G$ is said to be a minimum $u - v$ loss path (min$(u - v)$ $l$-path).

That is, a path is a max$(u - v)$ $g$-path if $g(P) > g(P')$ and is a min$(u - v)$ $l$-path if $l(P) < l(P')$, where $P'$ is any $u - v$ path in $G$.

Note that, a max$(u - v)$ $g$-path need not be a gain path and a min$(u - v)$ $l$-path need not be a loss path.

*Example 8.7.13* Consider the following example of a bipolar fuzzy graph $G$ with four vertices given in Fig. 8.3. The gain and loss of different paths are given in the table.

| Vertices | Max-gain | Max $g$-path | Min-loss | Min $l$-path |
|---|---|---|---|---|
| $a - b$ | 0.1 | Any path | 0.1 | $ab$ |
| $a - c$ | 0.1 | Any path | 0.3 | $adc$ |
| $a - d$ | 0.1 | Any path | 0.1 | $ad$ |
| $b - c$ | 0.3 | $bc, bdc$ | 0.3 | $badc$ |
| $b - d$ | 0.3 | $bd, bcd$ | 0.1 | $bad$ |
| $c - d$ | 0.4 | $cd$ | 0.3 | $cd$ |

Note that, $P_1$ : $abc$ is a loss path ($a - c$ loss path) because $l(P) = 0.5 > 0.1 = g(P)$ and edge $cd$ is a $c - d$ gain path because $g(cd) = 0.4 > 0.3 = l(cd)$.

**Definition 8.7.14** A $u - v$ path $P$ in a bipolar fuzzy graph is said to be **balanced** if $g(P) = l(P)$. Also, $P$ is said to be **optimal** if it is a max $(u - v)$ $g$-path and min $(u - v)$ $l$-path.

In the Example 8.7.13 (Fig. 8.3), edge $cd$ is an optimal $c - d$ path. $adc$ is an optimal $a - c$ path. Also, there are many balanced paths in $G$. For example, $bad$ is a balanced $b - d$ path.

**Definition 8.7.15** Let $G = (V', E')$ be a bipolar fuzzy graph and let $u, v \in V'$. The gain of $u$ and $v$, denoted by $G(u, v)$ is defined as the gain of a $\max(u - v)$ $g$-path and the loss of $u$ and $v$, denoted by $L(u, v)$ is the loss of a $\min(u - v)$ $l$-path. If $H$ is a bipolar fuzzy subgraph of $G$, then the gain of $u$ and $v$ in $H$ is the gain of a $\max(u - v)$ $g$-path strictly belonging to $H$ and is denoted by $G_H(u, v)$. The loss of $u$ and $v$ in $H$ is similarly defined. If there exists no $\max(u - v)$ $g$-path (or $\min(u - v)$ $l$-path) completely in $H$, we define $G_H(u, v) = 0$ (or $L_H(u, v) = 0$).

Next we have a trivial proposition.

**Proposition 8.7.16** *If $H$ be a subgraph of a bipolar fuzzy graph $G = (V', E')$, then $G_H(u, v) \leq G(u, v)$ and $L_H(u, v) \leq L(u, v)$ for all pairs of vertices $u$ and $v$.*

Next we introduce an important concept called the Gain-Loss Matrix (GLM) in bipolar fuzzy graphs.

**Definition 8.7.17** Let $G = (V', E')$ be a bipolar fuzzy graph with vertices $\{a_1, a_2, \ldots, a_n\}$. The Gain-Loss Matrix (GLM) of $G$ is defined as $M = [(G_{ij}, L_{ij})]$, where $G_{ij} = G(a_i, a_j)$ and $L_{ij} = L(a_i, a_j)$ for $i \neq j$, $(\mu_1^P(a_i), |\mu_1^N(a_i)|) = 0$ if $i = j$.

Consider the following example.

*Example 8.7.18* The GLM of the bipolar fuzzy graph in Example 8.7.5 (Fig. 8.1) is given below.

$$\text{GLM}(G_1) = \begin{bmatrix} (0.3, 0.5) & (0.3, 0.3) & (0.3, 0.3) \\ (0.3, 0.3) & (0.7, 0.5) & (0.4, 0.3) \\ (0.3, 0.3) & (0.4, 0.3) & (0.5, 0.5) \end{bmatrix}$$

Clearly, GLM of a bipolar is a symmetric matrix.

**Theorem 8.7.19** *In a complete bipolar fuzzy graph $G = (V', E')$, $G(u, v) = \mu_2^P(u, v)$ for all $u, v \in V$.*

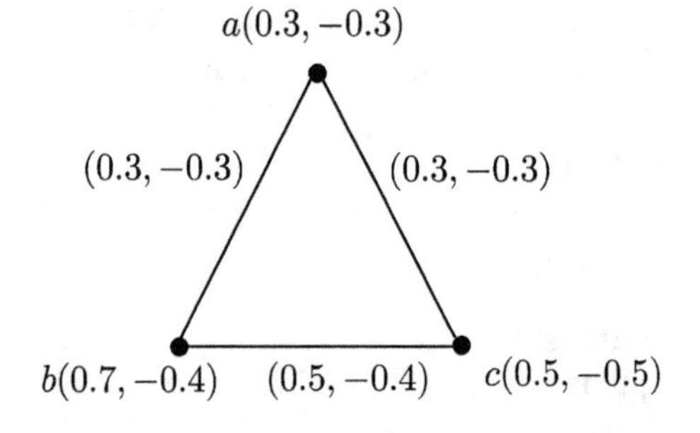


**Fig. 8.4** A complete bipolar fuzzy graph

*Proof* Consider a complete bipolar fuzzy graph $G = (V', E')$ with vertices $v_1, v_2, \ldots, v_n$. By definition, for all $v_i, v_j \in V$, we have

$$\mu_2^P(v_i v_j) = \mu_1^P(v_i) \wedge \mu_1^P(v_j).$$

Let $u, v \in V$ and let $P : u = u_1, u_2, \ldots, u_m = v$ be a $u - v$ path in $G$. Then

$$\begin{aligned} g(P) &= \mu_2^P(u_1u_2) \wedge \mu_2^P(u_2u_3) \wedge \cdots \wedge \mu_2^P(u_{m-1}u_m) \\ &\leq \mu_2^P(u_1u_2) \wedge \mu_2^P(u_{m-1}u_m) \\ &= \wedge\{\mu_1^P(u_1) \wedge \mu_1^P(u_2), \mu_1^P(u_{m-1}) \wedge \mu_1^P(u_m)\} \\ &\leq \mu_1^P(u_1) \wedge \mu_1^P(u_m) \\ &= \mu_1^P(u) \wedge \mu_1^P(v) \\ &= \mu_2^P(uv). \end{aligned}$$

Thus, $g(P) \leq \mu_2^P(uv)$ for any $u - v$ path $P$. In particular, the gain of edge $uv$ is $\mu_2^P(uv)$ and hence, $G(u, v) = \mu_2^P(uv)$. ■

**Note** In a complete bipolar fuzzy graph, $L(u, v)$ need not be equal to $|\mu_2^N(uv)|$ for all $u, v \in V$. For example, Consider the complete BPFG $G$ in Fig. 8.4. Here, $L(b, c) = 0.3$ and $|\mu_2^-(bc)| = 0.4$.

Now, we introduce bipolar fuzzy cutvertices and bridges. Three types of cutvertices are possible in a bipolar fuzzy graph, which are given below.

**Definition 8.7.20** Let $G = (V', E')$ be a bipolar fuzzy graph with bipolar functions, $\mu_1$ and $\mu_2$. A vertex $u \in V'$ is said to be a **bipolar fuzzy cutvertex** (**bf-cutvertex**) if there exist two vertices $x, y \in V'$, $x \neq u \neq v$ such that $G_{G-u}(x, y) < G_G(x, y)$ and $L_{G-u}(x, y) > L_G(x, y)$. A vertex in a bipolar fuzzy graph is called a **gain cutvertex** if the first condition is satisfied and a **loss cutvertex** if the second condition is satisfied.

Now, we characterize bipolar fuzzy cutvertices in the following theorem.

**Theorem 8.7.21** ([117]) *A vertex u in a bipolar fuzzy graph* $G = (V', E')$ *is a bipolar fuzzy cutvertex if and only if u is a vertex in every* $\max(x - y)$ *gain path and is in every* $\min(x - y)$ *loss path for some* $x, y \in V'$.

*Proof* Let $G = (V', E')$ be a bipolar fuzzy graph with bipolar functions, $\mu_1$ and $\mu_2$. Suppose that $u$ is a bipolar fuzzy cutvertex. By definition, there exist vertices $x, y$ in $G$ such that

$(i)$ $G_{G-u}(x, y) < G_G(x, y)$,

$(ii)$ $L_{G-u}(x, y) > L_G(x, y)$.

$(i)$ implies that the removal of $u$ from $G$ removes all $\max(x - y)$ gain paths and $(ii)$ implies that the removal of $u$ removes all $\min(x - y)$ loss paths. Thus, $u$ is in every $\max(x - y)$ gain path and is in every $\min(x - y)$ loss path.

Conversely, suppose that $u$ is in every $\max(x - y)$ gain path and is in every $\min(x - y)$ loss path. Then the removal of $u$ from $G$ results in the removal of all $\max(x - y)$ gain paths and $\min(x - y)$ loss paths. Hence, the gain will decrease and loss will increase between $x$ and $y$. Thus, $G_{G-u}(x, y) < G_G(x, y)$ and $L_{G-u}(x, y) > L_G(x, y)$. That is, $u$ is a bipolar fuzzy cutvertex. ■

Now, we give a characterization theorem for the other two types of cutvertices.

**Theorem 8.7.22** *Let* $G = (V', E')$ *be a bipolar fuzzy graph. A vertex u is a gain cutvertex (g-cutvertex) if and only if u is in every* $\max(x - y)$ *gain path for some vertices x and y such that* $x \neq y \neq u$ *and is a loss cutvertex (l-cutvertex) if and only if u is in every* $\min(s - t)$ *loss path for some vertices s and t such that* $s \neq t \neq u$.

**Definition 8.7.23** Let $G = (V', E')$ be a bipolar fuzzy graph with bipolar functions, $\mu_1$ and $\mu_2$. Let $e = xy$ be an edge in $G$. $e$ is said to be a **bipolar fuzzy bridge (bf-bridge)** if $G_{G-e}(x', y') < G_G(x', y')$ and $L_{G-e}(x', y') > L_G(x', y')$ for some $x', y' \in V'$. If at least one of $x'$ or $y'$ is different from $x$ and $y$, $e$ is said to be a bipolar **fuzzy bond** and a bipolar **fuzzy cutbond** if both $x'$ and $y'$ is different from $x$ and $y$.

Also, we can define gain bridges and loss bridges similar to their counterparts in vertices.

Similar to bipolar fuzzy cutvertices, we have a characterization for bipolar fuzzy bridges, which is stated below without proof.

**Theorem 8.7.24** *An edge* $e \in E$ *of a bipolar fuzzy graph* $G = (V', E')$ *is a bf-bridge if and only if it is in every* $\max(u - v)$ *gain path and in every* $\min(u - v)$ *loss path for some vertices* $u, v$ *in* $V'$.

Next we have an easy theorem to verify whether a particular edge is a bf-bridge or not.

**Theorem 8.7.25** *An edge xy is a bf-bridge if and only if* $G_{G-xy}(x, y) < \mu_2^P(xy)$ *and* $L_{G-xy}(x, y) > \left|\mu_2^N(xy)\right|$.

*Proof* Suppose $G = (V', E')$ is a bipolar fuzzy graph and $xy$ is an edge in $G$ in such that $G_{G-xy}(x, y) < \mu_2^P(xy)$ and $L_{G-xy}(x, y) > \left|\mu_2^N(xy)\right|$. Because $\mu_2^P(xy) \leq G(x, y)$ and $\left|\mu_2^N(xy)\right| > L(x, y)$, we have

$$G_{G-xy}(x, y) < G(x, y),$$
$$L_{G-xy}(x, y) > L(x, y).$$

It follows that $xy$ is a bipolar fuzzy bridge.

Assume, $xy$ is a bipolar fuzzy bridge. By Theorem 8.7.24, there exists a pair of vertices $s$ and $t$ in $V'$ such that $xy$ is present on every $\max(s - t)$ $g$-path and every $\min(s - t)$ $l$-path.

Suppose, $G_{G-xy}(x, y) \geq \mu_2^P(xy)$, Then $G_{G-xy}(x, y) = G(x, y)$. It follows that there is a $\max(x - y)$ $g$-path in $G$ (say, $P$) which is different from $xy$. Let $Q$ be a $\max(s - t)$ $g$-path in $G$. Replace $xy$ in $Q$ by $P$ to obtain an $s - t$ walk. This walk contains an $s-t$ path. The gain of this path is greater than or equal to $G_G(s, t)$ which is not possible. Therefore, $G_{G-xy}(x, y) < \mu_2^P(xy)$.

Assume, $L_{G-xy}(x, y) \leq \left|\mu_2^N(xy)\right|$. $L_{G-xy}(x, y) = L(x, y)$. It follows that there is a $\min(x-y)$ $l$-path in $G$ (say, $P'$) which is different from $xy$. Let $Q'$ be a $\min(s-t)$ $l$-path in $G$. Replace $xy$ in $Q'$ by $P'$ to obtain an $s - t$ walk. This walk contains an $s-t$ path. The loss of this path is less than or equal to $L_G(s, t)$ which is not possible. Therefore, $L_{G-xy}(x, y) > \left|\mu_2^N(xy)\right|$. ■

**Theorem 8.7.26** *An edge $e = xy$ of a bipolar fuzzy graph $G = (V', E')$, which is a cycle is a bf-bridge if and only if there exists edges $st$, $s't' \in E'$ such that $\mu_2^P(st) < \mu_2^P(xy)$ and $\left|\mu_2^N(s't')\right| > \left|\mu_2^N(xy)\right|$.*

*Proof* Let $e = xy$ be a bipolar fuzzy bridge. By definition, there exist two distinct vertices $x'$ and $y'$ such that $xy$ lies on every $\max(x' - y')$ $g$-path and on every $\min(x' - y')$ $l$-path. Because $G$ is a cycle, exactly one of the two $x' - y'$ paths (say, $P$) in $G$ contains $xy$ and is both the $\max(x' - y')$ $g$-path and $\min(x' - y')$ $l$-path. Let the other $x' - y'$ path be $Q$. Then

$$g(Q) < g(P) \leq \mu_2^P(xy),$$
$$l(Q) > l(P) \geq \left|\mu_2^N(xy)\right|.$$

Assume that, $xy$ is not a bipolar fuzzy bridge. Then at least one of the below conditions holds according to Theorem 8.7.25.

$(i)$ $G_{G-xy}(x, y) \geq \mu_2^P(xy)$

$(ii)$ $L_{G-xy}(x, y) \leq \left|\mu_2^N(xy)\right|$.

If $(i)$ is true, then the path $P$ in $G$ from $x$ to $y$ other than edge $xy$ has gain greater than or equal to $\mu_2^P(xy)$. It follows that for each edge $e \in E'$, $\mu_2^P(e) \geq \mu_2^P(xy)$.

If $(ii)$ is true, then the path $P$ in $G$ from $x$ to $y$ other than edge $xy$ has loss less than or equal to $\left|\mu_2^N(xy)\right|$. It follows that for each edge $e \in E'$, $\left|\mu_2^N(e)\right| \leq \left|\mu_2^N(xy)\right|$. ■

# Bibliography


1. Abiyev, R. H., Fuzzy wavelet neural network based on fuzzy clustering techniques for time series prediction, *Neural Comput. Appl.*, 20 (2011) 249–259.
2. Acharya, B. D. and Acharya, M., On self-antipodal graphs, *Nat. Acad. Sci. Lett.*, 8 (1985) 151–153.
3. Adamo, J. H., Fuzzy decision tress, *Fuzzy Sets and Systems*, 4 (1980) 207–219.
4. Ahmed, B. and Nagoor Gani, A., Perfect fuzzy graphs, *Bulletin or Pure and Applied Sciences*, 28 (2009) 83–90.
5. Akram, M., Bipolar fuzzy graphs, *Information Sciences*, 181 (2011) 5548–5564.
6. Akram, M., Bipolar fuzzy graphs with applications, *Knowledge-Based Systems*, 39(2013) 1–8.
7. Akram, M., Cofuzzy graphs, *J. Fuzzy Math*, 19 (2011) 1–12.
8. Akram, M., Fuzzy Lie ideals of Lie algebra with interval-valued membership functions, *Quasigroups and Related Systems*, 16 (2008) 1–12.
9. Akram, M., Interval-valued fuzzy line graphs, *Neural Comput & Applic*, 21 (2012) 145–150.
10. Akram, M., Intuitionistic $(S, T)$-fuzzy Lie ideals of Lie Algebra, *Quasigroups and Related Systems*, 15 (2007) 201–208.
11. Akram, M. and Al-Shehri, N. O., Intuitionistic fuzzy cycles and intuitionistic fuzzy trees, *Sci. World L.*, 2014(2014), Article ID 305836.
12. Akram, M. and Dar, K. H., Generalized Fuzzy $K$-Algebra, VMD Verlag, ISBN: 978-3-639-27095-2, 2010 pp. 288.
13. Akram, M. and Davvaz, B., Strong intuitionistic fuzzy graphs, *FILO-MAT*, 26(1)(2012) 177–196.
14. Akram, M. and Dudek, W. A., Interval-valued fuzzy graphs, *Computers and Mathematics with Applications*, 61(2)(2011) 289–299.
15. Akram, M. and Dudek, W. A., Intuitionistic fuzzy hypergraphs with applications. *Information Sciences*, 218 (2013) 182–193.
16. Akram, M. and Dudek, W. A., Regular bipolar fuzzy graphs, *Neural Computing and Applications*, 1 (2011) 1–9.
17. Akram, M. and Karunambihai, M. G., Metric in bipolar fuzzy graphs, *World Applied Science Journal*, 14 (2011) 1920–1927.
18. Akram, M. and Shun, K. P., Bifuzzy ideals of near rings, *Algebras, Groups Geom*, 24 (2007) 389–407.
19. Akram, M., Al-Shehri, N. O. and Dudek, W. A., Certain types of interval-valued fuzzy graphs, *J. Appl. Math.*, 2013(2013), Article ID 857070.




S. Mathew et al., *Fuzzy Graph Theory*, Studies in Fuzziness and Soft Computing 363, https://doi.org/10.1007/978-3-319-71407-3

20. Akram, M., Dar, K. H and Shum, K. P., Interval-valued $(\alpha, \beta)$-fuzzy $K$-algebras, *Applied Soft Computing Journal* , 11 (2011) 1213–1222.
21. Akram, M., Li, S.G. and Shum, K. P., Antipodal Bipolar Fuzzy Graphs, *Italian Journal of Pure and Applied Mathematics*, 31 (2013) 97–110.
22. Akram, A., Saeid, A. B., Shum, K. P. and Meng, B. L., Bipolar fuzzy K-algebras, *International Journal of Fuzzy Systems*, 10 (2010) 252–258.
23. Akram, M., Yousaf, M. M. and Dudek, W. A., Self centered interval-valued fuzzy graphs, *Afr. Mat.*, https://doi.org/10.1007/s13370-014-0256-9, 2014.
24. Alaoui, A., On fuzzification of some concepts of graphs, *Fuzzy Sets and Systems*, 101 (1999) 363–389.
25. Alavi, Y., Chartrand, G., Chung, F. R. K., Erdös, P., Grahm, R. L. and Oellermann, O. R., Highly irregular graphs, *Journal of Graph Theory*, 11 (1987) 235–249.
26. AL-Hawary, T., Complete fuzzy graphs, *International Journal of Mathematical Combinatorics*, 4 (2011) 26–34.
27. Alspach, B. and Mishna, M., Enumeration of Cayley graphs and digraphs, *Discrete Mathematics*, 256 (2002) 527–539.
28. Anjali, N. and Mathew, S., Critical blocks in fuzzy graphs, *The Journal of Fuzzy Mathematics*, 23(4) (2015) 907–916.
29. Anjali N. and Mathew, S., On blocks and stars in fuzzy graphs, *Journal of Intelligent and Fuzzy Systems*, 28(2015) 1659–1665.
30. Mathew, S., Anjali, N. and Mordeson, J. N., Transitive blocks and their applications in fuzzy interconnection networks, *Fuzzy Sets and Systems*, 2018 (In Press).
31. Aravamudhan, R. and Rajendran, B., On antipodal graphs, *Discrete Math.*, 49 (1984) 193–195.
32. Atanassov, K., Index matrix representation of the intuitionistic fuzzy graphs, Preprint MRL-MFAIS-10-94, Sifia, 1994, 36–41.
33. Atanassov, K., *Intuitionistic Fuzzy Sets*, Springer Physica-Verlag, Berlin, 1999.
34. Atanassov, K., *Intuitionistic Fuzzy Sets: Theory and Applications.* Studies in Fuzziness and Soft Computing, Physica-Verlag, Heidelberg, New York, 1999.
35. Atanassov, K., Pasi, G., Yager, R. and Atanassova, V., Intuitionistic fuzzy graph interpretations of multi-person multi-criteria decision making, www.eusflat.org/proceedings/EUSFLAT_2003papers/09/Atanassov/pdf.
36. Banerjee, S., An optimal algorithm to find degrees of connectedness in an undirected edge weighted graph, *Pattern Recognition Letters*, 12 (1991) 421–424.
37. Bhattacharya, P., Some remarks on fuzzy graphs, *Pattern Recognition Letters*, 6 (1987) 297–302.
38. Bhattacharya, P. and Suraweera, F., An algorithm to compute the supremum of max-min powers and a property of fuzzy graphs, *Pattern Recognition Letters*, 12 (1991) 413–420.
39. Bhragsam, S. G. and Ayyaswamy, S.K., Neighbourly irregular graphs, *Indian Journal of Pure and Applied Mathematics*, 35 (2004) 389–399.
40. Bhutani, K. R., On automorphisms of fuzzy graphs, Pattern Recognition Letters, 9 (1989) 159–162.
41. Bhutani, K. R. and Battou, A., On M-strong fuzzy graphs, *Information Sciences*, 155 (2003) 103–109.
42. Bhutani, K. R. and Rosenfeld, A., Fuzzy end vertices in fuzzy graphs, *Information Sciences*, 152 (2003) 323–326.
43. Bhutani, K.R and Rosenfeld, A., Geodesics in fuzzy graphs, *Electronic Notes in Discrete Mathematics*, 15 (2003) 51–54.
44. Bhutani, K.R and Rosenfeld, A., Strong arcs in fuzzy graphs, *Information Sciences*, 152 (2003) 319–322.
45. Bhutani, K.R., Mordeson, J.N. and Rosenfeld, A., On degrees of end vertices in fuzzy graphs, *Iranian Journal of Fuzzy Graphs*, 1(1) (2004) 57–64.
46. Bloch, I., Dilation and erosion of spatial bipolar fuzzy sets, *Lecture Notes in Artificial Intelligence*, (2007) 385–393.

47. Bloch, I., Geometry of spatial bipolar fuzzy sets based on bipolar fuzzy numbers and mathematical morphology, *Fuzzy Logic and Applications, Lecture Notes in Computer Science*, 5571 (2009) 237–245.
48. Biglarbegian, A., Sadeghian, W. and Melek, M., On the accessibility/controllability of fuzzy control systems, *Information Sciences*, 202(2012) 58–72.
49. Boulmakoul, A., Fuzzy graph modelling for HazMat telegeomonitoring, *European Journal of Operation Research*, 175 (2006) 1514–1525.
50. Buckley, F., Self-centered graphs, graphs theory and its applications: *East and West. Ann. New York Acad. Sci.*, 576 (1989) 71–78.
51. Buckley, F. and Harary, F., *Distance in graphs*, Addison-Wesley, 1990.
52. Bustince, H., Beliakov, G., Goswami, D., Mukherjee, U. and Pal, N., On averaging operators for Atanassov's intuitionistic fuzzy sets, *Information Sciences*, 181 (2011) 1116–1124.
53. Chanas, S. and Kolodziejczynk, W., Maximum flow in a network with fuzzy arc capacities, *Fuzzy Sets and Systems*, 8 (1982) 165–173.
54. Chartrand,G. and Zhang, P., Distance in graphs-Taking the long view, *AKCE Int. J. of Graphs Comb.*, 1(1) (2004) 1–13.
55. Changat, M., Klavzar, S. and Mulder, H.M., The all-paths transit functions of a graph, *Czechoslovak Mathematical Journal*, 51(2)(2001) 439–448.
56. Changat, M., Mulder, H.M. and Sierksma, G., Convexities related to path properties on graphs, *Discrete Mathematics*, 290(2)(2005) 17–131.
57. Chen, S.M., Interval-valued fuzzy hypergraph and fuzzy partition, *IEEE Transactions on Systems, Man, and Cybernetics* B, 27 (1997) 725–733.
58. Chen, J.C. and Tsai, C.H., Conditional edge-fault tolerant Hamiltonicity of dual cubes, *Information Sciences*, 181 (2011) 620–627.
59. Chen, D., Hu, Q. and Yang, Y., Parameterized attribute reduction with Gaussian kernel based fuzzy rough sets, *Information Sciences*, 181(23)(2011) 5169–5179.
60. Cheng, C.T., Ou, C.P. and Chau, K.W., Combining a fuzzy optimal model with a generic algorithm on solving multiobjective rainfall-runoff model calibration, *Journal of Hydrology*, 268 (2002) 72–86.
61. Craine, W. L., Characterization of fuzzy interval graphs, *Fuzzy Sets and Systems*, 68 (1994) 181–193.
62. Delgado, M., Verdegay, J. L. and Vila, M. A., On fuzzy tree definition, *European J. Operational Res.*, 22 (1985) 243–249.
63. Deo, N., *Graph Theory with Applications to Engineering and Computer Science*, Prentice Hall, Englewood Cliffs, NJ, USA, 1990.
64. Deschrijver, G. and Cornelis, C., Representability in interval–valued fuzzy set theory, *International Journal of Uncertainty, Fuzziness and Knowledge-Based Systems*, 15 (2007) 345–361.
65. Deschrijver, G. and Kerre, E. E., On the relationships between some extensions of fuzzy set theory, *Fuzzy Sets and Systems* , 133 (2003) 227–235.
66. Dhanyamol, M. V. and Mathew, S., Gates in fuzzy graphs, *New Mathematics and Natural Computation*, (In press).
67. Dhanyamol, M. V. and Mathew, S., Intervals in fuzzy graphs, *Fuzzy Information and Engineering*, (In press)
68. Diamond, P., A fuzzy max-flow min-cut theorem, *Fuzzy Sets and Systems*, 119 (2001) 139–148.
69. Dress, A. W. M. and Scharlau, R., Gated sets in metric spaces, *Aequationes Mathematicae*, 34(1987) 112–120.
70. Dubois, D. and Prade, H., *Fuzzy Sets and Systems*, Academic Press, New York 1980.
71. Dubios, D., Kaci, S. and Prade, H., Bipolarity in reasoning and decision, an introduction, in: Inf. Pro. Man. Unc. IPMU'04, 2002, pp. 959–966.
72. Erdos. P. and Gallai, A., On maximal paths and circuits in graphs, *Acta Math. Sci. Hung*, 10 (1959) 337–356.
73. Fang, J. F., The bipancycle-connectivity of hypercube, *Information Sciences*, 178 (2008) 4679–4687.

74. Fulkerson, D. R. and Gross, O. A., Incidence matrices and interval graphs, *Pacific J. Math.*, 5 (1965) 835–855.
75. Gacto, M. J., Alcala, R. and Herrera, F., Interpretability of Linguistic fuzzy rule-based systems: An overview of interpretability measures, *Information Sciences*, 181(20)(2011) 4340–4360.
76. Gani, A.L. and Radha, K., Isomorphism on fuzzy graph, *Int. J. Comput. Math. Sci.*, 2 (2008) 190–196.
77. Gani, A. N. and Latha, S. R., On irregular fuzzy graphs, *Applied Mathematical Sciences*, 6 (2012) 517–523.
78. Gilmore, P.C. and Hoffman, A. J., A characterization of comparability graphs and interval graphs, *Canad. J. Math.*, 16 (1964) 539–548.
79. Goddard, W., Ortrud, R. and Oellermann, Distance in graphs, *Structural Analysis of Complex Networks*, (2011) 49–72.
80. Goldman, A. J. and Witzgall, C. J., A localisation theorem for optimal facility location, *Transportation Science*, 4(1970) 406–409.
81. Gorzalczany, M. B., A method of inference in approximate reasoning based on interval-valued fuzzy sets, *Fuzzy Sets and Systems* , 21 (1987) 1–17.
82. Gorzalczany, M. B., An interval-valued fuzzy inference method some basic properties, *Fuzzy Sets and Systems*, 31 (1989) 243–251.
83. Harary, F., *Graph Theory*, Addison Wesley, Third Edition, October 1972.
84. Hongmei, J. and Lianhua, W., Interval-valued fuzzy subsemigroups and subgroups associated with interval-valued fuzzy graphs, in: 2009 WRI Global Congress on Intelligent Systems, 2009, pp. 484–487.
85. Homenda, W. and Pedrycz, W., Balanced fuzzy gates, RSCTC (2006) 107–116.
86. Huber, K. P. and Berthold, M. R., Application of fuzzy graphs for metamodeling, in: Proceedings, 1998. IEEE World Congress on Computational Intelligence. Anchorage, USA. 4–9 May 1998 (current version 6 Aug 2002).
87. Jicy, N. and Mathew, S., Connectivity analysis of cyclicaly balanced fuzzy graphs, *Fuzzy Information and Engineering*, 7 (2015) 245–255.
88. Ju, H. and Wang, L., Interval-valued fuzzy subsemigroups and subgroups associated by interval-valued fuzzy graphs, Proceedings of the WRI Global Congress on Intelligent Systems (GCIS'09), pp. 484–487, Xiamen China, May 2009.
89. Karunambigal, M. G., Rangasamy, P., Atanassov, K. T. and Palaniappan, N., An intuitionistic fuzzy graph method for finding the shortest paths in network, in O. Castill et al. (Eds.) Theor. Adv. and Appl. of Fuzzy logic, Vol. 42, ASC, 2007, pp. 3–10.
90. Kaufman, A., Introduction a la Theorie ds sons-ensembles flous, Vol. 1, Masson Paris, (1973) 41–189.
91. Kauffmann, A., *Introduction to the Theory of Fuzzy Sets* , Vol. 1, Academic Press, Inc., Orlando, Florida, 1973.
92. Kiss, A., An application of fuzzy graphs in database theory, PU. M. A. Ser. A, 1 (1990) 337–342.
93. Kiss, A. λ-decomposition of fuzzy relational databases, Annales Univ. Sci. Budapest, Sect. Comp. 12 (1991) 133–148.
94. Klein, C. M., Fuzzy shortest paths, *Fuzzy Sets and Systems*, 39 (1991) 27–41.
95. Koczy, L. T., Fuzzy graphs in the evaluation and optimization of networks, *Fuzzy Sets and Systems*, 46 (1992) 307–319.
96. Lam, Y. F. and Li, V., Reliability modeling and analysis of communication networks with dependent failures, *IEEE Transactions on Communications*, 34(1986) 82–84.
97. Lee, K. M., Bipolar-valued fuzzy sets and their basic operations, in: Proceedings of the International Conference, Bangkok, Thailand, 2000, pp. 307–317.
98. Lee, K. M., Comparison of interval-valued fuzzy sets, intuitionistic fuzzy sets and bipolar fuzzy sets, *J. Fuzzy Logic Intelligent Systems*, 14 (2004) 125–129.
99. Li, Y., Finite automata theory with membership values in lattices, *Information Sciences*, 181 (2011) 1003–1007.

100. Lin, J. Y., Cheng, C.T. and Chau, K.W., Using support vector machines for long-term discharge prediction, *Hydrological Sciences Journal*, 51 (2006) 599–612.
101. Linda, J. P and Sunitha, M. S., Fuzzy detour g-centre in fuzzy graphs, *Annals of Fuzzy Mathematics and Informatics*, 7(2)(2014) 219–228.
102. Linda, J. P and Sunitha, M. S., Fuzzy detour g-interior nodes and fuzzy detour g-boundary nodes of a fuzzy graph, *Journal of Intelligent and Fuzzy Systems*, 27(2014) 435–442.
103. Ma, X., Zhan, J., Davvaz, B. and Jun, Y.B., Some kinds of $(\epsilon, \epsilon \vee q)$-interval-valued fuzzy ideals of BCI algebra, *Information Sciences*, 178 (2008) 3738–3754.
104. Malik, D. S. and Mordeson, J. N., *Fuzzy Discrete Structures*, Studies in Fuzziness and Soft Computing, 58, Physica-Verlag 2000.
105. Marczewski, E., Sur deux Proprieties des Classes d'ensembles, *Fund. Math.*, 33 (1945) 303–307.
106. Mathew, J. K. and Mathew, S., On strong extremal problems in fuzzy graphs, *Journal of Intelligent and Fuzzy Systems*, 30 (2016) 2497–2503.
107. Mathew, J. K. and Mathew, S., Some special sequences in fuzzy graphs, *Fuzzy Information and Engineering*, 8 (2016) 31–40
108. Mathew, J. K., Yang, H. L. and Mathew, S., Saturation in fuzzy graphs, *New Mathematics and Natural Computation* (World Sci.) (In press).
109. Mathew, S., Anjali, N. and Mordeson, J.N., Transitive blocks and their applications in fuzzy interconnection networks, *Fuzzy Sets and Systems*, 2017, (preprint).
110. Mathew, S. and Sunitha, M. S., A characterization of blocks in fuzzy graphs, *The Journal of Fuzzy Mathematics*, 18(4) (2010) 999–1006.
111. Mathew, S. and Sunitha, M. S., Bonds in graphs and fuzzy graphs, *Advances in Fuzzy Sets and Systems*, 6(2) (2010) 101–119.
112. Mathew, S. and Sunitha, M. S., Cycle connectivity in fuzzy graphs, *Journal of Intelligent and Fuzzy Systems*, 24(2013) 549–554.
113. Mathew, S. and Sunitha, M. S., Menger's theorem for fuzzy graphs, *Information Sciences*, 222 (2013) 717–726.
114. Mathew, S. and Sunitha, M. S., Node connectivity and arc connectivity in fuzzy graphs, *Information Sciences*, 180(2010) 519–531.
115. Mathew, S. and Sunitha, M. S., Strongest strong cycles and $\theta$-fuzzy graphs, *IEEE Transactions on Fuzzy Systems*, 21(6) (2013) 1096–1104.
116. Mathew, S. and Sunitha, M. S., Types of arcs in a fuzzy graph, *Information Sciences*, 179(11) (2009) 1760–1768.
117. Mathew, S., Sunitha, M. S. and Anjali, N., Some connectivity concepts in bipolar fuzzy graphs, *Annals of Pure and Applied Mathematics*, 7 (2014) 98–108.
118. Matula, D. W., Cluster analysis via graph theoretic techniques, Proc. of Louisiana Conf. on Combinatorics, Graph Theory, and Computing, 199–212, March 1970.
119. McAllister, L. M. N., Fuzzy intersection graphs, *Intl. J. Computers in Mathematics with Applications*, 5 (1988) 871–886.
120. McKee, T. A., Foundation of intersection graph theory, *Utilitas Math.*, 40 (1991) 77–86.
121. Mendel, J. M., *Uncertainty Rule-Based Fuzzy Logic Systems: Introduction and New Directions,* Prentice-Hall, Upper Saddle River, New Jersey, 2001.
122. Mendal, J. M. and Gang, X., Fast computation on centroids for constant-width interval-valued fuzzy sets, Fuzzy Information Processing Society, NAFIPS (2006) 621–626.
123. Menger, K., Zur allgemeinen Kurventheorie, *Fund. Math.*, 10(1927), 96–115.
124. Miyamoto, S., *Fuzzy Sets in Information Retieval and Cluster Analysis, Theory and Decision Library*, Series D: System Theory, Knowledge Engineering and Problem Solving, Kluwer Academic Publishers, 1990.
125. Mordeson, J. N., Fuzzy line graphs, *Pattern Recognition Letters*, 14 (1993) 381–384.
126. Mordeson, J. N. and Nair, P. S., Arc disjoint fuzzy graphs, Proceedings of the 18th International Conference of the North American Fuzzy Information Processing Society, June 199, New York, https://doi.org/10.1109/NAFIPS.1999.781654.

127. Mordeson, J. N. and Nair, P. S., Cycles and cocycles of fuzzy graphs, *Information Sciences*, 90 (1996) 39–49.
128. Mordeson, J. N. and Nair, P. S., *Fuzzy Graphs and Fuzzy Hypergraphs*, 46, Physica-Verlag, 2000.
129. Mordeson, J. N. and Nair, P. S., *Fuzzy Mathematics: An Introduction for Engineers and Scientists*, 20, Physica-Verlag, 2001.
130. Mordeson, J. N. and Peng, C. S., Operations on fuzzy graphs, *Information. Sciences*, 79 (1994) 159–170.
131. Mordeson, J. N. and Yao, Y. Y., Fuzzy cycles and fuzzy trees, *The Journal of Fuzzy Mathematics*, 10 (2002) 189–202.
132. Mulder, H. M., *The interval function of a graph*, Mathematisch Centrum, Amsterdam, 1980.
133. Mulder, H. M. and Nebesky, L., Axiomatic characterization of the interval function of a graph, *European Journal of Combinatorics* , 30(2009) 1172–1185.
134. Muttill, N. and Chau, K. W., Neural network and genetic programming for modelling coastal algal blooms, *International Journal of Environment and Pollution,* 28 (2006) 223–238.
135. Nagoor Gani, A. and Radha, K., Isomorphism of fuzzy graphs, *International Journal of Computational and Mathematical Sciences*, 2 (2008) 190–196.
136. Nagoor Gani, A., Umamaheswari, J., Fuzzy detour $\mu$ center in fuzzy graphs, *International Journal of Algorithms, Computing and Mathematics*, 3(2) (2010) 57–63.
137. Nebesky, L., A characterization of the set of all shortest paths in a connected graph, *Mathematica Bohemica*, 119(1994) 15–20.
138. Nebesky, L., A characterization of the interval function of a connected graph, *Czechoslovak Mathematical Journal*, 44(1994) 173–178.
139. Nebesky, L., A characterization of the interval function of a (finite or infinite) connected graph, *Czechoslovak Mathematical Journal*, 51(2001) 635–642.
140. Nebesky, L., Characterizing the interval function of a connected graph, *Mathematica Bohemica*, 123(1998) 137–144.
141. Parvathi, R., Karunambigai, M. G. and Atanassov, K. T., Operations on intuitionistic fuzzy graphs, in Proceedings of the IEEE International Conference on Fuzzy Systems, 2009, pp. 1396–1401.
142. Pasi, G., Yager, R. and Atanassov, K. T., Intuitionistic fuzzy graph interpretations of multi-person multi-criteria decision making: generalized net approach, in: Intelligent Systems, Proceedings of the 2004 2nd International IEEE Conference, Vol. 2, 2004, pp. 434–439.
143. Pedrycz, W., *Fuzzy Sets Engineering*, CRC Press, Boca Raton, FL, 1995.
144. Pedrycz, W., Human centricity in computing with fuzzy sets: an interpretability quest for higher granular constructs, *J. Ambient Intelligence and Humanized Computing*, 1 (2010) 65–74.
145. Pedrycz, W. and Bargiela, A., Fuzzy clustering with semantically distinct families of variables; descriptive and predictive aspects, *Pattern Recognition Letters*, 31 (2010) 1952–1958.
146. Peng, C., Zhang, R., Zhang, Q. and Wang, J., Dominating sets in directed graphs, *Information Sciences*, 180 (2010) 3647–3652.
147. A. Perchant and I. Bloch, Fuzzy morphisms between graphs, *Fuzzy Sets and Systems*, 128 (2002) 149–168.
148. Hu, Q., Yu, D. and Guo, M., Fuzzy preference based rough sets, *Information Sciences*, 180(10)(2010) 2003–2022.
149. Hu, Q., Shuang, A., Yu, D., Soft fuzzy dependency for robust feature evaluation, *Information Sciences*, 180 (2010) 4384–4400.
150. Rashmanlou, H., Samanta, S., Pal, M. and Borzooei, R. A., A study on bipolar fuzzy graphs, to appear in *Journal of Intelligent and Fuzzy Systems*.
151. Rias, F. and Ali, K. M., Applications of graph theory in computer science, in: 2011 Third International Conference Computational Intelligence, Communication Systems and Networks (CICSyN), 2011, pp. 142–145.
152. Riff, M., Zúniga, M. and Montero, E., A graph-based immune-inspired constraint satisfaction search, *Neural Comput Appl*, 19 (2010) 1133–1142.

153. Roberts, F., *Discrete Mathematical Models*, Prentice Hall, Englewood Cliffs, New Jersey, 1976.
154. Rosenfeld, A., *Fuzzy graphs*, In: L. A. Zadeh, K. S. Fu and M. Shimura, Eds., Fuzzy Sets and Their Applications, Academic Press, New York, pp.77–95, 1975.
155. Roy, M. K. and Biswas, R., I-V fuzzy relations and Sanchez's approach to medical diagnosis, *Fuzzy Sets and Systems*, 47 (1992) 35–38.
156. Samanta, S. and Pal, M., Bipolar fuzzy hypergraphs, *International Journal of Fuzzy Logic Systems*, 2 (1) (2012) 17–28.
157. Samantha, S. and Pal, M., Fuzzy $k$-competition graphs and $p$ -competition graphs, *Fuzzy Inf. Eng.*, 5 (2013) 191–204.
158. Samantha, S. and Pal, M., Irregular bipolar fuzzy graphs, *Inter. J. Appl. Fuzzy Sets*, 2 (2012) 91–102.
159. Samanta, S. and Pal, M., Some more results on bipolar fuzzy sets and bipolar fuzzy intersection graphs, *The Journal of Fuzzy Mathematics*, 22 (2) (2014) 253–262.
160. Sameena, K. and Sunitha, M. S., A Characterization of g-selfcentered fuzzy graphs, *The Journal of Fuzzy Mathematics*, 16(4) (2008) 787–791.
161. Sameena, K. and Sunitha, M. S., On g-distance in fuzzy trees, *The Journal of Fuzzy Mathematics*, 19(4) (2011) 787–791.
162. Sameena, K. and Sunitha, M. S., Strong arcs and maximum spanning trees in fuzzy graphs, *International Journal of Mathematical Sciences*, (2006) 17–20.
163. Shahzamanian, M. H., Shirmohammadi, M. and Davvaz, N. B., Roughness in Caley graphs, *Information Sciences*, 180 (2010) 3362–3372.
164. Shannon, A. and Atanassov, K. T., A first step to a theory of the intuitionistic fuzzy graph, (in: D. Lakov Ed.), Proceeding of FUBEST, Sofia, Sept. 28–30, 1994, pp. 59–61.
165. Shannon, A. and Atanassov, K. T., Intuitionistic fuzzy graphs from $\alpha$-, $\beta$-, and $(\alpha, \beta)$-levels, *Notes on Intuitionistic Fuzzy Sets*, 1 (1995) 32–35.
166. Shrinivas, S. G., Vertrivel, S. and Elango, N. M., Appliations of graph theory in computer science an overview, *International Journal of Engineering Science and Technology*, 2 (9) (2010) 610–621.
167. Sunitha, M. S. and Vijayakumar, A., A characterization of fuzzy trees, *Information Sciences*, 113 (1999) 293–300.
168. Sunitha, M. S. and Vijayakumar, A., Blocks in fuzzy graphs, *The Journal of Fuzzy Mathematics*, 13(1)1 (2005) 13–23.
169. Sunitha, M. S. and Vijayakumar, A., Complement of a fuzzy graph, *Indian Journal of Pure and Applied Mathematics*, 33(2002) 1451–1464.
170. Sunitha, M. S. and Mathew, S., Fuzzy graph theory, a survey, *Annals of Pure and Applied Mathematics*, 4(1) (2013) 92–110.
171. Sunitha, M. S. and Vijayakumar, A., Some metric aspects of fuzzy graphs, In: Balakrishna, R., Mulder, H. M., Vijayakumar, A. (Eds), *Proceedings of the Conference on Graph Connections* CUSAT, Allied Publishers, Cochin, 1999, pp. 111–114.
172. Suvarna, N. T. and Sunitha, M. S., Convexity in fuzzy graphs, *Advances in Fuzzy Sets and Systems*, 5(1) (2010) 73–80.
173. Tamura, S., Higuchi, S. and Tanaka, K., Pattern classification based on fuzzy relations, *IEEE Tansactions SMC*-1, 61–66, 1971.
174. Tom, M. and Sunitha, M. S., On strongest paths, delta arcs and blocks in fuzzy graphs, *World Applied Science Journal*, 22 (2013) 10–17.
175. Tom, M. and Sunitha, M. S., Sum distance in fuzzy graphs, *Annals of Pure and Applied Mathematics*, 7(2) (2014) 73–89.
176. Tong, Z. and Zheng, D., An algorithm for finding the connectedness matrix of a fuzzy graph, *Congr. Numer.*, 120 (1996) 189–192.
177. Tuksen, I. B., Interval-valued fuzzy sets based on normal forms, *Fuzzy Sets and Systems*, 20 (1986) 191–210.
178. Wu, L., Shan, E. and Liu, Z., On the $k$-tuple domination of generalized de Brujin and Kautz digraphs, *Information Sciences*, 180 (2010) 4430–4435.

179. Wu, S.Y., The composition of fuzzy graphs, *Journal of Research in Education Sciences*, 31 (1986) 603–629.
180. Wu, C. L., Chau, K. W. and Li, Y. S., Predicting monthly stream flowusing data driven models coupled with data pre-processing techniques, Water resources research, Vol. 45, W08432, https://doi.org/10.1029/2007WR006737, 2009.
181. Xie, J. X., Cheng, C. T., Chau, K. W. and Pei, Y. Z., A hypbrid adaptive time-delayed neural network model for multi-step-ahead prediction of sunspot activity, *International Journal of Environment and Pollution*, 28 (2006) 364–381.
182. Xu, J., The use of fuzzy graphs in chemical structure research, in Fuzzy Logic in Chemistry, D. H Rouvry, Ed., Academic Press, 1997, pp. 249–282.
183. Yang, H. L., Li, S. G., Guo, Z. L. and Ma, C. H., Transformation of bipolar fuzzy rough set models, *Knowledge-Based Systems*, 27 (2012) 60–68.
184. Yang, H., Li, S., Yang, W. and Lu, Y., Notes on bipolar fuzzy graphs, *Information Sciences*, 242 (2013) 113–121.
185. Yamak, S., Kazanc, O. and Davaz, B., Normal fuzzy hyperideals in hypernear-rings, *Neural Comput Appl*, 20 (2010) 25–30.
186. Yeh, R. T. and Bang, S. Y., *Fuzzy graphs, fuzzy relations, and their applications to cluster analysis*, In: Zadeh, L. A., Fu, K. S. and Shimura, M. Eds., Fuzzy Sets and Their Applications, Academic Press, New York, 125–149, 1975.
187. Yener, B., Gunduz, C. and Gultekin, S. H., The cell graphs of cancer, *Bioinformatics*, 20(1) (2004) 145-151.
188. Yin, Y., Zhan, J. and Corsini, P., Fuzzy roughness of $n$-ary hypergroups based on a complete residuated lattice, *Neural Comput. Appl.*, 20 (2011) 41–57.
189. Zadeh, L. A., From precise to granular probabilities, *Fuzzy Sets and Systems*, 154 (2005) 370–374.
190. Zadeh, L. A., Fuzzy sets, *Inform. and Control* 8 (1965) 338–353.
191. Zadeh, L. A., Is there a need for fuzzy logic?, *Information Sciences*, 178(13) (2008) 2751–2779.
192. Zadeh, L. A., Similarity relations and fuzzy orderings, *Information Sciences*, 3 (1971) 177–200.
193. Zadeh, L. A., Toward a generalized theory of uncertainty (GTU) an outline, *Information Sciences* 172 (2005) 1–40.
194. Zadeh, L. A., The concept of a linguistic and application to approximate reasoning, *Information Sciences*, 8 (1975) 199–249.
195. Zhang, W. R., Bipolar fuzzy sets, Proceedings of FUZZ-IEEE (1998), 835–840.
196. Zhang, W. R., Bipolar fuzzy sets and relations: a computational framework for cognitive modeling and multiagent decision analysis, Proceedings of IEEE Conf., 1994, pp. 305–309.
197. Zhang, W. R., Yin Yang bipolar relativity, IGI Global 2011.
198. Zhang, J. and Chau, K. W., Multilayer ensemble pruning via novel multi-sub-swarm particle swarm optimization, *Journal of Universal Computer Science*, 15(4) (2009) 840–858.
199. Zhang, J. and Yang, X., Some properties of fuzzy reasoning in propositional fuzzy logic systems, *Information Sciences*, 180 (2010) 4661–4671.

# Index





S. Mathew et al., *Fuzzy Graph Theory*, Studies in Fuzziness
and Soft Computing 363, https://doi.org/10.1007/978-3-319-71407-3

**D**



**E**

CPSIA information can be obtained
at www.ICGtesting.com
Printed in the USA
/081409090619
638LV00008B/496/P